Botany: Principles and Applications

Botany: Principles and Applications

Editor: Allen Davies

STATES
ACADEMIC PRESS
www.statesacademicpress.com

States Academic Press,
109 South 5th Street,
Brooklyn, NY 11249, USA

Visit us on the World Wide Web at:
www.statesacademicpress.com

ISBN: 978-1-63989-082-8 (Hardback)

Cataloging-in-Publication Data

Botany : principles and applications / edited by Allen Davies.
p. cm.
Includes bibliographical references and index.
ISBN 978-1-63989-082-8
1. Botany. 2. Botany--Methodology. I. Davies, Allen.
QK45.2 .B68 2022
580--dc23

Table of Contents

Preface

Every book is initially just a concept; it takes months of research and hard work to give it the final shape in which the readers receive it. In its early stages, this book also went through rigorous reviewing. The notable contributions made by experts from across the globe were first molded into patterned chapters and then arranged in a sensibly sequential manner to bring out the best results.

Botany is a branch of biology that deals with the study of plants. It is an umbrella discipline which includes branches such as plant morphology, plant physiology, plant ecology, systematics, etc. Plant morphology studies the structure and form of plants. It also encompasses cytology, histology, anatomy and morphogenesis, besides others. The study of plant functions is essential for understanding the diversity of plant structures, which is addressed under the branch of plant physiology. The understanding of how the physical and natural environment comprising of the soil, climate and atmosphere affect the physiological functions of plants falls under the scope of plant ecology. Such studies have major implications in the provision of a number of products of human interest such as medicine, rubber, timber, fiber, oil, etc. Botanical research further aids in the synthesis of raw materials and chemicals for energy production and construction, environmental management and biodiversity conservation. This book brings forth some of the most innovative concepts and elucidates the unexplored aspects of botany. It elucidates new techniques and their applications in a multidisciplinary manner. The extensive content of this book provides the readers with a thorough understanding of the subject.

It has been my immense pleasure to be a part of this project and to contribute my years of learning in such a meaningful form. I would like to take this opportunity to thank all the people who have been associated with the completion of this book at any step.

Editor

1

Feedback regulation by trehalose 6-phosphate slows down starch mobilization below the rate that would exhaust starch reserves at dawn in Arabidopsis leaves

Letícia dos Anjos[1,2] | Prashant Kumar Pandey[1†] | Thiago Alexandre Moraes[1] | Regina Feil[1] | John E. Lunn[1] | Mark Stitt[1]

[1]Max Planck Institute of Molecular Plant Physiology, Potsdam, Golm, Germany

[2]Universidade Federal do Ceará, Fortaleza, Brazil

Correspondence
Mark Stitt, Max Planck Institute of Molecular Plant Physiology, Am Muehlenberg 1, Potsdam, Golm 14474, Germany.
Email: mstitt@mpimp-golm.mpg.de

Present address
Prashant Kumar Pandey, National Research Council Canada (NRC-CNRC), 110 Gymnasium Place, Saskatoon, Saskatchewan, S7N 0W9, Canada.

Funding information
Capes (Coordenação de Aperfeiçoamento de Nível Superior, Brazil); Max Planck Society and European Union, Grant/Award Number: 245143; Brazilian National Council for Scientific and Technological Development (CNPq), Grant/Award Number: 246681/2012-8

Abstract

Trehalose 6-phosphate (Tre6P), a sucrose signaling metabolite, inhibits transitory starch breakdown in Arabidopsis (*Arabidopsis thaliana*) leaves and potentially links starch turnover to leaf sucrose status and demand from sink organs (Plant Physiology, 163, 2013, 1142). To investigate this relationship further, we compared diel patterns of starch turnover in ethanol-inducible Tre6P synthase (iTPS) lines, which have high Tre6P and low sucrose after induction, with those in *sweet11;12* sucrose export mutants, which accumulate sucrose in their leaves and were predicted to have high Tre6P. Short-term changes in irradiance were used to investigate whether the strength of inhibition by Tre6P depends on starch levels. *sweet11;12* mutants had twofold higher levels of Tre6P and restricted starch mobilization. The relationship between Tre6P and starch mobilization was recapitulated in *iTPS* lines, pointing to a dominant role for Tre6P in feedback regulation of starch mobilization. Tre6P restricted mobilization across a wide range of conditions. However, there was no correlation between the level of Tre6P and the absolute rate of starch mobilization. Rather, Tre6P depressed the rate of mobilization below that required to exhaust starch at dawn, leading to incomplete use of starch. It is discussed how Tre6P interacts with the clock to set the rate of starch mobilization.

KEYWORDS
Arabidopsis, circadian clock, diel, starch, trehalose 6-phosphate

1 | INTRODUCTION

Plants use light energy to drive photosynthetic carbon (C) gain, metabolism, and growth, but at night depend on C reserves accumulated in previous light periods. In many species, including Arabidopsis, foliar starch is the major C reserve (Smith & Stitt, 2007). Diel regulation of starch turnover may depend on the conditions (Paul & Foyer, 2001). In source-limited plants, C is in short supply and it is crucial to manage C reserves to insure rapid investment in growth while avoiding C starvation at night (Scialdone and Howard, 2015; Smith & Stitt, 2007; Stitt & Zeeman, 2012). In sink-limited conditions, C regulation of metabolism and growth is relaxed (Baerenfaller et al., 2015; Sulpice et al., 2014) and starch often accumulates in leaves and other parts of the plant. This incomplete utilization of starch may be at least partly due to feedback inhibition of starch mobilization by the sucrose signal trehalose 6-phosphate (Figueroa and Lunn, 2016; Lunn, Delorge, Figueroa, Van Dijck, & Stitt, 2014; Martins et al., 2013). The following experiments provide further evidence that Tre6P plays a key role in the feedback regulation of starch mobilization. In particular, we ask whether feedback inhibition by Tre6P is minimized to allow full use of starch in conditions where C is in short supply, but operates effectively when C is in excess.

When Arabidopsis plants grow in conditions where less C is available per 24 hr cycle, they accumulate a larger proportion of their fixed C to starch in the daytime and slow down mobilization of starch during the night, compared to plants growing with a large C supply. As a result, starch reserves are almost but not completely exhausted at dawn, irrespective of the overall availability of C. This pattern of diel starch turnover maximizes growth in low C conditions by insuring that almost all of the fixed C are invested in growth within a 24 hr cycle, while avoiding a deleterious period of C starvation at the end of the night (EN; Geiger and Servaites 1994; Ishihara, Obata, Sulpice, Fernie, & Stitt, 2015; Smith & Stitt, 2007; Stitt & Zeeman, 2012). Starch turnover shows this diel profile in many species including important crops (Chatterton & Silvius 1979, 1980, 1981; Cheng, Moore, & Seemann, 1998; Matt et al., 2001; Mullen & Koller, 1988; Silvius, Chatterton, & Kremer, 1979; reviewed in Smith & Stitt, 2007).

Arabidopsis maintains this pattern of diel starch turnover across a wide range of growth conditions including different photoperiods (Gibon, Pyl, Sulpice, Höhne, & Stitt, 2009; Sulpice et al., 2014), light intensities (Mengin et al., 2017), and night temperatures (Pilkington et al., 2015; Pyl et al., 2012). In addition, and crucially, sudden perturbations in growth conditions trigger a change in the rate of starch mobilization such that starch reserves last until the coming dawn. Examples include slowing down of mobilization after a sudden early dusk (Feike et al., 2016; Graf, Schlereth, Stitt, & Smith, 2010; Martins et al., 2013; Scialdone et al., 2013) or a single low light (LL) day (Feike et al., 2016; Pilkington et al., 2015), and speeding up of mobilization after a sudden late dusk (Scialdone et al., 2013) or an interruption of the night with a short interval of light (Scialdone et al., 2013). These rapid responses will be important in optimizing growth in a fluctuating environment.

These observations prompted the idea that starch mobilization is timed to the next dawn by the circadian clock (Graf et al., 2010; Scialdone et al., 2013). There are several lines of evidence for involvement of the clock. Wild-type plants growing in a 17-hr or a 28-hr light-dark cycle exhaust their starch at about 24 hr after the previous dawn, reflecting the innate 24-hr periodicity of the wild-type clock. The *lhy cca1* mutant exhausts its starch about 18–20 hr after the previous dawn, matching the shortened period in this mutant (Graf et al., 2010; Scialdone et al., 2013). Several models have been proposed to explain how the clock might regulate starch turnover (Dodd, Dalchau, Gardner, Baek, & Webb, 2014; Pokhilko, Flis, Sulpice, Stitt, & Ebenhöh, 2014; Scialdone and Howard, 2015; Scialdone et al., 2013; Seki et al., 2017). Some models propose that metabolic signals related to C availability modulate clock gene expression or clock period (Seki et al., 2017; Shin et al., 2017) or clock output pathways (Pokhilko et al., 2014). The arithmetic division model (Scialdone et al., 2013) proposes that a clock-dependent mechanism measures time until the next dawn (*T*) and that this information is integrated with a measure of the starch content (*S*) to set the rate of starch mobilization ($R = S/T$). The robust pacing of starch mobilization to the coming dawn in the face of sudden perturbations (see above) is explained well by the molecular division model, but less easily by models that invoke changes in clock periodicity. However, it remains unclear which molecular outputs from the clock regulate the rate of starch mobilization, and how the amount of starch is measured. The biochemical mechanism that sets the rate of mobilization is also not known. There is some evidence that changes in the rate of starch mobilization may be linked with changes in phosphorylation of starch (Martins et al., 2013; Scialdone et al., 2013), and *early starvation1* (*esv1*) mutants were recently described in which premature exhaustion of starch is associated with altered granule structure and shape (Feike et al., 2016).

The widespread elevation of crop yield in Free Air CO_2 Enhancement studies points to C supply often being at least partly limiting for plant growth in the field (Long, Ainsworth, Leakey, Nösberger, & Ort, 2006). In some conditions, like nutrient-limiting conditions, C gain does not limit growth (Körner, 2006). Sink-limited plants do not fully exhaust their starch (Grimmer, Bachfischer, & Komor, 1999; Hädrich et al., 2012; Lawlor & Mitchell, 1991; Pilkington et al., 2015), implying that feedback mechanisms related to low demand for C depress the rate of starch mobilization below that which would be allowed by the clock. In some cases, starch is exhausted in source leaves but reserves remain elsewhere in the plant (Czedik-Eysenberg et al., 2016).

While it has been known for a long time that starch accumulates in sink-limited plants, an underlying mechanism was only recently uncovered. Trehalose 6-phosphate (Tre6P) has been proposed to act as a sucrose signal to regulate sucrose homeostasis (Figueroa and Lunn, 2016; Lunn et al., 2006, 2014; Yadav et al., 2014). Martins et al. (2013) used ethanol-inducible TPS-overexpressor (*iTPS*) lines to investigate whether Tre6P regulates the rate of mobilization of leaf starch. *iTPS* plants were grown in a 12-hr photoperiod for 3 weeks and then induced at the end of the day (ED) to increase Tre6P levels in the following night. This led to slower starch mobilization and strongly decreased levels of maltose (Martins et al., 2013), an early intermediate in foliar leaf mobilization (Lu & Sharkey, 2006; Niittylä et al., 2004; Stitt & Zeeman, 2012), as well as a large decrease in the levels of sucrose and reducing sugars (Martins et al., 2013). These results indicated that Tre6P exerts feedback regulation on starch mobilization that Tre6P acts at an early step in the mobilization pathway and that high Tre6P is able to override any other signals that might be generated by changes in sugar levels. However, the molecular mechanism by which Tre6P inhibits starch mobilization remains unclear. Furthermore, the action and efficacy of Tre6P has only been analyzed in *iTPS* lines and in one condition in which (see Sulpice et al., 2014) the plants were slightly source limited. We also lack studies that document the full chain of events in which a decreased demand or rate of export of sucrose acts via an increase in Tre6P to slow down starch mobilization.

Export of sucrose in Arabidopsis involves its release to the apoplast by the sucrose effluxers *SWEET11* and *SWEET12* (Chen et al.,

2012) followed by active uptake into the phloem complex by sucrose-H^+ cotransporters (Ayre, 2011; Lemoine et al., 2013; Riesmeier, Hirner, & Frommer, 1993). Mutants deficient in SWEET proteins show a reproducible but relatively small increase in leaf sucrose, compared to the drastic increases seen in mutants deficient in sucrose-H^+ cotransporters (Chen et al., 2012). They also bring the advantage that there is no accumulation of sucrose in the apoplast, which can trigger complex signaling and biotic resistance responses (Bolouri Moghaddam & Van den Ende, 2012; Herbers et al., 2000; Lemoine et al., 2013).

In the following experiments, we first asked whether the elevated levels of sucrose in leaves of *sweet* mutants lead to an increase in Tre6P and an inhibition of starch mobilization. We then present experiments with *sweet* mutants in conditions in which dusk starch levels were either low or high to learn whether the effectiveness of Tre6P depends on C status. In a parallel set of experiments with *iTPS* lines, we investigated the relationship between increased Tre6P levels and starch mobilization when sucrose and other sugars are falling (see above) rather than rising. The results strengthen the evidence that Tre6P plays a key role in the feedback regulation of starch mobilization. Rather surprisingly, they also reveal that there is no simple relationship between the Tre6P level and the rate of starch mobilization. Rather, Tre6P depresses the rate of mobilization below that required to exhaust starch at the coming dawn. This finding points to a close link between the mechanisms by which Tre6P and the clock act on starch mobilization.

2 | MATERIALS AND METHODS

2.1 | Plant material

Wild-type *Arabidopsis thaliana* (L.) Heyhn. accession Columbia-0 (Col-0) was provided by in-house collection (MPIMP, Golm). The Arabidopsis (Col-0) ethanol-inducible TPS-overexpressor (*iTPS*) and empty-vector control *AlcR* lines were the same as those described in (Martins et al., 2013).The Arabidopsis (Col-0) *sweet11*, *sweet12*, and *sweet11,12* mutants and wild-type segregant line (Chen et al., 2012) were kindly provided by Prof. Wolf Frommer (Heinrich Heine University, Düsseldorf, Germany).

2.2 | Plant growth conditions and ethanol induction

Seeds were sown in soil mixed with vermiculite (1:1) in 10-cm pots and placed in a phytotron under long-day (LD) conditions, 16 hr (20°C) light/8 hr (4°C) dark, with an irradiance of 160 μmol m^{-2} s^{-1}. After 1 week, pots were transferred directly to equinoctial conditions (12 hr light/12 hr dark) at constant 20°C, with an irradiance of 160 μmol m^{-2} s^{-1}. At 2 weeks after sowing, seedlings were transplanted into 10-cm pots (5 plants per pot). For the continuous light experiment, seeds were sown on soil in 10-cm pots and placed directly in a phytotron under continuous light with a constant irradiance of 160 μmol m^{-2} s^{-1} and constant temperature of 20°C. At 2 weeks after sowing, plants were thinned out inside the phytotron, leaving five plants per pot.

Short-term responses were evaluated at 21 days after sowing by submitting a batch of plants to different growth conditions. On the day of the experiment, some of the wild-type Col-0 and *sweet11;12* mutant plants was transferred at dawn to either a LD (15 hr light/9 hr dark, 160 μmol m^{-2} s^{-1}, 20°C), or high light (HL; 12 hr/12 hr dark, 320 μmol m^{-2} s^{-1},20°C) or to combined LD and HL (LD + HL; 15 hr light/9 hr dark, 320 μmol m^{-2} s^{-1}, 20°C). Similarly, on the day of the experiment, some of the empty-vector control *AlcR* and *iTPS* plants were transferred at dawn to HL; 12 hr (320 μmol m^{-2} s^{-1}, 20°C) or to LL (80 μmol m^{-2} s^{-1}, 20°C) Plants were induced by spraying the mutant and the empty-vector control *AlcR* 2 hr before dusk with 2% (v/v) ethanol.

For the continuous light experiment, empty-vector control *AlcR* and *iTPS* plants were germinated, thinned to four plants per pot, and grown in continuous light at 20°C for 3 weeks. On the day of the experiment, plants were sprayed with 2% (v/v) ethanol and then darkened 2 hr later.

In all experiments, plants were harvested just before darkening and at various times during the night, as indicated for each individual experiment. Whole rosettes were excised under the ambient conditions and immediately quenched in liquid nitrogen. For sampling in the light, care was taken not to shade the leaves at any time before quenching. Plants from one or two pots (i.e., five or 10 plants) were pooled to form one sample, ground to a fine powder at −70°C using a robotized ball mill (Smith & Stitt, 2007), subaliquoted in a cryorobot, and stored at −80°C until use.

2.3 | Extraction and measurement of metabolites

Starch, glucose, fructose, and sucrose were determined by enzymatic assays in ethanolic extracts of 20 mg frozen plant material as described by Cross et al. (2006). Assays were performed in 96-well microplates using a Janus pipetting robot (Perkin Elmer, Wellesley, MA, USA; www.perkinelmer.com). Absorbances were determined using a Synergy, an ELx800 or an ELx808 microplate reader (Bio-Tek, Bad Friedrichshall, Germany; www.bio-tek.com). For all the assays, two technical replicates were determined per biological replicate. Tre6P and ADP-glucose (ADPGlc) were extracted in chloroform-methanol and measured by high-performance anion-exchange chromatography coupled to tandem mass spectrometry (LC-MS/MS) as described by Lunn et al. (2006) with modifications as in Figueroa et al. (2016). Tre6P was quantified using enzymatically calibrated standards and a [$_2$H]Tre6P internal standard to correct for ion suppression and matrix effects (Lunn et al., 2006).

Transcript analysis was performed essentially as in Flis et al. (2015, 2016). mRNA was extracted from samples using an RNeasy Plant Mini kit (Qiagen; www.qiagen.com). cDNA was synthesized using a SuperScript III First-strand Synthesis System Kit (Invitrogen; www.thermofisher.com). Quantitative real-time PCR was performed using the SYBR Green PCR kit (Qiagen). Data were collected using SDS software (Applied Biosystems; www.appliedbiosystems.com). Concentration of target gene transcripts in the samples was calculated using standard curves generated for each sample, based on C_t

values and spike-in controls added to the samples before RNA extraction (Flis et al., 2015; Piques, Schulze, Gibon, Rohwer, & Stitt, 2009).

3 | RESULTS

3.1 | Kinetics of starch breakdown in the *sweet11;12* mutant

Wild-type Col-0, the *sweet11* and *sweet12* mutants, and the *sweet11;12* double mutant were grown in a 12-hr photoperiod with 160 μmol m^{-2} s^{-1} photon flux density (PFD) at rosette level for 30 days and then sampled at dusk and 4-hr intervals during the night (Figure 1, original data provided in Supporting Information Table S1). Two Col-0 wild-type lines were included; one from standard seed stock, and one that segregated from the cross between *sweet11* and *sweet12* (Col-$0_{(seg)}$).

In the wild-type lines, sucrose levels were fairly constant throughout the night. The levels (about 2 μmol/g FW) resembled those seen previously in our standard growth conditions (12-hr photoperiod, 160 μmol m^{-2} s^{-1} m 20°C; Annunziata et al., 2017; Flis et al., 2016; Mengin et al., 2017; Pal et al., 2013; Sulpice et al., 2014). Sucrose was not significantly changed in the single *sweet* mutants, but was elevated two- to threefold in the double mutant (Figure 1a; see also Chen et al., 2012). Glucose and fructose (Supporting Information Figure S1) were very low throughout the night in the wild-types, not significantly increased in the single mutants, and strongly increased in *sweet11;12*. Tre6P (Figure 1b) was unaltered in the single mutants, and 80%–150% increased in *sweet11;12* compared to wild-type lines. In wild-type lines, starch decreased during the night in a linear manner and was almost completely exhausted at dawn (Figure 1c), as seen in many previous studies (Annunziata et al., 2017; Flis et al., 2016; Gibon et al., 2004; Graf et al., 2010; Mengin et al., 2017; Pal et al., 2013; Scialdone et al., 2013). Starch mobilization in the single mutants resembled Col-0. At dusk, *sweet11;12* contained about 30% more starch than the wild-types or the single mutants. Unexpectedly, *sweet11;12* broke its starch down at almost the same rate as the other three genotypes. As a result of the high dusk starch content, *sweet11;12* had a large starch excess at dawn. Of the starch at dusk, <10% remained at dawn in wild-type lines and the single *sweet* mutants, but about 37% in *sweet11;12*.

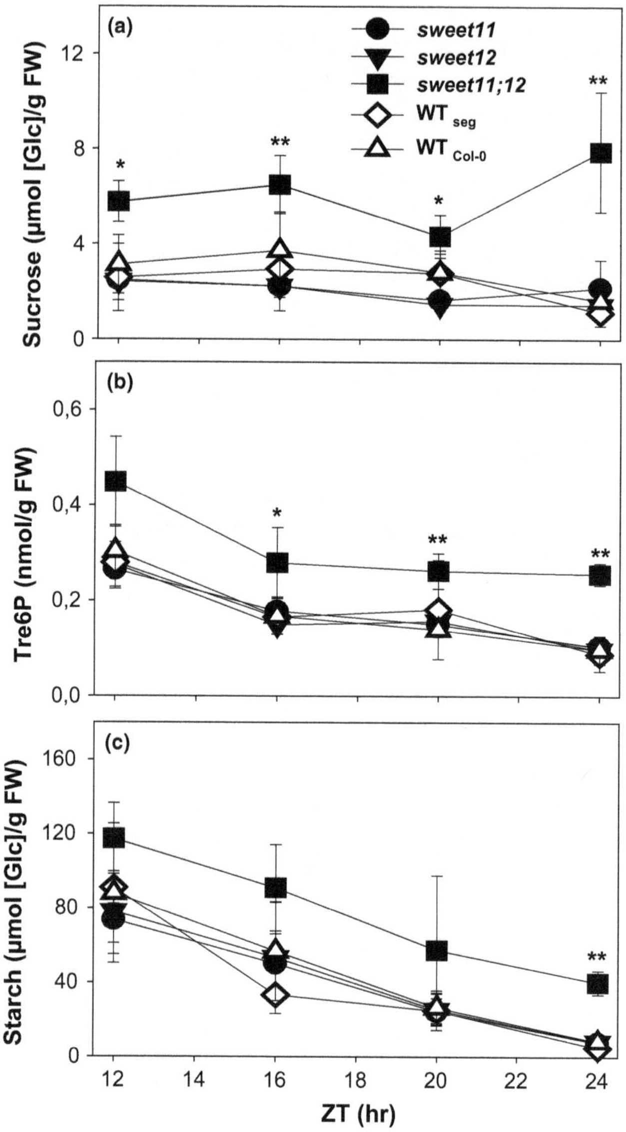

FIGURE 1 Night-time metabolite levels in *Arabidopsis thaliana sweet* mutants in standard growth conditions. Arabidopsis *sweet11*, *sweet12*, and *sweet11;12* mutants were grown in standard conditions (12-hr/12-hr light/dark cycle, 21°C, 160 μmol m^{-2} s^{-1} PFD) along with a wild-type segregant (WT_{seg} and wild-type Col-0, WT $_{Col0}$) plants, and rosettes were harvested at various times during the night. (a) Sucrose, (b) Tre6P, (c) Starch. Values shown are mean ± *SD* (n = 3). Significant differences between *sweet11;12* and WT_{Col0}, using Student's t test are indicated by asterisks: *$p < 0.05$, **$p < 0.01$, and ***$p < 0.001$. The original data are provided in Supporting Information Table S1; glucose and fructose are shown in Supporting Information Figure S1. ZT, zeitgeber time (hours after dawn)

3.2 | Effect of increased night-time sucrose and Tre6P levels on starch mobilization

We were curious whether the generally elevated levels of starch and unaltered rate of starch mobilization in *sweet11;12* were a consequence of the conditions used in the experiment of Figure 1, or a robust phenotype of this mutant. Wild-type plants and *sweet11;12* were grown in standard conditions (12-hr photoperiod, 21°C, PFD of 160 μmol m^{-2} s^{-1} at rosette level) for 21 days, and batches were then either left in standard conditions (control; Ctrl) or shifted for a single day to a 15-hr photoperiod (LD) with unaltered light intensity (160 μmol m^{-2} s^{-1}), or to high-light (HL, 320 μmol m^{-2} s^{-1} PFD, which is close to Arabidopsis light-saturation point, see Pilkington et al., 2015) in a 12-hr photoperiod, or to a combination of LD and HL (LD + HL). These treatments were chosen to generate different starch contents at dusk. Starch, sucrose, and Tre6P contents were investigated at the ED, at 3-hr intervals during the night, and at the EN (Figures 2 and 3, Supporting Information Figure S2; original data provided in Supporting Information Table S2).

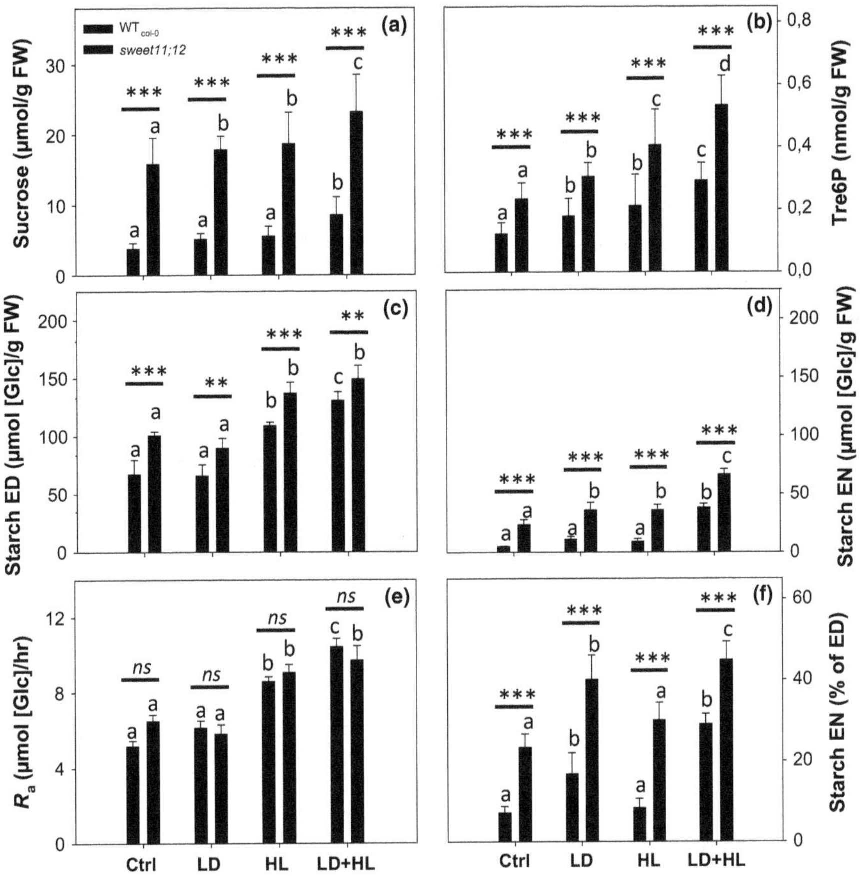

FIGURE 2 Effect of short-term treatments to alter dusk starch content on night-time levels of Tre6P and sucrose, starch content at dusk and dawn and starch mobilization in wild-type Col-0 and the *sweet11;12* mutant. Wild-type Col-0 plants (WT) and *sweet11;12* double mutants were grown in standard conditions (12-hr/12-hr light/dark cycle, 21°C, 160 μmol m^{-2} s^{-1} PFD). Three weeks after germination, batches of plants were left in standard conditions (Ctrl) or were shifted for a single day either to a long-day condition (LD, 15-hr photoperiod, 21°C, 160 μmol m^{-2} s^{-1} PFD), or to high-light condition (HL, 12-hr photoperiod, 21°C, 320 μmol m^{-2} s^{-1} PFD), or a combined LD and high-light condition (LD + HL). Samples were collected at the end of the light period (ED) after the shift, and throughout the following dark period at 3-hr intervals, up to the end of the night (EN) for metabolite analysis. (a) Night-time sucrose and (b) night-time Tre6P levels, averaged from ZT15 to ZT24, see Supporting Information Figure S3 for values at each time; (c) starch content at ED; (d) starch content at EN; (e) absolute rate of starch mobilization (R_a), estimated from the slope of the linear regression of starch contents measured along the night from dusk to dawn (see Figure 3); (f) starch content at EN expressed as a percentage of starch content at ED. Values are mean ± *SD* (*n* = 5). Asterisks indicate statistically significant genotype-dependent differences in a given treatment: **p* < 0.05, ***p* < 0.01, and ****p* < 0.001. Letters indicate significant treatment-dependent differences in a given genotype; blue letters indicate treatment differences within $WT_{Col\text{-}0}$, and red letters indicate differences within *sweet11;12* double mutant (one-way ANOVA, Holm-Sidak post hoc pairwise multiple comparison testing, *p* < 0.01). The original data are provided in Supporting Information Table S2, and the temporal changes of Tre6P, sucrose, and other selected metabolites are shown in Supporting Information Figure S2. A replicate experiment for the Crtl, LD, and HL treatments is shown in Supporting Information Figures S3 and S4. ZT, zeitgeber time (hours after dawn)

In wild-type plants, the average night-time levels of sucrose (Figure 2a) and Tre6P (Figure 2b) rose progressively between the Ctrl, LD, HL, and LD + HL treatments, with the increase being significant in LD + HL for sucrose and in all treatments for Tre6P. In all treatments, *sweet11;12* had higher average night-time sucrose and Tre6P levels than wild-type plants (*p* < 0.001 in all cases). Sucrose levels were fourfold higher in *sweet11;12* than wild-types in Ctrl, and about threefold higher in the LD, HL, or LD + HL treatments. Average night-time Tre6P levels in *sweet11;12* were about twofold higher than wild-type in all treatments. In wild-type plants, starch content at dusk (Figure 2c) was unaltered in LD and rose significantly in HL and LD + HL, compared to Ctrl. The lack of response of dusk starch in LD is because starch mobilization in the light accelerates with the duration of the light period (Fernandez et al., 2017). Starch in wild-type plants at dawn (Figure 1d) was very low in Ctrl, marginally higher in LD and HL and significantly higher in LD + HL. Starch at dawn was significantly higher in *sweet11;12* than the wild-type in all treatments. Relative to the wild-type value, the increase in *sweet11;12* was relatively small at dusk (14%–50%) and larger at dawn (76%–244%), with the relative increase being larger in Ctrl than in HL, LD, or LD + HL.

The first dedicated and regulated reaction in starch synthesis is catalyzed by ADPGlc pyrophosphorylase, leading to formation of ADPGlc. The higher rates of starch accumulation in HL and HL + LD

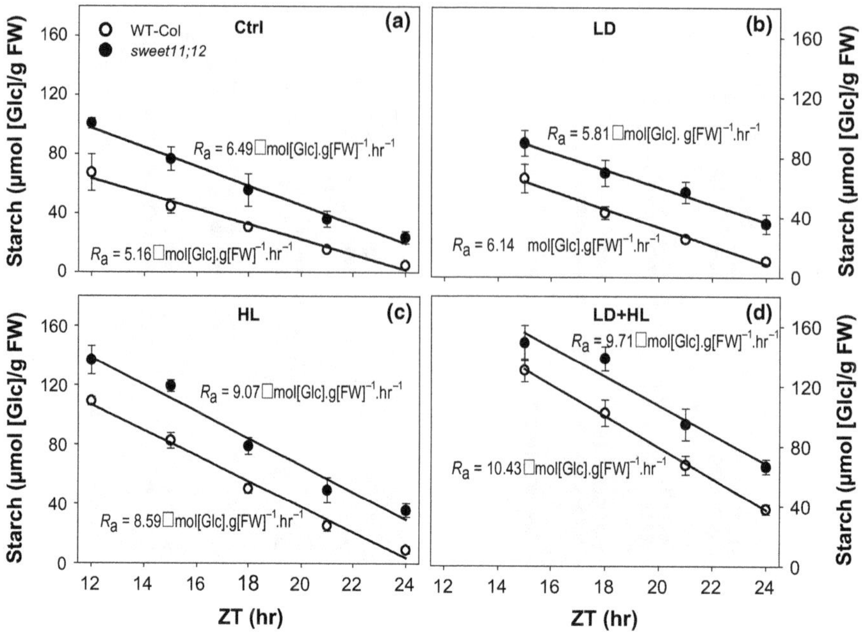

FIGURE 3 Effect of short-term treatments to alter dusk starch content on the temporal kinetics of starch mobilization in wild-type Col-0 and the *sweet11;12 double* mutant. This is a further analysis of data from the experiment shown in Figure 2. Wild-type Col-0 plants (WT) and *sweet11;12* double mutants were grown in standard conditions (12-hr/12-hr light/dark cycle, 21°C, 160 µmol m^{-2} s^{-1} PFD) for 3 weeks after germination. On the day of the experiment, plants were either left in the standard growth condition (a) or shifted to a new condition for a single day to increase starch content at dusk: (b) high-light (HL, 12-hr photoperiod, 21°C, 320 µmol m^{-2} s^{-1} PFD); (c) long-day (LD, 15-hr photoperiod, 21°C, 160 µmol m^{-2} s^{-1} PFD); or (d) long-day and high-light (LD + HL). Rosettes were harvested at dusk and at 3-hr intervals during the night to measure starch content. Values are means ± *SD* (*n* = 5). Lines show the linear regressions of starch content from which the absolute rate of starch mobilization (R_a) was calculated (numbers given in panel). ZT, zeitgeber time (hours after dawn)

were accompanied by a twofold increase in the dusk level of ADPGlc (Supporting Information Figure S2). The increase in Tre6P in *sweet11;12* did not lead to an increase in dusk levels of ADPGlc. This is consistent with the similar rates of starch accumulation between dawn and dusk in *sweet11;12* and wild-type plants. It is also consistent with the lack of effect of elevated Tre6P on ADPGlc levels and starch accumulation previously reported by Figueroa et al. (2017). ADPGlc levels were negligible in the dark in wild-type plants. They were also negligible in the *sweet11;12* mutant, despite its high sucrose levels during the night.

Figure 3 shows the kinetics of starch mobilization during the night. Both wild-type Col-0 and *sweet11;12* degraded their starch in a near-linear manner during the night (Figure 3). The slope of the linear regression on starch contents at ZT12, ZT16, ZT20, and ZT24 (ZT, zeitgeber time, i.e., hours since previous dawn) was used to estimate the absolute rate of starch mobilization (R_a; Figure 2e). Comparing across treatments, wild-type plants mobilized their starch slightly faster after the LD and considerable faster after the HL and LD + HL treatments than in the control (5.16, 6.14, 8.59, and 9.71 µmol [Glc] g^{-1} FW hr^{-1} in Crtl, LD, HL, and LD + HL, respectively). This increase in HL and LD + HL reflects the increased dusk starch content. Comparing across treatments, a higher dusk starch content led to a faster starch mobilization in *sweet11;12* (6.49, 5.81, 9.97, and 10.43 µmol [Glc] g^{-1} FW hr^{-1} in Crtl, LD, HL, and LD + HL, respectively. The key finding, however, is that in a given condition, the absolute rate of starch mobilization was not substantially or consistently slower in *sweet11;12* than in wild-type plants. Comparing *sweet11;12* to wild-type, mobilization was slightly faster in Ctrl (6.5 compared to 5.16 µmol [Glc] g^{-1} FW hr^{-1}, p = 0.05), marginally faster in HL and marginally slower in LD and HL + LD, withir none of differences being significant (p = 0.18, 0.65 and 0.34). Nevertheless, due to the differing starch content at dusk, *sweet11;12* plants did not completely degrade their starch reserves in any of the treatments whereas wild-type plants remobilized almost all their reserves by dawn in all treatments except LD + HL. The starch content at the EN, expressed as a percentage of the dusk starch content, was about twofold higher in *sweet11;12* (28% in Ctrl rising to 40%–43% in LD, HL, and LD + HL) than in wild-type plants (about 10% in Ctrl, rising to 20% in LD and HL, and 28% in LD + HL; Figure 2f). Thus, the increase in sucrose and Tre6P in *sweet11;12* did not strongly or consistently decrease the absolute rate of starch mobilization in the experiment of Figures 2 and 3; rather, it decreased the proportion of the starch that is degraded during the night.

In a further experiment with control, LD and HL treatments (Supporting Information Figure S3) in any given treatment sucrose was again about threefold (Supporting Information Figure S2A) and Tre6P about twofold (Supporting Information Figure S2B) higher in *sweet11;12* than Col-0. Compared to the experiment of Figures 2 and 3, wild-type plants contained more starch at dusk (Supporting Information Figure S3D). This may reflect small differences in the growth conditions affecting starch turnover; a 12-hr photoperiod with this irradiance and fertilization regime is close to the point at which growth of wild-type Col-0 switches from being source- to sink-limited (Sulpice et al., 2014). Compared to wild-type plants, starch levels in *sweet11;12* showed a trend to slightly higher levels at dusk (Supporting Information Figure S3C, significant in control conditions) and a large increase at dawn (Supporting Information Figure S3D, significant in all conditions). Compared to wild-type plants, R_a was slightly higher in *sweet11;12* in Ctrl (6.03 compared to 6.82 μmol [Glc] g^{-1} FW $hr^{-1,}$ respectively, p = 0.074) and lower in LD (8.77 compared to 6.95 μmol [Glc] g^{-1} FW hr^{-1}) and HL (9.03 compared to 6.56) μmol [Glc] g^{-1} FW hr^{-1}; Supporting Information Figure S4A–C), with none of these differences being significant (Supporting Information Figure S3E). As in the experiment of Figures 2 and 3, compared to Col-0, *sweet11;12* showed a large and highly significant twofold increase in the percentage of the dusk starch content that remained at dawn (Supporting Information Figure S3F). Thus, in this third independent experiment, an increase in sucrose and Tre6P again did not consistently slow starch mobilization but instead led to a decrease in the percentage of starch that was mobilized during the night.

3.3 | Effect of induced changes in Tre6P levels on starch mobilization rate in carbon-limiting and carbon-replete conditions

To further investigate whether the incomplete mobilization of starch in *sweet11;12* was associated with changes in Tre6P, we investigated starch mobilization in an ethanol-inducible TPS-overexpressor (*iTPS*) line (Figueroa et al., 2016; Martins et al., 2013). In this line, Tre6P levels can be inducibly increased within 2 hr in the absence of any increase in sucrose; indeed, the induced increase of Tre6P leads to a decrease of sucrose and reducing sugars (see Introduction, also below for more data). We investigated the response of starch mobilization to an increase in Tre6P in conditions where the plants were C-limited and the plants were C-replete.

iTPS plants were grown in parallel with the empty-vector control line, *AlcR*, in standard conditions for 4 weeks. One batch was then subjected to a single HL (320 μmol photons m^{-2} s^{-1}) and another batch to a single LL (80 μmol photons m^{-2} s^{-1}; not much above the light-compensation point of 30 μmol m^{-2} s^{-1}) light period. All plants were sprayed with 2% (v/v) ethanol 2 hrs before dusk to insure the inducible mutants would enter the night with increased Tre6P levels. Rosettes were harvested in the light just before the onset of darkness (ED) and at 2-hr intervals for the first 10 hr of the night to quantify starch, sugars, Tre6P, and other metabolites.

Figure 4 shows the average Tre6P (Figure 4a) and sucrose (Figure 4b) levels during the night, starch content at ZT12 (ED) and ZT20 (Figure 4c, d) as well as the estimated average rate of starch mobilization (Figure 4e) and the starch content at ZT20 as a percentage of that at dusk (Figure 4f). Time kinetics are also provided for the changes of Tre6P (Figure 5a, b), sucrose (Figure 5c, d), and starch (Figure 5e, f). The original data are provided in Supporting Information Table S3.

Night-time Tre6P levels were about twofold higher in *iTPS* than control *AlcR* plants (Figure 4a, b). This increase was highly significant ($p < 0.001$) when averaged over all times between ZT14 and ZT20 both in the LL and the HL treatment. Time resolved plots revealed that in the LL treatment (Figure 5a) Tre6P was low at dusk and remained low during the night in the control *AlcR* plants, whereas it rose steadily during the night in the induced *iTPS* line. In the HL treatment (Figure 5b), Tre6P was high at dusk and fell strongly in the first 4 hr of the night in the *AlcR* control, whereas it declined only slightly at 14 and 16 hr and then rose in *iTPS*. This resembles the response previously seen in plants grown in standard conditions and induced in the middle of the day or just before dusk (Martins et al., 2013).

Induction of *TPS* led to sucrose levels falling to under 20% and about 40% of the *AlcR* levels in the LL and HL treatment, respectively (Figure 4b). This decrease was significant within the first 2 hr of the night (Figure 5c, d). A large decrease in sucrose in the night after induction of *TPS* was seen previously (Martins et al., 2013) and attributed to an inhibition of starch mobilization by Tre6P. The decrease in sucrose was accompanied by a decrease of glucose and fructose (see Supporting Information Table S3).

Starch at dusk was nearly threefold higher in the HL than the LL treatment (Figure 4c). Starch was degraded in a near-linear manner after the HL and LL treatments both in *AlcR* and *iTPS* (Figure 5e, f). *AlcR* plants mobilized less of their starch after the HL treatment than the LL treatment, with 49% and 39% of the dusk content remaining at ZT20. Extrapolation indicated that after the HL and LL treatments about 27% and 8% of their starch would remain at dawn. The response of an induced increase in Tre6P differed between the HL and LL treatments. Compared to the *AlcR* control, starch mobilization in *iTPS* was weakly but non-significantly inhibited after LL, and significantly decreased after HL (to about 70% and 30% of the control rate, respectively, Figure 4e). The starch excess phenotype was prominent in *iTPS* in LL and very strong in HL with over 60% and 80%, respectively, of the starch remaining at ZT20 (Figure 4f). Extrapolation indicated that about 50% and 75% of the dusk starch would have remained at the next dawn. For comparison, in the study of Martins et al. (2013) using *iTPS* plants growing in similar conditions to those of Figure 4 but left in growth irradiance on the day on which *TPS* was induced, *iTPS* showed an approximately twofold increase in Tre6P compared to control plants after induction, and the starch content at dawn was about 60% of the starch content at dusk.

3.4 | Integration of the response to an increase of Tre6P in the *sweet11;12* mutant and induced *iTPS* lines

We next evaluated the impact of night-time Tre6P level on starch mobilization in a combined data set, which contained the experiments with *sweet11;12* and the experiments with *iTPS*. There was no correlation between night-time levels of Tre6P and the absolute rate of starch breakdown, R_a (Figure 6a).

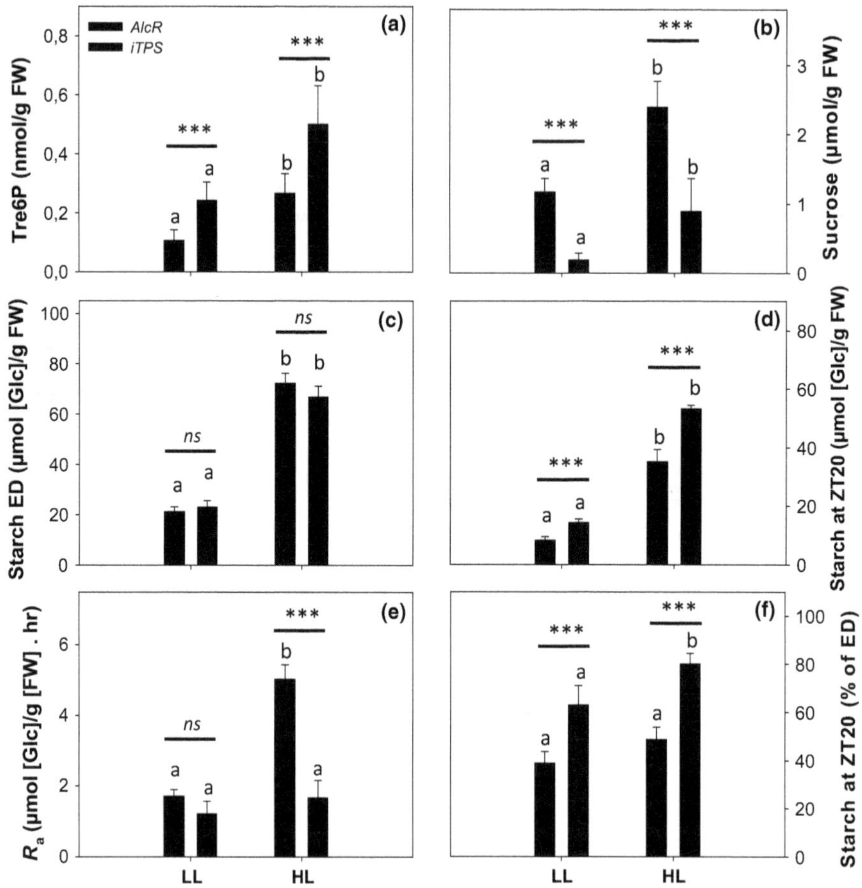

FIGURE 4 Effect of induced *TPS* overexpression on night-time metabolite levels after short-term perturbations to generate high or low starch content at dusk. The empty-vector control *AlcR* and *iTPS* plants were grown under standard conditions (12-hr/12-hr light/dark cycle, 21°C, 160 µmol m^{-2} s^{-1} PFD). Three weeks after germination, batches of plants were shifted for a single day either to low light (LL, 12-hr photoperiod, 21°C, 80 µmol m^{-2} s^{-1} PFD), or to high light (HL, 12-hr photoperiod, 21°C, 320 µmol m^{-2} s^{-1} PFD). Plants were induced 2 hr before dusk by spraying with 2% (v/v) ethanol. Samples were collected at the end of the day (ED) after the shift, and throughout the following dark period at 2-hr intervals, up to ZT20. (a) Night-time sucrose and (b) night-time Tre6P levels, averaged from ZT14 to ZT20, see Figure 5 for values at each time; (c) starch content at ED; (d) absolute rate of starch mobilization (R_a), estimated from the slope of the linear regression on starch content between ZT12 and ZT20 (see Figure 6); (e) starch content at ZT20; (f) starch content at ZT20 expressed as a percentage of the starch content at ED (ZT12). Values are means ± *SD* (*n* = 5). Asterisks indicate statistically significant genotype-dependent differences in a given treatment: $^*p < 0.05$, $^{**}p < 0.01$, and $^{***}p < 0.001$. Letters indicate significant treatment-dependent differences in a given genotype: blue letters indicate treatment differences within $WT_{Col\text{-}0}$, and red letters indicate differences within *sweet11;12* double mutant (one-way ANOVA, Holm-Sidak post hoc pairwise multiple comparison testing, $p < 0.01$). ZT, zeitgeber time (hours after dawn)

Irrespective of whether we used *sweet11;12* or *iTPS*, an increase in Tre6P was associated with a higher starch content toward the EN. This indicated that Tre6P might depress the rate of starch mobilization relative to the amount of starch available at dusk. We calculated, for each experiment and reatment, the maximum rate of starch mobilization (R_{max}) that could be supported by the dusk starch content under the provisos that breakdown is linear and that starch is not exhausted before dawn. R_{max} is equivalent to the slope of a linear regression through the measured starch content at ED and a hypothetical null starch content at EN (ZT24). We then estimated the extent to which starch mobilization was depressed relative to R_{max} ($R_r = R_a/R_{max}$) and compared R_r with the average night-time Tre6P level. As shown in Figure 6b, there was a strong and highly significant negative correlation between Tre6P and R_r ($R^2 = 0.72$, $p < 0.001$). This strong relationship was found even though the starch content at dusk and absolute rate of starch mobilization varied considerably between the various treatments. Importantly, a similar relationship was found for *sweet11;12* and *iTPS*, even though the former had elevated sucrose and the latter had lower sucrose than control plants. This consistent match across genotypes strongly indicates that the negative relationship in Figure 6b is driven by Tre6P.

3.5 | Impact of Tre6P on starch mobilization when the clock is desynchronized

The arithmetic division model (Scialdone et al., 2013) provides a simple conceptual framework to understand how starch breakdown is paced to dawn in a wide range of conditions including sudden

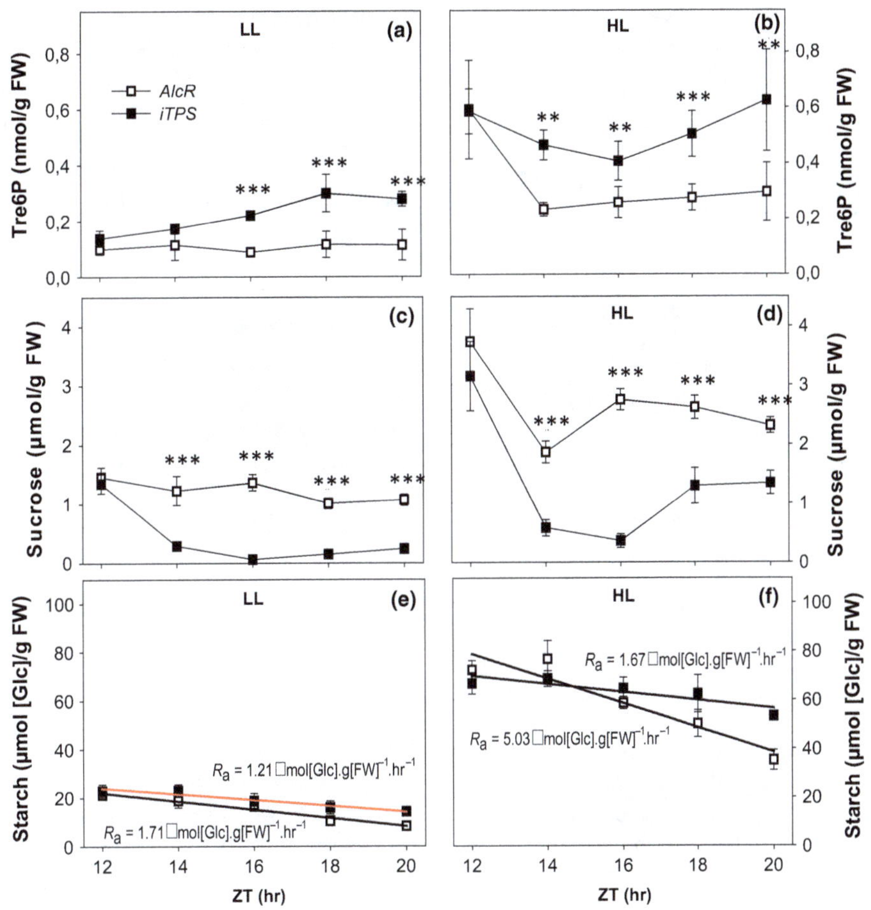

FIGURE 5 Effect of induced TPS overexpression on the temporal kinetics of Tre6P and sucrose levels and night-time starch mobilization after short-term perturbations to generate high or low starch content at dusk. This figure shows data from the same experiment as Figure 4. After one photoperiod in low light (LL) or high light (HL), empty-vector control *AlcR* and the inducible TPS-overexpressor line (*iTPS*) plants were induced 2 hr before dusk by spraying with 2% (v/v) ethanol. (a, b) Tre6P, (c, d) sucrose, and starch (d, e) in the LL treatment (a, c) or the HL treatment (b, d). Values are means ± *SD* (*n* = 5). In panels (e, f), lines represent the linear regressions of starch content from which the absolute rate of starch mobilization (R_a) was calculated (numbers shown in panel). Significant differences between *AlcR* and *iTPS*, using Student's *t* test, are indicated by asterisks: $*p < 0.05$,$**p < 0.01$, and $***p < 0.001$. ZT, zeitgeber time (hours after dawn)

perturbations (see Introduction). In this model, a clock-dependent mechanism measures time remaining until the next dawn ('*T*'). This information is integrated with a measure of the starch content ('*S*') to set the rate of starch mobilization ('*R*' = '*S*'/'*T*'). The term R_{max} in the calculations underlying Figure 6b is analogous to term '*R*' in the arithmetic division model. The negative correlation in Figure 6b is consistent with Tre6P depressing the rate of starch mobilization below that set by the clock.

We therefore asked whether the inhibition of starch mobilization by Tre6P depends on operation of the clock. To do this, we grew the *AlcR* and *iTPS* lines in continuous light and continuous temperature from germination onwards for 3 weeks and then turned the light off. Plants were induced with ethanol 2 hr prior to the onset of darkness. Rosettes were harvested at 4-hr intervals in the 24 hr prior to the onset of the darkness, and at 2-hr intervals for the first 12 hr after darkening.

We first checked whether the clock was desynchronized. To do this, we monitored transcript abundance of ten core clock genes (*LHY, CCA1, PRR9, PRR7, PRR5, GIGANTEA* (*GI*), *TIMING OF CAB EXPRESSION 1* (*TOC1*), *EARLY FLOWERING3* (*ELF3*), *ELF4*, and *LUX ARRHYTHMO* (*LUX*)) as well as two output genes (*PHYTOCHROME-INTERACTING FACTOR 4* (*PIF4*) and *PIF5*) in *AlcR* and *iTPS* lines in the last 24 hr of continuous light and the first 12 hr after darkening (Supporting Information Figure S5). The results are compared with oscillations of the same transcripts in 21-day-old wild-type Col-0 growing in a 12-hr photoperiod (Flis et al., 2016). The latter time series is plotted such that dusk coincides with the time at which the *AlcR* and *iTPS* plants were darkened. The strong oscillations of clock, *PIF4*, and *PIF5* transcripts in a light-dark cycle were abolished in continuous light, both in *AlcR* and in *iTPS*. In the first 12 hr after darkening, there was a small decrease of *PRR9*, *GI*, and *LUX* transcripts and a slow rise of *LHY*, *CCA1*, *PIF4*, and *PIF5* transcripts, but these changes were very small compared to the oscillations in plants grown in a light-dark cycle.

AlcR and *iTPS* had similar and nearly constant levels of starch, sucrose, and Tre6P in continuous light (Figure 7, original data

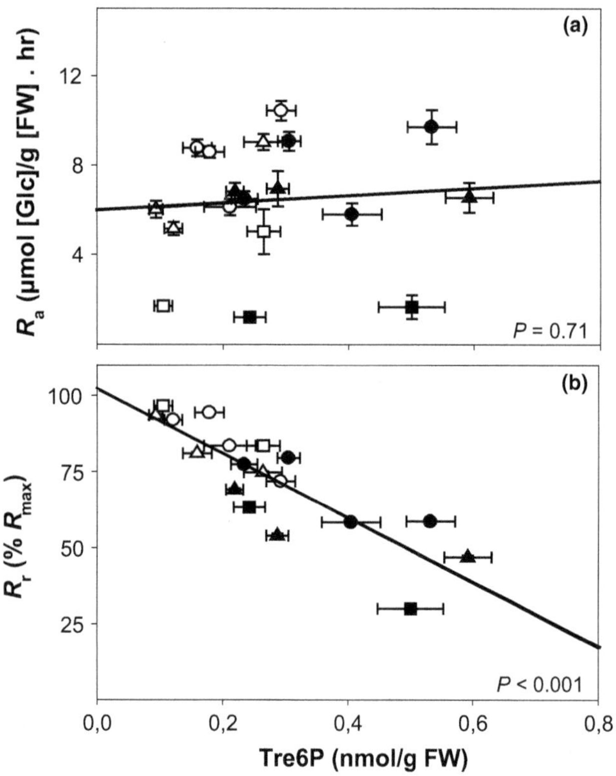

FIGURE 6 Correlation of average night-time Tre6P contents and starch mobilization in Arabidopsis rosettes. The plots show the relation between Tre6P levels and (a) the absolute starch mobilization rate (R_a), and (b) the relative starch mobilization rate (R_r) in wild-type Col-0, *sweet11;12* double mutant, *AlcR* empty-vector control and induced TPS-overexpressor (*iTPS*) line grown under standard conditions and subjected for the preceding light period to different light intensities or to different photoperiods (for details see legends of Figures 2 and 4). Values for Tre6P are the average values between ZT15 and ZT24 for experiments with wild-type Col-0 and *sweet11;12* mutant (see Figure 2b, Supporting Information Figure S3B) and between ZT14 and ZT20 for the experiment with *AlcR* and *iTPS* (Figure 4b). Values for R_a are derived from Figure 3 and Supporting Information Figure S4 for wild-type Col-0 and *sweet11;12*, and from Figure 6 for *AlcR* and *iTPS*. R_r is calculated as R_a/R_{max}, where R_{max} is the rate of starch mobilization that would exhaust starch at EN. R_r is also geometrically equivalent to the proportion of starch at ED that would remain at EN. The underlying calculations are provided in Supporting Information Table S4

provided in Supporting Information Table S5). It is noteworthy that the leaf starch content in continuous light (about 30 µmol [Glc] g^{-1} FW) was rather low compared to that found in wild-type, *AlcR*, or *iTPS* in a light-dark cycle (60–130 µmol [Glc] g^{-1} FW). After darkening, Tre6P declined in *AlcR* controls but remained high in induced *iTPS* plants (average levels in the first 12 hr after darkening were 0.09 ± 0.03 and 0.22 ± 0.07 µmol [Glc] g^{-1} FW, respectively, Figure 7a). Starch mobilization was faster in *AlcR* than *iTPS* (Figure 7b). At 8 hr after darkening, *AlcR* and *iTPS* had remobilized 77% and 33% of their starch, respectively. After this time, the rate of mobilization fell off in the *AlcR* plants. Estimation from the linear regression analysis in the linear phase during the first 8 hr in darkness (Supporting Information Figure S6) revealed a rate of starch mobilization of about 2.1 µmol [Glc] g^{-1} FW hr^{-1} in *AlcR* and 1.14 µmol [Glc] g^{-1} FW hr^{-1} in *iTPS*. We conclude that Tre6P can inhibit starch mobilization in the absence of a synchronized and entrained clock.

3.6 | Testing whether the clock overrides inhibition by Tre6P

As already mentioned, starch content typically decreases proportionally with the decrease in time to dawn, resulting in a linear depletion of starch and exhaustion of starch at a time around dawn. We reasoned that when Tre6P is inhibiting starch mobilization, the starch content at any given time would be higher than it would if there was no inhibition by Tre6P. Within the context of the molecular division model, this would represent a higher value for '*S*', and lead to a higher value for '*R*' (i.e., a higher rate of mobilization) that would partly compensate for the inhibition by Tre6P. We modeled this interaction, using a numerical Tre6P input that would lead to 0%, 30%, or 60% inhibition of the initial rate of starch mobilization (Supporting Information Figure S7). These values were chosen to span the inhibition seen in the experiments with *sweet11;12* (Figures 1–3) and the *iTPS* line (Figures 4 and 5) as well as in Figure 7. The predicted inhibition was slightly weaker when the action of Tre6P was modified by a downstream interaction with a clock output (Supporting Information Figure S7B) than in a scenario in which the rate of mobilization was independent of the clock and starch content (Supporting Information Figure S7A). However, the effect was too small to explain why Tre6P often has little or no effect on the absolute rate of starch mobilization (see Figures 1c, 3a–d, 4e, 5e).

4 | DISCUSSION

As outlined in the Introduction, in C-limiting conditions, the clock sets the rate of starch mobilization such that starch reserves are almost exhausted at dawn (Scialdone and Howard, 2015; Smith and Stitt 2007; Stitt & Zeeman, 2012), while in C-replete conditions, starch reserves are not fully exhausted at dawn. Experiments with *iTPS* lines in Martins et al. (2013) pointed to a role for the sucrose signal metabolite Tre6P in the feedback regulation of starch mobilization. Our results confirm and extend the idea that Tre6P regulates starch mobilization in C-replete conditions. First, impaired sucrose export in *sweet11;12* results in accumulation of sucrose, higher levels of Tre6P, and a restriction of starch mobilization. Second, the relation between Tre6P and starch mobilization in *sweet11;12* is largely recapitulated in *iTPS* lines, even though sucrose and other C-metabolites rose in the former and decreased in the latter treatment. This points to a dominant role for Tre6P in the feedback regulation of starch mobilization. Third, Tre6P restricts mobilization under a wide range of conditions. However, maybe surprisingly, there is no clear relationship between the level of Tre6P and the absolute rate of starch mobilization. Rather, Tre6P depresses the rate of mobilization below that required to exhaust starch at

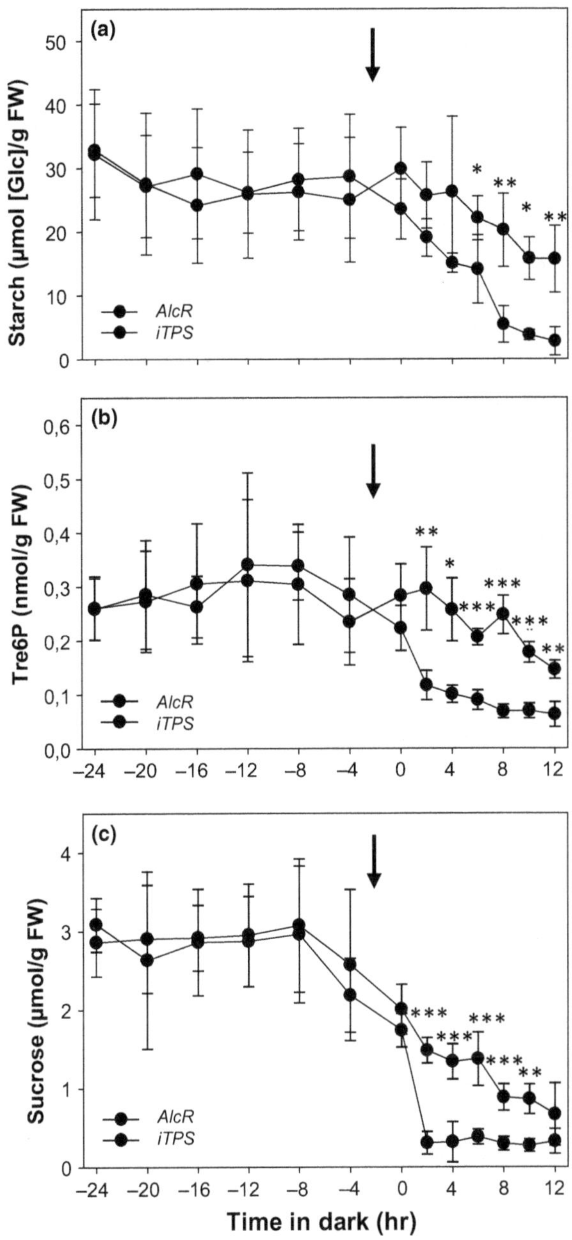

FIGURE 7 Effect of induced TPS overexpression on starch mobilization after darkening plants that were previously germinated and grown in continuous light. The empty-vector control *AlcR* and the inducible TPS-overexpressor line (*iTPS*) were grown from germination in continuous light and temperature (12-hr/12-hr light/dark cycle, 160 µmol m^{-2} s^{-1} PFD, 21°C). Three weeks after germination, plants were induced by spraying with 2% (v/v) ethanol (indicated by black arrow) and then darkened 2 hr later. Samples were harvested at 4-hr intervals during the 24 hr prior to darkening and the first 12 hrs after darkening for measurements of: (a) Tre6P, (b) starch, and (c) sucrose. Values are means ± *SD* (n = 4). Significant differences between *AlcR* and *iTPS*, using Student's t test, are indicated by asterisks: *$p < 0.05$, **$p < 0.01$, and ***$p < 0.001$. Expanded plots for starch levels in the first 8 hr after darkening and regressions to estimate the rate of starch mobilization are provided in Supporting Information Figure S6. Time series for clock gene transcripts from the same experiment are shown in Supporting Information Figure S5

dawn. As will be discussed, this finding implies that Tre6P interacts closely with the clock to set the rate of starch mobilization.

4.1 | Elevated Tre6P consistently leads to incomplete exhaustion of starch at dawn but has no consistent impact on the absolute rate of starch mobilization

Our experiments employed two complementary strategies to elevate Tre6P; *sweet11;12* mutants with constitutively higher levels of sucrose and reducing sugars, or transient induction of a bacterial *TPS*, in which case Tre6P levels increased while sucrose and reducing sugars levels fell (see also Martins et al., 2013). We investigated starch mobilization after short-term changes in the light regime to generate high or low starch levels at dusk. Irrespective of the time frame over which Tre6P levels were elevated and whether the elevation was associated with higher or lower levels of sugars, elevated Tre6P always led to a restriction of starch breakdown, in particular, incomplete mobilization of starch leaving a starch excess at dawn. Crucially, there was a robust negative relationship between the Tre6P level and R_r, the rate of starch mobilization expressed relative to the rate required to exhaust starch at dawn (Figure 6b).

In contrast, there was no consistent relation between Tre6P levels and R_a, the absolute rate of starch mobilization (Figure 6a). In experiments with the *sweet11;12* mutant, elevated Tre6P led to either a minor decrease or even a minor increase in the absolute rate of starch mobilization (Figures 2–4, Supporting Information Figure S3). In experiments with *iTPS* lines, elevated Tre6P did lead to a decrease in the rate of starch mobilization, but the inhibition was small when they entered the night with low levels of starch (Figure 4). Overall, elevated Tre6P decreased R_a when plants with elevated Tre6P and control plants entered the night with similar and high levels of starch, but it had only a small effect when they entered the night with similar but low levels of starch, and little or no effect when the plants with elevated Tre6P entered the night with higher levels of starch than control plants (see below for more discussion).

We conclude that Tre6P plays a major role in the feedback regulation of starch mobilization. In particular, elevated Tre6P prevents complete mobilization of starch. The similar relationship between Tre6P levels and R_r across genotypes underlines the importance of Tre6P in the feedback regulation of starch mobilization. Importantly, elevated Tre6P restricts starch mobilization, irrespective of whether sugars are rising or falling, implying that other C-signals play only a small role in the feedback regulation of starch mobilization.

4.2 | Masked relationship between the amount of starch and the rate of starch mobilization

Many earlier studies showed that the rate of starch mobilization in Arabidopsis is fairly constant during the night (i.e., starch content decreases in a linear manner; Flis et al., 2016; Gibon et al., 2004;

Graf et al., 2010; Mengin et al., 2017; Scialdone et al., 2013; Sulpice et al., 2014). This was also seen in our study. On the face of it, these observations imply that the rate of starch mobilization is independent of starch content until the latter falls to rather low values.

However, several aspects of our data indicate that a higher starch content does favor a higher absolute rate of starch mobilization. First, in wild-type plants, higher dusk starch levels typically led to a higher absolute rate of starch mobilization (see e.g., Figures 2c, e and 4c, e, see also Mengin et al., 2017; Scialdone et al., 2013; Pilkington et al., 2015). Second, as already mentioned, the only conditions in which elevated Tre6P led to a large decrease in the absolute rate of starch mobilization (R_a) were when starch levels at the start of the night were high and similar to those in the control (see Figures 6b and 7, Supporting Information Figure S6; also Martins et al., 2013). Third, incomplete mobilization of starch due to feedback inhibition by Tre6P in one night will leave a residue of non-mobilized starch at dawn (see Figure 4a, B) and, if the rate of starch accumulation during the next day remains unaltered, a higher starch content at the following dusk. This cycle will be repeated until a steady state is reached with a similar absolute rate of starch mobilization (R_a) in the Tre6P-inhibited case and the control. This new balance can be explained if the inhibition by Tre6P is offset by a higher basal rate of starch mobilization due to the higher starch content. This pattern is seen in the *sweet11;12* mutant in growth conditions (Figure 1d; Supporting Information Figure S4A). It was also seen after overexpressing an activated form of ADPGlc pyrophosphorylase to increase starch accumulation in the light (Hädrich et al., 2012). This response has been modeled more generally for scenarios in which starch mobilization is restricted by factors like decreased starch mobilization capacity or defects in circadian regulation (Seaton et al., 2015).

At first sight, dependence of the rate of mobilization on starch content ('*S*') appears to be incompatible with the observed linear decline in starch content during the night (see above). However, this inconsistency can be resolved if a further factor promotes starch mobilization as the 24-hr cycle progresses, that is, as the remaining time to dawn decreases. One simple and plausible factor is the parameter '*T*' as defined in the arithmetic division model (Scialdone et al., 2013).

4.3 | Interaction between Tre6P and circadian regulation of starch mobilization

In terms of the arithmetic division model, there are several possible explanations for the negative action of Tre6P on starch mobilization: (a) *Tre6P and the clock act independently*. This seems unlikely because the weak and inconsistent impact of Tre6P on the absolute rate of starch mobilization could not be mimicked in a simple model investigating how an independent inhibition by elevated Tre6P would impact on the rate of starch mobilization predicted by the arithmetic division model (Supporting Information Figure S6). This hypothesis also does not explain why elevated Tre6P in iTPS lines led to only a weak and non-significant decrease in the absolute rate of starch degradation when plants entered the night with low and similar levels of starch to those in control *AlcR* lines (Figures 4c and 5e) (b) *The clock overrides the inhibition by Tre6P*. This seems unlikely, because elevated Tre6P consistently led to an inhibition of starch mobilization (or at least incomplete exhaustion of starch) even under conditions in which dusk starch content was low and mobilization was already slow in the control and, by implication, strongly restricted by the clock. (c) *Tre6P overrides the clock*, that is, modifies '*R*'. This also seems unlikely, because the largest inhibition of the absolute rate was seen in a treatment (Figure 6b) in which dusk starch content was high in the control plants, and starch was only partly exhausted at dawn implying that regulation by the clock was rather relaxed. Further, Tre6P inhibited starch mobilization when plants were darkened after 21 days in continuous light, when their transcriptional clock was desynchronized and non-entrained (Figure 8). (d) *Tre6P interacts with the clock output 'T'*. This explanation appears unlikely, because Tre6P inhibited starch mobilization in plants with a desynchronized clock. However, it cannot be totally excluded that Tre6P acts on a default value for '*T*'. An interaction with a clock output is also a possible explanation in terms of models in which metabolic feedback alters clock period and delays the time at which dawn is anticipated (Seki et al., 2017). However, a short-term increase in Tre6P at the beginning of the night slows down starch mobilization early in the same night (Figure 6a, b, see also Martins et al., 2013). This rather fast response cannot be easily understood in terms of changes that occur due to C depletion at the following dawn.

The most plausible hypothesis may be that Tre6P interacts with the measure of starch content, '*S*'. This provides a simple explanation for the consistent relationship between the level of Tre6P and the proportion of starch that is predicted to be mobilized during the night (R_r, Figure 6b). It also provides a simple explanation why elevated Tre6P only depresses the absolute rate of starch mobilization (R_a) when dusk starch levels are the same in the control and treatments with elevated Tre6P. Furthermore, it is consistent with the finding that Tre6P inhibits starch mobilization in plants with a desynchronized and non-entrained clock, without requiring additional hypotheses that there is a default measure of time to dawn ('*T*') in such conditions. However, it should be noted that within the context of the arithmetic division model, modification of '*T*', and modification of '*S*' will be mathematically similar with respect to the effect on '*R*'.

Starch is located in the plastid, but TPS1 lacks a plastid transit sequence and non-aqueous fractionation indicates that much or even all the Tre6P in the Arabidopsis rosette is located outside the chloroplasts (Martins et al., 2013). The observed inhibition of starch mobilization by Tre6P implies either that some Tre6P does enter the plastid or that Tre6P acts on an extra-plastidic signaling network that generates signals that enter the plastid to regulate starch mobilization. The reported presence of two Tre6P phosphatase family members in the chloroplast (TPPD and TPPE; Krasensky, Broyart, Rabanal, & Jonak, 2014) means the former scenario deserves further attention, even though there is to date no evidence that Tre6P directly inhibits enzymes involved in starch mobilization (Martins et al., 2013). The core clock operates via transcriptional regulation of

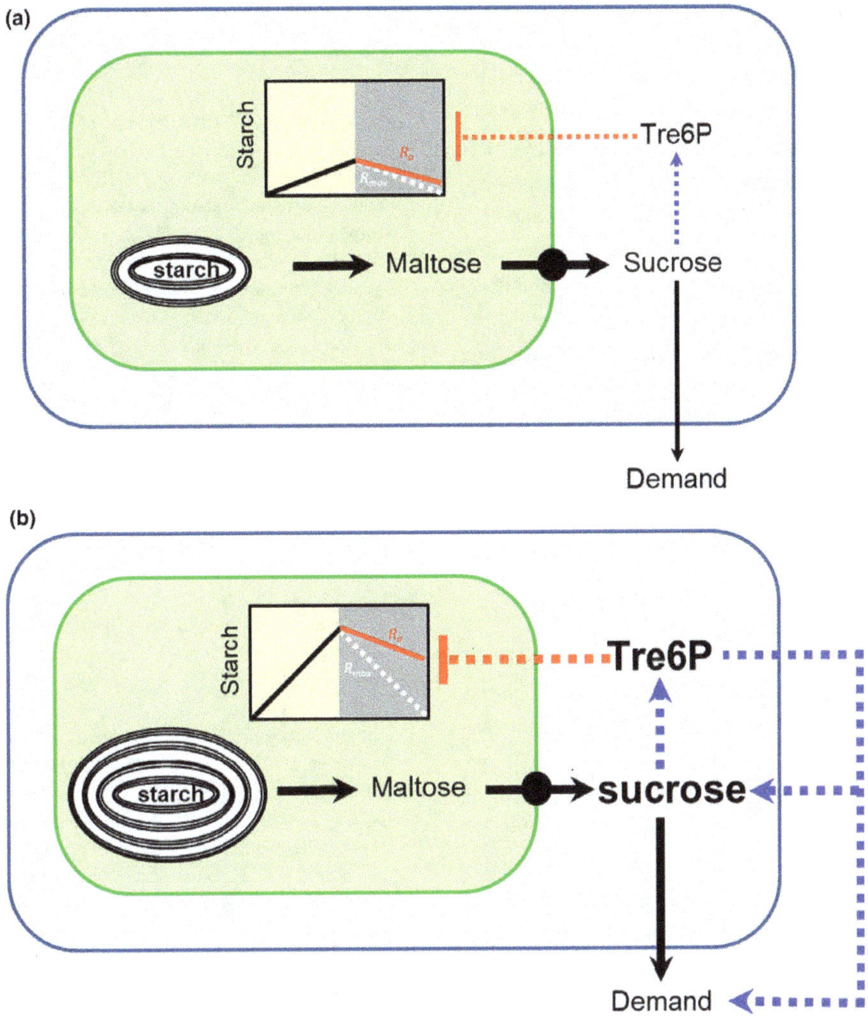

FIGURE 8 Schematic model to illustrate the inhibitory effect of Tre6P on night-time starch mobilization in C-replete and C-limiting scenarios. In conditions where a low amount of starch is available at dusk, such as in short photoperiod or low irradiance, the clock enforces a low rate of starch mobilization during the night, leading to low levels of sucrose and Tre6P. Tre6P only slightly depresses the relative rate of starch mobilization (R_r) below that permitted by the clock (R_{max}), and this small depression is further attenuated by the compensatory increase modeled in Supporting Information Figure S7. In conditions where ample starch is available at dusk and starch mobilization supplies more C than can be consumed for metabolism, maintenance and growth, sucrose will increase. The resulting increase in Tre6P will restrict starch mobilization leading to incomplete utilization of starch. High Tre6P (and other C-signals derived from high sucrose and other sugars) will also act to increase the rate of C utilization. The extent to which starch mobilization is slowed down will depend on the balance between negative feedback of starch mobilization by Tre6P and feedforward signaling by Tre6P and other C-signals to increase C utilization

gene expression in the nucleus; thus, involvement of the clock in the regulation of starch mobilization implies the movement of a signal from the nucleus or cytosol into the plastid. Changes in Tre6P might be integrated with the clock-derived signal in either the cytosol or the plastid. The biochemical mechanism that regulates starch mobilization still needs to be elucidated but appears to involve changes in the cycle of starch phosphorylation and dephosphorylation at the surface of the starch granule, irrespective of whether mobilization is responding to the clock (Scialdone et al., 2013) or Tre6P (Martins et al., 2013).

The interaction between the clock and Tre6P in regulating starch mobilization is summarized in Figure 8. In conditions where dusk starch levels are low and the rate of starch mobilization is restricted by the clock, sucrose and Tre6P levels are low at night. This occurs in plants growing in short photoperiods or low irradiance (Mengin et al., 2017; Sulpice et al., 2014) or after a sudden early dusk (Martins et al., 2013). Tre6P will only slightly depress the rate of starch mobilization below that permitted by the clock, and this small depression will be attenuated by the compensation modeled in Supporting Information Figure S7. In conditions where ample starch is available at dusk, starch is rapidly mobilized leading to high levels of sugars and thus Tre6P, which restricts mobilization below the rate that would be permitted by the clock. This occurs in wild-type plants after treatments to increase the starch content at dusk (see Figures 2a–c, f and 4a–c, f; see also Mengin et al., 2017; Pilkington et al., 2015; Sulpice et al., 2014) and when induced *iTPS* lines enter the night with high starch content (Figure 6). In this interaction, circadian regulation acts to maximize starch mobilization in each 24 hr cycle, while Tre6P restricts starch mobilization when it is supplying more C than can be consumed for metabolism. Importantly, Tre6P not only restricts starch mobilization and sucrose production in leaves. Tre6P also regulates sucrose consumption, including in growing sink organs (Figueroa and Lunn, 2016). For example, Tre6P stimulates utilization of C for amino acid synthesis (Figueroa et al., 2017). Tre6P also stimulates C utilization in the long term by promoting shoot branching and flowering (Fichtner et al., 2017; Wahl et al., 2013). The relative strengths of the inhibitory action of Tre6P on starch mobilization and the positive action of Tre6P and other C-signals on C utilization will determine to what extent allocation of C to growth is maximized and accumulation of foliar starch is minimized during the day in C-replete conditions. This balance may depend on the physiological state of the plant, the environment and, likely, the genetic background.

ACKNOWLEDGMENTS

L.d.A. was recipient of a scholarship from Capes (Coordenação de Aperfeiçoamento de Nível Superior, Brazil). Research was supported by the Max Planck Society and European Union (FP7 collaborative project TiMet, contract no. 245143). The authors thank Dr. Karin Köhl, Christin Abel, and the greenhouse staff for advice and help with plant cultivation, Dr. Danilo M. Daloso, and Dr. Carlos M. Figueroa for discussions and help with harvesting, Dr. Maria Grazia Annunziata, Dr. Armin Schlereth, Ursula Krause, and Melanie Höhne for help with metabolite extraction and measurement. T.A.M. was supported by the AM was supported by the Brazilian National Council for Scientific and Technological Development (CNPq).

AUTHOR CONTRIBUTIONS

LdA: Performed experiments and measurements, data analysis, contributed to writing the manuscript; PKP: Performed experiments and measurements, data analysis; TAM: Data analysis; RF: Performed measurements; JEL: Designed experiments, contributed to writing the manuscript; MS: Designed experiments, contributed to writing the manuscript.

ORCID

Letícia dos Anjos http://orcid.org/0000-0002-9313-7262
Prashant Kumar Pandey http://orcid.org/0000-0002-5647-6601
Thiago Alexandre Moraes http://orcid.org/0000-0002-7156-6947
Regina Feil http://orcid.org/0000-0002-9936-1337
John E. Lunn http://orcid.org/0000-0001-8533-3004
Mark Stitt http://orcid.org/0000-0002-4900-1763

REFERENCES

Annunziata, M. G., Apelt, F., Carillo, P., Krause, U., Feil, R., Mengin, V., ... Raines, C. (2017). Getting back to nature: A reality check for experiments in controlled environments. *Journal of Experimental Botany*, *68*, 4463–4477. https://doi.org/10.1093/jxb/erx220

Ayre, B. G. (2011). Membrane-transport systems for sucrose in relation to whole-plant carbon partitioning. *Molecular Plant*, *4*, 377–394. https://doi.org/10.1093/mp/ssr014

Baerenfaller, K., Massonnet, C., Hennig, L., Russenberger, D., Sulpice, R., Walsh, S., ... Gruissem, W. (2015). A long photoperiod relaxes energy management in Arabidopsis leaf six. *Current Plant Biology*, *2*, 34–45. https://doi.org/10.1016/j.cpb.2015.07.001

Bolouri Moghaddam, M. R., & Van den Ende, W. (2012). Sugars and plant innate immunity. *Journal of Experimental Botany*, *63*, 3989–3998. https://doi.org/10.1093/jxb/ers129

Chatterton, N. J., & Silvius, J. E. (1979). Photosynthate partitioning into starch in soybean leaves: I. Effects of photoperiod versus photosynthetic period duration. *Plant Physiology*, *64*, 749–753.

Chatterton, N. J., & Silvius, J. E. (1980). Acclimation of photosynthate partitioning and photosynthetic rates to changes in length of the daily photosynthetic period. *Annals of Botany*, *46*, 739–745.

Chatterton, N. J., & Silvius, J. E. (1981). Photosynthate partitioning into starch in soybean leaves. 2. Irradiance level and daily photosynthetic period duration effects. *Plant Physiology*, *67*, 257–260.

Chen, L. Q., Qu, X. Q., Hou, B. H., Sosso, D., Osorio, S., Fernie, A. R., & Frommer, W. B. (2012). Sucrose efflux mediated by SWEET proteins as a key step for phloem transport. *Science*, *335*, 207–211.

Cheng, S.-H., Moore, B., & Seemann, J. R. (1998). Effects of short- and long-term elevated CO_2 on the expression of ribulose-1,5-bisphosphate carboxylase/oxygenase genes and carbohydrate accumulation in leaves of *Arabidopsis thaliana* (L.) Heynh. 1. *Plant Physiology*, *116*, 715–723. https://doi.org/10.1104/pp.116.2.715

Cross, J. M., von Korff, M., Altmann, T., Bartzetko, L., Sulpice, R., Gibon, Y., ... Stitt, M. (2006). Variation of enzyme activities and metabolite levels in 24 Arabidopsis accessions growing in carbon-limited conditions. *Plant Physiology*, *142*, 1574–1588. https://doi.org/10.1104/pp.106.086629

Czedik-Eysenberg, A., Arrivault, S., Lohse, M. A., Feil, R., Krohn, N., Encke, B., ... Stitt, M. (2016). The interplay between carbon availability and growth in different zones of the growing maize leaf. *Plant Physiology*, *172*, 943–967.

Dodd, A. N., Dalchau, N., Gardner, M. J., Baek, S.-J., & Webb, A. A. R. (2014). The circadian clock has transient plasticity of period and is required for timing of nocturnal processes in Arabidopsis. *New Phytology*, *201*, 168–179.

Feike, D., Sueng, D., Graf, A., Bischof, S., Ellick, T., Coiro, M., ... Smith, A. M. (2016). The starch granule-associated protein EARLY STARVATION1 is required for the control of starch degradation in *Arabidopsis thaliana* leaves. *Plant Cell*, *28*, 1472–1489.

Fernandez, O., Ishihara, H., George, G. M., Mengin, V., Flis, A., Sumner, D., ... Stitt, M. (2017). Foliar starch turnover occurs in long days and in falling light at the end of the day. *Plant Physiology*, *174*, 2199–2212.

Figueroa, C. M., Feil, R., Ishihara, H., Watanabe, M., Kölling, K., Krause, U., ... Lunn, J. E. (2016). Trehalose 6-phosphate coordinates organic and amino acid metabolism with carbon availability. *Plant Journal*, *85*, 410–423. https://doi.org/10.1111/tpj.13114

Figueroa, C. M., & Lunn, J. E. (2016). A tale of two sugars: Trehalose 6-phosphate and sucrose. *Plant Physiology*, *172*, 7–27.

Flis, A., Fernández, A. P., Zielinski, T., Mengin, V., Sulpice, R., Hume, A., ... Millar, A. J. (2015). Robust regulation of the clock gene circuit identified from comprehensive timeseries data in a shared data infrastructure. *Open Biology*, *5*, 150042. https://doi.org/10.1098/rsob.150042

Flis, A., Sulpice, R., Seaton, D. D., Ivakov, A. A., Liput, M., Abel, C., ... Stitt, M. (2016). Photoperiod-dependent changes in the phase of core clock transcripts and global transcriptional outputs at dawn and dusk in Arabidopsis. *Plant, Cell and Environment*, *39*, 1955–1981. https://doi.org/10.1111/pce.12754

Geiger, D. R., & Servaites, J. C. (1994). Diurnal regulation of photosynthetic carbon metabolism in plants. *Annual Review of Plant Biology, 45*, 235–256.

Gibon, Y., Pyl, M.-T., Sulpice, R., Höhne, M., & Stitt, M. (2009). Adjustment of growth, starch turnover, protein content and central metabolism to a decrease of the carbon supply when Arabidopsis is grown in very short photoperiods. *Plant Cell and Environment, 32*, 859–874.

Gibon, Y., Bläsing, O. E., Palacios-Rojas, N., Pankovic, D., Hendriks, J. H., Fisahn, J., ... Stitt, M. (2004). Adjustment of diurnal starch turnover to short days: Depletion of sugar during the night leads to a temporary inhibition of carbohydrate utilization, accumulation of sugars and post-translational activation of ADP-glucose pyrophosphorylase in the following light period. *Plant Journal, 39*, 847–862.

Graf, A., Schlereth, A., Stitt, M., & Smith, A. M. (2010). Circadian control of carbohydrate availability for growth in Arabidopsis plants at night. *Proceedings of the National Academy of Sciences, 107*, 9458–9463. https://doi.org/10.1073/pnas.0914299107

Grimmer, C., Bachfischer, T., & Komor, E. (1999). Carbohydrate partitioning into starch in leaves of *Ricinus communis* L. grown under elevated CO_2 is controlled by sucrose. *Plant, Cell and Environment, 22*, 1275–1280. https://doi.org/10.1046/j.1365-3040.1999.00481.x

Hädrich, N., Hendriks, J. H. M., Kötting, O., Arrivault, S., Feil, R., Zeeman, S. C., ... Lunn, J. E. (2012). Mutagenesis of cysteine 81 prevents dimerization of the APS1 subunit of ADP-glucose pyrophosphorylase and alters diurnal starch turnover in *Arabidopsis thaliana* leaves. *Plant Journal, 70*, 231–242. https://doi.org/10.1111/j.1365-313X.2011.04860.x

Herbers, K., Takahata, Y., Melzer, M., Mock, H.-P., Hajirezaei, M., & Sonnewald, U. (2000). Regulation of carbohydrate partitioning during the interaction of potato virus Y with tobacco. *Molecular Plant Pathology, 1*, 51–59. https://doi.org/10.1046/j.1364-3703.2000.00007.x

Ishihara, H., Obata, T., Sulpice, R., Fernie, A. R., & Stitt, M. (2015). Quantitative determination of the rates of protein synthesis and turnover in Arabidopsis rosettes using dynamic $^{13}CO_2$ labelling and analysis of isotope enrichment of individual amino acids in free pools and in protein. *Plant Physiology, 168*, 74–93.

Körner, C. (2006). Plant CO_2 responses: An issue of definition, time and resource supply. *New Phytologist, 172*, 393–411. https://doi.org/10.1111/j.1469-8137.2006.01886.x

Krasensky, J., Broyart, C., Rabanal, F. A., & Jonak, C. (2014). The redox-sensitive chloroplast trehalose-6-phosphate phosphatase AtTPPD regulates salt stress tolerance. *Antioxidants & Redox Signaling, 21*, 1289–1304. https://doi.org/10.1089/ars.2013.5693

Lawlor, D. W., & Mitchell, R. A. C. (1991). The effects of increasing CO_2 on crop photosynthesis and productivity: A review of field studies. *Plant, Cell and Environment, 14*, 807–818. https://doi.org/10.1111/j.1365-3040.1991.tb01444.x

Lemoine, R., La Camera, S., Atanassova, R., Dédaldéchamp, F., Allario, T., Pourtau, N., ... Durand, M. (2013). Source-to-sink transport of sugar and regulation by environmental factors. *Frontiers in Plant Science, 4*, 272.

Long, S. P., Ainsworth, E. A., Leakey, A. D. B., Nösberger, J., & Ort, D. R. (2006). Food for thought: Lower-than-expected crop yield stimulation with rising CO_2 concentrations. *Science, 312*, 1918–1921. https://doi.org/10.1126/science.1114722

Lu, Y., & Sharkey, T. D. (2006). The importance of maltose in transitory starch breakdown. *Plant, Cell and Environment, 29*, 353–366. https://doi.org/10.1111/j.1365-3040.2005.01480.x

Lunn, J. E., Delorge, I., Figueroa, C. M., Van Dijck, P., & Stitt, M. (2014). Trehalose metabolism in plants. *Plant Journal, 79*, 544–567. https://doi.org/10.1111/tpj.12509

Lunn, J. E., Feil, R., Hendriks, J. H., Gibon, Y., Morcuende, R., Osuna, D., ... Stitt, M. (2006). Sugar-induced increases in trehalose 6-phosphate are correlated with redox activation of ADPglucose pyrophosphorylase and higher rates of starch synthesis in *Arabidopsis thaliana*. *Biochemical Journal, 397*, 139–148. https://doi.org/10.1042/BJ20060083

Martins, M. C. M., Hejazi, M., Fettke, J., Steup, M., Feil, R., Krause, U., ... Lunn, J. E. (2013). Feedback inhibition of starch degradation in Arabidopsis leaves mediated by trehalose 6-phosphate. *Plant Physiology, 163*, 1142–1163. https://doi.org/10.1104/pp.113.226787

Matt, P., Geiger, M., Walch-Liu, P., Engels, C., Krapp, A., & Stitt, M. (2001). Elevated carbon dioxide increases nitrate uptake and nitrate reductase activity when tobacco is growing on nitrate, but increases ammonium uptake and inhibits nitrate reductase activity when tobacco is growing on ammonium nitrate. *Plant, Cell and Environment, 24*, 1119–1137. https://doi.org/10.1046/j.1365-3040.2001.00771.x

Mengin, V., Pyl, E.-T., Moraes, T. A., Sulpice, R., Krohn, N., Encke, B., & Stitt, M. (2017). Photosynthate partitioning to starch in *Arabidopsis thaliana* is insensitive to light intensity but sensitive to photoperiod due to a restriction on growth in the light in short photoperiods. *Plant, Cell and Environment, 40*, 2608–2627. https://doi.org/10.1111/pce.13000

Mullen, J. A., & Koller, H. R. (1988). Daytime and nighttime carbon balance and assimilate export in soybean leaves at different photon flux densities. *Plant Physiology, 86*, 880–884. https://doi.org/10.1104/pp.86.3.880

Niittylä, T., Messerli, G., Trevisan, M., Chen, J., Smith, A. M., & Zeeman, S. C. (2004). A previously unknown maltose transporter essential for starch degradation in leaves. *Science, 303*, 87–89. https://doi.org/10.1126/science.1091811

Pal, S. K., Liput, M., Piques, M., Ishihara, H., Obata, T., Martins, M. C. M., ... Stitt, M. (2013). Diurnal changes of polysome loading track sucrose content in the rosette of wild-type arabidopsis and the starchless pgm mutant. *Plant Physiology, 162*, 1246–1265. https://doi.org/10.1104/pp.112.212258

Paul, M. J., & Foyer, C. H. (2001). Sink regulation of photosynthesis. *Journal of Experimental Botany, 52*, 1383–1400. https://doi.org/10.1093/jexbot/52.360.1383

Pilkington, S. M., Encke, B., Krohn, N., Höhne, M., Stitt, M., & Pyl, E. T. (2015). Relationship between starch degradation and carbon demand for maintenance and growth in *Arabidopsis thaliana* in different irradiance and temperature regimes. *Plant, Cell and Environment, 38*, 157–171.

Piques, M., Schulze, W., Gibon, Y., Rohwer, J., & Stitt, M. (2009). Quantitative analysis of ribosome recruitment and transcript occupancy by polysomes: The dynamics of protein turnover in central metabolism in Arabidopsis Rosettes. *Molecular Systems Biology, 5*, 314.

Pokhilko, A., Flis, A., Sulpice, R., Stitt, M., & Ebenhöh, O. (2014). Adjustment of carbon fluxes to light conditions regulates the daily turnover of starch in plants: A computational model. *Molecular BioSystems, 10*, 613–662.

Pyl, E.-T., Piques, M., Ivakov, A., Schulze, W. X., Stitt, M., & Sulpice, R. (2012). Metabolism and growth in Arabidopsis depend on the daytime temperature but are temperature-compensated against cool nights. *Plant Cell, 24*, 2443–2469.

Riesmeier, J. W., Hirner, B., & Frommer, W. B. (1993). Potato sucrose transporter expression in minor veins indicates a role in phloem loading. *Plant Cell, 5*, 1591–1598. https://doi.org/10.1105/tpc.5.11.1591

Scialdone, A., & Howard, M. (2015). How plants manage food reserves at night: Quantitative models and open questions. *Frontiers in Plant Science, 6*, 204.

Scialdone, A., Mugford, S. T., Feike, D., Skeffngton, A., Borrill, P., Graf, A., ... Howard, M. (2013). Arabidopsis plants perform arithmetic division to prevent starvation at night. *Elife, 2*, e00669. https://doi.org/10.7554/eLife. 00669

Seaton, D. D., Smith, R. W., Song, Y. H., MacGregor, D. R., Stewart, K., Steel, G., ... Halliday, K. J. (2015). Linked circadian outputs control elongation growth and flowering in response to photoperiod and temperature. *Molecular Systems Biology, 11*, 776. https://doi.org/10.15252/msb.20145766

Seki, M., Ohara, T., Hearn, T. J., Frank, A., da Silva, V. C. H., Caldana, C., ... Satake, A. (2017). Adjustment of the Arabidopsis circadian oscillator by sugar signalling dictates the regulation of starch metabolism. *Scientific Reports*, *7*, 8305. https://doi.org/10.1038/s41598-017-08325-y

Shin, J., Sánchez-Villarreal, A., Davis, A. M., Du, S., Berendzen, K. W., Koncz, C., … Davis, S. J. (2017). The metabolic sensor AKIN10 modulates the *Arabidopsis* circadian clock in a light-dependent manner. *Plant, Cell and Environment*, *40*, 997–1008. https://doi.org/10.1111/pce.12903

Silvius, J. E., Chatterton, N. J., & Kremer, D. F. (1979). Photosynthate partitioning in soybean leaves at two irradiance levels: Comparative responses of acclimated and unacclimated leaves. *Plant Physiology*, *64*, 749–753.

Smith, A. M., & Stitt, M. (2007). Coordination of carbon supply and plant growth. *Plant, Cell and Environment*, *30*, 1126–1149. https://doi.org/10.1111/j.1365-3040.2007.01708.x

Stitt, M., & Zeeman, S. C. (2012). Starch turnover: Pathways, regulation and role in growth. *Current Opinion in Plant Biology*, *15*, 282–292. https://doi.org/10.1016/j.pbi.2012.03.016

Sulpice, R., Flis, A., Ivakov, A. A., Apelt, F., Krohn, N., Encke, B., … Stitt, M. (2014). Arabidopsis coordinates the diurnal regulation of carbon allocation and growth across a wide range of photoperiods. *Molecular Plant*, *7*, 137–155.

Yadav, U., Ivakov, A., Feil, R., Duan, G. Y., Walther, D., Giavaliaco, P., … Lunn, J. E. (2014). The sucrose-trehalose-6-phosphate (Tre6P) nexus: Specificity and mechanisms of sucrose signalling by Tre6P. *Journal of Experimental Botany*, *65*, 1051–1068.

2

Regulation of cyclic electron flow by chloroplast NADPH-dependent thioredoxin system

Lauri Nikkanen | Jouni Toivola | Andrea Trotta | Manuel Guinea Diaz | Mikko Tikkanen | Eva-Mari Aro | Eevi Rintamäki*

Molecular Plant Biology, Department of Biochemistry, University of Turku, Turku, Finland

Correspondence
Eevi Rintamäki, Molecular Plant Biology, Department of Biochemistry, University of Turku, Turku, Finland.
Email: evirin@utu.fi

Funding information
This work was funded by the Academy of Finland grants 276392 (to E.R.) and 307335 (the Center of Excellence in Molecular Biology of Primary Producers to E-M.A.) and by the Doctoral Program in Molecular Life Sciences in the University of Turku Graduate School (to L.N.).

Abstract

Linear electron transport in the thylakoid membrane drives photosynthetic NADPH and ATP production, while cyclic electron flow (CEF) around photosystem I only promotes the translocation of protons from stroma to thylakoid lumen. The chloroplast NADH dehydrogenase-like complex (NDH) participates in one CEF route transferring electrons from ferredoxin back to the plastoquinone pool with concomitant proton pumping to the lumen. CEF has been proposed to balance the ratio of ATP/NADPH production and to control the redox poise particularly in fluctuating light conditions, but the mechanisms regulating the NDH complex remain unknown. We have investigated potential regulation of the CEF pathways by the chloroplast NADPH-thioredoxin reductase (NTRC) in vivo by using an Arabidopsis knockout line of *NTRC* as well as lines overexpressing NTRC. Here, we present biochemical and biophysical evidence showing that NTRC stimulates the activity of NDH-dependent CEF and is involved in the regulation of generation of proton motive force, thylakoid conductivity to protons, and redox balance between the thylakoid electron transfer chain and the stroma during changes in light conditions. Furthermore, protein–protein interaction assays suggest a putative thioredoxin-target site in close proximity to the ferredoxin-binding domain of NDH, thus providing a plausible mechanism for redox regulation of the NDH ferredoxin:plastoquinone oxidoreductase activity.

KEYWORDS
chloroplast, cyclic electron transfer, fluctuating light, NDH, NTRC, photosynthesis, thioredoxin

1 | INTRODUCTION

In their natural habitats, plants face constant fluctuation of light intensity, including both seasonal changes in photoperiod and daily fluctuations according to environmental conditions. Optimization of photosynthesis in plant leaves requires strict balancing between conversion of light energy to chemical energy in photosynthetic light reactions and the energy-consuming reactions of chloroplast metabolism. Multiple regulatory and photoprotective mechanisms have evolved in photosynthetic organisms to cope with fluctuating light conditions and to prevent the photodamage of both Photosystem (PS) II and PSI (Tikkanen & Aro, 2014; Tikkanen et al., 2012;

*The author responsible for distribution of materials integral to the findings presented in this article in accordance with the policy described in the Instructions for Authors (www.plantphysiol.org) is: Eevi Rintamäki (evirin@utu.fi).

Tiwari et al., 2016; Townsend, Ware, & Ruban, 2017). Regularly occurring light variations induce long-term acclimatory changes in the photosynthetic machinery via signaling mechanisms, while temporary fluctuation of light within a day transiently activates short-term regulatory mechanisms (Armbruster et al., 2014; Bailey, Walters, Jansson, & Horton, 2001; Grieco, Tikkanen, Paakkarinen, Kangasjärvi, & Aro, 2012; Kono & Terashima, 2014). The short-term mechanisms include nonphotochemical quenching (NPQ), photosynthetic control of electron flow between PSII and PSI, state transitions (ST), cyclic electron flow (CEF), and activation of photosynthetic enzymes both in light and carbon fixation (CBC) reactions (Balsera, Uberegui, Schürmann, & Buchanan, 2014; Demmig-Adams, Cohu, Muller, & Adams, 2012; Gollan, Lima-Melo, Tiwari, Tikkanen, & Aro, 2017; Tikkanen & Aro, 2014; Yamori, Makino, & Shikanai, 2016).

Light drives the electron flow from water through PSII, plastoquinone (PQ), cytochrome *b6f*, plastocyanin (PC), and PSI to ferredoxin and ultimately to $NADP^+$, producing NADPH. These photosynthetic electron transfer reactions are coupled with ATP synthesis via translocation of protons to the thylakoid lumen, generating a proton gradient over the thylakoid membrane (ΔpH), which together with membrane potential ($\Delta\Psi$) constitutes the proton motive force (*pmf*) (Armbruster, Correa Galvis, Kunz, & Strand, 2017; Hangarter & Good, 1982). ΔpH also contributes to induction of the energy-dependent qE component of NPQ, a photoprotective mechanism that dissipates excess excitation energy from the electron transfer chain (Niyogi & Truong, 2013; Ruban, 2016), and maintains photosynthetic control at Cyt *b6f* (Johnson, 2011; Joliot & Johnson, 2011). Other regulatory mechanisms include the reversible rearrangements of light-harvesting complexes to balance the excitation of PSII and PSI known as state transitions (Rochaix, 2011; Ruban & Johnson, 2009; Tikkanen et al., 2006) as well as cyclic electron flow around PSI (CEF), a process where electrons are transferred from ferredoxin back to the PQ pool. CEF contributes to the generation of *pmf* and therefore to the production of ATP, and has been suggested to adjust the ATP/NADPH ratio in chloroplasts according to the needs of the CBC (for a recent review, see Yamori & Shikanai, 2016). CEF also provides an alternative electron acceptor mechanism for PSI to relieve stromal overreduction, which is needed to protect the photosystems from damage during early developmental stages of chloroplasts (Allorent et al., 2015; Suorsa, 2015), and during excess illumination or fluctuating light conditions (Miyake, Shinzaki, Miyata, & Tomizawa, 2004; Suorsa et al., 2012; Yamori & Shikanai, 2016; Yamori et al., 2016). CEF has also been shown to be important for controlling the magnitude of the *pmf* (Shikanai & Yamamoto, 2017; Wang, Yamamoto, & Shikanai, 2015), and during induction of photosynthesis (Fan et al., 2007; Joliot & Joliot, 2002). Fan et al. (2007) calculated that CEF contributes a maximum of 68% of total electron flux after a 30-s illumination of spinach leaves with red and far red light.

Two distinct pathways of CEF have been suggested to exist in plant chloroplasts (Munekage et al., 2004). One CEF pathway involves the chloroplast NADH dehydrogenase-like complex (NDH), an ortholog of mitochondrial respiratory complex I (Peltier, Aro, & Shikanai, 2016; Shikanai, 2016). However, unlike complex I, which is reduced by NADH, the chloroplast NDH complex is reduced by ferredoxin (Yamamoto, Peng, Fukao, & Shikanai, 2011; Yamamoto & Shikanai, 2013). It has been suggested recently in several studies that CEF via the NDH complex is essential for photosynthesis in low light conditions (Kou, Takahashi, Fan, Badger, & Chow, 2015; Martin, Noarbe, Serrot, & Sabater, 2015; Yamori, Shikanai, & Makino, 2015) as well as for the tolerance of drought (Horvath et al., 2000) and low temperature (Yamori, Sakata, Suzuki, Shikanai, & Makino, 2011). The antimycin A-sensitive CEF pathway depends on the proteins PROTON GRADIENT REGULATION 5 (PGR5) (Munekage et al., 2002) and PGR5-LIKE 1 (PGRL1) (DalCorso et al., 2008), and has been suggested to constitute the hypothetical ferredoxin-plastoquinone reductase (FQR) (Hertle et al., 2013). However, controversy still exists over the molecular identity of FQR and the physiological function of PGR5 (Kanazawa et al., 2017; Leister & Shikanai, 2013; Tikkanen & Aro, 2014). The PGR- and NDH-dependent pathways differ in their energetic properties; two protons per electron are translocated to the lumen (by the Q-cycle) in the FQR-pathway, whereas the NDH complex functions as a proton pump and additionally transfers two protons per electron to the lumen (Strand, Fisher, & Kramer, 2017; Strand, Livingston, et al., 2017). A third CEF pathway involving transfer of electrons from ferredoxin or FNR to PQ via heme c_n in the Cyt b_6f complex has also been proposed (Hasan, Yamashita, Baniulis, & Cramer, 2013). In general, CEF activity is highly dependent on stromal redox state (Breyton, Nandha, Johnson, Joliot, & Finazzi, 2006), and both the PGR-dependent pathway (Hertle et al., 2013; Strand, Fisher, Davis, & Kramer, 2016) and the NDH pathway (Courteille et al., 2013) have been proposed to be subject to thiol regulation by chloroplast thioredoxins. The physiological roles of each CEF pathway and TRXs involved in the regulation are nevertheless still unclear.

In chloroplasts of Arabidopsis, two thioredoxin systems function in parallel. The ferredoxin–thioredoxin system depends on photosynthetically reduced ferredoxin to supply electrons to the ferredoxin–thioredoxin reductase (FTR), which in turn reduces several thioredoxins, namely TRX-*f*1 and *f*2, four isoforms of TRX-*m*, TRX-*x* as well as TRX-*y*1 and *y*2 (Schürmann & Buchanan, 2008; Yoshida & Hisabori, 2017). The other system consists of a single enzyme, NADPH-thioredoxin reductase (NTRC) that contains both a reductase and a thioredoxin domain (Serrato, Perez-Ruiz, Spinola, & Cejudo, 2004). NTRC is reduced by NADPH, which in addition to the light reactions, is also produced in the oxidative pentose phosphate pathway (OPPP) in darkness. Both chloroplast TRX systems are essential for normal development and growth of plants (Serrato et al., 2004; Wang et al., 2014). The *ntrc* knockout has a stunted and low chlorophyll phenotype, which is particularly severe in plants grown under short photoperiods (Lepistö et al., 2009, 2013; Pérez-Ruiz et al., 2006). The mutant suffers from an impaired ability to activate the ATP synthase and CBC enzymes as well as elevated nonphotochemical quenching (NPQ) (Carrillo, Froehlich, Cruz, Savage, & Kramer, 2016; Naranjo et al., 2016; Nikkanen, Toivola, & Rintamäki, 2016; Thormählen et al., 2017). In contrast, NTRC overexpression lines (OE-NTRC), with

15–20 times higher NTRC content compared to WT, show wild-type–like visible phenotype, enhanced vegetative growth, and increased activation of the ATP synthase and CBC enzymes, particularly in darkness and low light (Nikkanen et al., 2016; Toivola et al., 2013). NTRC has a less negative midpoint redox potential than FTR (Hirasawa et al., 1999; Yoshida & Hisabori, 2016) and plays an important regulatory role under low irradiance, while the FTR-dependent system probably requires more extensive illumination to be fully activated (Geigenberger, Thormählen, Daloso, & Fernie, 2017; Nikkanen et al., 2016; Thormählen et al., 2017). Recent studies have revealed a significant functional overlap and crosstalk between the two chloroplast TRX systems, and indicated that they cooperatively regulate ATP synthesis, the CBC, starch synthesis, and scavenging of reactive oxygen species (ROS) (Geigenberger et al., 2017; Nikkanen et al., 2016; Pérez-Ruiz, Naranjo, Ojeda, Guinea, & Cejudo, 2006; Thormählen et al., 2015).

Here, we have used NTRC overexpression lines as well as the *ntrc* knockout mutant of *Arabidopsis thaliana* to investigate the potential role of the NTRC system in regulating CEF. Our results emphasize the important role of thioredoxins in the chloroplast regulatory network, particularly controlling the photosynthetic redox balance under fluctuating light conditions. We suggest that NTRC plays a crucial role in the activation of the NDH-dependent electron flow in darkness (chlororespiration) and during dark to light transitions. Overexpression of NTRC, on the other hand, maintains constant NDH-CEF activity leading to elevated *pmf* and improved utilization of light energy under fluctuating light conditions. Our results also suggest that NTRC does not activate the PGR-dependent CEF, but contributes to the PGR5-dependent downregulation of thylakoid membrane proton conductivity upon transient exposure of leaves to high light intensity. Through control of both CEF and the activity of the ATP synthase, NTRC would play a pivotal role in adjusting the proton motive force and photosynthetic redox poise in Arabidopsis chloroplasts.

2 | RESULTS

2.1 | NTRC is an active reductant in darkness and low light conditions

The NADPH produced in the oxidative pentose phosphate pathway (OPPP) has been proposed to maintain the NTRC pool partially reduced, and thus active in darkness and when low irradiance limits photosynthesis (Geigenberger et al., 2017; Pérez-Ruiz et al., 2006). To test this hypothesis, we analyzed the in vivo redox state of NTRC by a mobility shift assay using the WT or OE-NTRC protein extracts alkylated with methoxypolyethylene glycol maleimide (MAL-PEG). The assays indicated that the redox state of the NTRC pool remains fairly constant in all light intensities and during dark-to-light transitions, with a significant proportion of the enzyme pool in fully or partially reduced form (Figure 1a,b). This is also the case in OE-NTRC, despite a 20-fold increase in the NTRC content of leaves (Figure 1 and Supporting Information Figure S1).

2.2 | NDH-dependent CEF is enhanced by overexpression of NTRC

In order to determine the effect of altered chloroplast thiol redox state on the activity of NDH-dependent CEF, we measured the post-illumination rise of chlorophyll a fluorescence (PIFR). The PIFR has been suggested to represent electron flow from stromal reductants via the NDH complex to the plastoquinone (PQ) pool upon cessation of illumination (Gotoh, Matsumoto, Ogawa, Kobayashi, & Tsuyama, 2010; Shikanai et al., 1998). The OE-NTRC line showed a significantly larger PIFR after preillumination with low intensity white light than WT, suggesting increased CEF activity (Figure 2a). No PIFR was detected in the *ndho* mutant, which is lacking a functional NDH complex (Rumeau et al., 2005). A diminished PIFR was also observed in the *pgr5* line, which is deficient in PGR-dependent CEF (Munekage et al., 2002) (Figure 2b), but this decrease in PIFR is likely due to the lower content of reduced ferredoxin in *pgr5* chloroplasts because of impaired electron transfer from PSI (Takagi & Miyake, 2018; Tiwari et al., 2016). These results suggest that NTRC contributes to the activation of NDH-dependent CEF. In order to confirm that the increased PIFR in OE-NTRC derives from the activity of the NDH complex, we generated an NTRC overexpression line in the *ndho* mutant background (OE-NTRC *ndho*), which indeed was fully missing the PIFR (Figure 2c). Additionally, the relaxation of fluorescence after saturating pulses during actinic illumination was delayed in OE-NTRC *ndho* (Figure 2c). The level of NTRC overexpression in OE-NTRC *ndho* plants was confirmed by immunoblotting and found to be in the same range as the OE-NTRC line (Supporting Information Figure S1B).

The *ntrc* knockout exhibited a slower initial PIFR response after illumination, but the PIFR did not decline after 15–20 s in darkness as in WT or OE-NTRC. Instead, the PIFR continued to rise throughout the duration of the dark phase of the experiment (Figure 2a). A brief pulse of far red (FR) light quenched the fluorescence, but after cessation of the FR light, fluorescence quickly rose back to its high pre-FR level. The F_0 level was elevated in dark-adapted *ntrc* leaves in comparison to WT, OE-NTRC, or other mutant lines (Figure 2a). The abnormal fluorescence pattern in Figure 2a may also be due to the highly pleiotropic phenotype of the *ntrc* mutant, particularly when grown under a short-day photoperiod. The *ntrc* knockout had high NPQ in the experimental conditions (Supporting Information Figure S2B), and its relaxation in darkness likely contributed to the PIFR. Moreover, the *ntrc* mutant has an impaired capacity to scavenge hydrogen peroxide (H_2O_2) (Kirchsteiger, Pulido, Gonzalez, & Javier Cejudo, 2009; Pulido et al., 2010), whose accumulation has been shown to cause an increase in NDH-dependent CEF (Strand et al., 2015; Strand, Livingston, et al., 2017). In order to clarify whether the differences in PIFR were caused indirectly by metabolic disturbances due to impaired growth in a short photoperiod and/or accumulation of H_2O_2, we estimated the levels of H_2O_2 in illuminated WT, *ntrc*, and OE-NTRC leaves by DAB staining, and repeated the PIFR-experiment with plants grown in a 12 hr/12 hr photoperiod. An increased

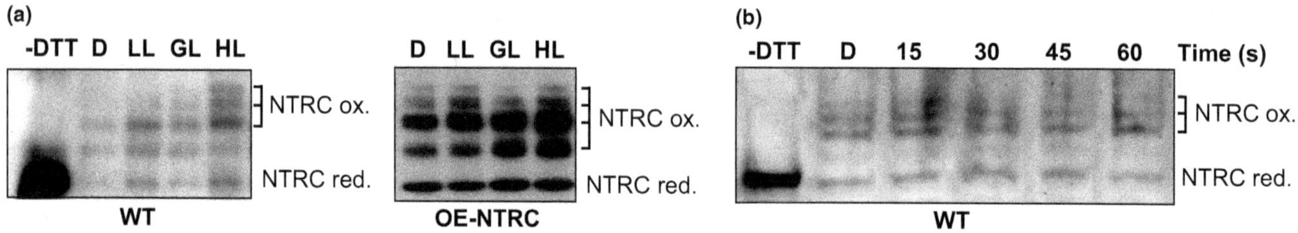

FIGURE 1 In vivo redox state of NTRC in dark-adapted and illuminated leaves. (a) Total protein extract was isolated from WT and OE-NTRC leaves incubated in darkness (D), or illuminated for 2 hr in low light (LL, 40 μmol photons m^{-2} s^{-1}), growth light (GL, 200 μmol photons m^{-2} s^{-1}) or high light (HL, 800 μmol photons m^{-2} s^{-1}). Free thiols of proteins were blocked with NEM, disulfides reduced with DTT and newly formed thiols alkylated with MAL-PEG. The in vivo-reduced form of NTRC therefore migrates faster in SDS-PAGE than the in vivo-oxidized forms. –DTT stands for the unlabeled control sample where DTT was not added after incubating the leaf extracts in a buffer containing NEM. Protein content of samples has been equalized only based on the amount of starting leaf material, and the apparent differences in band intensity should not be taken as an indication of differences in NTRC content between light treatments. For an analysis of the origin of different MAL-PEG–labeled bands see Supporting Information Figure S1. (b) NTRC redox state in WT during a transition from dark to growth light. Samples were taken from darkness (2 hr) (D) and 15, 30, 45, and 60 s after onset of illumination

amount of H_2O_2 was detected in both low light- and high light-treated *ntrc* leaves in comparison to WT (Supporting Information Figure S2C). No difference was observed in the amount of H_2O_2 between OE-NTRC and WT (Supporting Information Figure S2C), indicating that the elevation of PIFR in OE-NTRC (Figure 2a) is not caused by increased content of H_2O_2. Furthermore, the PIFR mostly disappeared in a 12-hr photoperiod-grown *ntrc*, but remained similar to 8-hr photoperiod-grown plants in WT and OE-NTRC (Supporting Information Figure S2A).

Altered PIFR responses in OE-NTRC and *ntrc* could hypothetically be caused by changes in the available amount of reduced ferredoxin (Fd^-), the substrate of the NDH complex (Yamamoto & Shikanai, 2013; Yamamoto et al., 2011). In order to investigate this possibility, we used the Dual/Klas-NIR spectrometer, which allows deconvolution of the Fd signal from plastocyanin (PC) and P700 signals (Klughammer & Schreiber, 2016; Schreiber, 2017; Schreiber & Klughammer, 2016), to measure the redox state of Fd in WT, *ntrc*, and OE-NTRC under similar light and postillumination conditions as used in the PIFR measurements. In OE-NTRC, reduced Fd was reoxidized slightly faster than in WT upon onset of illumination, but upon cessation of actinic illumination, there was no difference in the fraction of reduced Fd between WT and OE-NTRC (~11% in both lines, Figure 2d). In *ntrc* leaves, however, almost 40% of the Fd pool was reduced at the end of actinic illumination, and a significantly increased reduction of Fd also occurred in darkness after initial oxidation (Figure 2d). These results indicate that the elevated PIFR in OE-NTRC is not caused by increased accumulation of substrate of the NDH complex. In contrast, the PIFR signal in *ntrc* is very likely partially caused by an increase in the relative amount of reduced Fd at the end of illumination.

Differences in the rate of PQ reduction could also be caused by altered content of PSII or PSI complexes, the NDH complex, Cyt *b6f* or plastid terminal oxidase (PTOX). No statistically significant differences were detected in the amounts of the PSII core protein D1, Cyt *b6f* subunit Cyt *f*, or NDH subunits NhdS and NdhH between the studied lines, while a decrease in the amount of PGR5 and the PSI core protein PsaB, as well as elevated PTOX content were detected in *ntrc* (Figure 3).

2.3 | NTRC promotes dark reduction of the plastoquinone pool

To determine whether the higher NDH activity that was observed in darkness after illumination of OE-NTRC leaves (Figure 2a) alters the redox state of the PQ pool, we proceeded to analyze the phosphorylation level of LHCII proteins in dark-adapted and illuminated leaves. The reduction in the PQ pool induces phosphorylation of LHCII by activating the STN7 kinase through interaction with the Cyt *b6f* complex (Bellafiore, Barneche, Peltier, & Rochaix, 2005; Shapiguzov et al., 2016; Vener, VanKan, Rich, Ohad, & Andersson, 1997). In WT plants LHCII proteins were mostly nonphosphorylated in darkness, maximally phosphorylated in low light, moderately phosphorylated in growth light, and mostly dephosphorylated in high light (Figure 4h), in agreement with earlier studies (Rintamäki et al., 1997; Tikkanen, Grieco, Kangasjarvi, & Aro, 2010). The *stn7* mutant was unable to phosphorylate LHCII (Bellafiore et al., 2005). In contrast to WT, LHCII was phosphorylated in darkness in OE-NTRC (Figure 4h). As in WT, only a small amount of phosphorylated LHCII was present in thylakoids isolated from dark-adapted leaves of *ntrc*, whereas slightly higher accumulation of phosphorylated LHCII was observed in high light-illuminated *ntrc* (Figure 4h).

To further investigate the effect of NTRC on the reduction state of the PQ pool in darkness, we measured the kinetics of Chl a fluorescence OJIP transients in dark-adapted leaves and in leaves preilluminated with far red light (FR) to fully oxidize the PQ pool. The difference in F/Fm at the J phase of the transient (F_J, 3 ms after onset of illumination) between dark-adapted and preilluminated leaves is an indicator for the redox state of the PQ pool in darkness (Stirbet, Riznichenko, Rubin, & Govindjee, 2014; Toth, Schansker, & Strasser, 2007). As demonstrated in Figure 4, the OE-NTRC line had a significantly larger proportion of reduced PQ in darkness when compared to WT. The *ndho* mutant had a more oxidized PQ pool in

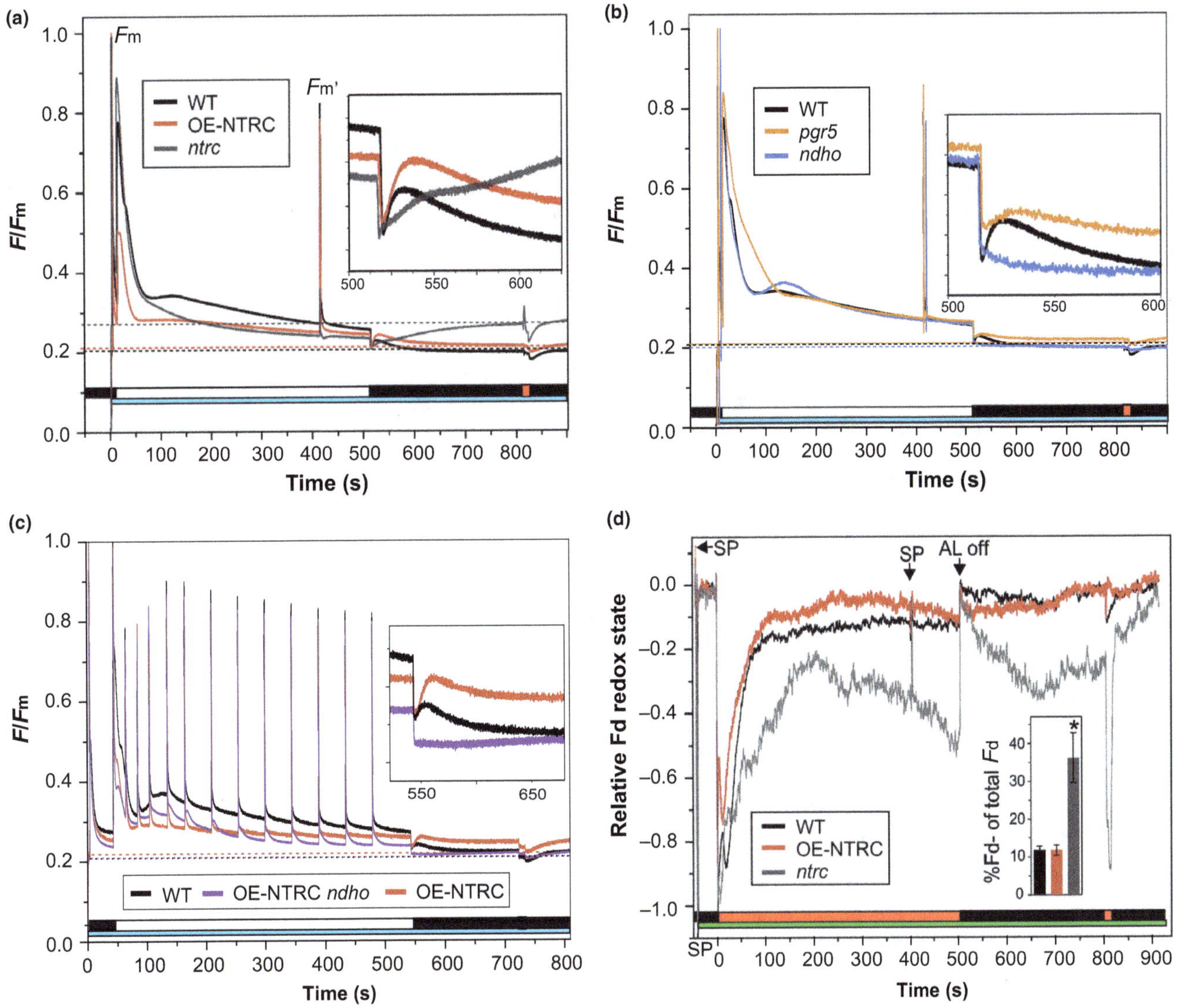

FIGURE 2 Postillumination fluorescence rise (PIFR) in dark-adapted leaves. (a) and (b) PIFR was measured from WT, OE-NTRC, *ntrc* (a), *pgr5* and *ndho* (b) leaves. The smaller windows show magnifications of the ~100 s of the PIFR. The cyan bars indicate exposure to a 480 nm measuring light of 0.28 μmol photons m^{-2} s^{-1}, the white bar depicts illumination with 67 μmol photons m^{-2} s^{-1} white light and the red bar shows the duration of a pulse of far red light. The dashed lines indicate the F_O values of the lines. The curves are averages of measurements from three to seven individual leaves, and are normalized to Fm. (c) PIFR in WT, OE-NTRC, and OE-NTRC *ndho*. PIFR was measured as in (a) and (b), but saturating pulses were administered with 20-s intervals for the first minute, 30-s interval for the second minute, and 45-s intervals for the rest of the experiment. Representative curves of three to four measurements of individual leaves are shown. (d) Fd redox changes during and after illumination of dark-adapted WT, *ntrc*, and OE-NTRC leaves. The Fd redox changes were deconvoluted from four near-infrared absorbance differences measured with a Dual/Klas-NIR spectrometer according to Klughammer and Schreiber (2016). Leaves were illuminated at actinic red light (630 nm) of 61 μmol photons m^{-2} s^{-1}. The red bar shows the duration of a pulse of far red light, the green bar the duration of the four measuring beams, the black bar the duration of the dark period after illumination, SP means saturating pulse. Averages of the relative amount of reduced Fd at the end of actinic illumination ± *SE* in three to seven individual leaves are shown in the column chart inside the figure. Representative curves of WT, OE-NTRC, and *ntrc* are shown in the figure

darkness than WT, while there was no significant difference between *pgr5* and WT (Figure 4g), suggesting that the NDH complex is the main CEF pathway contributing to dark-reduction of the PQ pool. The differences in dark-redox state of PQ (Figure 4g) were also supported by an elevated dark-adapted apparent F_O value, a steeper initial slope of the transient, and a decreased area above the O-J phase of the OJIP transient in both *ntrc* and OE-NTRC when compared to WT (Supporting Information Table S1) (Toth et al., 2007). Attribution of the elevated dark-reduction of PQ in OE-NTRC to enhanced activity of the NDH complex was further supported by the observation that the redox state of the PQ pool in darkness was more oxidized in the OE-NTRC *ndho* line in comparison to WT and OE-NTRC (Figure 4g). OJIP transients also showed higher dark-reduction of PQ in *ntrc* mutant when compared to WT (Figure 4), but it must be noted that the overall kinetics of the OJIP transient in *ntrc* differed considerably from the other lines.

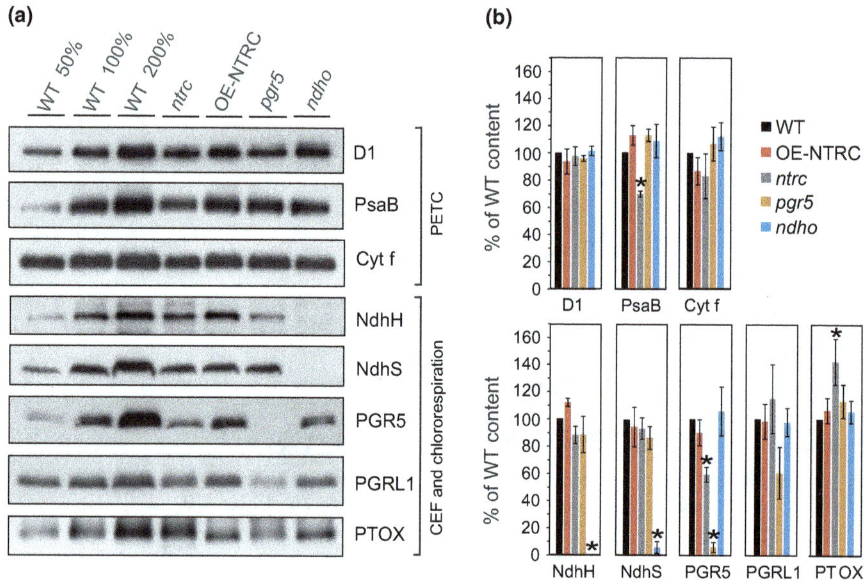

FIGURE 3 Content of proteins functioning in the photosynthetic electron transfer chain (PETC), cyclic electron flow (CEF), and chlororespiration in WT, *ntrc*, OE-NTRC, *pgr5*, and *ndho*. (a) Representative immunoblots showing the content of D1, PsaB, Cyt *f*, NdhH, NdhS, PGR5, PGRL1, and PTOX. Appropriate amount of thylakoid extract (based on protein content) was separated with SDS-PAGE and probed with specific antibodies. Equal loading was confirmed by protein staining with Li-Cor Revert Total Protein Stain. (b) Relative content of proteins in mutant lines as percentage of WT. The numbers represent the average protein content ±*SE* in three to five biological replicates. The quantified values were normalized to the total protein content in the sample determined with Li-Cor Revert Total Protein Stain. Statistically significant differences to WT according to Student's *t* tests ($p < 0.05$) are marked with *

2.4 | NTRC enhances the generation of proton motive force during dark-to-light transitions

Reduction of plastoquinone by the thylakoid NDH complex is known to be coupled with the translocation of protons to the lumen, which contributes to the formation of *pmf* and enhances ATP synthesis (Strand, Fisher, & Kramer, 2017). We therefore investigated whether the generation of *pmf* during dark-to-light transitions and in plants illuminated with different light conditions is affected by deficiency or overexpression of NTRC. The *pmf*, conductivity of the thylakoid membrane to protons (g_{H+}), and proportions of the *pmf* components ΔpH and ΔΨ were determined by measuring a difference in absorbance at 550 and 515 nm, also known as the electrochromic shift (ECS) (Cruz et al., 2005).

In WT, *pmf* was transiently elevated upon onset of illumination at growth light intensity, peaking after 15–20 s (Figure 5a) and coinciding with a decrease in g_{H+} (Figure 5c). The initial decrease in g_{H+} occurred despite rapid reduction of the gamma subunit of the ATP synthase ($CF_1\gamma$), as after 20 s under growth light $CF_1\gamma$ was already fully reduced (Figure 5f). Under growth light intensity, another slight rise in *pmf* was observed after ca. 30–60 s in light (Figure 5a), coinciding with P700 oxidation (Figure 6b). Subsequently *pmf* slowly decreased to a steady-state value.

In OE-NTRC, the initial *pmf* increase occurred already in a few seconds after the onset of illumination, reached a higher level and decreased more slowly than in WT (Figure 5a). While g_{H+} was drastically elevated in dark-adapted OE-NTRC leaves, it rapidly decreased to a level comparable to WT (Figure 5c). Both total *pmf* and the contribution of ΔpH to it were higher in OE-NTRC in all light intensities when compared to WT (Figure 5e). There was no significant difference between OE-NTRC and WT in g_{H+} under growth light illumination apart from the enhanced conductivity in dark-adapted OE-NTRC leaves (Figure 5c). PSII quantum yield (Figure 6a) increased only slightly during early photosynthetic induction and slightly but insignificantly decreased at steady-state illumination in comparison to WT, while P700 oxidation was significantly faster (Figure 6b). These results suggested that the increase in *pmf* in OE-NTRC derives from CEF, which is also supported by the nonlinear relationship between thylakoid proton flux (v_{H+}) and quantum yield of PSII in plants with modified NTRC content (Supporting Information Figure S3).

In *ndho*, generation of *pmf* followed similar kinetics as in WT, but its level remained lower (Figure 5a). In comparison to WT, thylakoid proton conductivity was increased during the 30–60-s time period (Figure 5c). In OE-NTRC *ndho*, however, an elevation of *pmf* occurred during transitions from dark to growth light similarly to OE-NTRC, except for a time period between 15–40 s after onset of illumination, where *pmf* was lowered in OE-NTRC *ndho* in comparison to OE-NTRC (Figure 5a). Upon dark-to-light transitions, the high initial thylakoid proton conductivity in OE-NTRC *ndho* decreased as in OE-NTRC, but after 10 s of illumination g_{H+} again rose more rapidly than in OE-NTRC (Figure 5c). In the absence of the NDH complex, NTRC overexpression was also unable to enhance P700 oxidation during dark-to-growth light transitions (Figure 6b), and OE-NTRC *ndho* suffered from increased PSI acceptor side limitation in comparison to WT (Figure 6c). These data indicate that the enhanced

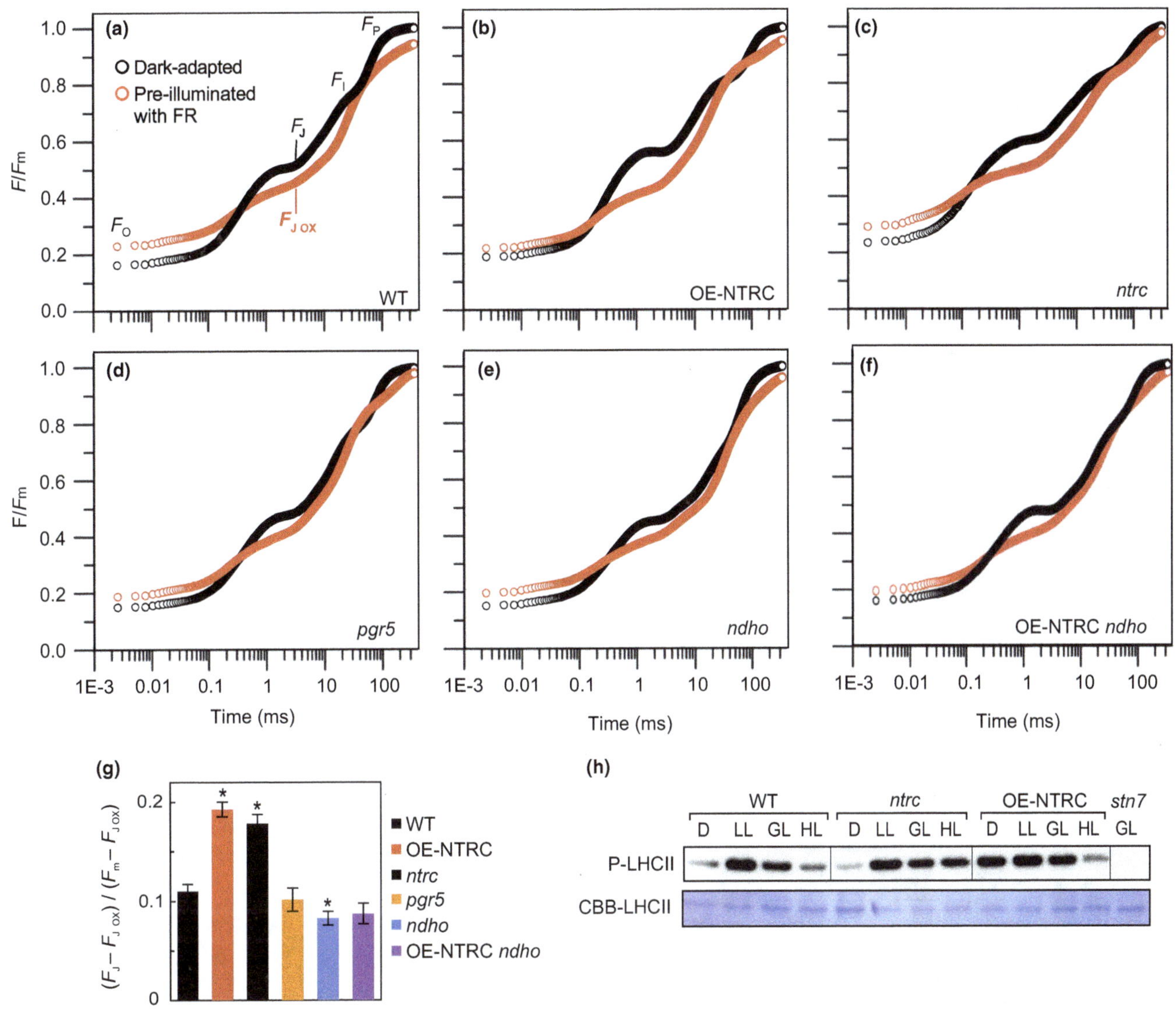

FIGURE 4 Redox state of the PQ pool in dark-adapted leaves. (a–f) PQ pool redox state in darkness was determined from measurement of the OJIP transients in dark-adapted leaves (black) and leaves preilluminated with far red light (red). Data are presented as averaged curves from measurements of 5 to 10 individual leaves on a logarithmic time scale from WT (a), OE-NTRC (b), *ntrc* (c), *pgr5* (d), *ndho* (e), and OE-NTRC *ndho* (f). The slight rise in F_O values after FR-illumination is likely due to the small actinic effect of FR on PSII, as discussed by Schansker and Strasser (2005). (g) Proportions of reduced Q_A calculated as $(F_J\text{-}F_{J\,ox})/(F_m\text{-}F_{J\,ox})$. Values are averages of measurements from 5 to 10 individual leaves ± *SE*. * indicates statistically significant difference to WT according to Student's *t* test ($p < 0.05$). All values are normalized to dark-adapted Fm. (h) Determination of phosphorylation status of LHCII proteins in WT, *ntrc*, OE-NTRC, and *stn7* after 2 hr of darkness and in low light (40 μmol photons $m^{-2}\ s^{-1}$), growth light (200 μmol photons $m^{-2}\ s^{-1}$) and high light (600 μmol photons $m^{-2}\ s^{-1}$). Thylakoid extracts containing 0.4 μg chlorophyll were separated with SDS-PAGE and phosphoproteins were detected with a Phosphothreonine-specific antibody. Coomassie Brilliant Blue staining of LHCII on the membrane (CBB-LHCII) was used as loading control

capacity of the stroma in OE-NTRC to pull electrons from PSI during dark-to-light transitions is dependent on the NDH complex.

Increased activation of the NDH complex is not, however, sufficient to fully explain the elevated *pmf* in OE-NTRC, especially immediately after dark-to-light transitions and at steady state. Therefore, we also generated an NTRC overexpression line in the *pgr5* mutant background (OE-NTRC *pgr5*) whose NTRC expression level was comparable to OE-NTRC (Supporting Information Figure S1B). In the *pgr5* mutant, *pmf* generation at the onset of illumination was impaired and the steady-state level of *pmf* was lower than in WT (Figure 5b), in part due to increased g_{H+} (Figure 5d). The absence of PGR5 did not alter the kinetics of *pmf* induction in plants overexpressing NTRC, but the magnitude of *pmf* remained lower (Figure 5b). During early photosynthetic induction, proton conductivity of the thylakoid membrane in OE-NTRC *pgr5* was high, like in OE-NTRC, but dropped less than in OE-NTRC, and at steady state, stayed higher even in comparison to *pgr5* (Figure 5d). Moreover, PSI acceptor side limitation was only slightly alleviated in OE-NTRC *pgr5* in comparison to *pgr5* (Figure 6c). Interestingly, a high initial g_{H+} value and rapid generation of a *pmf* peak after onset of illumination

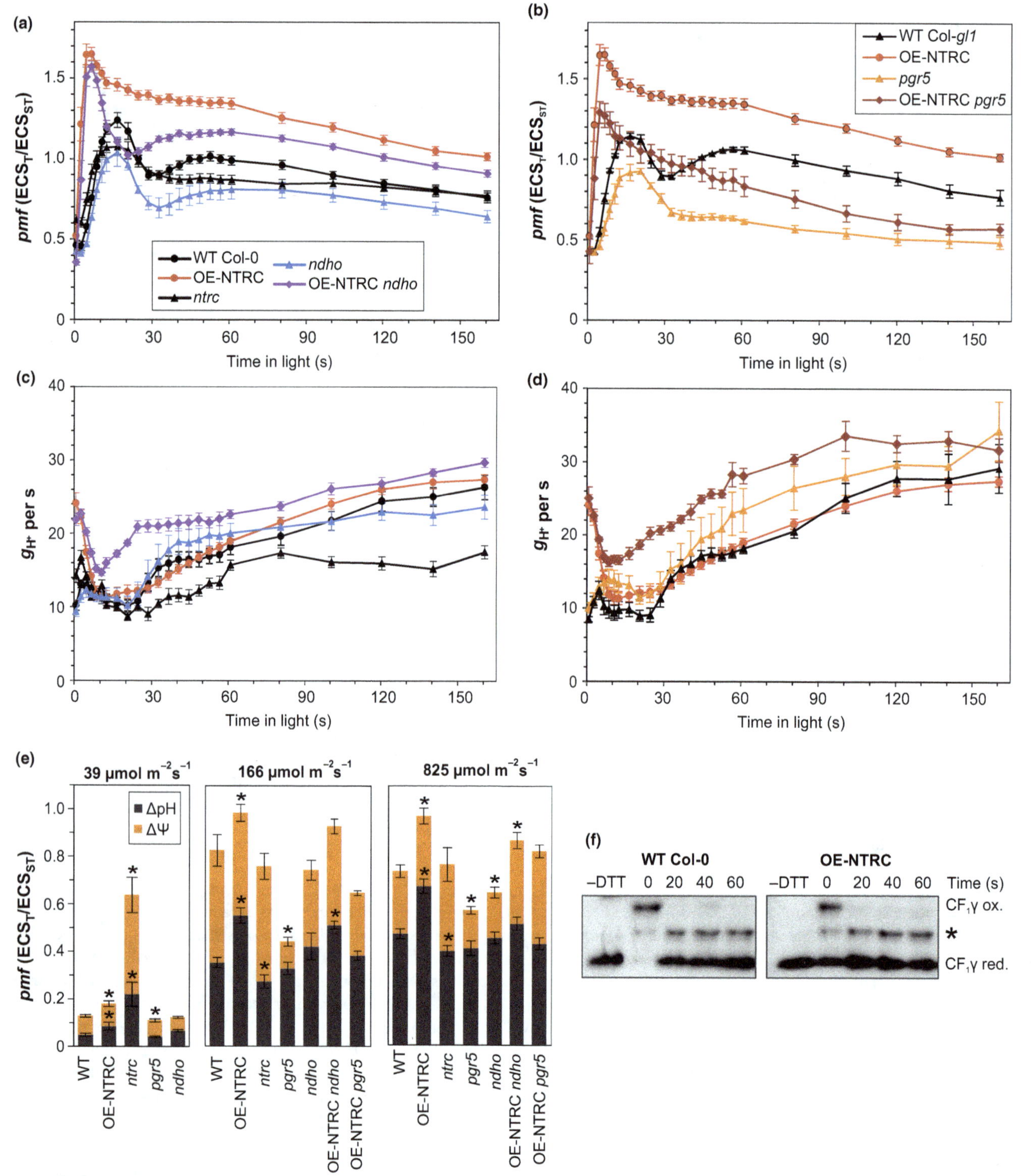

FIGURE 5 Generation of the proton gradient during dark-to-light transitions. (a, b) Proton motive force (*pmf*) at specific time points during transitions from dark to actinic light of 166 μmol photons m^{-2} s^{-1} (AL) in dark-adapted leaves of WT Col-0, OE-NTRC, *ntrc*, *ndho*, and OE-NTRC *ndho* (a) and in WT Col-*gl1*, OE-NTRC, *pgr5*, and OE-NTRC *pgr5* (b). The *pmf* was measured as light-induced change in the ECS signal (ECS_T) and normalized with the magnitude of ECS induced by a 20-μs saturating single-turnover flash administered prior to the onset of AL (ECS_{ST}). Values are averages of measurements from 4 to 16 individual leaves ±*SE*. (c, d) Conductivity of the thylakoid membrane to protons (g_{H+}), calculated as the inverse of the time constant of a first-order fit to the decay of ECS during 250 ms dark intervals. (e) Total *pmf* and its partitioning to ΔpH and ΔΨ after 3 min illumination with low, growth or high light. Values are averages of measurements from 3 to 6 (for 39 μmol m^{-2} s^{-1}), 5 to 10 (for 166 μmol m^{-2} s^{-1}), and 8 to 15 (for 825 μmol m^{-2} s^{-1}) individual leaves ±*SE*. Statistically significant differences to WT according to Student's *t* tests are marked with *. (f) Mobility shift assays with MAL-PEG labeled protein extracts to determine the in vivo redox state of $CF_{1}\gamma$ subunit of the ATP synthase during first 60 s of dark-to-growth light transitions. * marks an unspecific band of unknown origin

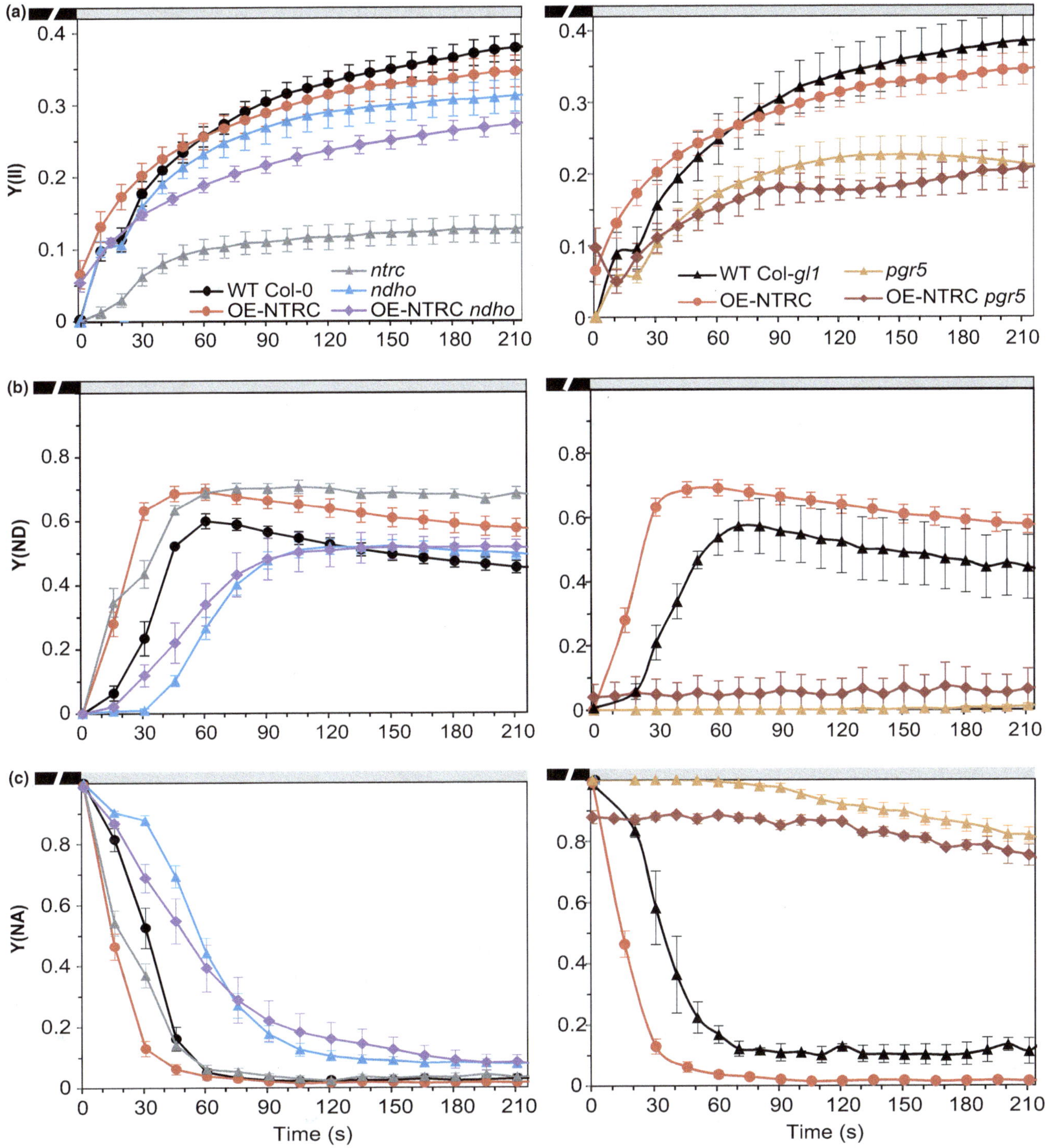

FIGURE 6 Photosynthetic parameters measured from leaves during dark-to-light transitions. (a–c) Dark-adapted WT Col-0, *ntrc*, OE-NTRC, *ndho*, OE-NTRC *ndho*, WT Col-*gl1*, *pgr5*, and OE-NTRC *pgr5* leaves were shifted from darkness to the actinic light of 166 μmol photons m^{-2} s^{-1} and illuminated for 210 s. Photosystem II quantum yield (a), was calculated from Chl a fluorescence. P700 oxidation (Y(ND)) (b) and PSI acceptor side limitation (Y(NA)) (c) were calculated from absorbance difference at 875 and 830 nm. The graphs are averages of measurements from four to nine individual leaves ±*SE*

were observed in all plants overexpressing NTRC (OE-NTRC, OE-NTRC *ndho*, and OE-NTRC *pgr5*) but missing in WT, *ndho*, and *pgr5* plants (Figure 5a,b).

In *ntrc*, the initial *pmf* increase occurred with similar kinetics to WT upon onset of illumination at growth light intensity, but had a lesser magnitude (Figure 5a). The secondary increase in *pmf* was absent in *ntrc* leaves. In growth light, the steady-state *pmf* was comparable to WT, but contribution of ΔpH to total *pmf* was slightly diminished (Figure 5e). Decreased thylakoid conductivity to protons was observed in growth light intensity in *ntrc* (Figure 5c). However,

$CF_1\gamma$ is reduced normally in growth light in *ntrc* (Nikkanen et al., 2016), implying that thylakoid proton conductivity is inhibited by other means. A high donor side limitation of PSI was measured in *ntrc* under growth light (Figure 6b) despite high excitation pressure in the PQ pool (Supporting Information Figure S4B) and lowered PsaB content (Figure 3), suggesting inhibition of electron transfer between the PQ pool and PSI.

2.5 | In fluctuating light, NTRC overexpression enhances PSI yield in low light and represses thylakoid conductivity to protons upon transitions from low to high irradiance

Due to the recent suggestions implicating a particularly important role for NTRC under low and fluctuating light conditions (Carrillo et al., 2016; Nikkanen et al., 2016; Thormählen et al., 2017), we proceeded to investigate *pmf* generation and photosynthetic electron transfer in OE-NTRC under these conditions. During transitions from dark to low light intensity, increased *pmf* formation was again observed in OE-NTRC, but the difference to WT was less dramatic than in growth light (Figure 7a). PSII yield was enhanced in OE-NTRC during the transition from dark to low light (Figure 8a), also contributing to the *pmf* increase. P700 oxidation was also enhanced during dark-to-low light transitions (Figure 8c), while NPQ was decreased (Supporting Information Figure S5A), despite higher ΔpH (Figure 5e). A high PSI yield was maintained throughout the low light periods in OE-NTRC due to low acceptor side limitation (Figure 8b). Notably, the overexpression of NTRC in *ndho* background reverted these changes observed in OE-NTRC plants to levels comparable to or even more severe than in *ndho* knockout plants, except for the increased PSII yield during dark-to-low light transitions (Figures 7a,b and 8a), suggesting that enhanced activation of NDH-mediated CEF contributes to photosynthetic performance of OE-NTRC during dark-to-light transitions and under low light.

During transitions from dark to low light, the *pgr5* and *ndho* mutants both showed impaired *pmf* generation (Figure 7). In contrast, *pmf* generation in OE-NTRC *pgr5* during dark-to-low and low-to-high light transitions was recovered to WT levels (Figure 7c) despite elevated g_{H+} especially in high light (Figure 7d). The recovery of *pmf* most likely occurred due to enhanced activity of NDH-CEF by increased NTRC content. Slightly improved tolerance to light fluctuation was also observed in OE-NTRC *pgr5* in comparison to *pgr5* as PSI yield was better maintained in low light following the high light periods (Figure 8b). Importantly, overexpression of NTRC improved the ability to oxidize PSI in low light even in the *pgr5* background (Figure 8c).

In all lines a transient *pmf* spike was observed upon the switch from low to high light intensity (Figure 7). In WT, the proton conductivity of the thylakoid membrane decreased gradually (Figure 7b), but the decrease in g_H+ was not due to oxidation of $CF_1\gamma$, as it remained fully reduced in high light conditions (Supporting Information Figure S6B). The decrease in g_{H+} was even stronger in OE-NTRC upon the shift from low to high light (Figure 7b). Overexpression of NTRC in *ndho* background decreased *pmf* generation during transitions from low to high light in comparison to the OE-NTRC line (Figure 7a), suggesting that enhanced activation of NDH-mediated CEF contributes to the high *pmf* in OE-NTRC in these conditions. OE-NTRC *ndho* showed increased steady-state g_{H+} under high irradiance similarly to *ndho* (Figure 7b).

The *pgr5* mutant was unable to oxidize P700 in high light (Figure 8c) or to decrease proton efflux from the lumen (Figure 7d), resulting in a loss of *pmf* (Figure 7c). The strong decrease in g_{H+} observed in OE-NTRC in high light disappeared in OE-NTRC *pgr5*, which lacks PGR5 (Figure 7d).

In *ntrc*, high steady-state *pmf* under low light intensity (Figures 5e, and 7a) was likely caused by impaired activation of the chloroplast ATP synthase and the Calvin–Benson cycle as previously reported (Carrillo et al., 2016; Nikkanen et al., 2016). Furthermore, exceptionally high NPQ was recorded in the *ntrc* line, especially at low light (Supporting Information Figure S5A).

Concluding from Figures 5–8, it is evident that both the knockout and overexpression of NTRC had a distinct influence on the formation of *pmf* during transitions from dark to light and from low to high light through regulation of the activities of CEF and ATP synthase.

2.6 | Interaction of CEF-related proteins with NTRC

Distinct effects of NTRC overexpression or deficiency on the postillumination fluorescence rise (Figure 2), the dark-reduction level of the PQ pool (Figure 4), and generation of *pmf* during dark/light transitions and low/high light transitions (Figures 5 and 7) suggested that NTRC may either directly or indirectly regulate CEF. In order to screen for potential targets of NTRC-mediated regulation, we performed coimmunoprecipitation (Co-IP) assays with an antibody against NTRC, and analyzed eluates from WT, *ntrc*, and OE-NTRC total leaf extracts by mass spectrometry (MS). A full list of identified peptides is provided in Supporting Information Dataset S1, while Supporting Information Table S2 lists 100 chloroplast proteins

FIGURE 7 Formation and regulation of the proton motive force during changes in light conditions. (a, b) The *pmf* (a) and proton conductivity of the thylakoid membrane (g_{H+}) (b) at specific time points during transitions from darkness to low actinic light (39 μmol photons m^{-2} s^{-1}) and from low to high light (825 μmol photons m^{-2} s^{-1}) in dark-adapted leaves of WT Col-0, OE-NTRC, *ntrc*, *ndho*, and OE-NTRC *ndho*. The *pmf* and g_{H+} were measured and calculated as explained in the legend for Figure 5. The graphs shown are averages of measurements from 4 to13 individual leaves ±*SE*. (c, d) The *pmf* (c) and g_{H+} (d) in dark-adapted leaves of WT Col-*gl1*, OE-NTRC, *pgr5*, and OE-NTRC *pgr5* during changes in light conditions. The experiment was performed as explained in the figure legend for (a, b). The graphs shown are averages of measurements from 3 to 13 individual leaves ±*SE*

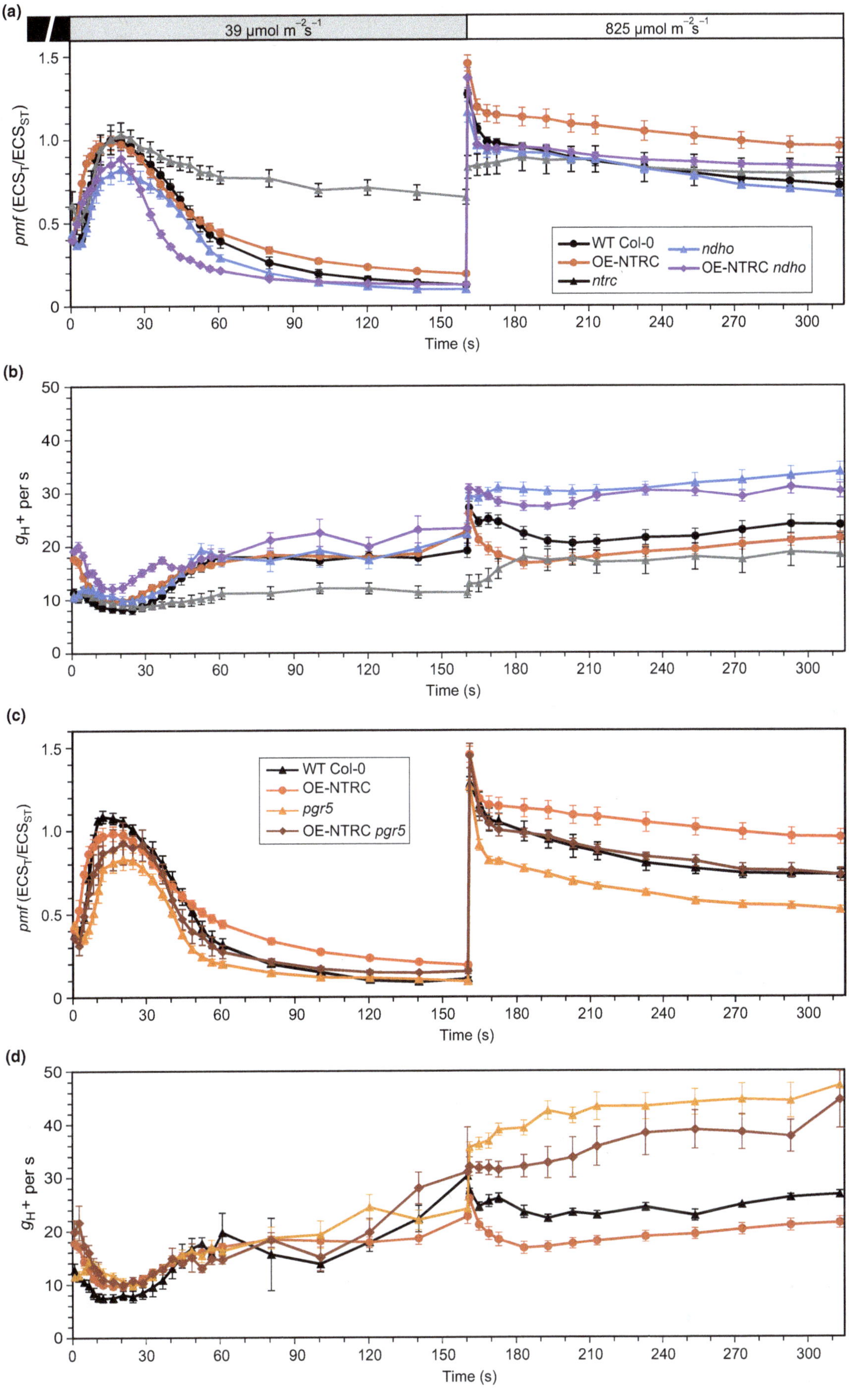
(a)
39 µmol m⁻² s⁻¹
825 µmol m⁻² s⁻¹
pmf (ECS_T/ECS_ST)
WT Col-0
OE-NTRC
ntrc
ndho
OE-NTRC ndho
Time (s)
(b)
g_H+ per s
Time (s)
(c)
pmf (ECS_T/ECS_ST)
WT Col-0
OE-NTRC
pgr5
OE-NTRC pgr5
Time (s)
(d)
g_H+ per s
Time (s)

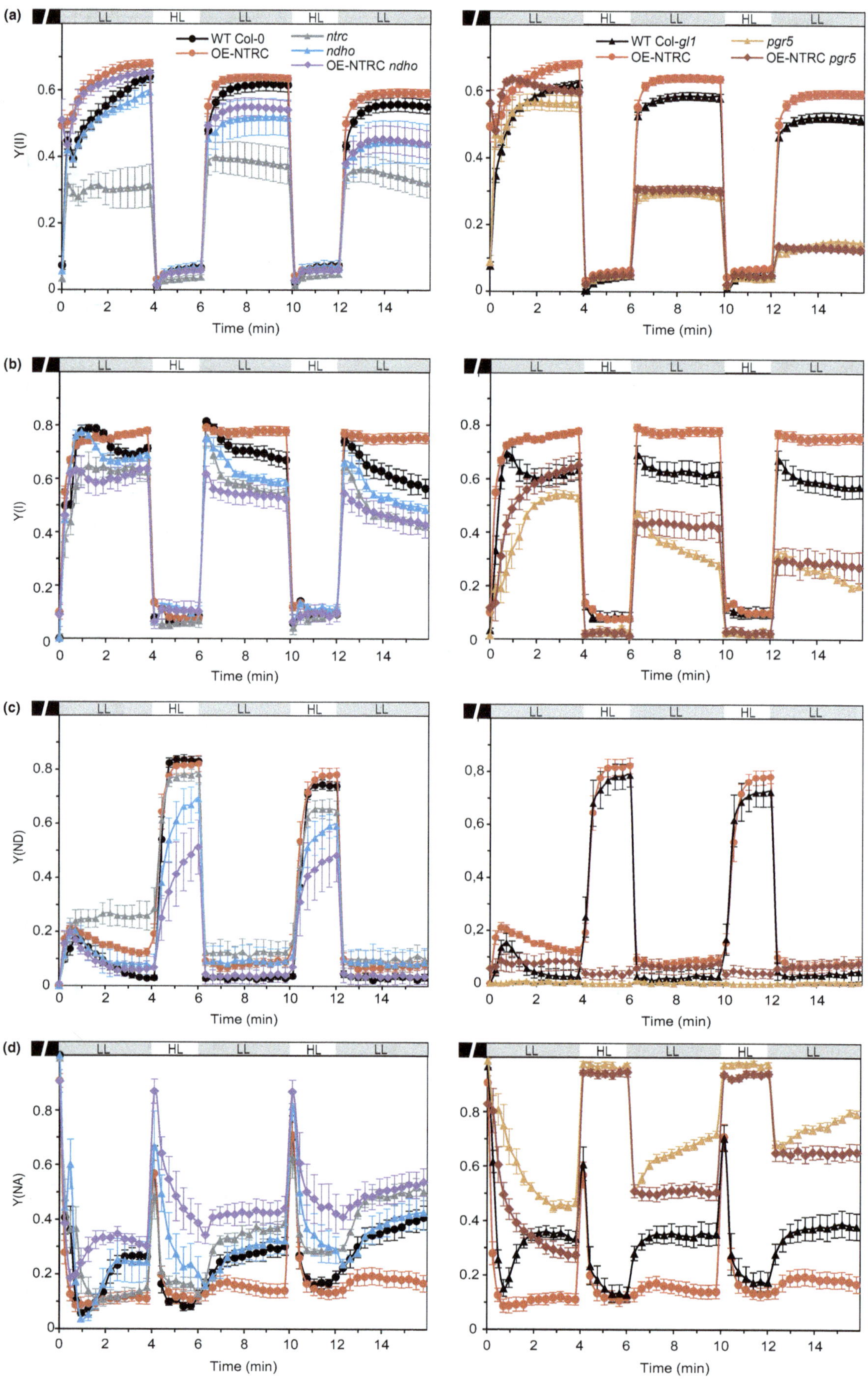
(a)
LL
HL
LL
HL
LL
WT Col-0
OE-NTRC
ntrc
ndho
OE-NTRC ndho
Y(II)
Time (min)
WT Col-gl1
OE-NTRC
pgr5
OE-NTRC pgr5
(b)
Y(I)
(c)
Y(ND)
(d)
Y(NA)

FIGURE 8 Analysis of Chlorophyll a fluorescence and P700 oxidation in fluctuating light. (a) PSII yield (Y(II)) in light conditions fluctuating between periods of low actinic light (LL, 39 μmol photons m^{-2} s^{-1}) and high light (HL, 825 μmol photons m^{-2} s^{-1}) in WT Col-0, OE-NTRC, *ntrc*, *ndho*, OE-NTRC *ndho*, WT Col-*gl1*, *pgr5*, and OE-NTRC *pgr5*. Five-week-old plants were dark-adapted for 30 min before measuring fluorescence from detached leaves. All values are averages of measurements from 3 to 10 individual leaves ±*SE*. (b–d) PSI yield (Y(I)) (b), P700 oxidation (Y (ND)) (c), and PSI acceptor side limitation (Y(NA)) (d) in light conditions fluctuating between periods of low actinic light (LL, 39 μmol photons m^{-2} s^{-1}) and high light (HL, 825 μmol photons m^{-2} s^{-1}) in WT Col-0, OE-NTRC, *ntrc*, *ndho*, OE-NTRC *ndho*, WT Col-*gl1*, *pgr5*, and OE-NTRC *pgr5*. Five-week-old plants were dark-adapted for 30 min before measuring fluorescence from detached leaves. All values are averages of measurements from 3 to 10 individual leaves ±*SE*

in order of their abundance in WT and OE-NTRC eluates but missing in *ntrc* eluates. Among these proteins were several TRX interactors identified in previous studies (Balmer et al., 2003; Marchand et al., 2006; Hall et al., 2010), as well as established NTRC target proteins such as 2-cysteine peroxiredoxins (Pérez-Ruiz et al., 2006) and enzymes involved in chlorophyll biosynthesis (Richter et al., 2013). Relevantly in the current context, several proteins involved in CEF around PSI were identified by the Co-IP/MS screening (Table 1). Most notably, five subunits of the thylakoid NDH complex, NdhH, Ndh48, NdhS, NdhU, and NdhO, (in order of abundance in the Co-IP eluates) as well as PGR5 were identified as potential NTRC interactors. Intriguingly, all of the NDH subunits identified are located in close proximity to the proposed ferredoxin binding and oxidation site on the stromal side of the NDH complex (Peltier et al., 2016; Shikanai, 2016; Yamamoto & Shikanai, 2013; Yamamoto et al., 2011).

The potential interactions of NTRC with NdhS and PGR5 were further supported by positive results in bimolecular fluorescence complementation tests (BiFC). Coexpression of NTRC with both NdhS and PGR5 in *Nicotiana benthamiana* leaves resulted in YFP fluorescence that was strictly colocalized with chlorophyll autofluorescence, suggesting that it originated from the thylakoid membrane (Figure 9). TRX-*m*1 interacted with PGRL1, while no interaction capability was detected between PGRL1 and NTRC

TABLE 1 Screening of putative NTRC target proteins in CEF pathways by Co-IP/MS

AGI code	Description	*ntrc*	WT	OE-NTRC	MW (kDa)	#Cys
ATCG01110.1	NdhH	-	+	+	45.5	3
AT1G15980.1	Ndh48 (NDF1)	-	+	+	51.0	3
AT4G23890.1	NdhS (CRR31)	-	+	+	27.7	2
AT1G74880.1	NdhO	-	+	+	17.6	0
AT5G21430.2	NdhU	-	+	-	24.3	0
AT2G05620.1	PGR5	-	+	-	14.3	1

Notes. Only proteins of which at least two unique peptides were detected in WT/OE-NTRC eluates (+), and which were absent from *ntrc* eluates are included in the table. MW (kDa) indicates molecular weight of the protein, and #Cys the number of conserved cysteine residues (see Supporting Information Tables S3–S5). For a description of experimental procedures see Materials and Methods. For a list of 100 chloroplast proteins detected in WT and/or OE-NTRC but not in *ntrc* eluates, see Supporting Information Table 2, and for a full list of detected peptides see Supporting Information Dataset S1.

(Figure 9). Neither NdhS nor PGR5 interacted with TRX-*x* in BiFC, which was used as a control in the test (Supporting Information Figure S7).

To assess the potential of the CEF-related proteins identified by Co-IP/MS (Table 1) and BiFC (Figure 9) to be targets of redox regulation by TRXs, their amino acid sequences were analyzed for conserved cysteine residues. Of the NDH subunits identified as putative NTRC targets by Co-IP/MS, NdhS, NdhH, and Ndh48 contain cysteine residues that are conserved in angiosperms (Table 1, Supporting Information Tables S3–S5), and therefore could in theory be subject to thiol regulation. NdhO and NdhU do not contain conserved cysteine residues, but they likely coprecipitate with the NTRC antibody because of their interactions with NdhH and NdhS subunits of the NDH complex, respectively (Shikanai, 2016).

PGR5 has been shown to form redox-dependent heterodimers with PGRL1 which have been proposed to be required for acceptance of electrons from ferredoxin and for reduction of PGRL1 (Hertle et al., 2013; Leister & Shikanai, 2013). The mature PGR5 polypeptide contains a single highly conserved cysteine residue (Munekage et al., 2002), which could hypothetically form an intermolecular disulfide with PGRL1 or some other partner, or be a target for S-nitrosylation or glutathionylation (Couturier, Chibani, Jacquot, & Rouhier, 2013; Zaffagnini et al., 2016). As a direct determination of PGR5 redox state with the alkylation method was not feasible, we investigated if the redox state of PGRL1 is affected by NTRC deficiency or overexpression. PGRL1 contains six conserved thiol-sensitive cysteine residues that form inter- and intramolecular disulfides (Hertle et al., 2013; Petroutsos et al., 2009). We observed that PGRL1 was mostly oxidized in dark-adapted leaves, but underwent a transient reduction during approximately 60 s of illumination with growth light (Supporting Information Figure S6A). This corresponds with the timescale of NPQ induction (Supporting Information Figure S5A), as well as with the transient increase in *pmf* and decrease in g_{H+} during dark-to-light transitions (Figure 5a–d). No significant difference to WT in PGRL1 reduction or protein content was detected in OE-NTRC (Supporting Information Figure S6A, Figure 3). PGR5 content of thylakoid membranes was, however, decreased in *ntrc* by 40% in comparison to WT (Figure 3).

3 | DISCUSSION

The role of CEF around PSI in the response of plants to fluctuating light conditions has attracted great attention during the past

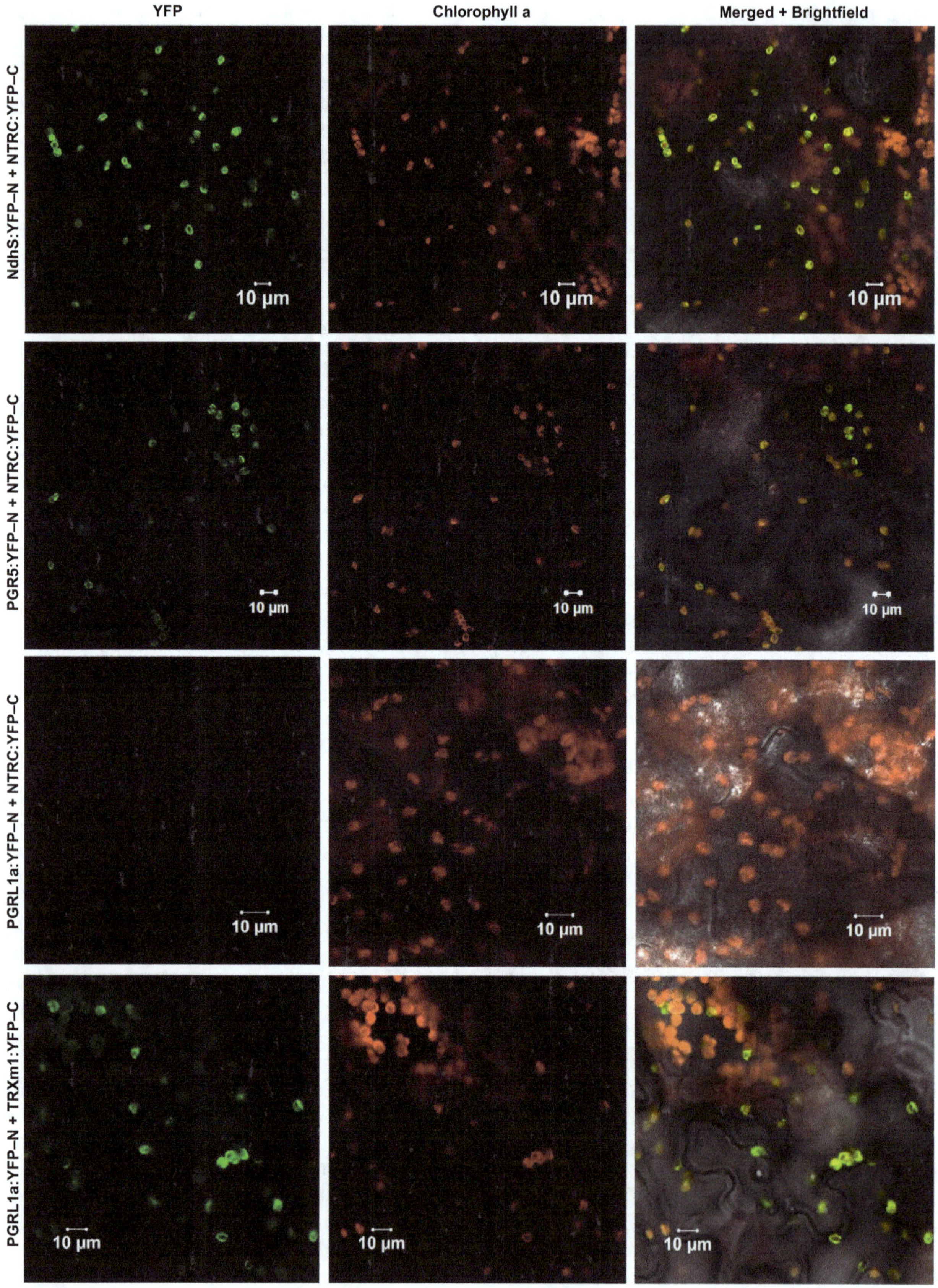

FIGURE 9 Bimolecular fluorescence complementation (BiFC) tests of *in planta* interactions between chloroplast TRXs and potential target proteins in cyclic electron flow. The left panel shows yellow fluorescent protein (YFP) fluorescence in green, the middle panel Chlorophyll a autofluorescence in red and the right panel, a merged image of YFP, chlorophyll, and brightfield images. YFP-N and YFP-C indicate expression of a fusion proteins including the N-terminal and C-terminal parts of YFP, respectively, in tobacco (*Nicotiana benthamiana*) leaves

10 years. Importance of CEF likely relies in its capacity to maintain redox balance in chloroplasts upon fluctuations in light intensity and during dark-to-light transitions (Strand, Fisher, & Kramer, 2016; Suorsa et al., 2016; Yamori et al., 2016). Plastidial thioredoxin systems, on the other hand, are crucial regulators of chloroplast metabolic reactions in the stroma. It has, however, remained unclear whether thioredoxin-related redox regulation is also involved in the attainment of CEF-mediated redox balance upon exposure of plants to changing light intensities. To this end, we applied here an *in vivo* approach to investigate, whether NTRC contributes to regulation of CEF pathways in chloroplasts.

3.1 | Regulation of photosynthesis is altered in the pleiotropic *ntrc* mutant

While the *ntrc* knockout plants are plain green and stunted (Lepistö et al., 2009; Pérez-Ruiz et al., 2006; Serrato et al., 2004), the rosettes of OE-NTRC lines are healthy and vigorous (Nikkanen et al., 2016; Toivola et al., 2013). Accordingly, the photosynthetic parameters measured for the *ntrc* knockout plants were not always in line with the results obtained with NTRC-overexpressing plants (Figures 2 and 4). In a short photoperiod, the *ntrc* plants have a highly pleiotropic phenotype (Supporting Information Figure S2A,B) (Kirchsteiger et al., 2009; Naranjo et al., 2016; Nikkanen et al., 2016; Pérez-Ruiz et al., 2006 Pulido et al., 2010; Thormählen et al., 2015) that complicates the interpretation of results from the *ntrc* line. The high, slow-rising PIFR and increased dark-reduction of PQ in short-day grown *ntrc* (Figures 2a and 4g) were likely caused by a high NPQ (Supporting Information Figures S2B and S5A) and its relaxation in darkness, lowered PSI content (Figure 3) (Thormählen et al., 2015), and impaired stromal redox metabolism (Nikkanen et al., 2016; Pérez-Ruiz et al., 2006). Furthermore, we observed increased accumulation of reduced Fd, the substrate of the NDH complex, in *ntrc* (Figure 2d), which may be due to lower activity of CBC enzymes and consequent lower consumption of NADPH in carbon fixation (Nikkanen et al., 2016). Accordingly, a high $NADPH/NADP^+$ ratio has been shown to activate CEF (Breyton et al., 2006; Okegawa, Kagawa, Kobayashi, & Shikanai, 2008). A high NADPH/NADP+ ratio together with high accumulation of reduced Fd may also explain the activation of NDH-dependent CEF by H_2O_2 (Strand et al., 2015; Strand, Fisher, & Kramer, 2017; Strand, Livingston, et al., 2017), since oxidative treatment decreases the activation of redox-regulated CBC enzymes, and consequently the consumption of NADPH in illuminated chloroplasts. Accordingly, the *ntrc* mutant has a higher accumulation of H_2O_2 than WT (Supporting Information Figure S2C and Pulido et al., 2010), a higher NADPH/NADP+ ratio (Thormählen et al., 2015), and lower activation states of CBC enzymes (Nikkanen et al., 2016; Pérez-Ruiz et al., 2006), which likely all contribute to the altered PIFR observed in Figure 2a. The PIFR was significantly diminished in *ntrc* plants grown in a longer photoperiod, where NPQ is lower (Supporting Information Figure S2A) and the phenotype of *ntrc* has been shown to be less severe in terms of growth, chlorophyll content, and efficiency of photochemistry (Lepistö et al., 2009, 2013). In *ntrc*, electrons also accumulate in the PQ pool upon increases in light intensity (Supporting Information Figure S5), although PSI is limited on the donor side (Figures 6 and 8), indicating that electron transfer is limited between the PQ pool and PSI. Higher content of PTOX (Figure 3) may assist relaxation of excitation pressure in the PQ pool of *ntrc* plants.

The pleiotropy described above for *ntrc* complicates the general interpretation of the results and makes it difficult to assess the contribution of NTRC to the regulation of photosynthesis when using the *ntrc* line alone in the experiments. To avoid these obstacles we have constructed the OE-NTRC line, whose phenotype and development are not considerably dissimilar to WT (Nikkanen et al., 2016; Toivola et al., 2013). OE-NTRC in different backgrounds (*ntrc*, *ndho* and *pgr5*) provides a more reliable platform to examine the direct effects of NTRC on specific plastidial processes.

3.2 | NTRC in regulation of CEF

In-depth analysis of NTRC overexpression lines with respect to thylakoid functional properties provided evidence that NTRC is indeed involved in regulation of CEF in the thylakoid membrane. Support for NTRC-induced activation of the NDH complex was obtained by analyzing thylakoid CEF-related functions in NTRC overexpression lines made on the backgrounds of *ndho* (OE-NTRC *ndho*) and *pgr5* (OE-NTRC *pgr5*), incapable of performing NDH- and PGR-dependent CEF, respectively. A distinct effect of NTRC overexpression on the postillumination rise of chlorophyll a fluorescence (Figure 2), the redox state of the plastoquinone pool in darkness (Figure 4) as well as on the generation of the *pmf* (Figures 5 and 7) and oxidation of P700 (Figures 6b and 8c) upon dark-to-light transitions and sudden increases in light intensity demonstrated the activating effect of NTRC on NDH-dependent CEF. Higher activity of NDH-dependent CEF in plants overexpressing NTRC was shown not to be due to increased accumulation of the NDH complex or its substrate, because no increase in the accumulation of either NDH subunits or reduced Fd was detected in OE-NTRC plants in comparison to WT (Figures 2d and 3). Furthermore, identification of NDH subunits in close proximity of the ferredoxin binding site as potential NTRC interactors supports the proposition that NTRC may have a direct effect on the activation of NDH (Table 1, Figure 9). Although several NDH subunits were detected by Co-IP/MS, most likely only one or few of these subunits are genuine NTRC targets. The others likely coprecipitate with the NTRC antibody due to reciprocal interactions of the NDH subunits on the stromal side of the thylakoid membrane (Peltier et al., 2016; Shikanai, 2016). Existence of a thiol-regulated component in the ferredoxin binding site would provide a mechanism for dynamic control of the ferredoxin:plastoquinone oxidoreductase activity of the complex in response to fluctuations in light conditions. Redox regulation of the NDH complex would allow rapid adjustment of *pmf* and nonphotochemical quenching as well as the maintenance of a redox balance between the electron transfer chain and the electron sink capacity of stromal acceptors, most importantly the CBC. In high light, less active NDH could prevent the reverse function of the complex (i.e., oxidization of PQ to reduce ferredoxin and transfer of protons from lumen to stroma) in conditions of high *pmf* and a reduced PQ pool (Strand, Fisher, & Kramer, 2017).

Notably, overexpression of NTRC also affects the function of PGR5. Nevertheless, as discussed below, the effect is not necessarily related to the putative role of PGR5 in CEF. Our results, more likely, support the hypothesis (Kanazawa et al., 2017; Tikkanen, Rantala, & Aro, 2015) that PGR5 is involved in controlling the proton conductivity of the thylakoid membrane, which, consequently, affects the generation of *pmf*.

3.3 | Regulation of *pmf* and the thylakoid redox balance via NDH and NTRC during changes in light conditions

Overexpression of NTRC caused elevated *pmf* under all light conditions, while no significant changes were observed in PSII quantum yield or thylakoid proton conductivity in comparison to WT (Figures 5, 6a, 7 and 8a). These results suggest that the elevation of *pmf* derives from enhanced CEF. Increased P700 oxidation during dark-to-light transitions in OE-NTRC was reverted in OE-NTRC *ndho* (Figure 6b), and a lack of NDH also delayed the ability to oxidize P700 during the high-light phases in fluctuating light (Figure 8c). It is therefore evident that the NDH complex controls the strength of the trans-thylakoid *pmf* as well as the redox balance between the electron transfer chain and the stroma, and that this regulation is under the control of the stromal TRX systems, with our results suggesting a specific role for the NTRC system.

While our results demonstrate enhancement of the NDH-dependent CEF by NTRC overexpression (Figure 2), earlier studies have revealed an inhibition of NDH-dependent CEF upon TRX-*m4* overexpression and, conversely, an enhancement in *trxm4* mutants (Courteille et al., 2013). Thus, it is conceivable that the two chloroplast TRX systems regulate CEF in an antagonistic way, although it remains to be elucidated how such a regulation might be mechanistically accomplished. We propose that in low light and upon sudden changes in the light intensity, NTRC is crucial for activation of the NDH-dependent CEF, while the TRX-*m4*-dependent inhibition of NDH-CEF requires higher light intensity or longer duration of illumination. Moderate to high light illumination is required to fully activate the FTR-dependent TRX system (reviewed in Geigenberger et al., 2017), which possibly contributes to the downregulation of NDH. In OE-NTRC, the NDH-dependent CEF is constitutively active in light, which contributes to elevated *pmf* in all light intensities. Upon transition from dark to low light, there is less difference between OE-NTRC and WT in terms of *pmf* formation (Figure 7a), because in those conditions the NTRC-mediated activation of NDH occurs similarly in WT and OE-NTRC.

The NDH complex translocates protons from the stroma to the lumen not only via Cyt *b6f* but also itself functions as a proton pump with a 2 H^+/e^- stoichiometry (Strand, Fisher, & Kramer, 2017). NDH-mediated CEF therefore contributes relatively more to *pmf* generation and consequently to ATP synthesis and NPQ induction than the PGR-dependent pathway. It has been postulated that the NDH complex is unlikely responsible for CEF during the early induction phase of photosynthesis, due to a low concentration of the complex in thylakoids in relation to the total PSI content (Joliot & Joliot, 2002). However, the NDH complex forms functional CEF-supercomplexes with PSI in stroma thylakoids (Peng et al., 2008) and recently it has been shown that a single NDH complex can bind up to six PSI complexes (Yadav et al., 2017). Specific association of NDH complex with several PSI complexes indicate that even a relatively low NDH content may have a significant impact on *pmf* generation. Moreover, in vivo results in the current paper as well as in another recent study (Shimakawa & Miyake, 2018) demonstrate the importance of the NDH complex for the redox poise of the photosynthetic electron transfer chain during dark-to-light transitions as well as during fast fluctuations in light intensity.

3.4 | Contribution of TRX systems to PGR-related CEF pathway

Increased activation of NDH-CEF alone is not sufficient to explain all observed changes of *pmf* in OE-NTRC plants. When compared to WT, *pmf* remained elevated in OE-NTRC *ndho* during the first seconds of photosynthetic induction and at steady state in growth and high light (Figures 5a and 7a). These results could be explained by activation of PGR-dependent CEF as well in plants overexpressing NTRC. Stromal thiol redox state has been previously suggested to control PGR-dependent CEF by a component that has a midpoint redox potential of −310 mV (Hertle et al., 2013; Strand, Fisher, Davis, et al., 2016). It has also been proposed that *m*-type TRXs, with redox potentials between −357 and −312 mV (Collin et al., 2003; Yoshida, Hara, & Hisabori, 2015), reduce an intermolecular disulfide in PGRL1 homodimers, and subsequently, the released monomeric PGRL1 may function as the ferredoxin-plastoquinone reductase (Hertle et al., 2013). Here, we confirm the previously reported transient reduction of PGRL1 during dark-to-light transitions (Hertle et al., 2013), but NTRC overexpression does not intervene in the reduction (Supporting Information Figure S6A). Moreover, TRX-*m1* but not NTRC interacts with PGRL1 in BiFC (Figure 9). Our results thus support the hypothesis that TRX-*m* is a primary reductant of PGRL1. Crosstalk between NTRC and FTR-dependent systems (Nikkanen et al., 2016; Thormählen et al., 2015; Toivola et al., 2013), and the interaction of NTRC with TRX-*m1* in BiFC (Nikkanen et al., 2016), further support the interpretation that the activation of PGR-dependent CEF is indirectly increased in NTRC-overexpressing plants through enhancement of TRX-*m* reduction. This would also be in line with the steady-state *pmf* increase observed in OE-NTRC *ndho* in comparison to WT (Figures 5a and 7a).

The NTRC overexpression may also affect the function of PGR5 in a way that is independent of its involvement in CEF. Redox regulation of PGR5 may occur to control its association with the ATP synthase during dark-to-light and low-to-high light transitions, and thereby allow inhibition of the ATP synthase in an unknown mechanism, as suggested earlier (Kanazawa et al., 2017; Tikkanen et al., 2015). Such a mechanism would result in acidification of the lumen and induction of NPQ, allowing dissipation of excess excitation energy from the electron transfer chain until CBC is activated. This hypothesis is supported by the impaired abilities of *pgr5* and OE-NTRC *pgr5* to control the activity of the ATP synthase at early stages of dark-light transitions and upon transitions to high light intensities (Figures 5d and 7d, Avenson, Cruz, Kanazawa, & Kramer, 2005). Furthermore, the elevated NTRC content in leaves caused decreased thylakoid proton conductivity upon increases in light intensity (Figure 7b,d), suggesting that NTRC controls the PGR5-dependent downregulation of proton efflux from the lumen. This is supported by the identification of PGR5 as a potential NTRC interactor (Table 1, Figures 3 and 9).

The *pgr5* mutant is known to suffer from an impaired ability to induce ΔpH-dependent photosynthetic control at Cyt b_6f upon shifts to high irradiance, possibly due to loss of the *pmf* (Suorsa et al., 2013; Takagi & Miyake, 2018). Interestingly, recovery of *pmf* in high light through NTRC overexpression (Figure 7) was not sufficient to prevent excess electron flow to PSI in *pgr5* background (Figure 8). This supports the hypothesis that the PGR5 protein is directly required to control linear electron flow (Suorsa et al., 2013; Takagi & Miyake, 2018; Tikkanen et al., 2015).

The initial strong *pmf* increase in OE-NTRC after onset of growth light illumination was evident also in both the OE-NTRC *ndho* and OE-NTRC *pgr5* plants (Figure 5a,b) indicating that this *pmf* peak is not caused by CEF. More likely, the initial *pmf* results from dark-activation of the CBC enzymes in plants overexpressing NTRC (Nikkanen et al., 2016), which provides an enhanced ability of the stroma to accept electrons from the PETC upon dark-to-light transition and consequently enhances proton pumping to the lumen.

3.5 | Cooperative regulation of photosynthetic electron transfer and carbon fixation by chloroplast thioredoxin systems

Light-dependent reductive activation of the ATP synthase, the CBC, and the NADP-malate dehydrogenase (NADP-MDH) by TRXs has been well established for several decades (reviewed in Buchanan, 2016). More recently, knowledge of TRX-mediated control has been extended to various regulatory and photoprotective mechanisms of photosynthesis, including regulation of state transitions (Rintamäki, Martinsuo, Pursiheimo, & Aro, 2000; Shapiguzov et al., 2016), NPQ (Brooks, Sylak-Glassman, Fleming, & Niyogi, 2013; Da et al., 2017; Hall et al., 2010; Naranjo et al., 2016) and CEF (Courteille et al., 2013; Hertle et al., 2013; Strand, Fisher, Davis, et al., 2016). We propose here a model, comprising a cooperative function of the two chloroplast TRX systems with distinct reductants and redox potentials that allows the maintenance of redox balance between the two photosystems and stromal metabolism during fluctuations in light conditions. This is achieved through dynamic regulation of the activities of the ATP synthase, NPQ, the NDH complex, PGRL1/PGR5 as well as the LHCII kinase STN7 by reversible thiol modifications. We propose a specific role for NTRC in regulating NDH-CEF, the ATP synthase and CBC enzymes in low light, dark-to-light transitions, and during sudden increases in light intensity, as schematically depicted in Figure 10.

In darkness, a proportion of NDH complexes in the thylakoid membrane is activated by NTRC, and moderate chlororespiration from NDH to PTOX occurs. Due to an inactive ATP synthase and proton pumping activity of NDH, a weak proton gradient over the thylakoid membrane is established. Redox-regulated CBC enzymes are inactive, causing PC and P700 to be reduced (Schreiber, 2017) due to lack of electron acceptors in the stroma. In OE-NTRC, chlororespiration via NDH to the PQ pool is elevated due to an increased amount of active NDH complexes. This leads to increased protonation of the lumen and higher reduction of the PQ pool. Proportions of the ATP synthase and CBC enzyme pools are also activated due to high NTRC content.

Upon transition from dark to low light, the ATP synthase pool is fully reduced and the CBC enzyme pool partially reduced by NTRC in WT plants. Delay in activation of the CBC enzymes causes, however, a reduction of the PQ pool due to scarcity of stromal acceptors limiting electron transfer. NTRC contributes to the activation of NDH-dependent CEF, which alleviates electron pressure at PSI and transiently increases *pmf* and induces NPQ. In OE-NTRC, P700 and PC are effectively oxidized upon onset of low illumination, as the acceptor side limitation is negligible due to fully active NDH-dependent CEF, ATP synthase, and redox-activated CBC enzymes. This results in an elevated ΔpH and faster induction of NPQ in comparison to WT at the initial phase of illumination. At steady state, NPQ is lower than in WT despite high ΔpH, suggesting downregulation of NPQ by thioredoxin via a ΔpH-independent mechanism, as reported previously by Brooks et al. (2013).

When a leaf is shifted from low to high irradiance, both TRX systems become fully active, and the CBC enzymes as well as the PGR-dependent CEF are fully activated. NTRC affects PGR5-dependent inhibition of the ATP synthase, which contributes to the accumulation of protons in the lumen. Consequently, NPQ and downregulation of electron transfer at Cyt *b6f* are induced. Electrons are effectively pulled from PSI, and the donor side is limiting electron transfer. In OE-NTRC, increased reduction in PGR5 likely leads to stronger downregulation of the ATP synthase. This, together with proton pumping by constantly active NDH and possibly through increased TRX-*m*-mediated activation of the PGR-dependent pathway, results in high *pmf*. NPQ is, however, lower than in WT due to ΔpH-independent downregulation of NPQ by overexpressed NTRC.

4 | MATERIALS AND METHODS

4.1 | Plant material and growth conditions

Experiments have been carried out with *Arabidopsis thaliana* wild-type (WT) lines of the Columbia ecotype (Col-0 and Col-*gl1*), and with the following transgenic lines: NTRC overexpression line (Toivola et al., 2013), T-DNA knockout mutants of *NTRC* (At2g41680, SALK_096776) (Lepistö et al., 2009), *NDH-O* (At1g74880, SALK_068922) (Rumeau et al., 2005) and *STN7* (At1g68830,

SALK_073254) (Bellafiore et al., 2005) as well as the *pgr5* mutant (At2g05620) (Munekage et al., 2002). The plants were grown in a photoperiod of 8 hr light/16 hr darkness at 23°C under 200 μmol of photons $m^{-2}s^{-1}$ for all experiments except for the measurements shown in Supporting Information Figure S2A, for which plants were grown in a 12 hr/12 hr photoperiod under 130 μmol $m^{-2}s^{-1}$. Wild-type tobacco (*Nicotiana benthamiana*) plants used in BiFC tests were grown under 130 μmol photons $m^{-2}s^{-1}$ at 23°C in a 16 hr light/8 hr dark photoperiod. The OE-NTRC *ndho* and OE-NTRC *pgr5* lines were generated by *Agrobacterium tumefaciens* and floral dipping-mediated transformation of the *ndho* knockout and *pgr5* mutant lines, respectively, with the NTRC overexpression construct as described previously (Toivola et al., 2013). The OE-NTRC *ndho* and OE-NTRC *pgr5* plants used in the experiments were heterozygous T2 generation plants that were selected on agar plates with 0.5X Murashige–Skoog medium (MS) (Murashige & Skoog, 1962) and 50 μg/ml kanamycin. The plants were subsequently transferred to soil and grown in an 8 hr light/16 hr darkness photoperiod at 23°C under 200 μmol of photons $m^{-2}s^{-1}$ for 4 weeks before usage in the experiments. As control, OE-NTRC plants were similarly selected on kanamycin-containing plates while WT Col-0 and WT Col-*gl1* (ecotype of the *pgr5* mutant) plants were grown on 0.5X MS-agar plates without antibiotics for an equivalent time.

4.2 | Determination of H_2O_2 content in leaves

The hydrogen peroxide content in leaves was estimated by staining with diaminobenzidine (DAB), as previously described in Lepistö et al. (2013). Detached leaves from 4-week-old WT, *ntrc*, and OE-NTRC plants were incubated overnight in darkness in 0.1 mg ml^{-1} solution of diaminobenzidine (DAB; Sigma-Aldrich) (pH 3.8), after which the leaves were illuminated with either 40 or 200 μmol photons $m^{-2}s^{-1}$ for 1 hr. Chlorophyll was then bleached by incubating the leaves in ethanol and subsequently photographed. Image J software (Schneider, Rasband, & Eliceiri, 2012) was used to quantify the intensity of the staining.

4.3 | Measurement of chlorophyll a fluorescence and P700 and Fd redox changes

The postillumination chlorophyll a fluorescence rise (PIFR) was measured from detached leaves with the Multicolor-PAM fluorometer (Walz). A 480-nm measuring beam at an intensity of 0.2 μmol photons $m^{-2}s^{-1}$ was used to measure fluorescence changes after illumination of dark-adapted (30 min) leaves with 67 μmol photons m^{-2} s^{-1} of white actinic light for 500 s, with saturating pulses of 800 ms (10,000 μmol photons $m^{-2}s^{-1}$) in the beginning and at 400 s to determine Fm and Fm'. For Figure 2c, saturating pulses were administered at 20-s intervals for the first minute, 30-s intervals for the second minute, and at 45-s intervals for the rest of the experiment in order to more closely follow photosynthetic parameters such as NPQ and Y(II). The actinic light was then switched off and the changes in chlorophyll a fluorescence in the dark were observed for 300 s. A 10-s pulse of far red light was then given to fully oxidize

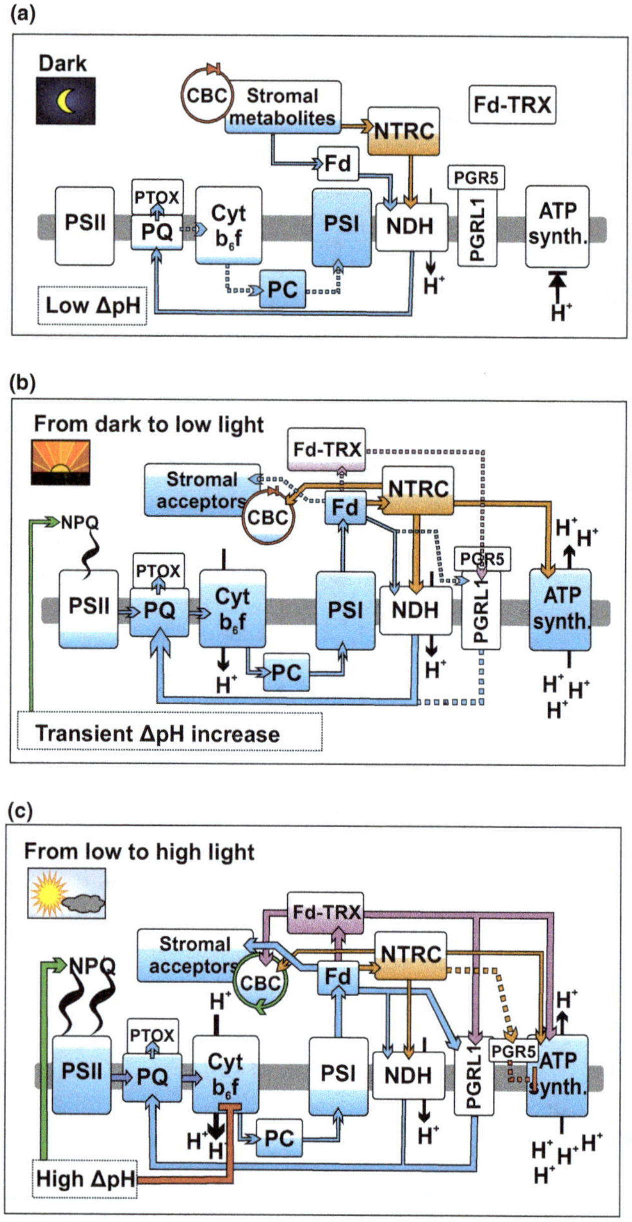

FIGURE 10 A schematic model of the role of chloroplast TRX systems in regulating CEF and the *pmf* during dark-to-light transitions and fluctuations in light intensity. (a–c) Dark-adapted leaves (a), transition from dark to low light (b) and transition from low to high light (c). Blue color indicates the approximate reduction levels of different photosynthetic redox components based on data in the current paper (Figures 1, 2, 8, and 9) as well as other reports (Yoshida, Hara and Hisabori, 2015; Nikkanen et al., 2016; Schreiber, 2017). Green and red arrows indicate activating and inhibitory effects, respectively, while orange color represents the thiol regulation by NTRC and purple by the Fd-TRX system. Thicker lines depict stronger effect than thin and dotted lines. For details see the text. The arrow to NPQ refers to the induction of the qE component of NPQ due to acidification of the lumen (Demmig-Adams et al., 2012). The arrows representing reduction of PQ by the NDH complex and PGRL1 have been drawn through the lumen only to increase clarity of the illustration

the PQ pool, and the subsequent rereduction of the PQ pool was detected through a rise in Chl fluorescence. All values were normalized to Fm.

The OJIP transients were recorded with the Multicolor-PAM from dark-adapted (30 min) leaves and from leaves preilluminated with far red light (intensity setting 15) for 6 s, according to the method described by (Toth et al., 2007). A saturating pulse of 3,000 μmol photons $m^{-2} s^{-1}$ and measuring light at 440 nm were used in the measurements. The initial slopes of the transients in Supporting Information Table S1 were calculated from F/Fm values between 50 and 150 μs.

The Dual-PAM-100 (Walz) was used to simultaneously record the Chl a fluorescence and P700-dependent difference in absorbance at 875 and 830 nm during transitions from dark to 166 μmol photons $m^{-2} s^{-1}$ (Figure 6) and during a light regime where a 620 nm AL fluctuates between 39 and 825 μmol photons $m^{-2} s^{-1}$ (Figure 8). Saturating pulses were administered at 10- or 15-s intervals for the measurements in Figure 6 and at 15-s intervals for the first minute after onset of illumination, and at 20-s intervals thereafter for Figure 8. Because *ntrc* leaves are very small in size and low in chlorophyll content, it was in some cases necessary to record from two or three leaves simultaneously to obtain a P700 signal of sufficient quality. The parameters shown were calculated with the Dual-PAM-100 software as described by Bilger and Björkman (1990), Klughammer and Schreiber (2008a,b) and Kramer, Johnson, Kiirats, and Edwards (2004).

For determination of Fd redox state, the Dual/Klas-NIR (Walz) spectrometer was used to record the four absorbance differences between 785 and 840, 810 and 870, 870 and 970, as well as 795 and 970 nm, from which the redox changes in P700, PC, and Fd were deconvoluted as described by Klughammer and Schreiber (2016) and Schreiber (2017). A similar illumination and postillumination regime was used as described above for the measurement of PIFR, with the exception that dark-adapted leaves were illuminated with 61 μmol photons $m^{-2} s^{-1}$ of 630 nm instead of white actinic light. Measured Fd redox changes were then normalized to the maximum level of Fd reduction, which was determined according to Schreiber and Klughammer (2016).

4.4 | Measurement of electrochromic shift (ECS)

In order to measure the magnitude and kinetics of *pmf* formation, changes in the electrochromic shift (ECS, P515) signal were recorded with the Dual-PAM-100 and the P515/535 accessory module (Walz) (Klughammer, Siebke, & Schreiber, 2013; Schreiber & Klughammer, 2008). A dual beam difference signal between 550 and 515 nm was used to avoid distortion of results by scattering effects. A measuring light at a 2,000-Hz pulse frequency was used in all ECS measurements. For the dark-to-light and low-to-high light transition measurements in Figures 5 and 7, plants were first dark-adapted for 30 min. A single-turnover saturating flash (20 μs) of 14,000 μmol photons $m^{-2} s^{-1}$ was then applied to obtain ECS_{ST}, a maximum absorbance change value that was used to normalize all results to account for differences in leaf thickness and chlorophyll content between individual leaves and lines (Kramer & Crofts, 1989). The obtained values of ECS_{ST} were in good correlation with the differences in chlorophyll content in OE-NTRC and *ntrc* lines reported previously (Toivola et al., 2013). In order to distinguish the light-induced ECS change (ECS_T) from signal drift and baseline change, dark intervals of 250 ms were applied at the following time points after the onset of AL illumination: 0.8; 2.7; 4.7; 6.7; 8.7; 10.7; 12.7; 16.7; 20.7; 24.7; 28.7; 32.7; 36.7; 40.6; 44.7; 48.7; 52.7; 56.7; 60.7; 80.6; 100.5; 120.5; 140.5, and 160.5 s after onset of illumination. Additionally, during the shift from low to high irradiance (Figure 7), dark intervals were applied at 1.1; 5.1; 9.1; 13.1; 23.1; 33.1; 43.1; 53.1; 73.1; 93.1; 113.1; 133.1, and 153.1 s after the increase in light intensity. ECS_T was calculated as the difference between total ECS in light and an Y_0 value obtained from the first-order exponential fit to the decay kinetics of the ECS signal during a dark interval. Total *pmf* was then calculated as ECS_T/ECS_{ST}. The g_H+ parameter, describing thylakoid membrane conductivity to protons, was calculated as the inverse of the time constant of a first-order exponential fit to ECS decay kinetics during a dark interval (Avenson et al., 2005; Cruz, Sacksteder, Kanazawa, & Kramer, 2001; Cruz et al., 2005). The v_{H+} parameter indicating the rate of proton flux over the thylakoid membrane and corresponding to the initial slope of the decay of the ECS signal upon cessation of illumination was calculated as *pmf* x g_{H+} (Cruz et al., 2005). Partitioning of total *pmf* to its components ΔpH and ΔΨ was determined from the light-off response of the ECS signal (Cruz et al., 2001) after 3 min of illumination, also using the Dual-PAM ECS module as described by Schreiber and Klughammer (2008). Same settings were used for the determination of *pmf* partitioning as for the dark-to-light and low-to-high light transition measurements.

4.5 | Protein extraction, alkylation of thiols, and SDS-PAGE

Proteins and thylakoids were isolated as previously described (Lepistö et al., 2009), while chlorophyll content was determined according to (Porra, Thompson, & Kriedemann, 1989) and protein content with the Bio-Rad Protein Assay kit. For determination of the redox states of TRX-regulated proteins, leaf proteins were precipitated with trichloroacetic acid (TCA) and free thiols in proteins alkylated with N-ethylmaleimide (NEM, Sigma-Aldrich). After alkylation protein disulfides were reduced with dithiothreitol (DTT, Sigma-Aldrich) and subsequently produced thiols were alkylated with methoxypolyethylene glycol maleimide M_n 5000 (MAL-PEG, Sigma-Aldrich) as described earlier (Nikkanen et al., 2016). Sodium dodecyl sulfate polyacrylamide gel electrophoresis (SDS-PAGE) and immunoblotting was performed as reported in (Nikkanen et al., 2016). For running the MAL-PEG samples precast 4%–20% Mini-PROTEAN TGX gels (Bio-Rad) were used, except for the gel in Figure 1a,b, and Supporting Information Figure S1A, where a 12% polyacrylamide gel was used. PVDF membranes were probed with antibodies raised against NTRC (Lepistö et al., 2009), D1 (Research Genetics, Inc (Thermo Fisher)), PsaB (Agrisera, AS10 695), Cyt *f* (kindly provided by L. Zhang), PTOX (kindly provided by M. Kuntz), NdhH (Agrisera), NdhS (Agrisera), $CF_1\gamma$ (Agrisera, AS08 312), PGRL1 (Agrisera, AS10 725), PGR5 (Agrisera) or phosphothreonine (P-Thr)

(New England Biolabs). Membranes were then treated with a horseradish peroxidase (HRP)-conjugated goat anti-rabbit secondary antibody (Agrisera, AS09 602) for 2 hr. All immunoblots shown are representative of at least three biological replicates. Quantifications of protein content shown in Figure 3b were performed using the ImageJ software (Schneider et al., 2012) and normalized according to the intensity of Li-Cor Revert Total Protein Stain. Statistical significance was determined using two-tailed Student's *t* tests for unequal variances with *p*-values below 0.05 interpreted as statistically significant.

4.6 | Coimmunoprecipitation and Mass spectrometry

For coimmunoprecipitation (Co-IP), WT, *ntrc*, and OE-NTRC leaves were frozen in liquid N_2, lysed in Pierce IP Lysis buffer containing 1% NP-40 detergent (Thermo-Fisher), and immunoprecipitated in a resin containing NTRC-specific antibody using the Pierce Co-IP kit (Thermo-Fisher) with an affinity-purified NTRC-specific antibody, as described previously (Nikkanen et al., 2016). Co-IP eluates were denatured and purified by SDS-PAGE in a 6% acrylamide gel with 6 M urea, subjected to in-gel tryptic digestion and the extracted peptides analyzed with the Q Exactive Hybrid Quadruple-Orbitrap mass spectrometer (Thermo-Fisher Scientific) in DDA mode as previously described (Trotta, Suorsa, Rantala, Lundin, & Aro, 2016). MS/MS spectra were analyzed with an in-house installation of Mascot (v.2.4) (Matrix Science) search engine and analyzed with Proteome Discoverer (v.1.4) Software (Thermo Scientific), restricting the search to the nonredundant database TAIR10 supplemented with most common laboratory contaminants (Trotta et al., 2016). Peptides were validated by Decoy Database Search, with target false discovery rates (FDR) set to be below 0.01 (strict) or below 0.05 (relaxed).

4.7 | BiFC tests

Bimolecular fluorescence complementation tests (BiFC) were performed as described in (Nikkanen et al., 2016). For the current study, coding sequences of PGR5, PGRL1a, and NdhS obtained from Arabidopsis Biological Resource Center (ABRC) were cloned into pSPYNE-35S and pSPYCE-35S binary vectors (Walter et al., 2004), and the resulting constructs were checked by sequencing. Primer sequences used for cloning are listed in Supporting Information Table S6. Imaging of YFP and chlorophyll autofluorescence from *N. benthamiana* leaves infiltrated with *Agrobacterium tumefaciens* strain GV3101 carrying the appropriate binary vectors was performed with a Zeiss LSM780 laser scanning confocal microscope 3 days after infiltration. The negative result between PGRL1:YFP-N and NTRC:YFP-C also serves as a negative control.

4.8 | Multiple alignment of amino acid sequences

Amino acid sequences of NdhH, Ndh48, NdhS, NdhJ, and Ndh45 in *Arabidopsis thaliana* and, as available, in *Populus trichocarpa, Vitis vinifera, Glycine max, Solanum lycopersicum, Oryza sativa, Sorghum bicolor, Brachypodium distachion, Physcomitrella patens, Selaginella moellendorffii*, and *Synechocystis PCC 6803* were obtained from the UniProtKB database and aligned with the Clustal Omega 1.2.4 online alignment tool (Sievers et al., 2011) using default settings.

ACKNOWLEDGMENTS

We thank Jesse Ojala for assistance with the experiments, Esa Tyystjärvi for the advice for PIFR measurements, Sari Järvi and Marjaana Suorsa for advice on the use of antibodies, Alexandrina Stirbet for expert advice on measurement of OJIP transients, and Mika Keränen, Kurt Ståhle, and Tapio Ronkainen for technical assistance.

AUTHOR CONTRIBUTIONS

L.N. and E.R. designed the research, L.N., J.T., and A.T. performed the research, L.N., J.T., A.T., M.T., and E.R. analyzed the data, L.N. and E.R. wrote the article with input from E.-M.A., M.T., A.T., M.G.D., and J.T.

ACCESSION NUMBERS

The Arabidopsis Genome Initiative locus identifiers (AGI) used in this paper are listed in Table 1, Supporting Information Table S2–S6 and Dataset S1.

REFERENCES

Allorent, G., Osorio, S., Vu, J. L., Falconet, D., Jouhet, J., Kuntz, M., … Finazzi, G. (2015). Adjustments of embryonic photosynthetic activity modulate seed fitness in Arabidopsis thaliana. *New Phytologist, 205*, 707–719. https://doi.org/10.1111/nph.13044

Armbruster, U., Carrillo, L. R., Venema, K., Pavlovic, L., Schmidtmann, E., Kornfeld, A., … Jonikas, M. C. (2014). Ion antiport accelerates photosynthetic acclimation in fluctuating light environments. *Nature Communications, 5*, 5439. https://doi.org/10.1038/ncomms6439

Armbruster, U., Correa Galvis, V., Kunz, H. H., & Strand, D. D. (2017). The regulation of the chloroplast proton motive force plays a key role for photosynthesis in fluctuating light. *Current Opinion in Plant Biology, 37*, 56–62. https://doi.org/10.1016/j.pbi.2017.03.012

Avenson, T., Cruz, J., Kanazawa, A., & Kramer, D. (2005). Regulating the proton budget of higher plant photosynthesis. *Proceedings of the National Academy of Sciences of the United States of America, 102*, 9709–9713. https://doi.org/10.1073/pnas.0503952102

Bailey, S., Walters, R. G., Jansson, S., & Horton, P. (2001). Acclimation of *Arabidopsis thaliana* to the light environment: the existence of separate low light and high light responses. *Planta, 213*, 794–801. https://doi.org/10.1007/s004250100556

Balmer, Y., Koller, A., del Val, G., Manieri, W., Schurmann, P., & Buchanan, B. (2003). Proteomics gives insight into the regulatory function of chloroplast thioredoxins. *Proc. Natl. Acad. Sci. U. S. A., 100*, 370–375.

Balsera, M., Uberegui, E., Schürmann, P., & Buchanan, B. B. (2014). Evolutionary development of redox regulation in chloroplasts. *Antioxidants and Redox Signaling, 21*, 1327–1355. https://doi.org/10.1089/ars.2013.5817

Bellafiore, S., Barneche, F., Peltier, G., & Rochaix, J. (2005). State transitions and light adaptation require chloroplast thylakoid protein kinase STN7. *Nature, 433*, 892–895. https://doi.org/10.1038/nature03286

Bilger, W., & Björkman, O. (1990). Role of the xanthophyll cycle in photoprotection elucidated by measurements of light-induced absorbance changes, fluorescence and photosynthesis in leaves of *Hedera canariensis*. *Photosynthesis Research*, *25*, 173–185. https://doi.org/10.1007/BF00033159

Breyton, C., Nandha, B., Johnson, G. N., Joliot, P., & Finazzi, G. (2006). Redox modulation of cyclic electron flow around photosystem I in C3 plants. *Biochemistry (New York)*, *45*, 13465–13475. https://doi.org/10.1021/bi061439s

Brooks, M. D., Sylak-Glassman, E. J., Fleming, G. R., & Niyogi, K. K. (2013). A thioredoxin-like/beta-propeller protein maintains the efficiency of light harvesting in Arabidopsis. *Proceedings of the National Academy of Sciences of the United States of America*, *110*, 2733–2740. https://doi.org/10.1073/pnas.1305443110

Buchanan, B. B. (2016). The path to thioredoxin and redox regulation in chloroplasts. *Annual Review of Plant Biology*, *67*, 1–24. https://doi.org/10.1146/annurev-arplant-043015-111949

Carrillo, L. R., Froehlich, J. E., Cruz, J. A., Savage, L. J., & Kramer, D. M. (2016). Multi-level regulation of the chloroplast ATP synthase: The chloroplast NADPH thioredoxin reductase C (NTRC) is required for redox modulation specifically under low irradiance. *The Plant Journal*, *87*, 654–663. https://doi.org/10.1111/tpj.13226

Collin, V., Issakidis-Bourguet, E., Marchand, C., Hirasawa, M., Lancelin, J., Knaff, D., & Miginiac-Maslow, M. (2003). The Arabidopsis plastidial thioredoxins - New functions and new insights into specificity. *Journal of Biological Chemistry*, *278*, 23747–23752. https://doi.org/10.1074/jbc.M302077200

Courteille, A., Vesa, S., Sanz-Barrio, R., Cazale, A., Becuwe-Linka, N., Farran, I., … Rumeau, D. (2013). Thioredoxin m4 controls photosynthetic alternative electron pathways in *Arabidopsis*. *Plant Physiology*, *161*, 508–520. https://doi.org/10.1104/pp.112.207019

Couturier, J., Chibani, K., Jacquot, J., & Rouhier, N. (2013). Cysteine-based redox regulation and signaling in plants. *Frontiers in Plant Science*, *4*, 105.

Cruz, J., Avenson, T., Kanazawa, A., Takizawa, K., Edwards, G., & Kramer, D. (2005). Plasticity in light reactions of photosynthesis for energy production and photoprotection. *Journal of Experimental Botany*, *56*, 395–406.

Cruz, J., Sacksteder, C., Kanazawa, A., & Kramer, D. (2001). Contribution of electric field (Delta psi) to steady-state transthylakoid proton motive force (pmf) in vitro and in vivo. Control of pmf parsing into Delta psi and Delta pH by ionic strength. *Biochemistry (New York)*, *40*, 1226–1237.

Da, Q., Sun, T., Wang, M., Jin, H., Li, M., Feng, D., … Liu, B. (2017). M-type thioredoxins are involved in the xanthophyll cycle and proton motive force to alter NPQ under low-light conditions in *Arabidopsis*. *Plant Cell Reports*, *37*, 279–291.

DalCorso, G., Pesaresi, P., Masiero, S., Aseeva, E., Schuenemann, D., Finazzi, G., … Leister, D. (2008). A complex containing PGRL1 and PGR5 is involved in the switch between linear and cyclic electron flow in *Arabidopsis*. *Cell*, *132*, 273–285. https://doi.org/10.1016/j.cell.2007.12.028

Demmig-Adams, B., Cohu, C. M., Muller, O., & Adams, W. W. 3rd (2012). Modulation of photosynthetic energy conversion efficiency in nature: From seconds to seasons. *Photosynthesis Research*, *113*, 75–88. https://doi.org/10.1007/s11120-012-9761-6

Fan, D., Nie, Q., Hope, A. B., Hillier, W., Pogson, B. J., & Chow, W. S. (2007). Quantification of cyclic electron flow around Photosystem I in spinach leaves during photosynthetic induction. *Photosynthesis Research*, *94*, 347–357. https://doi.org/10.1007/s11120-006-9127-z

Geigenberger, P., Thormählen, I., Daloso, D. M., & Fernie, A. R. (2017). The unprecedented versatility of the plant thioredoxin system. *Trends in Plant Science*, *22*, 249–262. https://doi.org/10.1016/j.tplants.2016.12.008

Gollan, P. J., Lima-Melo, Y., Tiwari, A., Tikkanen, M., & Aro, E. (2017). Interaction between photosynthetic electron transport and chloroplast sinks triggers protection and signalling important for plant productivity. *Philosophical Transactions of the Royal Society of London. Series B, Biological sciences*, *372*, 20160390. https://doi.org/10.1098/rstb.2016.0390

Gotoh, E., Matsumoto, M., Ogawa, K., Kobayashi, Y., & Tsuyama, M. (2010). A qualitative analysis of the regulation of cyclic electron flow around photosystem I from the post-illumination chlorophyll fluorescence transient in *Arabidopsis*: A new platform for the in vivo investigation of the chloroplast redox state. *Photosynthesis Research*, *103*, 111–123. https://doi.org/10.1007/s11120-009-9525-0

Grieco, M., Tikkanen, M., Paakkarinen, V., Kangasjärvi, S., & Aro, E. M. (2012). Steady-state phosphorylation of light-harvesting complex II proteins preserves photosystem I under fluctuating white light. *Plant Physiology*, *160*, 1896–1910. https://doi.org/10.1104/pp.112.206466

Hall, M., Mata-Cabana, A., Åkerlund, H. E., Florencio, F. J., Schroder, W. P., Lindahl, M., & Kieselbach, T. (2010). Thioredoxin targets of the plant chloroplast lumen and their implications for plastid function. *Proteomics*, *10*, 987–1001.

Hangarter, R. P., & Good, N. E. (1982). Energy thresholds for ATP synthesis in chloroplasts. *Biochimica et Biophysica Acta*, *681*, 397–404. https://doi.org/10.1016/0005-2728(82)90181-5

Hasan, S. S., Yamashita, E., Baniulis, D., & Cramer, W. A. (2013). Quinone-dependent proton transfer pathways in the photosynthetic cytochrome b_6f complex. *Proceedings of the National Academy of Sciences of the United States of America*, *110*, 4297–4302. https://doi.org/10.1073/pnas.1222248110

Hertle, A. P., Blunder, T., Wunder, T., Pesaresi, P., Pribil, M., Armbruster, U., & Leister, D. (2013). PGRL1 is the elusive ferredoxin-plastoquinone reductase in photosynthetic cyclic electron flow. *Molecular Cell*, *49*, 511–523. https://doi.org/10.1016/j.molcel.2012.11.030

Hirasawa, M., Schurmann, P., Jacquot, J., Manieri, W., Jacquot, P., Keryer, E.,… Knaff, D. (1999). Oxidation-reduction properties of chloroplast thioredoxins, Ferredoxin: Thioredoxin reductase, and thioredoxin f-regulated enzymes. *Biochemistry (New York)*, *38*, 5200–5205.

Horvath, E., Peter, S., Joet, T., Rumeau, D., Cournac, L., Horvath, G., … Medgyesy, P. (2000). Targeted inactivation of the plastid ndhB gene in tobacco results in an enhanced sensitivity of photosynthesis to moderate stomatal closure. *Plant Physiology*, *123*, 1337–1349. https://doi.org/10.1104/pp.123.4.1337

Johnson, G. N. (2011). Physiology of PSI cyclic electron transport in higher plants. *Biochimica et Biophysica Acta*, *1807*, 384–389. https://doi.org/10.1016/j.bbabio.2010.11.009

Joliot, P., & Johnson, G. N. (2011). Regulation of cyclic and linear electron flow in higher plants. *Proceedings of the National Academy of Sciences of the United States of America*, *108*, 13317–13322. https://doi.org/10.1073/pnas.1110189108

Joliot, P., & Joliot, A. (2002). Cyclic electron transfer in plant leaf. *Proceedings of the National Academy of Sciences of the United States of America*, *99*, 10209–10214. https://doi.org/10.1073/pnas.102306999

Kanazawa, A., Ostendorf, E., Kohzuma, K., Hoh, D., Strand, D. D., Sato-Cruz, M., … Kramer, D. M. (2017). Chloroplast ATP synthase modulation of the thylakoid proton motive force: Implications for photosystem I and photosystem II photoprotection. *Frontiers in Plant Science*, *8*, 719. https://doi.org/10.3389/fpls.2017.00719

Kirchsteiger, K., Pulido, P., Gonzalez, M., & Javier Cejudo, F. (2009). NADPH thioredoxin reductase C controls the redox status of chloroplast 2-Cys peroxiredoxins in *Arabidopsis thaliana*. *Molecular Plant*, *2*, 298–307. https://doi.org/10.1093/mp/ssn082

Klughammer, C., & Schreiber, U. (2008a). Complementary PSII quantum yields calculated from simple fluorescence parameters measured by PAM fluorometry and the Saturation Pulse method. *PAM Application Notes*, *1*, 27–35.

Klughammer, C., & Schreiber, U. (2008b). Saturation Pulse method for assessment of energy conversion in PSI. *PAM Application Notes*, *1*, 11–14.

Klughammer, C., & Schreiber, U. (2016). Deconvolution of ferredoxin, plastocyanin, and P700 transmittance changes in intact leaves with a new type of kinetic LED array spectrophotometer. *Photosynthesis Research, 128*, 195–214. https://doi.org/10.1007/s11120-016-0219-0

Klughammer, C., Siebke, K., & Schreiber, U. (2013). Continuous ECS-indicated recording of the proton-motive charge flux in leaves. *Photosynthesis Research, 117*, 471–487. https://doi.org/10.1007/s11120-013-9884-4

Kono, M., & Terashima, I. (2014). Long-term and short-term responses of the photosynthetic electron transport to fluctuating light. *Journal of Photochemistry and Photobiology B: Biology, 137*, 89–99. https://doi.org/10.1016/j.jphotobiol.2014.02.016

Kou, J., Takahashi, S., Fan, D., Badger, M. R., & Chow, W. S. (2015). Partially dissecting the steady-state electron fluxes in Photosystem I in wild-type and pgr5 and ndh mutants of Arabidopsis. *Frontiers in Plant Science, 6*, 758.

Kramer, D. M., & Crofts, A. R. (1989). Activation of the chloroplast atpase measured by the electrochromic change in leaves of intact plants. *Biochimica et Biophysica Acta, 976*, 28–41. https://doi.org/10.1016/S0005-2728(89)80186-0

Kramer, D. M., Johnson, G., Kiirats, O., & Edwards, G. E. (2004). New fluorescence parameters for the determination of Q(A) redox state and excitation energy fluxes. *Photosynthesis Research, 79*, 209–218. https://doi.org/10.1023/B:PRES.0000015391.99477.0d

Leister, D., & Shikanai, T. (2013). Complexities and protein complexes in the antimycin A-sensitive pathway of cyclic electron flow in plants. *Frontiers in Plant Science, 4*, 161.

Lepistö, A., Kangasjärvi, S., Luomala, E., Brader, G., Sipari, N., Keränen, M., ... Rintamäki, E. (2009). Chloroplast NADPH-thioredoxin reductase interacts with photoperiodic development in *Arabidopsis*. *Plant Physiology, 149*, 1261–1276. https://doi.org/10.1104/pp.108.133777

Lepistö, A., Pakula, E., Toivola, J., Krieger-Liszkay, A., Vignols, F., & Rintamäki, E. (2013). Deletion of chloroplast NADPH-dependent thioredoxin reductase results in inability to regulate starch synthesis and causes stunted growth under short-day photoperiods. *Journal of Experimental Botany, 64*, 3843–3854. https://doi.org/10.1093/jxb/ert216

Marchand, C., Le Marechal, P., Meyer, Y., & Decottignies, P. (2006). Comparative proteomic approaches for the isolation of proteins interacting with thioredoxin. *Proteomics, 6*, 6528–6537.

Martin, M., Noarbe, D. M., Serrot, P. H., & Sabater, B. (2015). The rise of the photosynthetic rate when light intensity increases is delayed in ndh gene-defective tobacco at high but not at low CO2 concentrations. *Frontiers in Plant Science, 6*, 34.

Miyake, C., Shinzaki, Y., Miyata, M., & Tomizawa, K. (2004). Enhancement of cyclic electron flow around PSI at high light and its contribution to the induction of non-photochemical quenching of chl fluorescence in intact leaves of tobacco plants. *Plant and Cell Physiology, 45*, 1426–1433. https://doi.org/10.1093/pcp/pch163

Munekage, Y., Hashimoto, M., Miyaka, C., Tomizawa, K., Endo, T., Tasaka, M., & Shikanai, T. (2004). Cyclic electron flow around photosystem I is essential for photosynthesis. *Nature, 429*, 579–582. https://doi.org/10.1038/nature02598

Munekage, Y., Hojo, M., Meurer, J., Endo, T., Tasaka, M., & Shikanai, T. (2002). PGR5 is involved in cyclic electron flow around photosystem I and is essential for photoprotection in Arabidopsis. *Cell, 110*, 361–371. https://doi.org/10.1016/S0092-8674(02)00867-X

Murashige, T., & Skoog, F. (1962). A revised medium for rapid growth and bio assays with tobacco tissue cultures. *Physiologia Plantarum, 15*, 473–497. https://doi.org/10.1111/j.1399-3054.1962.tb08052.x

Naranjo, B., Mignee, C., Krieger-Liszkay, A., Hornero-Mendez, D., Gallardo-Guerrero, L., Javier Cejudo, F., & Lindahl, M. (2016). The chloroplast NADPH thioredoxin reductase C, NTRC, controls non-photochemical quenching of light energy and photosynthetic electron transport in Arabidopsis. *Plant, Cell and Environment, 39*, 804–822. https://doi.org/10.1111/pce.12652

Nikkanen, L., Toivola, J., & Rintamäki, E. (2016). Crosstalk between chloroplast thioredoxin systems in regulation of photosynthesis. *Plant, Cell and Environment, 39*, 1691–1705. https://doi.org/10.1111/pce.12718

Niyogi, K. K., & Truong, T. B. (2013). Evolution of flexible non-photochemical quenching mechanisms that regulate light harvesting in oxygenic photosynthesis. *Current Opinion in Plant Biology, 16*, 307–314. https://doi.org/10.1016/j.pbi.2013.03.011

Okegawa, Y., Kagawa, Y., Kobayashi, Y., & Shikanai, T. (2008). Characterization of factors affecting the activity of photosystem I cyclic electron transport in chloroplasts. *Plant and Cell Physiology, 49*, 825–834. https://doi.org/10.1093/pcp/pcn055

Peltier, G., Aro, E., & Shikanai, T. (2016). NDH-1 and NDH-2 plastoquinone reductases in oxygenic photosynthesis. Annual review of plant biology. *Annual Review of Plant Biology, 67*, 55–80. https://doi.org/10.1146/annurev-arplant-043014-114752

Peng, L., Shimizu, H., & Shikanai, T. (2008). The Chloroplast NAD(P)H Dehydrogenase Complex Interacts with Photosystem I in Arabidopsis. *J. Biol. Chem., 283*, 34873–34879.

Pérez-Ruiz, J. M., Naranjo, B., Ojeda, V., Guinea, M., & Cejudo, F. J. (2006). NTRC-dependent redox balance of 2-Cys peroxiredoxins is needed for optimal function of the photosynthetic apparatus. *Proceedings of the National Academy of Sciences of the United States of America, 114*, 12069–12074. https://doi.org/10.1073/pnas.1706003114

Pérez-Ruiz, J. M., Spinola, M. C., Kirchsteiger, K., Moreno, J., Sahrawy, M., & Cejudo, F. J. (2006). Rice NTRC is a high-efficiency redox system for chloroplast protection against oxidative damage. *Plant Cell, 18*, 2356–2368. https://doi.org/10.1105/tpc.106.041541

Petroutsos, D., Terauchi, A. M., Busch, A., Hirschmann, I., Merchant, S. S., Finazzi, G., & Hippler, M. (2009). PGRL1 participates in iron-induced remodeling of the photosynthetic apparatus and in energy metabolism in *Chlamydomonas reinhardtii*. *Journal of Biological Chemistry, 284*, 32770–32781. https://doi.org/10.1074/jbc.M109.050468

Porra, R., Thompson, W., & Kriedemann, P. (1989). Determination of accurate extinction coefficients and simultaneous-equations for assaying chlorophyll-a and chlorophyll-B extracted with 4 different solvents - verification of the concentration of chlorophyll standards by atomic-absorption spectroscopy. *Biochimica et Biophysica Acta, 975*, 384–394. https://doi.org/10.1016/S0005-2728(89)80347-0

Pulido, P., Spinola, M. C., Kirchsteiger, K., Guinea, M., Belen Pascual, M., Sahrawy, M., ... Cejudo, F. J. (2010). Functional analysis of the pathways for 2-Cys peroxiredoxin reduction in *Arabidopsis thaliana* chloroplasts. *Journal of Experimental Botany, 61*, 4043–4054. https://doi.org/10.1093/jxb/erq218

Richter, A. S., Peter, E., Rothbart, M., Schlicke, H., Toivola, J., Rintamäki, E., & Grimm, B. (2013). Posttranslational influence of NADPH-dependent thioredoxin reductase C on enzymes in tetrapyrrole synthesis. *Plant Physiology, 162*, 63–73. https://doi.org/10.1104/pp.113.217141

Rintamäki, E., Martinsuo, P., Pursiheimo, S., & Aro, E. (2000). Cooperative regulation of light-harvesting complex II phosphorylation via the plastoquinol and ferredoxin-thioredoxin system in chloroplasts. *Proceedings of the National Academy of Sciences of the United States of America, 97*, 11644–11649. https://doi.org/10.1073/pnas.180054297

Rintamäki, E., Salonen, M., Suoranta, U., Carlberg, I., Andersson, B., & Aro, E. (1997). Phosphorylation of light-harvesting complex II and photosystem II core proteins shows different irradiance-dependent regulation in vivo - Application of phosphothreonine antibodies to analysis of thylakoid phosphoproteins. *Journal of Biological Chemistry, 272*, 30476–30482. https://doi.org/10.1074/jbc.272.48.30476

Rochaix, J. (2011). Regulation of photosynthetic electron transport. *Biochimica et Biophysica Acta-Bioenergetics, 1807*, 375–383. https://doi.org/10.1016/j.bbabio.2010.11.010

Ruban, A. V. (2016). Nonphotochemical chlorophyll fluorescence quenching: Mechanism and effectiveness in protecting plants from photo damage. *Plant Physiology*, *170*, 1903–1916. https://doi.org/10.1104/pp.15.01935

Ruban, A. V., & Johnson, M. P. (2009). Dynamics of higher plant photosystem cross-section associated with state transitions. *Photosynthesis Research*, *99*, 173–183. https://doi.org/10.1007/s11120-008-9387-x

Rumeau, D., Becuwe-Linka, N., Beyly, A., Louwagie, M., Garin, J., & Peltier, G. (2005). New subunits NDH-M, -N, and -O, encoded by nuclear genes, are essential for plastid Ndh complex functioning in higher plants. *Plant Cell*, *17*, 219–232. https://doi.org/10.1105/tpc.104.028282

Schansker, G., & Strasser, R. J. (2005). Quantificaiton of non-Q_B-reducing centers in leaves using far-red pre-illumination. *Photosynthesis Research*, *84*, 145–151. https://doi.org/10.1007/s11120-004-7156-z

Schneider, C. A., Rasband, W. S., & Eliceiri, K. W. (2012). NIH Image to ImageJ: 25 years of image analysis. *Nature Methods*, *9*, 671–675. https://doi.org/10.1038/nmeth.2089

Schreiber, U. (2017). Redox changes of ferredoxin, P700 and plastocyanin measured simultaneously in intact leaves. *Photosynthesis Research*, *134*, 343–360. https://doi.org/10.1007/s11120-017-0394-7

Schreiber, U., & Klughammer, C. (2008). New accessory for the DUAL-PAM-100: The P515/535 module and examples of its application. *PAM Application Notes*, *1*, 1–10.

Schreiber, U., & Klughammer, C. (2016). Analysis of Photosystem I donor and acceptor sides with a new type of online-deconvoluting kinetic LED-array spectrophotometer. *Plant and Cell Physiology*, *57*, 1454–1467.

Schürmann, P., & Buchanan, B. B. (2008). The ferredoxin/thioredoxin system of oxygenic photosynthesis. *Antioxidants and Redox Signaling*, *10*, 1235–1273. https://doi.org/10.1089/ars.2007.1931

Serrato, A. J., Perez-Ruiz, J. M., Spinola, M. C., & Cejudo, F. J. (2004). A novel NADPH thioredoxin reductase, localized in the chloroplast, which deficiency causes hypersensitivity to abiotic stress in Arabidopsis thaliana. *Journal of Biological Chemistry*, *279*, 43821–43827. https://doi.org/10.1074/jbc.M404696200

Shapiguzov, A., Chai, X., Fucile, G., Longoni, P., Zhang, L., & Rochaix, J. (2016). Activation of the Stt7/STN7 kinase through dynamic interactions with the cytochrome b(6)f complex. *Plant Physiology*, *171*, 1533–1533.

Shikanai, T. (2016). Chloroplast NDH: A different enzyme with a structure similar to that of respiratory NADH dehydrogenase. *Biochimica et Biophysica Acta-Bioenergetics*, *1857*, 1015–1022. https://doi.org/10.1016/j.bbabio.2015.10.013

Shikanai, T., Endo, T., Hashimoto, T., Yamada, Y., Asada, K., & Yokota, A. (1998). Directed disruption of the tobacco ndhB gene impairs cyclic electron flow around photosystem I. *Proceedings of the National Academy of Sciences of the United States of America*, *95*, 9705–9709. https://doi.org/10.1073/pnas.95.16.9705

Shikanai, T., & Yamamoto, H. (2017). Contribution of cyclic and pseudo-cyclic electron transport to the formation of proton motive force in chloroplasts. *Molecular Plant*, *10*, 20–29. https://doi.org/10.1016/j.molp.2016.08.004

Shimakawa, G., & Miyake, C. (2018). Changing frequency of fluctuating light reveals the molecular mechanism for P700 oxidation in plant leaves. *Plant Direct*, *2*, 1–11.

Sievers, F., Wilm, A., Dineen, D., Gibson, T. J., Karplus, K., Li, W., … Higgins, D. G. (2011). Fast, scalable generation of high-quality protein multiple sequence alignments using Clustal Omega. *Molecular Systems Biology*, *7*, 539.

Stirbet, A., Riznichenko, G. Y., Rubin, A. B., & Govindjee, (2014). Modeling chlorophyll a fluorescence transient: Relation to photosynthesis. *Biochemistry (Moscow)*, *79*, 291–323. https://doi.org/10.1134/S0006297914040014

Strand, D. D., Fisher, N., Davis, G. A., & Kramer, D. M. (2016). Redox regulation of the antimycin A sensitive pathway of cyclic electron flow around photosystem I in higher plant thylakoids. *Biochimica et Biophysica Acta-Bioenergetics*, *1857*, 1–6. https://doi.org/10.1016/j.bbabio.2015.07.012

Strand, D. D., Fisher, N., & Kramer, D. M. (2016). Distinct energetics and regulatory functions of the two major cyclic electron flow pathways in chloroplasts. In H. Kirchhoff (Ed.), *Chloroplasts: Current research and future trends* (pp. 89–100). Wymondham, UK: Caister Academic Press.

Strand, D. D., Fisher, N., & Kramer, D. M. (2017). The higher plant plastid NAD(P)H dehydrogenase-like complex (NDH) is a high efficiency proton pump that increases ATP production by cyclic electron flow. *Journal of Biological Chemistry*, *292*, 11850–11860. https://doi.org/10.1074/jbc.M116.770792

Strand, D. D., Livingston, A. K., Satoh-Cruz, M., Froehlich, J. E., Maurino, V. G., & Kramer, D. M. (2015). Activation of cyclic electron flow by hydrogen peroxide in vivo. *Proceedings of the National Academy of Sciences of the United States of America*, *112*, 5539–5544. https://doi.org/10.1073/pnas.1418223112

Strand, D. D., Livingston, A. K., Satoh-Cruz, M., Koepke, T., Enlow, H. M., Fisher, N., … Kramer, D. M. (2017). Defects in the expression of chloroplast proteins leads to H2O2 accumulation and activation of cyclic electron flow around photosystem I. *Frontiers in Plant Science*, *7*, 2073.

Suorsa, M. (2015). Cyclic electron flow provides acclimatory plasticity for the photosynthetic machinery under various environmental conditions and developmental stages. *Frontiers in Plant Science*, *6*, 800.

Suorsa, M., Grieco, M., Järvi, S., Gollan, P. J., Kangasjärvi, S., Tikkanen, M., & Aro, E. (2013). PGR5 ensures photosynthetic control to safeguard photosystem I under fluctuating light conditions. *Plant Signaling and Behavior*, *8*, e22714.

Suorsa, M., Jarvi, S., Grieco, M., Nurmi, M., Pietrzykowska, M., Rantala, M., … Aro, E. (2012). Proton gradient regulation 5 is essential for proper acclimation of arabidopsis photosystem I to naturally and artificially fluctuating light conditions. *Plant Cell*, *24*, 2934–2948. https://doi.org/10.1105/tpc.112.097162

Suorsa, M., Rossi, F., Tadini, L., Labs, M., Colombo, M., Jahns, P., … Pesaresi, P. (2016). PGR5-PGRL1-dependent cyclic electron transport modulates linear electron transport rate in *Arabidopsis thaliana*. *Molecular Plant*, *9*, 271–288. https://doi.org/10.1016/j.molp.2015.12.001

Takagi, D., & Miyake, C. (2018). Proton gradient regulation 5 supports linear electron flow to oxidize photosystem I. *Physiologia Plantarum*, *164*, 337–348. https://doi.org/10.1111/ppl.12723

Thormählen, I., Meitzel, T., Groysman, J., Ochsner, A. B., von Roepenack-Lahaye, E., Naranjo, B., … Geigenberger, P. (2015). Thioredoxin f1 and NADPH-dependent thioredoxin reductase C have overlapping functions in regulating photosynthetic metabolism and plant growth in response to varying light conditions. *Plant Physiology*, *169*, 1766–1786.

Thormählen, I., Zupok, A., Rescher, J., Leger, J., Weissenberger, S., Groysman, J., … Geigenberger, P. (2017). Thioredoxins play a crucial role in dynamic acclimation of photosynthesis in fluctuating light. *Molecular Plant*, *10*, 168–182. https://doi.org/10.1016/j.molp.2016.11.012

Tikkanen, M., & Aro, E. M. (2014). Integrative regulatory network of plant thylakoid energy transduction. *Trends in Plant Science*, *19*, 10–17. https://doi.org/10.1016/j.tplants.2013.09.003

Tikkanen, M., Grieco, M., Kangasjarvi, S., & Aro, E. (2010). Thylakoid protein phosphorylation in higher plant chloroplasts optimizes electron transfer under fluctuating light. *Plant Physiology*, *152*, 723–735. https://doi.org/10.1104/pp.109.150250

Tikkanen, M., Grieco, M., Nurmi, M., Rantala, M., Suorsa, M., & Aro, E. (2012). Regulation of the photosynthetic apparatus under fluctuating growth light. *Philosophical Transactions of the Royal Society B: Biological Sciences*, *367*, 3486–3493. https://doi.org/10.1098/rstb.2012.0067

Tikkanen, M., Pippo, M., Suorsa, M., Sirpio, S., Mulo, P., Vainonen, J., … Aro, E. M. (2006). State transitions revisited - a buffering system for dynamic low light acclimation of Arabidopsis. *Plant Molecular Biology, 62*, 779–793.

Tikkanen, M., Rantala, S., & Aro, E. (2015). Electron flow from PSII to PSI under high light is controlled by PGR5 but not by PSBS. *Frontiers in Plant Science, 6*, 521.

Tiwari, A., Mamedov, F., Grieco, M., Suorsa, M., Jajoo, A., Styring, S., … Aro, E. (2016). Photodamage of iron-sulphur clusters in photosystem I induces non-photochemical energy dissipation. *Nature Plants, 2*, 16035. https://doi.org/10.1038/nplants.2016.35

Toivola, J., Nikkanen, L., Dahlström, K. M., Salminen, T. A., Lepistö, A., Vignols, F., & Rintamäki, E. (2013). Overexpression of chloroplast NADPH-dependent thioredoxin reductase in Arabidopsis enhances leaf growth and elucidates in vivo function of reductase and thioredoxin domains. *Frontiers in Plant Science, 4*, 389.

Toth, S. Z., Schansker, G., & Strasser, R. J. (2007). A non-invasive assay of the plastoquinone pool redox state based on the OJIP-transient. *Photosynthesis Research, 93*, 193–203. https://doi.org/10.1007/s11120-007-9179-8

Townsend, A. J., Ware, M. A., & Ruban, A. V. (2017). Dynamic interplay between photodamage and photoprotection in photosystem II. *Plant, Cell and Environment, 41*, 1098–1112.

Trotta, A., Suorsa, M., Rantala, M., Lundin, B., & Aro, E. (2016). Serine and threonine residues of plant STN7 kinase are differentially phosphorylated upon changing light conditions and specifically influence the activity and stability of the kinase. *The Plant Journal, 87*, 484–494. https://doi.org/10.1111/tpj.13213

Vener, A., VanKan, P., Rich, P., Ohad, I., & Andersson, B. (1997). Plastoquinol at the quinol oxidation site of reduced cytochrome bf mediates signal transduction between light and protein phosphorylation: Thylakoid protein kinase deactivation by a single-turnover flash. *Proceedings of the National Academy of Sciences of the United States of America, 94*, 1585–1590. https://doi.org/10.1073/pnas.94.4.1585

Walter, M., Chaban, C., Schutze, K., Batistic, O., Weckermann, K., Nake, C., … Kudla, J. (2004). Visualization of protein interactions in living plant cells using bimolecular fluorescence complementation. *The Plant Journal, 40*, 428–438. https://doi.org/10.1111/j.1365-313X.2004.02219.x

Wang, P., Liu, J., Liu, B., Da, Q., Feng, D., Su, J., … Wang, H. (2014). Ferredoxin: Thioredoxin reductase is required for proper chloroplast development and is involved in the regulation of plastid gene expression in *Arabidopsis thaliana*. *Molecular Plant, 7*, 1586–1590. https://doi.org/10.1093/mp/ssu069

Wang, C., Yamamoto, H., & Shikanai, T. (2015). Role of cyclic electron transport around photosystem I in regulating proton motive force. *Biochimica et Biophysica Acta-Bioenergetics, 1847*, 931–938. https://doi.org/10.1016/j.bbabio.2014.11.013

Yadav, K.N.S., Semchonok, D.A., Nosek, L., Kouřil, R., Fucile, G., Boekema, E.J., & Eichacker, L.A. (2017). Supercomplexes of plant photosystem I with cytochrome b6f, light-harvesting complex II and NDH. *Biochim. Biophys. Acta-Bioenerg. 1858*, 12–20, doi/10.1016/j.bbabio.2016.10.006.

Yamamoto, H., Peng, L., Fukao, Y., & Shikanai, T. (2011). An Src homology 3 domain-like fold protein forms a ferredoxin binding site for the chloroplast NADH dehydrogenase-like complex in *Arabidopsis*. *Plant Cell, 23*, 1480–1493. https://doi.org/10.1105/tpc.110.080291

Yamamoto, H., & Shikanai, T. (2013). In planta mutagenesis of Src homology 3 domain-like fold of NdhS, a ferredoxin-binding subunit of the chloroplast NADH dehydrogenase-like complex in *Arabidopsis* A conserved Arg-193 playsa critical role in ferredoxin binding. *Journal of Biological Chemistry, 288*, 36328–36337. https://doi.org/10.1074/jbc.M113.511584

Yamori, W., Makino, A., & Shikanai, T. (2016). A physiological role of cyclic electron transport around photosystem I in sustaining photosynthesis under fluctuating light in rice. *Scientific Reports, 6*, 20147. https://doi.org/10.1038/srep20147

Yamori, W., Sakata, N., Suzuki, Y., Shikanai, T., & Makino, A. (2011). Cyclic electron flow around photosystem I via chloroplast NAD(P)H dehydrogenase (NDH) complex performs a significant physiological role during photosynthesis and plant growth at low temperature in rice. *The Plant Journal, 68*, 966–976. https://doi.org/10.1111/j.1365-313X.2011.04747.x

Yamori, W., & Shikanai, T. (2016). Physiological functions of cyclic electron transport around photosystem I in sustaining photosynthesis and plant growth. *Annual Review of Plant Biology, 67*, 81–106. https://doi.org/10.1146/annurev-arplant-043015-112002

Yamori, W., Shikanai, T., & Makino, A. (2015). Photosystem I cyclic electron flow via chloroplast NADH dehydrogenase-like complex performs a physiological role for photosynthesis at low light. *Scientific Reports, 5*, 15593. https://doi.org/10.1038/srep15593

Yoshida, K., Hara, S., & Hisabori, T. (2015). Thioredoxin selectivity for thiol-based redox regulation of target proteins in chloroplasts. *Journal of Biological Chemistry, 290*, 19540. https://doi.org/10.1074/jbc.A115.647545

Yoshida, K., & Hisabori, T. (2016). Two distinct redox cascades cooperatively regulate chloroplast functions and sustain plant viability. *Proceedings of the National Academy of Sciences of the United States of America, 113*, 3967–3976. https://doi.org/10.1073/pnas.1604101113

Yoshida, K., & Hisabori, T. (2017). Distinct electron transfer from ferredoxin-thioredoxin reductase to multiple thioredoxin isoforms in chloroplasts. *Biochemical Journal, 474*, 1347–1360. doi/10.1042/BCJ20161089.

Zaffagnini, M., De Mia, M., Morisse, S., Di Giacinto, N., Marchand, C. H., Maes, A., … Trost, P. (2016). Protein S-nitrosylation in photosynthetic organisms: A comprehensive overview with future perspectives. *Biochimica Biophysica Acta-Proteomics, 1864*, 952–966.

3

ABA-mediated regulation of stomatal density is OST1-independent

Pirko Jalakas[1] | Ebe Merilo[1] | Hannes Kollist[1] | Mikael Brosché[1,2]

[1]Institute of Technology, University of Tartu, Tartu, Estonia

[2]Viikki Plant Science Centre, Organismal and Evolutionary Biology Research Programme, Faculty of Biological and Environmental Sciences, University of Helsinki, Helsinki, Finland

Correspondence
Mikael Brosché, Viikki Plant Science Centre, Organismal and Evolutionary Biology Research Programme, Faculty of Biological and Environmental Sciences, University of Helsinki, P.O. Box 65 (Viikinkaari 1), FI-00014 Helsinki, Finland.
Email: mikael.brosche@helsinki.fi

Funding information
This work was supported by the Estonian Ministry of Science and Education (IUT2-21 to H.K. and PUT-1133 to E.M.), the European Regional Development Fund (Center of Excellence in Molecular Cell Engineering CEMCE to H.K.), the Academy of Finland (grant number #307335, Center of Excellence in Molecular Biology of Primary Producers 2014-2019 to M.B.).

Abstract

Stomata, small pores on the surfaces of leaves formed by a pair of guard cells, adapt rapidly to changes in the environment by adjusting the aperture width. As a long-term response, the number of stomata is regulated during stomatal development. The hormone abscisic acid (ABA) regulates both processes. In ABA mediated guard cell signaling the protein kinase OPEN STOMATA1 (OST1) has a central role, as stomatal closure in the *ost1* mutant is impaired in response to ABA and to different environmental stimuli. We aimed to dissect the contribution of different ABA-related regulatory mechanisms in determining stomatal conductance, a combination of stomatal density and aperture width, and crossed the *ost1* mutant with mutants that either decreased (*aba3*) or increased (*cyp707a1/a3*) the concentration of ABA in plants. The double mutant *ost1 aba3* had higher stomatal conductance than either parent due to a combination of increased stomatal aperture width and higher stomatal density. In the triple mutant *ost1 cyp707a1/a3*, stomatal conductance was significantly lower compared to *ost1-3* due to lower stomatal density. Further characterization of the single, double and triple mutants showed that responses to treatments that lead to stomatal closure were impaired in *ost1* as well as *ost1 aba3* and *ost1 cyp707a1/a3* mutants, supporting a critical role for OST1 in stomatal aperture regulation. On the basis of our results, we suggest that two signaling pathways regulate water flux from leaves, that is, stomatal conductance: an ABA-dependent pathway that determines stomatal density independent of OST1; and an OST1-dependent pathway that regulates rapid changes in stomatal aperture.

KEYWORDS
abscisic acid, aperture width, OST1, signaling pathways, Stomatal density

1 | INTRODUCTION

Stomata, formed by a pair of guard cells, are small pores responsible for gas exchange in leaves. They allow CO_2 uptake for photosynthesis, with the accompanying loss of water. In addition, air pollutants and some pathogens enter the plant through stomata. Hence, accurate adjustment of the stomatal pore is required for the plant to successfully thrive in a changing environment. Not only the width of the stomatal aperture, but also the number of stomata is regulated by environmental signals and influences plant gas exchange (Hetherington & Woodward, 2003). Several studies have shown that doubling of ambient CO_2 concentration leads to a reduction in stomatal density in different plant species and *Arabidopsis* accessions (Woodward & Kelly, 1995; Woodward, Lake, & Quick, 2002). In contrast, higher light intensity significantly increases stomatal density (Casson et al., 2009). The main difference between the adjustments of the stomatal aperture versus the number of stomata is the scale of time, stomatal aperture can change in minutes, whereas changes in stomatal density are fixed during leaf development.

The plant hormone abscisic acid (ABA) plays a central role in the regulation of guard cell function (Kim, Böhmer, Hu, Nishimura, & Schroeder, 2010; Kollist, Nuhkat, & Roelfsema, 2014). ABA-induced stomatal closure is initiated by binding of the hormone to PYR/RCAR receptors that leads to the inactivation of type 2C protein phosphatases (PP2Cs), which in turn releases SNF-related protein kinases (SnRK2s) such as OPEN STOMATA1 (OST1) to activate guard cell ion channels including SLOW ANION CHANNEL1 (SLAC1). This leads to the efflux of anions, followed by potassium and water efflux and stomatal closure (Kim et al., 2010; Kollist et al., 2014).

As ABA is the central regulatory molecule of stomatal function, fine-tuning of ABA levels and signaling is of utmost importance during acclimation to abiotic stress, for example, drought. Guard cell ABA levels are regulated by de novo biosynthesis, catabolism, and transport from other plant tissues (Merilo et al., 2015, 2018; Nambara & Marion-Poll, 2005). OST1 appears to have a critical role in ABA signaling. Stomatal closure induced by ABA or environmental factors is strongly impaired in *ost1* mutants (Merilo et al., 2013; Mustilli, Merlot, Vavasseur, Fenzi, & Giraudat, 2002). In addition to stomatal regulation, ABA affects stomatal development, which is also controlled by environmental factors such as light and the level of CO_2 (Casson & Hetherington, 2010; Chater et al., 2015). ABA-deficient mutants have increased stomatal densities compared to wild-type (Chater et al., 2015; Tanaka, Nose, Jikumaru, & Kamiya, 2013), whereas the ABA over-accumulating *cyp707a1/a3* double mutant had significantly lower stomatal density than wild-type (Tanaka et al., 2013), supporting the role of ABA in stomatal development. ABA3 encodes a molybdenum cofactor sulfurase required by an abscisic aldehyde oxidase to catalyze the conversion of abscisic aldehyde to ABA; its expression level increases in response to drought and ABA treatment (Xiong, Ishitani, Lee, & Zhu, 2001). In non-stressed conditions, the concentration of leaf ABA is approximately 45% of wild-type ABA in *aba3-1* (Merilo et al., 2018). The predominant ABA catabolic pathway, ABA 8′-hydroxylation, is mediated by four members of the CYP707A gene family and their transcription levels increase in response to salt and drought stress as well as ABA (Saito et al., 2004). CYP707A1 and CYP707A3 are important for post-germination growth, since seedling growth by exogenous ABA was inhibited more effectively in *cyp707a1* and *cyp707a3* mutants and was more pronounced in the double mutant that also contained higher concentration of ABA compared to the single mutants (Okamoto et al., 2006). Both *cyp707a1* and *cyp707a3* loss-of-function mutants showed reduced stomatal conductance, which was more pronounced in *cyp707a3* (Merilo et al., 2013).

Here, we report that while OST1 is required in rapid stomatal responses to several environmental conditions: reduced air humidity, darkness, elevated CO_2 concentration and exogenous ABA, OST1 is not required for regulation of stomatal density. These two pathways, OST1-dependent signaling that regulates stomatal aperture width and OST1–independent signaling that regulates stomatal density, coordinate the overall water flux through stomata.

2 | MATERIALS AND METHODS

2.1 | Plant material, growth and gas-exchange measurements

Col-0, *aba3-1* and *ost1-3* (*srk2e*, SALK_008068) were from the European Arabidopsis Stock Centre (www.arabidopsis.info). The *cyp707a1 cyp707a3* double mutant was a gift from Eiji Nambara (Okamoto et al., 2006). Double mutants and other crosses were made through standard techniques and genotyped with PCR-based markers.

aba3-1	aba3-1-left	TCATTCTTTCTACTGCTCCTGATTT	dCAPS marker, digest with MnlI
	aba3-1-right	GGTGAAGCAAATGAACTTATGATG	
ost1-3	SALK_008068 for	CCTCTGATGTCTTGGTGTCG	
	SALK_008068 rev	TGGAAGAAAAACCTCGCCTA	
cyp707a1	SALK_069127 for	CATGAACGTATTGGGTTTTGG	
	SALK_069127 rev	TCCTGATATTGAATCCATCGC	
cyp707a3	SALK_078173C for	GTTCCTGGAAGATTAATCGGC	
	SALK_078173C rev	ACGTGCTCTCGTCACTCTCTC	
SALK Lbb		GCGTGGACCGCTTGCTGCAACT	

Plants for gas-exchange measurements were sown into 2:1 (v:v) peat:vermiculite mixture and grown through a hole in a glass plate covering the pot as described in Kollist et al. (2007). Plants were grown in growth chambers (AR-66LX, Percival Scientific, IA, USA and Snijders Scientific, Drogenbos, Belgia) with 12 hr photoperiod, 23/18°C day/night temperature, 150 μmol m^{-2} s^{-1} light and 70% relative humidity. Plants were 24–30 days old during gas-exchange experiments.

Stomatal conductance of intact plants was measured using a rapid-response gas-exchange measurement device similar to the one described by Kollist et al. (2007) consisting of eight thermostated flow-through whole-rosette cuvettes. Plants were inserted into the measurement chambers and after stomatal conductance had stabilized, the following stimuli were applied: reduction in air humidity (decrease from 65%–75% to 30%–40%), darkness (decrease from 150 μmol m^{-2} s^{-1} to 0 light), CO_2 (increase from 400 ppm to 800 ppm) and spraying plants with 0 μM, 5 μM or 50 μM ABA solution with 0.012% Silwet L-77 (Duchefa) and 0.05% ethanol. ABA-induced stomatal closure experiments were carried out as described previously (Merilo et al., 2018). Initial changes in stomatal conductance were calculated as gs18-gs0, where gs0 is the pretreatment stomatal conductance and gs18 is the value of stomatal conductance 18 min after factor application; 16 min in case of ABA spraying.

2.2 | Measurement of stomatal aperture and density

Epidermal peels were stripped from 4-week-old plants grown in growth chambers as described above and incubated in resting buffer (containing 10 mM MES-KOH pH6.2) for 2.5 hr. Images of stomata were taken with a Zeiss Axio Examiner.D1 microscope. Images were taken of 15 stomata per leaf and averaged to characterize the stomatal aperture of each plant. Six plants per genotype were analyzed. Stomatal aperture width was measured using the image processing software ImageJ 1.51k (National Institutes of Health, USA).

For stomatal density (SD) measurements, leaves of 5 week-old plants grown as described above, one leaf per plant, were excised and the abaxial side was covered with dental resin (Xantropen VL Plus, Heraeus Kulzer, Germany). The hardened resin impressions were covered with transparent nail varnish. The dried nail varnish imprints were attached to a microscope glass slide with a transparent tape and images were taken with a Zeiss SteREO Discovery.V20 stereomicroscope. Twenty-four plants per genotype were analyzed. SD was determined from an image with an area of ~0.12 mm^2, taken from the middle of the leaf, close to the middle vein and calculated as: *SD* = number of stomata/area of the image.

2.3 | Statistical analysis

One-way ANOVA was used to compare the effect of genotype on the values of stomatal conductance, aperture, density and initial change in stomatal conductance. Comparisons between individual means were done with Tukey or Tukey unequal N HSD *post hoc* tests as indicated in figure legends. Stomatal conductance values before and after application of ABA were compared by repeated measures ANOVA with Tukey *post hoc* test. All effects were considered significant at $p < 0.05$. Statistical analyses were performed with Statistica, version 7.1 (StatSoft Inc., Tulsa, OK, USA).

2.4 | Accession numbers

ABA3—AT1G16540; OST1—AT4G33950; CYP707A1—AT4G19230; CYP707A3—AT5G45340.

3 | RESULTS

We crossed *ost1-3* into an ABA biosynthesis mutant (*aba3-1*) and to *cyp707a1 cyp707a3* (here abbreviated as *cyp707a1/a3*) that lacks two proteins involved in ABA catabolism. By doing so, we generated plants where strong ABA-insensitivity caused by impaired OST1 was combined with defective ABA biosynthesis or breakdown (Figure 1). Steady-state stomatal conductance and rapid stomatal responses to various closure-inducing stimuli were measured in intact plants with a custom-made gas-exchange device as described before (Kollist et al., 2007). Our results showed that the *aba3-1* mutant displayed higher stomatal conductance, whereas *cyp707a1/a3* had reduced stomatal conductance compared to Col-0 wild-type (Figure 2a), as can be expected on the basis of the ABA concentrations in these plants (Merilo et al., 2018; Okamoto et al., 2006). The double mutant *ost1 aba3* had higher stomatal conductance than either parent (Figure 2a) and the triple mutant *ost1 cyp707a1/a3* displayed lower stomatal conductance than the single *ost1-3* (Figure 2a).

As altered stomatal conductance can result from a change in stomatal aperture width or in stomatal density, we next determined the stomatal apertures of the mutants. There were no differences in aperture widths between *cyp707a1/a3* and wild-type (Figure 2b). Compared to wild-type, stomata of *aba3-1* and *ost1-3* single mutants had significantly wider apertures (Figure 2b). Aperture of *ost1 cyp707a1/a3* was similar to *ost1-3*, whereas *ost1 aba3* had significantly wider aperture compared to the single mutants (Figure 2b). These results suggest that ABA-deficiency leads to wider stomatal apertures, whereas over-accumulation of ABA seems to have no effect on aperture width. However, *cyp707a1/a3* (compared to

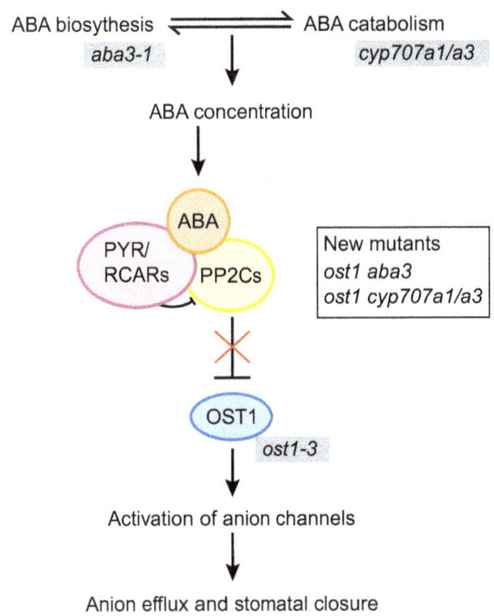

FIGURE 1 Schematic overview of ABA concentration determined by ABA biosynthesis and catabolism, followed by the core components in ABA signaling leading to stomatal closure. Mutants used in this study are indicated in gray background. New double and triple mutants generated for this study are *ost1 aba3* and *ost1 cyp707a1/a3*

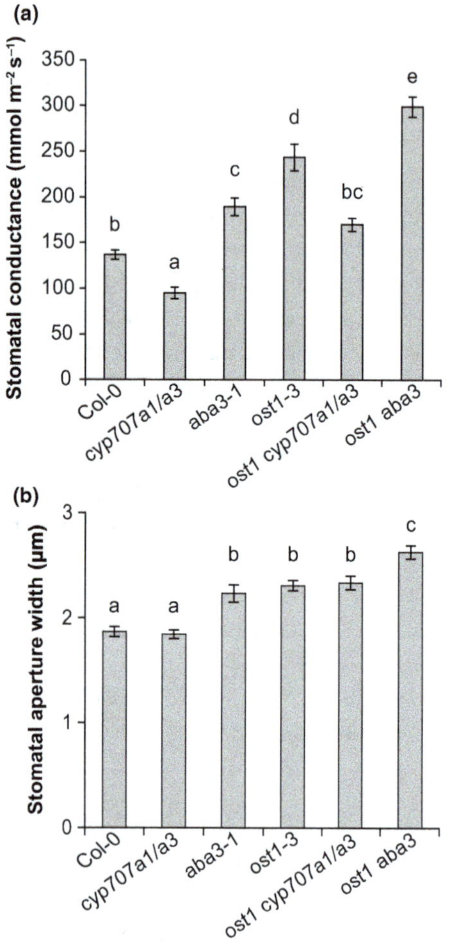

FIGURE 2 (a) Whole-plant steady-state stomatal conductance (gs) of 3- to 4-week-old plants. The ABA biosynthesis mutant *aba3-1* and catabolism double mutant *cyp707a1/a3* were crossed to *ost1-3* to genetically reduce or increase the ABA concentration in *ost1-3* background. Letters denote statistically significant differences between lines (ANOVA with Tukey unequal N HSD *post hoc* test, $p < 0.05$; $n = 8–13$). (b) Stomatal aperture measured on epidermal peels of 4-week-old plants. Letters denote statistically significant differences between lines (ANOVA with Tukey *post hoc* test, $p < 0.05$; $n = 6$)

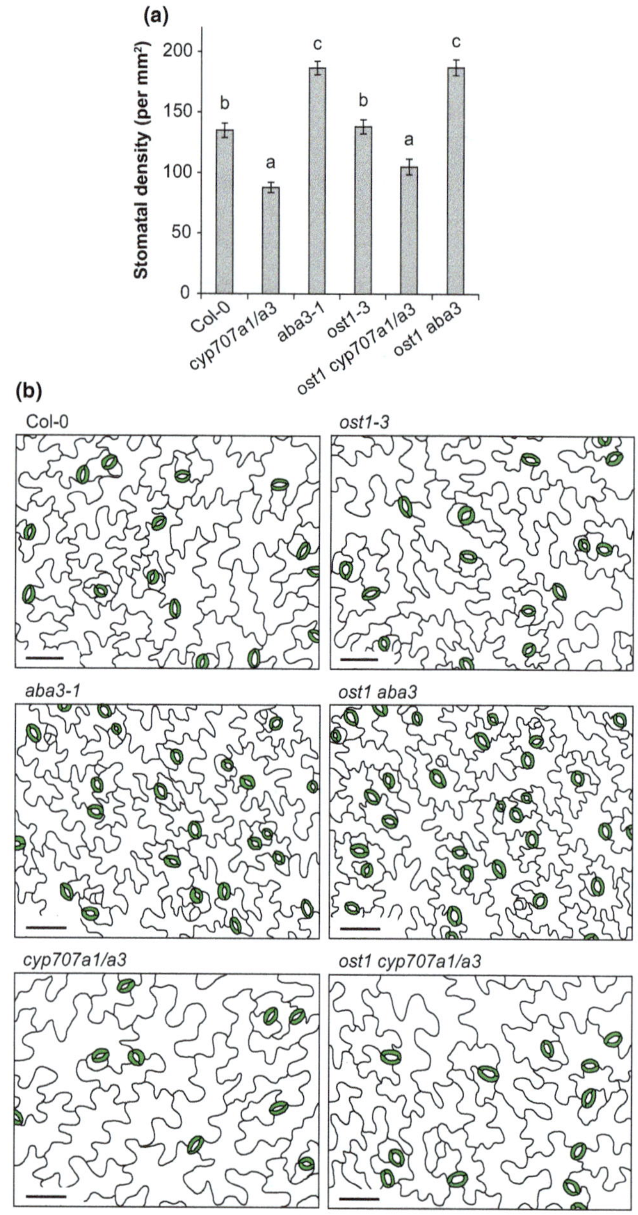

FIGURE 3 (a) Stomatal density of 5-week-old plants. Letters denote statistically significant differences between lines (ANOVA with Tukey *post hoc* test, $p < 0.05$; $n = 24$). (b) Tracing of epidermal impressions to illustrate the differences in stomatal densities between lines. The scale bar represents 50 μm

Col-0) and *ost1 cyp707a1/a3* (compared to *ost1*) showed differences in stomatal conductance but not in aperture widths, indicating that some other trait besides aperture is involved in determining stomatal conductance.

To test whether the differences in stomatal conductance in the studied mutants were associated with altered stomatal density, we measured stomatal density using leaf impressions. Consistent with already published results (Chater et al., 2015; Tanaka et al., 2013), *aba3-1* had higher and *cyp707a1/a3* lower stomatal density compared to wild-type in our experiment (Figure 3a,b). The stomatal density of *ost1-3* was similar to wild-type, but through genetically altering the ABA concentration in the *ost1-3* mutant, we could affect the stomatal development. Compared to the single *ost1-3* mutant, *ost1 aba3* and *ost1 cyp707a1/a3* had significantly higher or lower stomatal density, respectively (Figure 3a,b). Taken together, stomatal conductance, aperture and density results show that *ost1-3* has higher stomatal conductance due to more open stomata (Figures 2-3; Mustilli et al., 2002). However, ABA concentration is a crucial signal for stomatal development, which was apparently regulated by an OST1-independent mechanism.

To further characterize the role of ABA levels and OST1 in stomatal regulation, we tested the responses of single mutants, *ost1 aba3* and *ost1 cyp707a1/a3* to closure-inducing factors (Figure 4a–d). The closure induced by all stimuli was significantly impaired in *ost1-3*, whereas *aba3-1* plants showed wild-type-like closure or, in the case of reduced air humidity and ABA, even a hypersensitive response. In

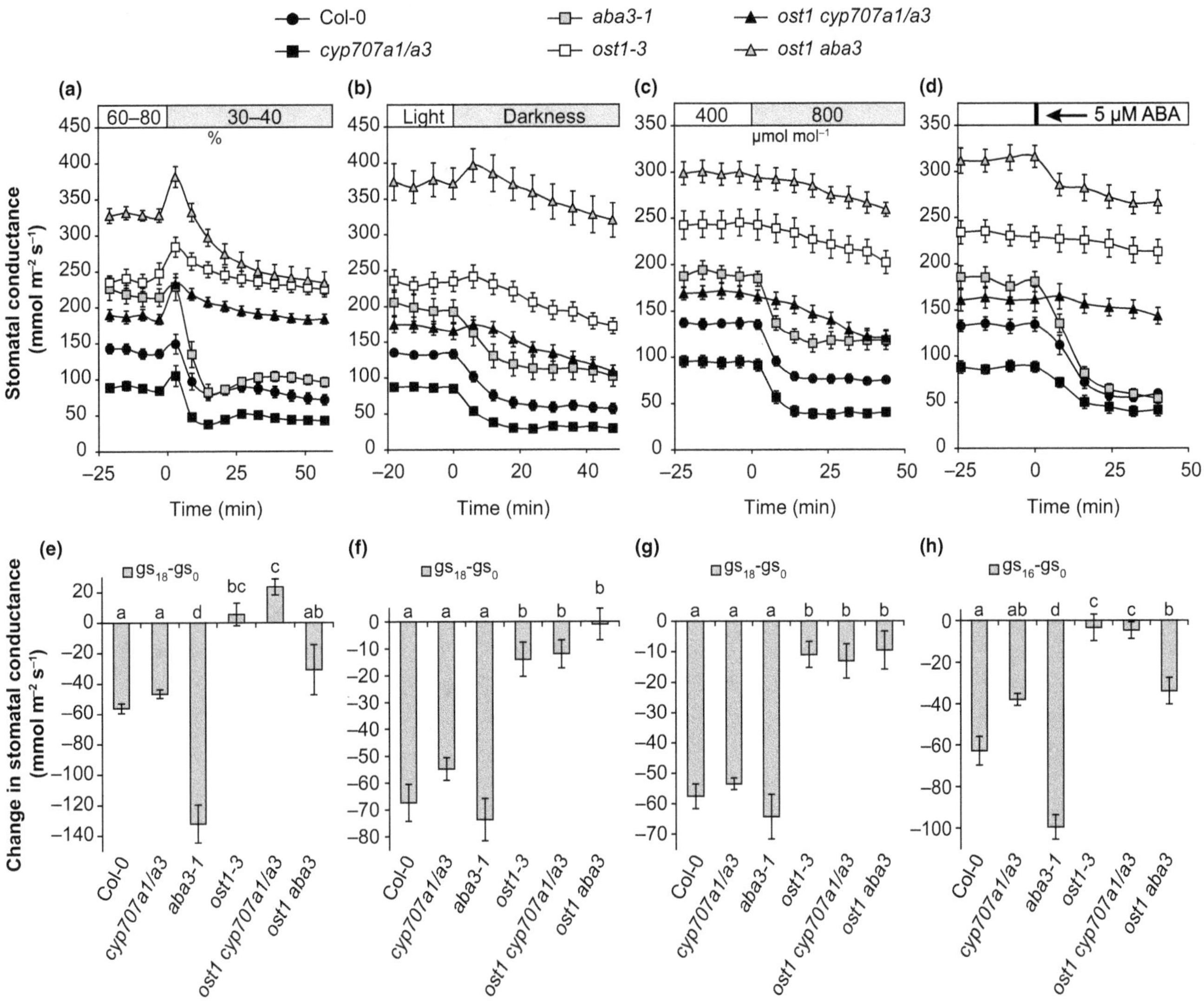

FIGURE 4 (a–d) Time courses of stomatal conductances in response to reduced air humidity (a), darkness (b), elevated CO_2 (c) and ABA treatment (d). (e–h) Changes in stomatal conductance during the first 18 min (first 16 min in the ABA treatment). Letters denote statistically significant differences between lines (ANOVA with Tukey unequal N HSD *post hoc* test, $p < 0.05$; $n = 8–15$)

response to darkness, reduced air humidity and elevated CO_2, the behavior of *ost1 aba3* and *ost1 cyp707a1/a3* mutants was not significantly different from *ost1-3* single mutant (Figure 4a–d, e–h). In response to ABA, *ost1 aba3* plants regained a small response that was larger compared to *ost1-3*, but reduced compared to wild-type (Figure 4d,h). Nevertheless, *ost1 aba3* and *ost1 cyp707a1/a3* were clearly impaired in rapid stomatal responses, supporting the critical role of OST1 in the regulation of stomatal aperture to sudden changes in the environment.

4 | DISCUSSION

Understanding stomatal function is critical for breeding plants with improved properties in water limiting conditions. As increased water loss from plants could be the result of either more open stomata or an increased number of stomata, the regulatory interplay between these traits is an important issue to be resolved. Changes in stomatal aperture width and density resulted in altered stomatal conductance, as suggested by our results (Figures 2, 3a). Through genetic manipulation of ABA levels, measurements of stomatal conductance, aperture, density, and responses to various treatments, we propose that the overall water flux through stomata is the sum of two signaling pathways: OST1-independent regulation of stomatal density; and an OST1-dependent pathway that regulates rapid changes of stomatal aperture. This conclusion is supported by our results which show significant differences in stomatal conductance due to differences in stomatal density and aperture between the single *ost1-3* mutant and its double and triple mutants, where ABA levels are genetically reduced (*ost1 aba3*) and increased (*ost1 cyp707a1/a3*), respectively, while responses to various environmental stimuli are impaired in these mutants. It remains to be clarified whether other SnRKs besides OST1 (SnRK2.6) are involved in the ABA-dependent signaling involved in stomatal development. Alternatively, stomatal density could be determined by a SnRK-independent mechanism. However, the recent findings that a mutant lacking six ABA receptors and the PP2C mutants *abi1-1* and *abi2-1* also showed higher stomatal densities (Merilo et al., 2018; Tanaka et al., 2013) indicate that the canonical ABA signaling pathway starting with ABA receptors (Figure 1) is involved in the regulation of stomatal development.

The change in stomatal conductance of mutants with altered concentrations of ABA appears to result from a change in stomatal density (Figure 3a), aperture width or both, as in *aba3-1* (Figure 2). The *aba3-1* mutant is relatively mildly impaired in ABA biosynthesis and still contains approximately 45% of wild-type ABA levels (Merilo et al., 2018). Mutants with more severely impaired ABA biosynthesis including *aba2-11* or *nced3 nced5* have considerably higher stomatal conductance than *aba3-1* (Merilo et al., 2018). Thus, the influence of ABA on stomatal aperture or density might become more prominent in plant lines where ABA concentrations are more severely reduced. Using stronger ABA-deficient lines including *aba2-11* or *nced3 nced5* (Merilo et al., 2018) or growing plants in water deficit conditions and measuring aperture and density may help to understand the balance between stomatal density and aperture in determining water flux through plants, that is, stomatal conductance. Stomatal density appeared to be more sensitive to reduced ABA levels than to increased levels as *cyp707a1/a3* showed only reduced density compared to wild-type (Figure 2b). To breed crops for future climate, we need to understand the contribution of both stomatal density and stomatal aperture to plant water relations (see also Hughes et al., 2017). The latter is subjected to a rapid and up-to-date environmental control, whereas the former is fixed during plant development.

The triple mutant *snrk2.2 snrk2.3 snrk2.6* is completely impaired in ABA responses, including seed germination and gene expression (Fujii & Zhu, 2009; Umezawa et al., 2009). Thus, the OST1-independent mechanism regulating stomatal density could be genetically redundant among this group of SnRKs. Unfortunately the severe developmental defects of the *snrk* triple mutant (Fujii & Zhu, 2009) make it difficult to directly test this hypothesis. The *ost1* mutant was previously shown to completely lack stomatal responses to ABA (Mustilli et al., 2002), and applied at 5 μM, the *ost1* mutant is unresponsive to ABA (Figure 4d). In an attempt to clarify if there is genetic redundancy among the SnRKs also in stomatal function, we treated *ost1-3* plants with very high 50 μM ABA, which induced a partial stomatal closure (Figure S1). This supports earlier findings that besides OST1 there are other components, including SNRK2.2 and SNRK2.3 and other possible kinases, such as calcium dependent protein kinases (Brandt et al., 2015) or GHR1 (GUARD CELL HYDROGEN PEROXIDE RESISTANT1) (Hua et al., 2012), that might contribute to ABA-induced stomatal closure. Other kinases may also explain the increased aperture width of *ost1 aba3* double mutant compared to single mutants. Several mitogen-activated protein kinases (MAPK) are involved in stomatal development (Wang, Ngwenyama, Liu, Walker, & Zhang, 2007) and ABA signaling (Jammes et al., 2009), indicating that MPKs might also contribute in the regulation of stomatal development and aperture in an ABA-dependent manner. Therefore, there are several candidate kinases in addition to OST1 in regulating stomatal responses to ABA.

Our results presented here show that it is possible to separate the regulation of stomatal aperture versus stomatal development. This information can be useful to breed separately for these traits to obtain plants suited for either rapidly changing environmental conditions or for conditions characterized by long-term drought.

ACKNOWLEDGMENTS

We would like to thank Hanna Hõrak for comments on the manuscript.

CONFLICT OF INTEREST

No conflict of interest declared.

AUTHOR CONTRIBUTIONS

M.B. conceived the project with help from H.K. and E.M., P.J. performed most of the experiments with assistance from E.M. and M.B., E.M. and M.B. supervised the experiments. P.J., E.M., H.K., and M.B. analyzed the data. P.J. and M.B. wrote the article with input from all authors.

REFERENCES

Brandt, B., Munemasa, S., Wang, C., Nguyen, D., Yong, T., Yang, P. G., … Schroeder, J. I. (2015). Calcium specificity signaling mechanisms in abscisic acid signal transduction in Arabidopsis guard cells. *Elife*, *4*, e10328.

Casson, S. A., Franklin, K. A., Gray, J. E., Grierson, C. S., Whitelam, G. C., & Hetherington, A. M. (2009). Phytochrome B and PIF4 regulate stomatal development in response to light quantity. *Current Biology*, *19*, 229–234.

Casson, S. A., & Hetherington, A. M. (2010). Environmental regulation of stomatal development. *Current Opinion in Plant Biology*, *13*, 90–95.

Chater, C., Peng, K., Movahedi, M., Dunn, J. A., Walker, H. J., Liang, Y.-K., … Neill, S. J. (2015). Elevated CO_2-induced responses in stomata require ABA and ABA signaling. *Current Biology*, *25*, 2709–2716.

Fujii, H., & Zhu, J.-K. (2009). Arabidopsis mutant deficient in 3 abscisic acid-activated protein kinases reveals critical roles in growth, reproduction, and stress. *Proceedings of the National Academy of Sciences of the United States of America*, *106*, 8380.

Hetherington, A. M., & Woodward, F. I. (2003). The role of stomata in sensing and driving environmental change. *Nature*, *424*, 901–908.

Hua, D., Wang, C., He, J., Liao, H., Duan, Y., Zhu, Z., … Gong, Z. (2012). A plasma membrane receptor kinase, GHR1, mediates abscisic acid- and hydrogen peroxide-regulated stomatal movement in Arabidopsis. *Plant Cell*, *24*, 2546–2561.

Hughes, J., Hepworth, C., Dutton, C., Dunn, J. A., Hunt, L., Stephens, J., … Gray, J. E. (2017). Reducing stomatal density in barley improves drought tolerance without impacting on yield. *Plant Physiology*, *174*, 776.

Jammes, F., Song, C., Shin, D., Munemasa, S., Takeda, K., Gu, D., … Leonhardt, N. (2009). MAP kinases MPK9 and MPK12 are preferentially expressed in guard cells and positively regulate ROS-mediated ABA signaling. *Proceedings of the National Academy of Sciences of the United States of America*, *106*, 20520–20525.

Kim, T.-H., Böhmer, M., Hu, H., Nishimura, N., & Schroeder, J. I. (2010). Guard cell signal transduction network: Advances in understanding abscisic acid, CO_2, and Ca^{2+} signaling. *Annual Review of Plant Biology*, *61*, 561–591.

Kollist, T., Moldau, H., Rasulov, B., Oja, V., Rämma, H., Hüve, K., … Kollist, H. (2007). A novel device detects a rapid ozone-induced transient stomatal closure in intact Arabidopsis and its absence in abi2 mutant. *Physiologia Plantarum*, *129*, 796–803.

Kollist, H., Nuhkat, M., & Roelfsema, M. R. G. (2014). Closing gaps: Linking elements that control stomatal movement. *New Phytologist*, *203*, 44–62.

Merilo, E., Jalakas, P., Laanemets, K., Mohammadi, O., Horak, H., Kollist, H., & Brosche, M. (2015). Abscisic acid transport and homeostasis in the context of stomatal regulation. *Molecular Plant*, *8*, 1321–1333.

Merilo, E., Laanemets, K., Hu, H., Xue, S., Jakobson, L., Tulva, I., … Kollist, H. (2013). PYR/RCAR receptors contribute to ozone-, reduced air humidity-, darkness-, and CO_2-induced stomatal regulation. *Plant Physiology*, *162*, 1652–1668.

Merilo, E., Yarmolinsky, D., Jalakas, P., Parik, H., Tulva, I., Rasulov, B., … Kollist, H. (2018). Stomatal VPD response: There is more to the story than ABA. *Plant Physiology*, *176*, 851–864.

Mustilli, A.-C., Merlot, S., Vavasseur, A., Fenzi, F., & Giraudat, J. (2002). Arabidopsis OST1 protein kinase mediates the regulation of stomatal aperture by abscisic acid and acts upstream of reactive oxygen species production. *Plant Cell*, *14*, 3089–3099.

Nambara, E., & Marion-Poll, A. (2005). Abscisic acid biosynthesis and catabolism. *Annual Review of Plant Biology*, *56*, 165–185.

Okamoto, M., Kuwahara, A., Seo, M., Kushiro, T., Asami, T., Hirai, N., … Nambara, E. (2006). CYP707A1 and CYP707A2, which encode abscisic acid 8′-hydroxylases, are indispensable for proper control of seed dormancy and germination in arabidopsis. *Plant Physiology*, *141*, 97–107.

Saito, S., Hirai, N., Matsumoto, C., Ohigashi, H., Ohta, D., Sakata, K., & Mizutani, M. (2004). Arabidopsis CYP707As encode (+)-Abscisic Acid 8′-Hydroxylase, a key enzyme in the oxidative catabolism of abscisic acid. *Plant Physiology*, *134*, 1439–1449.

Tanaka, Y., Nose, T., Jikumaru, Y., & Kamiya, Y. (2013). ABA inhibits entry into stomatal-lineage development in Arabidopsis leaves. *The Plant Journal*, *74*, 448–457.

Umezawa, T., Sugiyama, N., Mizoguchi, M., Hayashi, S., Myouga, F., Yamaguchi-Shinozaki, K., … Shinozaki, K. (2009). Type 2C protein phosphatases directly regulate abscisic acid-activated protein kinases in Arabidopsis. *Proceedings of the National Academy of Sciences of the United States of America*, *106*, 17588.

Wang, H., Ngwenyama, N., Liu, Y., Walker, J. C., & Zhang, S. (2007). Stomatal development and patterning are regulated by environmentally responsive mitogen-activated protein kinases in Arabidopsis. *Plant Cell*, *19*, 63–73.

Woodward, F. I., & Kelly, C. K. (1995). The influence of CO_2 concentration on stomatal density. *New Phytologist*, *131*, 311–327.

Woodward, F. I., Lake, J. A., & Quick, W. P. (2002). Stomatal development and CO_2: Ecological consequences. *New Phytologist*, *153*, 477–484.

Xiong, L., Ishitani, M., Lee, H., & Zhu, J.-K. (2001). The Arabidopsis LOS5/ABA3 locus encodes a molybdenum cofactor sulfurase and modulates cold stress– and osmotic stress-responsive gene expression. *Plant Cell*, *13*, 2063–2084.

4

Transcriptomic analysis of sweet potato under dehydration stress identifies candidate genes for drought tolerance

Kin H. Lau[1] | María del Rosario Herrera[2] | Emily Crisovan[1] | Shan Wu[3] | Zhangjun Fei[3] | Muhammad Awais Khan[2] | Carol Robin Buell[1,4] | Dorcus C. Gemenet[2]

[1]Department of Plant Biology, Michigan State University, East Lansing, Michigan
[2]International Potato Center, Lima, Peru
[3]Boyce Thompson Institute, Cornell University, Ithaca, New York
[4]Plant Resilience Institute, Michigan State University, East Lansing, Michigan

Correspondence
C. Robin Buell and Kin H. Lau, Department of Plant Biology, Michigan State University, East Lansing, MI.
Emails: buell@msu.edu and laukin3@msu.edu

Present address
Muhammad Awais Khan, Plant Pathology and Plant-Microbe Biology Section, Cornell University, Geneva, New York.

Funding information
Bill & Melinda Gates Foundation, Grant/Award Number: OPP1052983

Abstract

Sweet potato (*Ipomoea batatas* [L.] Lam.) is an important subsistence crop in Sub-Saharan Africa, yet as for many crops, yield can be severely impacted by drought stress. Understanding the genetic mechanisms that control drought tolerance can facilitate the development of drought-tolerant sweet potato cultivars. Here, we report an expression profiling study using the US-bred cultivar, Beauregard, and a Ugandan landrace, Tanzania, treated with polyethylene glycol (PEG) to simulate drought and sampled at 24 and 48 hr after stress. At each time-point, between 4,000 to 6,000 genes in leaf tissue were differentially expressed in each cultivar. Approximately half of these differentially expressed genes were common between the two cultivars and were enriched for Gene Ontology terms associated with drought response. Three hundred orthologs of drought tolerance genes reported in model species were identified in the *Ipomoea trifida* reference genome, of which 122 were differentially expressed under at least one experimental condition, constituting a list of drought tolerance candidate genes. A subset of genes was differentially regulated between Beauregard and Tanzania, representing genotype-specific responses to drought stress. The data analyzed and reported here provide a resource for geneticists and breeders toward identifying and utilizing drought tolerance genes in sweet potato.

KEYWORDS
drought, expression, *Ipomoea batatas*, polyethylene glycol, RNA-Seq, sweet potato

1 | INTRODUCTION

Sweet potato (*Ipomoea batatas* [L.] Lam.) is grown worldwide for its nutrient-rich storage roots. In Sub-Saharan African (SSA), sweet potato is a subsistence crop for small holder farmers and is a major source of energy and nutrition; as a consequence, it is an important food security crop (Low, Mwanga, Andrade, Carey, & Ball, 2017; Low et al., 2009; Mwanga & Ssemakula, 2011). Drought is a major concern for agriculture in many regions and is a particularly important issue among poor smallholder farmers who often lack the capacity to irrigate their crops, including sweet potato (Low et al., 2009). Screens for drought tolerance among sweet potato clones that grow well in SSA environments such as Mozambique (Maquia et al., 2013) and Kenya (Kivuva, Githiri, Yencho, & Sibiya, 2015) have shown that drought tolerance varies greatly in sweet potato germplasm, with the relative performance of sweet potato varieties in terms of yield shifting substantially between drought and non-drought conditions. The presence of genetic variation for drought tolerance among sweet potato germplasm suggests that drought tolerance loci can be introgressed into drought-sensitive cultivars. Thus, understanding the genetic and molecular basis of drought tolerance will aid in breeding for high-performing, drought tolerant sweet potato cultivars.

Drought stress induces a set of physiological responses with stomatal closure to reduce water loss through transpiration being the most immediate and important (reviewed in Osakabe, Osakabe, Shinozaki, & Tran, 2014). At the onset of drought, abscisic acid (ABA) levels are elevated, initiating signaling pathways that modulate guard cell osmolarity to achieve stomatal closure. As a result, components of the ABA signaling pathway as well as guard cell ion pumps are essential for proper stomatal closure during drought, leading to drought tolerance. However, control of stomatal closure during drought involves compromising between limiting water loss and maintaining CO_2 uptake for photosynthesis, which ultimately impacts crop yield. Thus, understanding the intricate processes that modulate drought responses such as stomatal closure and resiliency of photosynthetic rates during drought are crucial aspects of optimizing drought tolerance in crop plants.

Beauregard is a US-bred variety with low dry matter and orange flesh due to high levels of carotenoids (Rolston et al., 1987), while Tanzania is a popular Ugandan landrace with high dry matter and cream flesh (Mwanga et al., 2001). Though a Tanzania × Beauregard population (Cervantes-Flores, Yencho, Kriegner, et al., 2008) has been used to map quantitative trait loci (QTL) for root-knot nematode resistance, dry matter, starch, and β-carotene content (Cervantes-Flores, Yencho, Pecota, Sosinski, & Mwanga, 2008; Cervantes-Flores et al., 2011), drought tolerance traits have not yet been investigated in this population. However, we have an ongoing project to map QTL for drought-related traits using a population derived from the reciprocal cross, Beauregard × Tanzania. Promisingly, Kivuva et al. (2015) in a study involving 84 total genotypes showed that Beauregard differs from Tanzania in several drought response traits. Greenhouse-grown plantlets derived from cuttings of Tanzania stay green longer under drought conditions than those from Beauregard by 6–7 days, suggesting greater drought tolerance in Tanzania (Supporting Information Figure S1a). Mature plants grown at two different sites in Kenya under irrigated and rain-fed conditions resulted in a larger decrease in chlorophyll content in Tanzania compared to Beauregard, yet the number of storage roots and the fresh storage root weight were more sensitive in Beauregard than in Tanzania (Supporting Information Figure S1b–d).

A few studies have examined global transcriptomic responses for drought-related traits in different *Ipomoea* species. Yang et al. (2018) recently reported a three-way combined *de novo* transcriptome for whole plants of PEG-treated *Ipomoea triloba*, *I. batatas* (cv. Kokei No. 14) and a somatic hybrid between the two species, finding genes that are differentially regulated in the *I. batatas* parent compared to either *I. triloba* or the hybrid, which are more drought tolerant. Other de novo transcriptome approaches have been reported for hexaploid *Ipomoea trifida* and *Ipomoea imperati* for drought and salt stress, respectively (Peng et al., 2017; Solis, Baisakh, Brandt, Villordon, & Bonte, 2016). In this study, global gene expression patterns in leaf tissue of two varieties of *I. batatas*, Beauregard, and Tanzania, during PEG-simulated drought stress were characterized with RNA-Sequencing (RNA-Seq); diploid *I. trifida*, which is closely related to *I. batatas* and is a putative progenitor species (Austin, 1988; Kobayashi, 1983), was used as the reference genome sequence (http://sweetpotato.plantbiology.msu.edu/gt4sp_download.shtml). Differentially expressed genes (DEGs) were identified, revealing candidate genes for improving drought tolerance in sweet potato. In addition, genes with unique expression patterns in the two cultivars were classified into co-expression clusters showing enrichment for biological processes that could contribute to phenotypic differences in their drought responses. Data from this study will be useful in identifying candidate genes within drought tolerance QTL and improving drought tolerance in this key food security crop.

2 | MATERIALS AND METHODS

2.1 | Determination of optimal conditions to induce dehydration stress

In vitro plants of Beauregard (CIP440132) and Tanzania (CIP440166) were obtained from the International Potato Center (CIP) Genebank and nodes were cultured to form plantlets on MPB medium (Murashige and Skoog salts, 3% sucrose, 2 mg/L calcium pantothenate, 100 mg/L L-arginine, 200 mg/L ascorbic acid, 20 mg/L putrescine-HCl, 10 mg/L GA_3, 0.3% Phytagel, pH 5.7, autoclaved at 121°C for 15 min as recommended by the supplier) in culture tubes (three per tube). Shoot cultures were kept in a growth room at 27 ± 2°C, 3,000 lx, and 12 hr light/12 hr dark.

Polyethylene glycol (PEG 6000, Sigma) concentrations were chosen to create osmotic pressure levels between field capacity (−0.033 mpa) and permanent wilting (−1.5 mpa), as estimated based on osmotic pressure values reported for different PEG 6000 concentrations (http://www.plantstress.com/methods/peg.htm; Michel & Kaufmann, 1973). After 21 days of growth in solid MPB media, the plants were transferred to 35 ml of 15%, 20%, and 25% PEG6000 liquid MPB media, representing osmotic pressure values of −0.295, −0.491, and −0.735 mpa, respectively. Three replicates were used. Cultures were maintained in a growth chamber under the conditions described above and stress symptoms were evaluated at 4, 24, 48 hr, and 21 days after stress application.

2.2 | Plant growth conditions for RNA-Seq

For both varieties, one node was cut from the donor plant and placed in a 16 × 125 mm tube containing solid MPB media. Only the internode was submerged into the solid media to allow the node to grow. Fourteen days after planting, six plantlets with similar size were selected and transferred to another vessel containing liquid MPB media. A thin plate of acrylic plastic containing small holes was used to support plantlets and to allow leaf expansion. Three biological replicates and three controls were used. Cultures were maintained in the growth chamber under the same conditions as described above. At 21 days after planting, the initial liquid media was removed and replaced with liquid media containing 25% PEG in liquid MPB medium. Leaf samples were collected for RNA extraction at 24 and 48 hr after stress (HAS) imposition, flash frozen, and stored at −80°C. RNA extraction was performed with 200 mg of leaf tissue for each sample using TRIzol (Invitrogen) according to supplier instructions.

2.3 | Library construction, sequencing, and expression estimation

Strand-specific RNA-Seq libraries (one library for each of three biological replicates per experimental condition; Supporting Information Table S1) were constructed using the protocol described in Zhong et al. (2011) and sequenced on an Illumina HiSeq 2500 platform. Reads were trimmed using CutAdapt (v1.8.3; Martin, 2011), removing up to two adapter sequences from each read, retaining only reads with a minimum read length of 31 nt and trimming the 3′ end with a quality cut-off of 10. Reads were trimmed to 100 nt using FASTX (v0.0.14; http://hannonlab.cshl.edu/fastx_toolkit/) as a subset of libraries were sequenced to a read length of 151 nt instead of 101 nt. Reads flagged by the Illumina pipeline as low quality with "Y" were removed. Mononucleotide repeats of T and A 10-mers were removed up to 10 times using CutAdapt (v1.8.3), retaining only reads with a minimum read length of 31 nt. Libraries were run on multiple lanes (technical replicates) and were mapped independently to the *I. trifida* genome assembly v3 (http://sweetpotato.plantbiology.msu.edu/gt4sp_download.shtml) using TopHat2 (v2.1.1; Kim et al., 2013) implementing Bowtie2 (v2.2.9; Langmead & Salzberg, 2012) with the options: –min-intron-length 10, –max-intron-length 5000, –max-multihits 20, –library-type fr-firststrand, –min-segment-intron 10, –max-segment-intron 5000, –b2-N 1, –read-mismatches 3, and –read-edit-dist 3.

Cufflinks (v2.2.1; Trapnell et al., 2010) was used to calculate fragments per kilobase exon model per million mapped reads (FPKMs) for the set of high-confidence *I. trifida* gene models (v3) using the options: –multi-read-correct, –min-intron-length 10, –max-intron-length 5000, –library-type fr-first strand. One technical replicate was removed because of a markedly lower alignment rate compared to other technical replicates. Transcripts *itf01g31740.t1*, *itf02g16070.t2*, *itf03g01940.t1* were removed because these had identical coordinates as another isoform of the same gene. Correlations between different sequencing runs of the same library were quantified using Pearson Correlation Coefficients from FPKMs in R (pairwise.complete.obs, method = pearson). Technical replicates were highly correlated, with most correlation coefficients above 0.95 and others above 0.90. The alignment files for technical replicates were merged using the MergeSamFiles tool in PicardTools (v2.1.1; https://broadinstitute.github.io/picard/), and used for calculating FPKMs for each biological replicate (Supporting Information Table S2) and for differential expression analysis.

2.4 | Identification of differentially expressed genes

Gene annotations (v3; http://sweetpotato.plantbiology.msu.edu/gt4sp_download.shtml) were converted from GFF3 to GTF format using the "gffread" script included with Cufflinks (v2.2.2). Uniquely mapping reads overlapping gene models were counted using htseq-count in HTSeq (v0.6.1p1; Anders, Pyl, & Huber, 2015) with the options: –stranded = reverse –minaqual = 10 –type = exon –mode = union.

Read counts were used to detect differential expression with DESeq2 v1.16.1 (Love, Huber, & Anders, 2014). For differential expression between control and PEG treatment at each time-point in each variety, unique combinations of treatment, time-points, and variety were treated as a single factor, and the contrasts function in DESeq2 was used to test whether log2 fold change was equal to 0 for each pair of control and PEG treatment levels. An adjusted *p*-value cut-off of 0.05 and a log2 fold-change threshold of 1.5 were used. The "lfcShrink" function was used to restrain high log2 fold changes for genes with low expression levels.

To detect genes that responded differentially to PEG between Beauregard and Tanzania at either time-point or both time-points, the full model, ~treatment + variety + time_point + variety:time_point + treatment: time_pt + treatment:variety:time_point, was tested against a reduced model of, ~treatment + variety + time_point + variety:time_point + treatment:time_pt using a likelihood ratio test implemented in DESeq2 and an adjusted *p*-value cut-off of 0.01. For K-means clustering, log2 fold changes calculated by DESeq2 were obtained for significant genes and mean-centered by subtracting the arithmetic mean for each gene. K-means clustering was conducted using the "kmeans" function in R with the settings, centers = 16, nstart = 25, iter.max = 100, algorithm = "MacQueen" (MacQueen, 1967). A K of 16 was chosen because relatively little reduction in total within-cluster variation was observed for larger values.

2.5 | Gene ontology term enrichment analysis

Groups of DEGs or clusters with similar expression patterns were tested for enriched gene ontology (GO) terms using the "weight01" algorithm and Fisher's exact test implemented in topGO v2.28.0 (Alexa & Rahnenfuhrer, 2016) using all genes with GO terms as the background. *p*-values were corrected for multiple testing using the "p.adjust" function in R to implement the FDR method (Benjamini & Hochberg, 1995), and filtered at the 0.05 level.

2.6 | Orthology analysis

OrthoFinder v1.1.3 (Emms & Kelly, 2014) was run with the longest peptide isoforms from the predicted proteomes of *Amborella trichopoda* v1.0, *Arabidopsis thaliana* TAIR10, *I. trifida* v3, *Oryza sativa* v7, and *Solanum lycopersicum* iTAG2.4, obtained from Phytozome v12.1.5 (Goodstein et al., 2012).

3 | RESULTS AND DISCUSSION

3.1 | Establishing effective dehydration conditions

Drought conditions were simulated by adding PEG to the media that plantlets were grown in. No stress symptoms were observed up to 6 HAS. However, wilted and chlorotic leaves were observed in PEG-treated plantlets of both varieties by 24 HAS, and the severity increased at 48 HAS (Figure 1a), with plants dying after 21 days. In contrast, leaves on plantlets not treated with PEG remained green and turgid. No visible differences between the stress phenotypes of Beauregard and Tanzania were observed at these plantlet stages. An osmotic pressure of −0.735 mpa (25% PEG 6000) was chosen for the actual stress experiment for RNA-Seq analysis as it elicited strong differences between treated and untreated plants.

3.2 | Global gene expression profiles are conserved between Beauregard and Tanzania

To assess the quality of the biological replicates and the extent of gene expression variation explained by the experimental variables, a principal component analysis (PCA) was conducted (Figure 1b). Tight clustering of biological replicates indicated that the majority of the gene expression variation observed was due to the experimental conditions. Greater than 75% of variation was explained by the first principal component, which corresponded predominantly to the addition of PEG. In comparison, varietal effects of Beauregard and Tanzania, and the time after PEG application contributed to a much smaller portion of the expression variation. In support of the PCA results, expression fold changes of mock treatment vs. PEG treatment were significantly correlated between Beauregard and Tanzania at both 24 and 48 HAS ($p < 2.2e-16$ for both; Figure 2a). Indeed, genes with the largest fold change between mock and PEG treatments were differentially expressed in the same direction and at similar magnitude in Beauregard and Tanzania (Figure 2a). Conversely, a large number of genes with less extreme fold changes were differentially expressed in one variety but not the other, indicating that a substantial portion of dehydration-responsive genes are unique to each genotype.

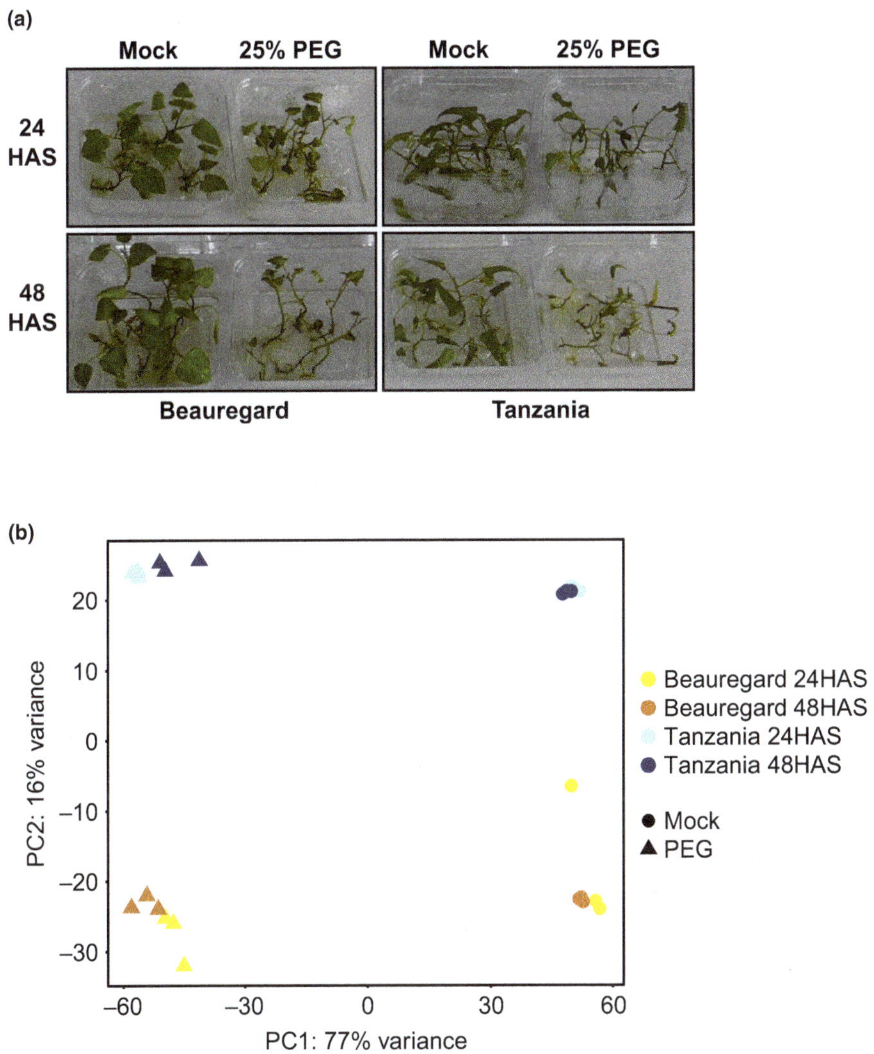

FIGURE 1 Phenotypic and genetic response of Beauregard and Tanzania sweetpotato varieties to the application of polyethylene glycol (PEG). (a) Plantlets incubated in untreated media or 25% PEG media at 24 and 48 hr after stress (HAS). (b) Principal component analysis of individual biological replicates using variance stabilizing transformed read counts. Each point represents a biological replicate

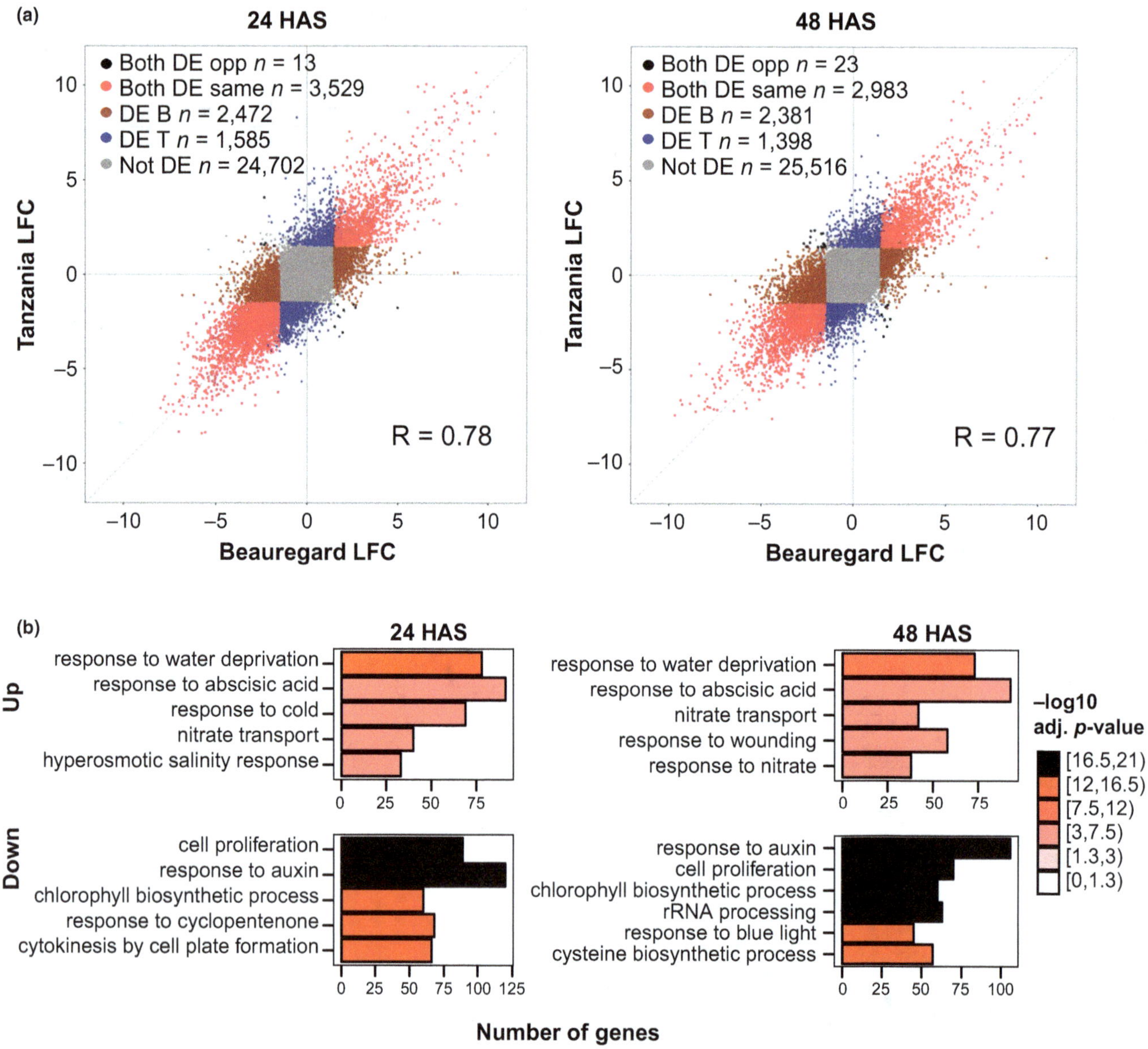

FIGURE 2 Conserved expression responses to polyethylene glycol in Beauregard and Tanzania. (a) Log2 fold change (LFC) of gene expression abundances in Beauregard and Tanzania at 24 and 48 hr after stress (HAS). "DE B" and "DE T" indicate genes differentially expressed in only Beauregard or Tanzania, respectively. (b) The most significantly enriched biological process gene ontology terms among genes differentially expressed in both Beauregard and Tanzania at 24 and 48 HAS

Considering that Beauregard is one of the predominant cultivars for large, commercial-scale cultivation in the U.S. while Tanzania is a landrace commonly grown by smallholder farmers in Uganda and other East African countries (Carpena, 2009), these two varieties are likely adapted to different environmental conditions, and shared genetic responses to PEG may be involved in critical drought tolerance or survival mechanisms. At 24 HAS, 2,216 and 1,809 genes were upregulated while 3,798 and 3,318 genes were downregulated in Beauregard and Tanzania, respectively. At 48 HAS, numbers of upregulated genes were comparable to 24 HAS, but downregulated genes decreased to 3,285 and 2,482. Among genes differentially expressed in the same direction in both genotypes, similar GO terms were enriched at 24 and 48 HAS. Consistent with a drought response, "water deprivation," "response to ABA," and "nitrate transport" were among the top biological processes (Figure 2b). Enrichment for the "response to ABA" GO term is expected because drought utilizes ABA signaling to induce stomatal closure in order to reduce water loss via transpiration (reviewed in Osakabe et al., 2014). An important part of this ABA signaling pathway is the transport of ABA to stomatal guard cells by ABA transporter proteins, such as NITRATE TRANSPORTER 1.2 (NRT1.2; Kanno et al., 2012), explaining the "nitrate transport" GO term enrichment. Indeed, upregulated genes enriched for the nitrate transport GO term included *NRT1.2* homologs, *NRT1.7* and – at 48 HAS only – *NRT1.5*, and an unrelated nitrate transporter, *NRT2.7* (Dechorgnat et al., 2011; Supporting Information Dataset S1). Alternatively, the "nitrate transport" GO term may be involved in the accumulation of free amino acids such as proline that are hypothesized to help maintain cell turgor by decreasing water potential inside cells (Stewart & Lee, 1974). Upregulated genes enriched for the "water deprivation" GO term (Supporting Information Dataset S1) included well-known drought tolerance genes such as *DEHYDRATION-RESPONSIVE ELEMENT-BINDING PROTEIN 2A* (*DREB2A*) which encodes a drought-responsive transcription factor (Liu et al., 1998; Sakuma et al., 2006) and *LATE EMBRYOGENESIS ABUNDANT* (*LEA*), a family of stress-inducible genes that encode proteins that have been implicated in diverse drought tolerance mechanisms such as preventing protein aggregation and membrane protection during dehydration (Goyal, Walton, & Tunnacliffe, 2005; Tolleter et al., 2007).

For downregulated genes, the most significant GO terms were primarily involved in cell division, response to auxin and chlorophyll biosynthesis (Figure 2b). Stomatal closure leads to reduced flux through the Calvin cycle, limiting the supply of $NADP^+$ electron acceptors to the electron transport chain in the chloroplasts causing an overaccumulation of electrons (Noctor, Mhamdi, & Foyer, 2014; Osmond & Grace, 1995). Thus, downregulation of genes involved in chlorophyll biosynthesis may represent an effort to avoid overaccumulation of electrons by reducing photosynthetic capacity. Regardless, reduced chlorophyll content due to drought has been reported in multiple plant species (Jiang & Huang, 2001; Li, Pezeshki, & Goodwin, 2004; Sarker, Rahman, & Paul, 1999) and as a consequence of decreased photosynthesis, less resources are available to support tissue growth thereby explaining the downregulated GO terms of "cell proliferation" and the "response to auxin," an important regulator of cell and tissue growth. Leaves, the tissue that RNA was extracted from for this study, are often more susceptible to reduced growth rate compared to roots and it is hypothesized that reduced leaf area helps limit water loss due to transpiration (Aguirrezabal et al., 2006; Pace, Cralle, El-halawany, Cothren, & Senseman, 1999; Xu et al., 2015). These overall expression patterns are consistent with a drought response and indicate that dehydration stress was successfully imposed by the PEG treatment.

3.3 | A group of leucine-rich kinases are downregulated at 24 HAS

For both directions of regulation, hundreds of DEGs were unique to a particular time-point in each variety, indicating that dynamic regulatory changes take place between 24 HAS and 48 HAS (Figure 3a,b). To examine their biological significance, GO term enrichment was tested for the DEGs that were specific to a particular time-point and common to both varieties (Figure 3b). While only weakly significant GO terms were observed for the other three classes of time-specific DEGs, genes downregulated at 24 HAS only were strongly enriched for protein phosphorylation and plasma membrane GO terms (Figure 3c). Out of the 39 genes downregulated at only 24 HAS that had a protein phosphorylation GO term, 24 were annotated as leucine-rich repeat protein kinases (LRR-PK) or leucine-rich receptor-like kinases (LRR-RLK) and five were other types of kinases (Supporting Information Dataset S2). The involvement of these genes in PEG response in sweet potato is consistent with previous transcriptomic analyses that have shown many LRR-RLK genes in Arabidopsis to be responsive (both down and upregulated) to osmotic stress (Chae, Sudat, Dudoit, Zhu, & Luan, 2009; Lehti-Shiu, Zou, Hanada, & Shiu, 2009).

3.4 | Identification and analysis of orthologs of drought tolerance candidate genes

DroughtDB (Alter et al., 2015) is a manually curated database of ~200 drought tolerance candidate genes compiled mainly from knockout, knockdown, and overexpression experiments in 38 different plant systems. Orthogroups were identified for the predicted proteomes of *I. trifida*, *A. trichopoda* (representing the most basal clade among angiosperms) and three highly represented species in DroughtDB, *A. thaliana*, *O. sativa*, and *S. lycopersicum*. Eighty-three percent of orthogroups containing candidate drought tolerance genes were conserved by at least four species, consistent with important biological roles for these genes (Figure 4a). In total, there were 300 DroughtDB orthologs in *I. trifida*, and 122 of these were differentially expressed in at least one variety-time point combination (Figure 4b).

3.4.1 | Upregulated candidate genes include a LEA gene and ABA metabolism genes

We focused on the 21 upregulated (*itf12g21010* was downregulated at 24 HAS in Beauregard) and 16 downregulated drought candidate genes that were differentially expressed at all four variety-time point combinations, reasoning that these genes are strong candidates to be involved in drought stress responses (Figure 4b; Supporting Information Figure S2). A *TAS14* homolog (*itf03g07310*) was upregulated by >8 log2 fold change across all four variety-time points combinations, substantially higher than any other drought tolerance candidate gene (Figure 4b). TAS14 is a LATE EMBRYOGENESIS ABUNDANT (LEA) protein of the dehydrin subfamily and its expression is induced by salt, ABA or mannitol treatment in tomato (Godoy, Pardo, & Pintor-Toro, 1990). Overexpression of *TAS14* in tomato under drought stress increases fruit biomass several fold compared to wild-type plants while producing no discernable deleterious effects under unstressed conditions (Muñoz-Mayor et al., 2012). The mechanism of TAS14 action may be related to an increase in soluble sugars in the leaves that coincides with higher water potential and cell turgor in drought-stressed plants overexpressing *TAS14* (Muñoz-Mayor et al., 2012). Other candidate genes upregulated in all four experimental conditions include the drought-inducible DREB2A transcription factor (Liu et al., 1998) and several components of the ABA signaling pathway. These include the ABA and drought-inducible bZIP transcription factor ABF3/ABF4/AREB1/OsbZIP23 that activates genes with ABRE (ABA responsive element) promoter elements (Kang, Choi, Im, & Kim, 2002; Uno et al., 2000; Xiang, Tang, Du, Ye, & Xiong, 2008), and the ATP-binding

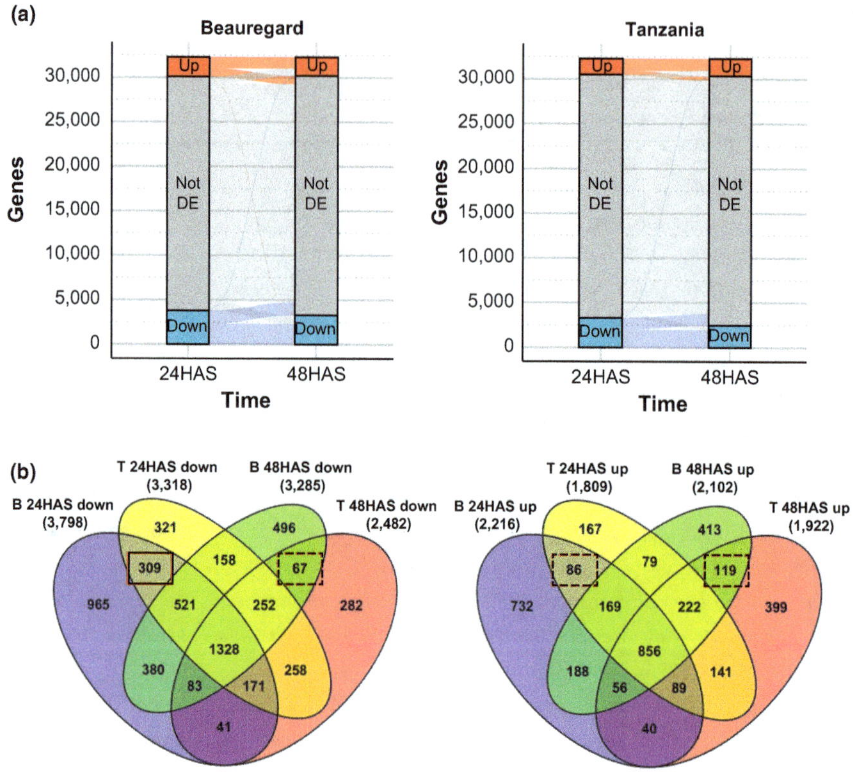

(c)

GO type	GO ID	GO term	Gene count	Adjusted *p*-value
CC	GO:0005886	plasma membrane	103	3.70E−10
BP	GO:0006468	protein phosphorylation	39	7.20E−10
BP	GO:0007169	transmembrane receptor protein tyrosine kinase signaling pathway	18	8.30E−07
MF	GO:0005524	ATP binding	24	5.00E−04
MF	GO:0004674	protein serine/threonine kinase activity	26	5.00E−04

FIGURE 3 Time-point specific differentially expressed genes (a) Differentially expressed genes at 24 and 48 hr after stress (HAS) in Beauregard and Tanzania. Genes labeled "up" or "down" were significantly upregulated or downregulated, respectively. (b) Overlap between genes significantly downregulated or upregulated in each variety (B: Beauregard, T: Tanzania) and at each time-point. Red rectangles indicate genes tested for GO term enrichment and dashed outlines indicate weak or insignificant enrichment. (c) The most significant GO terms for genes downregulated in both Beauregard and Tanzania at 24 HAS but not 48 HAS (solid red rectangle in panel b)

cassette (ABC) ABA transporters, ABCG22, 25 and 40 (Kang et al., 2010; Kuromori, Sugimoto, & Shinozaki, 2011; Kuromori et al., 2010). Besides the transport of ABA, its metabolism was also implicated in dehydration response in sweet potato. *NCED* encodes 9-cis-epoxycarotenoid dioxygenase, an ABA biosynthesis enzyme whose induction correlates with increases in ABA during drought stress (Iuchi et al., 2001). Conversely, CYP707A1 and 3 are ABA catabolic enzymes that are induced by both dehydration and rehydration in Arabidopsis, allowing controlled ABA responses and rapid disengagement of ABA responses, respectively (Kushiro et al., 2004; Saito et al., 2004; Umezawa et al., 2006).

3.4.2 | LHCSB6 and SLAC1 are unexpectedly downregulated during water stress

Surprisingly, some drought tolerance candidate genes that improve drought tolerance when overexpressed or functional in other species were downregulated in both varieties at both time-points. *SLOW ANION CHANNEL-ASSOCIATED 1* (*SLAC1*; Negi et al., 2008; Vahisalu et al., 2008) and *LIGHT-HARVESTING CHLOROPHYLL A/B-BINDING6* (*LHCB6*; Jansson, 1999) are two such genes and, intriguingly, these two genes lack paralogs in *I. trifida*, suggesting that their down-regulation during dehydration stress cannot be compensated by the activity of other genes (Figure 4c). KAT2 was another gene downregulated at all four variety-time point combinations that has no paralogs, but the downregulation of this gene, encoding an inwardly rectifying K^+ channel, is consistent with the need for net K^+ efflux from guard cells during stomatal closure (Schroeder, Hedrich, & Fernandez, 1984).

In contrast, the downregulation of *LHCB6* and *SLAC1* may have detrimental effects on stomatal closure during drought, and restoring their functions by overexpression may improve drought tolerance in sweet potato. *LHCSB6* encodes a protein associated with photosystem II (Jansson, 1999) that, in Arabidopsis, inhibits ABA-mediated stomatal closure when expression is knocked down and enhances ABA-mediated stomatal closure when overexpressed (Xu et al., 2012). Due to its role in mediating stomatal closure, the knockdown mutant of *LHCSB6* has increased drought sensitivity, with visibly more severe reductions in leaf size and greenness compared to

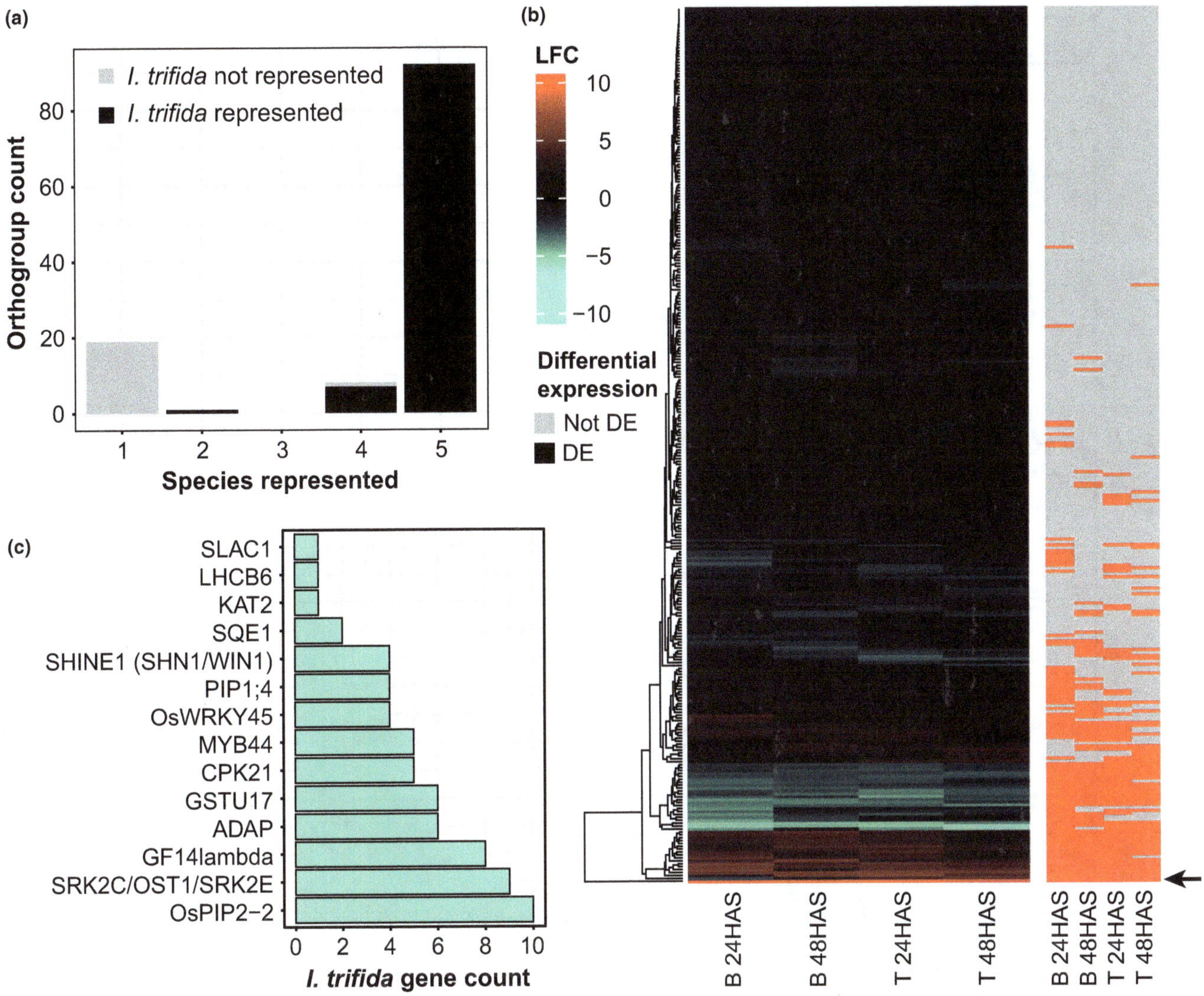

FIGURE 4 *Ipomoea trifida* orthologs of drought tolerance candidate genes identified in Arabidopsis, rice, and tomato. (a) Species representation of orthogroups containing DroughtDB genes. (b) Log2 fold changes (left; polyethylene glycol/control) and differential expression state (right; grey: not differentially expressed; red: differentially expressed). The arrow indicates itf03g07310 (*TAS14*). (c) Counts of *I. trifida* genes in orthogroups containing drought tolerance candidate genes significantly downregulated in all four variety-time point combinations

wildtype (Xu et al., 2012). SLAC1 is an anion efflux channel protein that is essential for proper stomatal closure in response to many factors, including ABA (Negi et al., 2008; Vahisalu et al., 2008). In Arabidopsis, leaf expression of *SLAC1* is restricted mainly to guard cells (Negi et al., 2008; Vahisalu et al., 2008) and is upregulated 1 to 1.6 log2 fold change by drought (Wang et al., 2016). In contrast, we observed a strong downregulation (−2.8 to −3.6 log2 fold change) of *SLAC1* in sweet potato leaves in response to PEG (Supporting Information Figure S2; Supporting Information Table S3).

3.5 | Varietal differences in differentially expressed genes under dehydration stress

Differential gene expression analysis for each individual variety-time point combination suggested that there are ~2,000 PEG-responsive genes unique to each variety (Figure 2a). We hypothesize that expression differences in some of these genes contribute to the differences in days to permanent wilting point, and drought-induced reductions in chlorophyll content and storage roots observed between the two varieties (Supporting Information Figure S1). To statistically identify genes that are regulated differentially between Beauregard and Tanzania in response to PEG, genes were tested for significant interaction effects between PEG treatment and variety, either as two-way interactions or as three-way interactions with time. The significant genes were classified into co-regulated clusters using a K-means approach in order to uncover expression patterns. In this way, 3,824 genes with significant interaction effects were identified and grouped into 16 clusters ranging from 60 to 500 genes in size (Supporting Information Figure S3; Supporting Information Table S4). Clusters 4, 5 and 6, encompassing 1,322 genes, have relatively mild differences between Beauregard and Tanzania (Figure 5a). Conversely, the other clusters, which compose the remaining

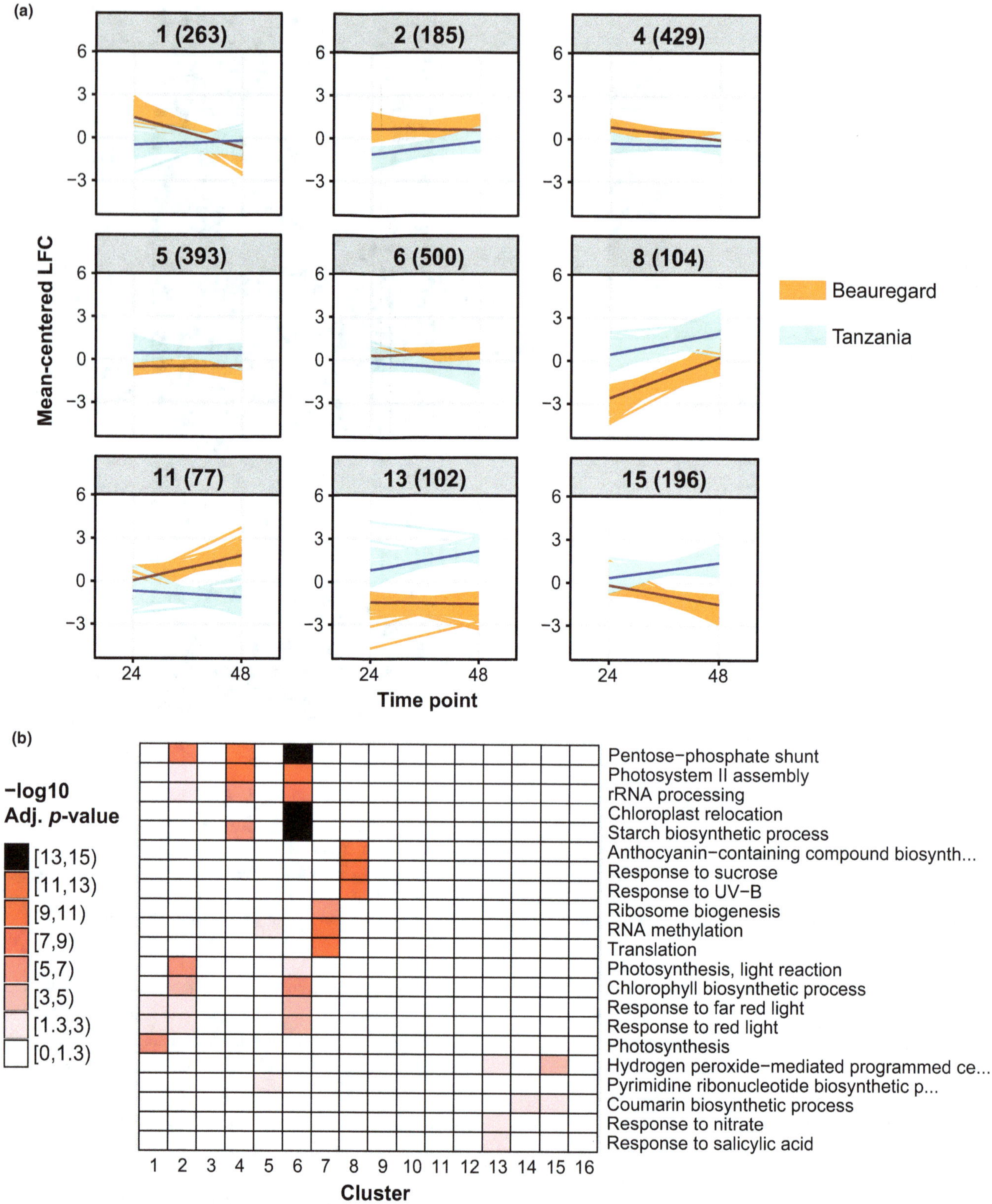

FIGURE 5 Genes differentially regulated between Beauregard and Tanzania in response to polyethylene glycol (PEG). (a) Representative K-means clusters for mean-centered (per gene) log2 fold changes (PEG/control) for Beauregard and Tanzania at 24 and 48 hr after stress. The number of genes in each cluster is indicated in parentheses. (b) Top three most significant biological processes gene ontology terms for each K-means cluster

2,502 genes used for clustering, had marked differences in expression pattern between the two varieties consistent with distinct genetic responses to dehydration stress.

Additionally, these expression differences between the two varieties were dramatically different at the two time-points in clusters 11, 13, and 15, emphasizing again that dynamic regulatory changes occur between the two time-points. Cluster 11 was not significantly enriched for any biological process GO term, but contained an ortholog of *OST2*, which encodes an H^+-ATPase proton pump that is involved in ABA-mediated stomatal closure in response to drought (Merlot et al., 2007; Figure 5b; Supporting Information Table S3). Three *OST2* orthologs were significant for interaction between treatment and variety effects (the other two were in clusters 3 and 6) and except for 24 HAS for *itf03g23220*, higher expression was observed for Beauregard for all data points suggesting that reduced *OST2* activity in Tanzania may contribute to its higher days to permanent wilting (Figure 1a, Supporting Information Figure S4). In contrast to cluster 11, clusters 13, and 15, which had increased expression in Tanzania at 48 HAS compared to 24 HAS but had either similar or weakened expression at 48 HAS compared to 24 HAS in Beauregard, were weakly significant for the biological process GO term "hydrogen peroxide-mediated programmed cell death" (Figure 5b; Supporting Information Dataset S3). This suggests that drought-induced hydrogen peroxide (Noctor, Veljovic-Jovanovic, Driscoll, Novitskaya, & Foyer, 2002) causes more cell death in Tanzania. However, the overall effect of hydrogen peroxide in Tanzania could potentially not be detrimental because hydrogen peroxide is an important signal molecule for proper ABA-mediated stomatal closure (Pei et al., 2000).

Clusters 1, 2, 4, and 6 were enriched for photosynthesis-related GO terms and contained genes that were, in general, expressed higher in Beauregard or at comparable levels in both varieties (Figure 5). Cellular compartment GO term analysis supported an enrichment for genes involved in photosynthesis in these clusters, with significant GO terms for chloroplast thylakoid membrane, stroma and envelope (Supporting Information Figure S5a). Clusters 2, 4, and 6 were also enriched for the pentose-phosphate shunt biological process GO term (Figure 5b). One of the products of the pentose-phosphate shunt is NADPH (Kruger & von Schaewen, 2003), which is required for the anti-oxidative function of glutathione reductase that may help quench reactive oxygen species (ROS) produced during drought that could damage chloroplast membranes (Das & Roychoudhury, 2014). Together, the higher expression of photosynthesis and pentose-phosphate shunt genes in clusters 1, 2, 4, and 6 in Beauregard may contribute to its smaller reduction in chlorophyll content compared to Tanzania plants in response to drought (Supporting Information Figure S1b).

Cluster 8 genes were on average regulated at ~3 log2 fold higher in Tanzania than in Beauregard at both 24 and 48 HAS (Figure 5a). Interestingly, cluster 8 was most significant for enrichment of the GO term, "anthocyanin-containing compound biosynthesis" (Figure 5b). Anthocyanin production is triggered by various biotic and abiotic stresses, including drought stress (Kovinich et al., 2014). While the function of stress-induced anthocyanin is still debated, overexpression of anthocyanin biosynthesis improves drought resistance and the leading hypothesis is that it is involved in removing ROS (Gould, McKelvie, & Markham, 2002; Kovinich, Kayanja, Chanoca, Otegui, & Grotewold, 2015; Nakabayashi et al., 2014). Because ROS overaccumulation can damage not only cellular membranes but also nucleic acids and proteins, the increased expression of anthocyanin-containing biosynthesis genes in Tanzania may contribute to its extended days to permanent wilting point compared to Beauregard (Supporting Information Figure S1a) by helping to limit levels of damaging ROS during drought stress.

4 | CONCLUSIONS

We present here an expression profiling resource for drought stress experiments in sweet potato and have demonstrated its utility for generating hypotheses for functional genomics. Overall expression patterns were conserved between Beauregard and Tanzania, and consistent with drought responses reported in the literature, with a general upregulation of ABA signaling components and downregulation of tissue growth processes. These observations corroborate the quality of our data and give weight to the other aspects of our analysis, including several interesting findings. We have identified a group of LRR kinases that are down-regulated at 24 HAS but not at 48 HAS suggesting their reduced activities at 24 HAS have important roles in dehydration response. Interestingly, we observed down-regulation of single-copy *LHCSB6* and *SLAC1* orthologs in both varieties at both time-points. These two genes encode effector proteins, a chlorophyll-binding component of PS II and a guard cell anion efflux protein, respectively, that have been shown to be highly important for stomatal closure during drought stress in Arabidopsis and we present these as strong candidates for overexpression experiments in sweet potato. Co-regulated clusters of genes involved in photosynthesis and the pentose-phosphate pathway that is expressed higher in Beauregard may contribute to its chlorophyll content being more resilient to drought than Tanzania. On the other hand, a separate co-regulated gene cluster involved in the production of anthocyanin-containing molecules may help increase the number of days to permanent wilting in Tanzania.

ACKNOWLEDGMENTS

This research was supported by a grant from the Bill & Melinda Gates Foundation (OPP1052983). The authors thank John P. Hamilton for bioinformatics support. Work at CIP was carried out as part of the CRP-RTB program of the CGIAR.

AUTHOR CONTRIBUTIONS

C.R.B., Z.F., and M.A.K. designed the experiment. M.A.K., M.R.H, S.W., and Z.F. performed the research. K.H.L., E.C., D.C.G., and C.R.B. analyzed and interpreted the data. K.H.L. and C.R.B. wrote the manuscript. All authors read and approved the manuscript.

REFERENCES

Aguirrezabal, L., Bouchier-Combaud, S., Radziejwoski, A., Dauzat, M., Cookson, S. J., & Granier, C. (2006). Plasticity to soil water deficit in *Arabidopsis thaliana*: Dissection of leaf development into underlying growth dynamic and cellular variables reveals invisible phenotypes. *Plant, Cell and Environment*, *29*, 2216–2227. https://doi.org/10.1111/j.1365-3040.2006.01595.x

Alexa, A., & Rahnenfuhrer, J. (2016). *topGO: Enrichment Analysis for Gene Ontology*.

Alter, S., Bader, K. C., Spannagl, M., Wang, Y., Bauer, E., Schön, C.-C., & Mayer, K. F. X. (2015). DroughtDB: An expert-curated compilation of plant drought stress genes and their homologs in nine species. *Database*, *2015*, bav046.

Anders, S., Pyl, P. T., & Huber, W. (2015). HTSeq—A Python framework to work with high-throughput sequencing data. *Bioinformatics*, *31*, 166–169. https://doi.org/10.1093/bioinformatics/btu638

Austin, D. F. (1988). The taxonomy, evolution and genetic diversity of sweet potatoes and related wild species. In P. Gregory (Ed.), *Exploration, maintenance and utilization of sweet potato genetic resources, Report of the 1st Sweet Potato Planning Conference 1987* (pp. 27–59). Lima, Peru: International Potato Center.

Benjamini, Y., & Hochberg, Y. (1995). Controlling the false discovery rate: A practical and powerful approach to multiple testing. *Journal of the Royal Statistical Society. Series B Methodological*, *57*, 289–300.

Carpena, A. L. (2009). Important cultivars, varieties, and hybrids. In G. Loebenstein & G. Thottappilly (Eds.), *The sweetpotato* (pp. 27–40). Dordrecht, The Netherlands: Springer. https://doi.org/10.1007/978-1-4020-9475-0

Cervantes-Flores, J. C., Sosinski, B., Pecota, K. V., Mwanga, R. O. M., Catignani, G. L., Truong, V. D., ... Yencho, G. C. (2011). Identification of quantitative trait loci for dry-matter, starch, and β-carotene content in sweetpotato. *Molecular Breeding*, *28*, 201–216. https://doi.org/10.1007/s11032-010-9474-5

Cervantes-Flores, J. C., Yencho, G. C., Kriegner, A., Pecota, K. V., Faulk, M. A., Mwanga, R. O. M., & Sosinski, B. R. (2008). Development of a genetic linkage map and identification of homologous linkage groups in sweetpotato using multiple-dose AFLP markers. *Molecular Breeding*, *21*, 511–532. https://doi.org/10.1007/s11032-007-9150-6

Cervantes-Flores, J. C., Yencho, G. C., Pecota, K. V., Sosinski, B., & Mwanga, R. O. M. (2008). Detection of quantitative trait loci and inheritance of root-knot nematode resistance in sweetpotato. *Journal of the American Society for Horticultural Science*, *133*, 844–851.

Chae, L., Sudat, S., Dudoit, S., Zhu, T., & Luan, S. (2009). Diverse transcriptional programs associated with environmental stress and hormones in the *Arabidopsis* receptor-like kinase gene family. *Molecular Plant*, *2*, 84–107. https://doi.org/10.1093/mp/ssn083

Das, K., & Roychoudhury, A. (2014). Reactive oxygen species (ROS) and response of antioxidants as ROS-scavengers during environmental stress in plants. *Frontiers in Environmental Science*, *2*, 1–13.

Dechorgnat, J., Nguyen, C. T., Armengaud, P., Jossier, M., Diatloff, E., Filleur, S., & Daniel-Vedele, F. (2011). From the soil to the seeds: The long journey of nitrate in plants. *Journal of Experimental Botany*, *62*, 1349–1359. https://doi.org/10.1093/jxb/erq409

Emms, D., & Kelly, S. (2014). OrthoFinder: Solving fundamental biases in whole genome comparisons dramatically improves orthogroup inference accuracy. *Genome Research*, *24*, 475–486.

Godoy, J. A., Pardo, J. M., & Pintor-Toro, J. A. (1990). A tomato cDNA inducible by salt stress and abscisic acid: Nucleotide sequence and expression pattern. *Plant Molecular Biology*, *15*, 695–705. https://doi.org/10.1007/BF00016120

Goodstein, D. M., Shu, S., Howson, R., Neupane, R., Hayes, R. D., Fazo, J., ... Rokhsar, D. S. (2012). Phytozome: A comparative platform for green plant genomics. *Nucleic Acids Research*, *40*, D1178–D1186. https://doi.org/10.1093/nar/gkr944

Gould, K. S., McKelvie, J., & Markham, K. R. (2002). Do anthocyanins function as antioxidants in leaves? Imaging of H_2O_2 in red and green leaves after mechanical injury. *Plant, Cell and Environment*, *25*, 1261–1269. https://doi.org/10.1046/j.1365-3040.2002.00905.x

Goyal, K., Walton, L. J., & Tunnacliffe, A. (2005). LEA proteins prevent protein aggregation due to water stress. *Biochemical Journal*, *388*, 151–157. https://doi.org/10.1042/BJ20041931

Iuchi, S., Kobayashi, M., Taji, T., Naramoto, M., Seki, M., Kato, T., ... Shinozaki, K. (2001). Regulation of drought tolerance by gene manipulation of 9-cis-epoxycarotenoid dioxygenase, a key enzyme in abscisic acid biosynthesis in *Arabidopsis*. *The Plant Journal*, *27*, 325–333. https://doi.org/10.1046/j.1365-313x.2001.01096.x

Jansson, S. (1999). A guide to the Lhc genes and their relatives in *Arabidopsis*. *Trends in Plant Science*, *4*, 236–240. https://doi.org/10.1016/S1360-1385(99)01419-3

Jiang, Y., & Huang, B. (2001). Drought and heat stress injury to two cool-season turfgrasses in relation to antioxidant metabolism and lipid peroxidation. *Crop Science*, *41*, 436–442. https://doi.org/10.2135/cropsci2001.412436x

Kang, J., Choi, H., Im, M., & Kim, S. Y. (2002). *Arabidopsis* basic leucine zipper proteins that mediate stress-responsive abscisic acid signaling. *The Plant Cell*, *14*, 343–357. https://doi.org/10.1105/tpc.010362

Kang, J., Hwang, J.-U., Lee, M., Kim, Y.-Y., Assmann, S. M., Martinoia, E., & Lee, Y. (2010). PDR-type ABC transporter mediates cellular uptake of the phytohormone abscisic acid. *Proceedings of the National Academy of Sciences of the United States of America*, *107*, 2355–2360. https://doi.org/10.1073/pnas.0909222107

Kanno, Y., Hanada, A., Chiba, Y., Ichikawa, T., Nakazawa, M., Matsui, M., ... Seo, M. (2012). Identification of an abscisic acid transporter by functional screening using the receptor complex as a sensor. *Proceedings of the National Academy of Sciences of the United States of America*, *109*, 9653–9658. https://doi.org/10.1073/pnas.1203567109

Kim, D., Pertea, G., Trapnell, C., Pimentel, H., Kelley, R., & Salzberg, S. L. (2013). TopHat2: Accurate alignment of transcriptomes in the presence of insertions, deletions and gene fusions. *Genome Biology*, *14*, R36. https://doi.org/10.1186/gb-2013-14-4-r36

Kivuva, B. M., Githiri, S. M., Yencho, G. C., & Sibiya, J. (2015). Screening sweetpotato genotypes for tolerance to drought stress. *Field Crops Research*, *171*, 11–22. https://doi.org/10.1016/j.fcr.2014.10.018

Kobayashi, M. (1983). The *Ipomoea trifida* complex closely related to sweet potato. In S. F. Shideler, & H. Rincon (Eds.), *Proceedings of the Sixth Symposium of the International Society of Tropical Root Crops* (pp. 561–568). Lima, Peru: International Potato Center (CIP).

Kovinich, N., Kayanja, G., Chanoca, A., Otegui, M. S., & Grotewold, E. (2015). Abiotic stresses induce different localizations of anthocyanins in *Arabidopsis*. *Plant Signaling and Behavior*, *10*, e1027850.

Kovinich, N., Kayanja, G., Chanoca, A., Riedl, K., Otegui, M. S., & Grotewold, E. (2014). Not all anthocyanins are born equal: Distinct patterns induced by stress in *Arabidopsis*. *Planta*, *240*, 931–940. https://doi.org/10.1007/s00425-014-2079-1

Kruger, N. J., & von Schaewen, A. (2003). The oxidative pentose phosphate pathway: Structure and organisation. *Current Opinion in Plant Biology*, *6*, 236–246. https://doi.org/10.1016/S1369-5266(03)00039-6

Kuromori, T., Miyaji, T., Yabuuchi, H., Shimizu, H., Sugimoto, E., Kamiya, A., ... Shinozaki, K. (2010). ABC transporter AtABCG25 is involved in abscisic acid transport and responses. *Proceedings of the National Academy of Sciences of the United States of America*, *107*, 2361–2366. https://doi.org/10.1073/pnas.0912516107

Kuromori, T., Sugimoto, E., & Shinozaki, K. (2011). Arabidopsis mutants of AtABCG22, an ABC transporter gene, increase water transpiration and drought susceptibility. *The Plant Journal*, *67*, 885–894. https://doi.org/10.1111/j.1365-313X.2011.04641.x

Kushiro, T., Okamoto, M., Nakabayashi, K., Yamagishi, K., Kitamura, S., Asami, T., ... Nambara, E. (2004). The *Arabidopsis* cytochrome P450 CYP707A encodes ABA 8′-hydroxylases: Key enzymes in ABA catabolism. *EMBO Journal, 23*, 1647–1656. https://doi.org/10.1038/sj.emboj.7600121

Langmead, B., & Salzberg, S. L. (2012). Fast gapped-read alignment with Bowtie 2. *Nature Methods, 9*, 357–359. https://doi.org/10.1038/nmeth.1923

Lehti-Shiu, M. D., Zou, C., Hanada, K., & Shiu, S.-H. (2009). Evolutionary history and stress regulation of plant receptor-like kinase/pelle genes. *Plant Physiology, 150*, 12–26. https://doi.org/10.1104/pp.108.134353

Li, S., Pezeshki, S. R., & Goodwin, S. (2004). Effects of soil moisture regimes on photosynthesis and growth in cattail (*Typha latifolia*). *Acta Oecologica, 25*, 17–22. https://doi.org/10.1016/j.actao.2003.10.004

Liu, Q., Kasuga, M., Sakuma, Y., Abe, H., Miura, S., Yamaguchi-Shinozaki, K., & Shinozaki, K. (1998). Two transcription factors, DREB1 and DREB2, with an EREBP/AP2 DNA binding domain separate two cellular signal transduction pathways in drought- and low-temperature-responsive gene expression, respectively, in *Arabidopsis*. *The Plant Cell, 10*, 1391–1406. https://doi.org/10.1105/tpc.10.8.1391

Love, M. I., Huber, W., & Anders, S. (2014). Moderated estimation of fold change and dispersion for RNA-seq data with DESeq2. *Genome Biology, 15*, 550. https://doi.org/10.1186/s13059-014-0550-8

Low, J., Lynam, J., Lemaga, B., Crissman, C., Barker, I., Thiele, G., ... Andrade, M. (2009). Sweetpotato in Sub-Saharan Africa. In G. Loebenstein & G. Thottappilly (Eds.), *The sweetpotato* (pp. 359–390). Dordrecht, The Netherlands: Springer. https://doi.org/10.1007/978-1-4020-9475-0

Low, J. W., Mwanga, R. O. M., Andrade, M., Carey, E., & Ball, A.-M. (2017). Tackling vitamin A deficiency with biofortified sweetpotato in sub-Saharan Africa. *Global Food Security, 14*, 23–30. https://doi.org/10.1016/j.gfs.2017.01.004

MacQueen, J. (1967). Some methods for classification and analysis of multivariate observations. *Proceedings of the Fifth Berkeley Symposium on Mathematical Statistics and Probability, 1*, 281–297.

Maquia, I., Muocha, I., Naico, A., Martins, N., Gouveia, M., Andrade, I., ... Ribeiro, A. I. (2013). Molecular, morphological and agronomic characterization of the sweet potato (*Ipomoea batatas* L.) germplasm collection from Mozambique: Genotype selection for drought prone regions. *South African Journal of Botany, 88*, 142–151. https://doi.org/10.1016/j.sajb.2013.07.008

Martin, M. (2011). Cutadapt removes adapter sequences from high-throughput sequencing reads. *EMBnet.journal, 17*, 10–12. https://doi.org/10.14806/ej.17.1.200

Merlot, S., Leonhardt, N., Fenzi, F., Valon, C., Costa, M., Piette, L., ... Leung, L. (2007). Constitutive activation of a plasma membrane H(+)-ATPase prevents abscisic acid-mediated stomatal closure. *EMBO Journal, 26*, 3216–3226.

Michel, B. E., & Kaufmann, M. R. (1973). The osmotic potential of polyethylene glycol 6000. *Plant Physiology, 51*, 914–916. https://doi.org/10.1104/pp.51.5.914

Muñoz-Mayor, A., Pineda, B., Garcia-Abellán, J. O., Antón, T., Garcia-Sogo, B., Sanchez-Bel, P., ... Bolarin, M. C. (2012). Overexpression of dehydrin tas14 gene improves the osmotic stress imposed by drought and salinity in tomato. *Journal of Plant Physiology, 169*, 459–468. https://doi.org/10.1016/j.jplph.2011.11.018

Mwanga, R. O. M., Odongo, B., Ocitti p'Obwoya, C., Gibson, R. W., Smit, N. E. J. M., & Carey, E. E. (2001). Release of five sweetpotato cultivars in Uganda. *HortScience, 36*, 385–386.

Mwanga, R. O. M., & Ssemakula, G. (2011). Orange-fleshed sweetpotatoes for food, health and wealth in Uganda. *International Journal of Agricultural Sustainability, 9*, 42–49. https://doi.org/10.3763/ijas.2010.0546

Nakabayashi, R., Yonekura-Sakakibara, K., Urano, K., Suzuki, M., Yamada, Y., Nishizawa, T., ... Saito, K. (2014). Enhancement of oxidative and drought tolerance in *Arabidopsis* by overaccumulation of antioxidant flavonoids. *The Plant Journal, 77*, 367–379. https://doi.org/10.1111/tpj.12388

Negi, J., Matsuda, O., Nagasawa, T., Oba, Y., Takahashi, H., Kawai-Yamada, M., ... Iba, K. (2008). CO_2 regulator SLAC1 and its homologues are essential for anion homeostasis in plant cells. *Nature, 452*, 483–486. https://doi.org/10.1038/nature06720

Noctor, G., Mhamdi, A., & Foyer, C. H. (2014). The roles of reactive oxygen metabolism in drought: Not so cut and dried. *Plant Physiology, 164*, 1636–1648. https://doi.org/10.1104/pp.113.233478

Noctor, G., Veljovic-Jovanovic, S., Driscoll, S., Novitskaya, L., & Foyer, C. H. (2002). Drought and oxidative load in the leaves of C3 plants: A predominant role for photorespiration? *Annals of Botany, 89*, 841–850. https://doi.org/10.1093/aob/mcf096

Osakabe, Y., Osakabe, K., Shinozaki, K., & Tran, L.-S. P. (2014). Response of plants to water stress. *Frontiers in Plant Science, 5*, 86.

Osmond, C. B., & Grace, S. C. (1995). Perspectives on photoinhibition and photorespiration in the field: Quintessential inefficiencies of the light and dark reactions of photosynthesis? *Journal of Experimental Botany, 46*, 1351–1362. https://doi.org/10.1093/jxb/46.special_issue.1351

Pace, P. F., Cralle, H. T., El-halawany, S. H. M., Cothren, J. T., & Senseman, S. A. (1999). Drought-induced changes in shoot and root growth of young cotton plants. *Journal of Cotton Science, 3*, 183–187.

Pei, Z.-M., Murata, Y., Benning, G., Thomine, S., Klüsener, B., Allen, G. J., ... Schroeder, J. I. (2000). Calcium channels activated by hydrogen peroxide mediate abscisic acid signalling in guard cells. *Nature, 406*, 731–734. https://doi.org/10.1038/35021067

Peng, M., Lin, F., Feng, J., Zhang, C., Yang, S., He, B., ... Li, M. (2017). Comparative transcriptome to reveal the drought tolerance mechanism in hexaploid *Ipomoea trifida*. *Molecular Plant Breeding, 8*, 100–112.

Rolston, L. H., Clark, C. A., Cannon, J. M., Randle, W. M., Riley, E. G., Wilson, P. W., & Robbins, M. L. (1987). Beauregard sweet potato. *HortScience, 22*, 1338–1339.

Saito, S., Hirai, N., Matsumoto, C., Ohigashi, H., Ohta, D., Sakata, K., & Mizutani, M. (2004). *Arabidopsis* CYP707As encode (+)-abscisic acid 8′-hydroxylase, a key enzyme in the oxidative catabolism of abscisic acid. *Plant Physiology, 134*, 1439–1449. https://doi.org/10.1104/pp.103.037614

Sakuma, Y., Maruyama, K., Osakabe, Y., Qin, F., Seki, M., Shinozaki, K., & Yamaguchi-Shinozaki, K. (2006). Functional analysis of an *Arabidopsis* transcription factor, DREB2A, involved in drought-responsive gene expression. *The Plant Cell, 18*, 1292–1309. https://doi.org/10.1105/tpc.105.035881

Sarker, A. M., Rahman, M. S., & Paul, N. K. (1999). Effect of soil moisture on relative leaf water content, chlorophyll, proline and sugar accumulation in wheat. *Journal of Agronomy and Crop Science, 183*, 225–229. https://doi.org/10.1046/j.1439-037x.1999.00339.x

Schroeder, J. I., Hedrich, R., & Fernandez, J. M. (1984). Potassium-selective single channels in guard cell protoplasts of *Vicia faba*. *Nature, 312*, 361–362. https://doi.org/10.1038/312361a0

Solis, J., Baisakh, N., Brandt, S. R., Villordon, A., & Bonte, D. L. (2016). Transcriptome profiling of beach morning glory (*Ipomoea imperati*) under salinity and its comparative analysis with sweetpotato. *PLoS One, 11*, e0147398. https://doi.org/10.1371/journal.pone.0147398

Stewart, G. R., & Lee, J. A. (1974). The role of proline accumulation in halophytes. *Planta, 120*, 279–289. https://doi.org/10.1007/BF00390296

Tolleter, D., Jaquinod, M., Mangavel, C., Passirani, C., Saulnier, P., Manon, S., ... Macherel, D. (2007). Structure and function of a mitochondrial late embryogenesis abundant protein are revealed by desiccation. *The Plant Cell, 19*, 1580–1589. https://doi.org/10.1105/tpc.107.050104

Trapnell, C., Williams, B. A., Pertea, G., Mortazavi, A., Kwan, G., van Baren, M. J., ... Pachter, L. (2010). Transcript assembly and quantification by RNA-Seq reveals unannotated transcripts and isoform switching during cell differentiation. *Nature Biotechnology, 28*, 511–515. https://doi.org/10.1038/nbt.1621

Umezawa, T., Okamoto, M., Kushiro, T., Nambara, E., Oono, Y., Seki, M., … Shinozaki, K. (2006). CYP707A3, a major ABA 8′-hydroxylase involved in dehydration and rehydration response in *Arabidopsis thaliana*. *The Plant Journal*, *46*, 171–182. https://doi.org/10.1111/j.1365-313X.2006.02683.x

Uno, Y., Furihata, T., Abe, H., Yoshida, R., Shinozaki, K., & Yamaguchi-Shinozaki, K. (2000). *Arabidopsis* basic leucine zipper transcription factors involved in an abscisic acid-dependent signal transduction pathway under drought and high-salinity conditions. *Proceedings of the National Academy of Sciences of the United States of America*, *97*, 11632–11637. https://doi.org/10.1073/pnas.190309197

Vahisalu, T., Kollist, H., Wang, Y.-F., Nishimura, N., Chan, W.-Y., Valerio, G., … Kangasjärvi, J. (2008). SLAC1 is required for plant guard cell S-type anion channel function in stomatal signalling. *Nature*, *452*, 487–491. https://doi.org/10.1038/nature06608

Wang, Y.-F., Zhang, A., Ren, H.-M., Tan, Y.-Q., Qi, G.-N., Yao, F.-Y., … Sun, S.-J. (2016). S-type anion channels SLAC1 and SLAH3 function as essential negative regulators for K^+ channel KAT1 and stomatal opening in *Arabidopsis*. *The Plant Cell*, *28*, 949–955.

Xiang, Y., Tang, N., Du, H., Ye, H., & Xiong, L. (2008). Characterization of OsbZIP23 as a key player of the basic leucine zipper transcription factor family for conferring abscisic acid sensitivity and salinity and drought tolerance in rice. *Plant Physiology*, *148*, 1938–1952. https://doi.org/10.1104/pp.108.128199

Xu, W., Cui, K., Xu, A., Nie, L., Huang, J., & Peng, S. (2015). Drought stress condition increases root to shoot ratio via alteration of carbohydrate partitioning and enzymatic activity in rice seedlings. *Acta Physiologiae Plantarum*, *37*, 9. https://doi.org/10.1007/s11738-014-1760-0

Xu, Y.-H., Liu, R., Yan, L., Liu, Z.-Q., Jiang, S.-C., Shen, Y.-Y., … Zhang, D.-P. (2012). Light-harvesting chlorophyll a/b-binding proteins are required for stomatal response to abscisic acid in *Arabidopsis*. *Journal of Experimental Botany*, *63*, 1095–1106. https://doi.org/10.1093/jxb/err315

Yang, Y., Wang, Y., Jia, L., Yang, G., Xu, X., Zhai, H., … Liu, Q. (2018). Involvement of an ABI-like protein and a Ca^{2+}-ATPase in drought tolerance as revealed by transcript profiling of a sweetpotato somatic hybrid and its parents *Ipomoea batatas* (L.) Lam. and *I. triloba* L. *PLoS One*, *13*, e0193193. https://doi.org/10.1371/journal.pone.0193193

Zhong, S., Joung, J.-G., Zheng, Y., Chen, Y., Liu, B., Shao, Y., … Giovannoni, J. J. (2011). High-throughput illumina strand-specific RNA sequencing library preparation. *Cold Spring Harbor Protocols*, *2011*, 940–949.

5

Field-grown soybean transcriptome shows diurnal patterns in photosynthesis-related processes

Anna M. Locke[1,2] | Rebecca A. Slattery[3,4] | Donald R. Ort[3,4,5]

[1]Soybean and Nitrogen Fixation Research Unit, USDA-ARS, Raleigh, North Carolina

[2]Department of Crop and Soil Sciences, North Carolina State University, Raleigh, North Carolina

[3]Carl R. Woese Institute for Genomic Biology, University of Illinois, Urbana, Illinois

[4]Global Change and Photosynthesis Research Unit, USDA-ARS, Urbana, Illinois

[5]Department of Plant Biology, University of Illinois, Urbana, Illinois

Correspondence
Anna M. Locke, Soybean and Nitrogen Fixation Research Unit, USDA-ARS, Raleigh, NC.
Email: Anna.Locke@ars.usda.gov

Abstract

Many plant physiological processes have diurnal patterns regulated by diurnal environmental changes and circadian rhythms, but the transcriptional underpinnings of many of these cycles have not been studied in major crop species under field conditions. Here, we monitored the transcriptome of field-grown soybean (*Glycine max*) during daylight hours in the middle of the growing season with RNA-seq. The analysis revealed 21% of soybean genes were differentially expressed over the course of the day. Expression of some circadian-related genes in field-grown soybean differed from previously reported expression patterns measured in controlled environments. Many genes in functional groups contributing to and/or depending on photosynthesis showed differential expression, with patterns particularly evident in the chlorophyll synthesis pathway. Gene regulatory network inference also revealed seven diurnally sensitive gene nodes involved with circadian rhythm, transcription regulation, cellular processes, and water transport. This study provides a diurnal overview of the transcriptome for an economically important field-grown crop and a basis for identifying pathways that could eventually be tailored to optimize diurnal regulation of carbon gain.

KEYWORDS

circadian rhythm, *Glycine max*, photosynthesis, RNA-seq, soybean, transcriptome

1 | INTRODUCTION

Diurnal changes in plant physiological responses and gene expression are governed by diurnally changing environmental conditions and endogenous circadian rhythms. Coordinating physiological functions with the 24-h clock allows plants to balance energy needs and resources (Dodd et al., 2005) and creates temporal compartments to prevent futile biochemical cycles. In C_3 plants, ATP and NADPH are produced through the light reactions to fuel carbon assimilation in the C_3 cycle only during daylight. Stomatal opening is stimulated by light to allow CO_2 to enter the leaf (Kinoshita et al., 2001), which means transpiration occurs almost entirely in the light, driving water flow through the plant. Water flow into the roots in turn drives nutrient uptake from the soil. During the daytime, extra carbon must be assimilated to fuel respiration for cellular processes in the dark. This extra carbon is typically stored as starch in the chloroplast, resulting in diurnal cycles of starch accumulation and depletion (Stitt & Zeeman, 2012; Zeeman, Smith, & Smith, 2007).

Environmental conditions change greatly even within the light period, as plants in the field experience fluctuations in light quality, light intensity, temperature, wind, and vapor pressure deficit on a daily basis. Photosynthetic processes continually acclimate throughout the day to optimize light use efficiency and to maintain a balance between energy from the light reactions and substrates for the carbon reactions

(Geiger & Servaites, 1994; Reinbothe & Reinbothe, 1996). Because light and temperature tend to peak in the middle of the day, it is likely that expression of photosynthesis-related genes follows characteristic patterns over the course of the light period. Some photosynthesis-related processes have been linked to expression of circadian rhythm-related genes. For example, altered expression of circadian-related genes in Arabidopsis (*Arabidopsis thaliana*) and maize (*Zea mays*) has been linked to increased rates of photosynthesis in hybrids (Bendix, Marshall, & Harmon, 2015; Ko et al., 2016).

Soybean (*Glycine max*) typically shows strong diurnal variations in photosynthesis, carbohydrate metabolism, and nitrogen assimilation (Bernacchi et al., 2006; Delhon, Gojon, Tillard, & Passama, 1995; Rogers et al., 2004; Upmeyer & Koller, 1973). Because carbon accumulation depends not just on peak midday photosynthesis, but on the daily integral of photosynthesis, the regulation of diurnal metabolic fluctuations impacts season-long yield. Understanding the transcriptional basis of diurnal photosynthetic cycles could provide insight into how carbon assimilation is, or could be, maximized over the entire day, rather than just at peak photosynthesis. Based on known fluctuations in photosynthesis, sugar metabolism, and environmental conditions, we hypothesized that genes related to photosynthetic function and sugar metabolism would likely show diurnal changes in transcription.

Using RNA-seq, this study examined diurnal changes in the leaf transcriptome of field-grown soybean. We first compared circadian-related gene expression in the field setting to previously measured expression patterns in controlled environments. We then identified diurnally regulated genes involved in various aspects of photosynthesis. Finally, we identified soybean genes integral to diurnal function in soybean.

2 | MATERIALS AND METHODS

2.1 | Plant material and growth conditions

For RNA-seq analyses, soybean (*Glycine max*) cv. 93B15 (Pioneer Hi-Bred, Johnston, IA) was planted on 27 May 2010 at the SoyFACE research farm near Champaign, Illinois (40.042° N, 88.237° W). Four plots in the field were the four biological replicates. Each replicate was sampled at four time points on the same day, and each replicate was sampled on a different day: 14, 16, 18, or 24 August 2010. All plants were in early to mid-seed fill (R5-R6) during this period. Day length during this time period ranged from 13:23 to 13:47 h. This experimental design accounted for environmental variation among sampling days with the replicate in the statistical model, distributed equally across time points. All three leaflets from an uppermost, fully expanded leaf were detached and immediately flash-frozen in liquid nitrogen at 8:00 (zeitgeber time [ZT] 2:20), 11:00 (ZT5:20), 14:00 (ZT8:20), and 17:00 (ZT11:20), and subsequently stored at −80°C. The three leaflets were combined to create one sample per plot.

For diurnal measurements of chlorophyll content, leaf disks from 10 randomly selected plants were harvested at 8:00/ZT2:20, 11:00/ZT5:20, 14:00/ZT8:20, and 17:00/ZT11:20. These samples were harvested at SoyFACE on 16 August 2013 from cv. 93Y40 (Pioneer Hi-Bred, Johnston, IA). Samples were cut from the uppermost, fully expanded leaf and immediately flash-frozen in liquid nitrogen.

Climate data (temperature, relative humidity [RH], and photosynthetic photon flux density [PPFD]) for 30 min intervals throughout each sampling day were obtained from the University of Illinois Energy Farm weather station located 3.7 km northeast of the SoyFACE site.

2.2 | RNA extraction and sequencing

Total RNA was isolated from whole leaflets with a method developed specifically for field-grown soybean (Bilgin, DeLucia, & Clough, 2009), and RNA was treated with the DNA-free kit (Ambion, Inc., Austin, TX). cDNA libraries were constructed and indexed with the TruSeq RNA Sample Preparation kit, and the average insert size was 361 bp (Illumina, Inc., San Diego, CA). Sixteen samples (four biological replicates at four time points) were sequenced on two randomly assigned lanes with the HiSeq2000 (Illumina, Inc., San Diego, CA). On average, 51.2 million paired-end 100 nt reads were generated for each sample. We previously reported transcript abundance for 34 soybean aquaporin genes from a separate analysis of these reads (Locke & Ort, 2015).

2.3 | Sequence alignment and processing

Raw reads were trimmed to remove adapter sequence contamination with Trimmomatic (Bolger, Lohse, & Usadel, 2014) and aligned to the soybean genome (Williams 82 assembly version 2 annotation version 1; (Schmutz et al., 2010)) in TopHat version 2.1.1 (Kim et al., 2013). Mapped reads per gene were counted with HTSeq-count (www-huber.embl.de/users/anders/HTSeq/; (Anders, Pyl, & Huber, 2015)).

2.4 | RNA-seq statistical analyses

Genes with very low or no expression (<8 counts for all biological replicates at every time point) were filtered from the data set prior to statistical analysis. Several normalization strategies were tested, of which log_2-transformed trimmed mean of M-values (TMM) (Robinson & Oshlack, 2010) most effectively normalized the data for parametric analysis. TMM correction factors were calculated using the edgeR package (McCarthy, Chen, & Smyth, 2012; Robinson, McCarthy, & Smyth, 2009). The log_2(TMM)-normalized expression was analyzed by repeated-measures ANOVA using PROC MIXED (SAS 9.4, SAS Institute, Cary, NC), in which time point was the main effect and plot was the repeated measures subject. This program was chosen because the repeated measures structure of the time course data set is not readily modeled by specialized statistical packages for gene expression. The residuals for each gene were tested for normality, a requirement for the validity of parametric testing; those genes with non-normal residual error distributions were instead analyzed in a repeated-measures ANOVA using simulated, bootstrapped "F"-distributions. In total, 357 genes fell into this nonparametric testing category; these are noted in Supporting information Dataset S1. *p*-values were corrected for false discovery rate (fdr) among all comparisons, and genes with fdr-corrected *p*-values below 0.05 and $\log_2$(fold change) ≥1 between at least two time points, which corresponds to a doubling or halving of transcription, were considered to be differentially expressed (DE). Log_2(fold changes) for DE genes were calculated from $\log_2$(TMM) normalized counts for each time point (11:00/ZT5:20, 14:00/ZT8:20, 17:00/ZT11:20) relative to 8:00/ZT2:20.

Gene ontology (GO) annotations were downloaded from SoyBase (https://soybase.org/; (Grant, Nelson, Cannon, & Shoemaker, 2009)), and to identify GO terms that were overrepresented among DE genes, Fisher's exact test with a Bonferroni *p*-value correction was applied through the tool on SoyBase (Morales et al., 2013). Dominant diurnal transcription patterns among DE genes were identified with *k*-means clustering of $\log_2$(TMM) values in SAS PROC FASTCLUS.

To examine the differences in variance that could indicate differential environmental effects on diurnal transcription patterns, the coefficient of variation (CV) was calculated for each gene/time point (absolute value[*SD*/mean]). The CVs for the four time points were then averaged, and the genes were ranked based on average CV with the highest average CV indicating the most variability in transcript abundance. Tukey's fences were used to identify genes that were moderate outliers (mean CV > Q3 + [1.5 × (Q3-Q1)]) and extreme outliers (mean CV > Q3 + [3 × (Q3-Q1)]) for average CV. Additionally, the equality of CVs among time points was tested for each gene with an asymptotic test (Feltz & Miller, 1996) using the R package "cvequality"(Marwick & Krishnamoorthy, 2018). In this analysis, each time point (8:00/ZT2:20, 11:00/ZT5:20, 14:00/ZT8:20, and 17:00/ZT11:20) was treated as a group, and $\log_2$(TMM), the same variable used to test the equality of means in the repeated-measures model, was the variable for which CV was calculated.

2.5 | Chlorophyll analysis

Chlorophyll was extracted from tissue samples, and total chlorophyll content and chlorophyll *a*/*b* ratios were determined according to Porra, Thompson, and Kriedemann (1989) and Lichtenthaler (1987). Data were analyzed by repeated-measures ANOVA in PROC MIXED (SAS 9.4, SAS Institute, Cary, NC).

2.6 | GRN inference

A dynamic gene regulatory network (GRN) was inferred using a graphical Gaussian model with the GeneNet package v. 1.2.13 in R (Opgen-Rhein & Strimmer, 2007; Schaefer, Opgen-Rhein, & Strimmer, 2015). A very strict threshold was used to filter the DE gene list for GRN inference (fdr-corrected $p < 0.0001$, $\log_2$(fold change) ≥ 2), and 1,199 strongly DE genes were included in the analysis. Significant edges and directions with fdr-corrected $p < 0.0001$ were included in the GRN.

2.7 | Accession numbers

Sequence data and read counts from this study can be found at the NCBI GEO data repository under accession number GSE114878.

3 | RESULTS AND DISCUSSION

3.1 | RNA-seq output and clustering

mRNA sequencing yielded over 820 million reads from 16 samples. Sufficient reads were counted for 36,059 genes to include in the analysis (Supporting information Dataset S1), which represents 64% of the 56,044 protein-coding genes predicted in the soybean genome (Schmutz et al., 2010). Of these, 11,984 genes were found to be differentially expressed [DE; fdr-corrected $p < 0.05$, $\log_2$(fold change ≥ 1)] between time points over the course of the day. The percentage of genes that were DE, approximately 21%, is smaller than the proportion of circadian-regulated genes found in the Arabidopsis transcriptome (Blasing et al., 2005; Covington, Maloof, Straume, Kay, & Harmer, 2008), which probably reflects the shorter time course examined here. Additionally, for the 357 genes that had to be analyzed using nonparametric tests, there was a higher Type II error rate than among the 35,702 genes analyzed with parametric tests, due to the lower statistical power of the nonparametric tests. The analysis method for each gene, parametric or non-parametric, is noted in Supporting information Dataset S1.

Varying environmental conditions over the four sampling days have the potential to introduce substantial variability in transcription. A microarray study comparing field and chamber-grown rice (*Oryza sativa*) found that over 7,000 genes required a term to be included in a model that summarized environmental variables in order for the model to accurately predict transcription (Nagano et al., 2012). For the four RNA sampling days in the present study, temperature, PPFD, and RH were remarkably consistent (Supporting information Figure S1). RH showed the greatest variability, with early-morning values ranging from <75% to >85%. Nonetheless, that this extensive DE gene list was able to be statistically resolved despite environmental variation among days implies a very robust diurnal response in these genes, especially considering the high potential for biological variation among the four replicates, which were spatially segregated field plots sampled on different days. Thus, the DE gene list presented and discussed in this study likely represents genes with the most consistent, environmentally stable diurnal responses.

Diurnal transcription patterns were examined by clustering DE genes based on $\log_2$(TMM) values (Figure 1). Each cluster contained between 692 and 3141 genes. Cluster analysis revealed seven distinct diurnal expression patterns (Figures 1, Supporting information Figure S2). Cluster 1 showed relatively low levels of expression in the morning that declined further throughout the day, whereas transcript abundance in cluster 2 was very low in the morning and increased through the day. Clusters 3 and 4 showed declining expression patterns throughout the day, although the overall magnitude of expression was greater in cluster 4. Clusters 5, 6, and 7 all

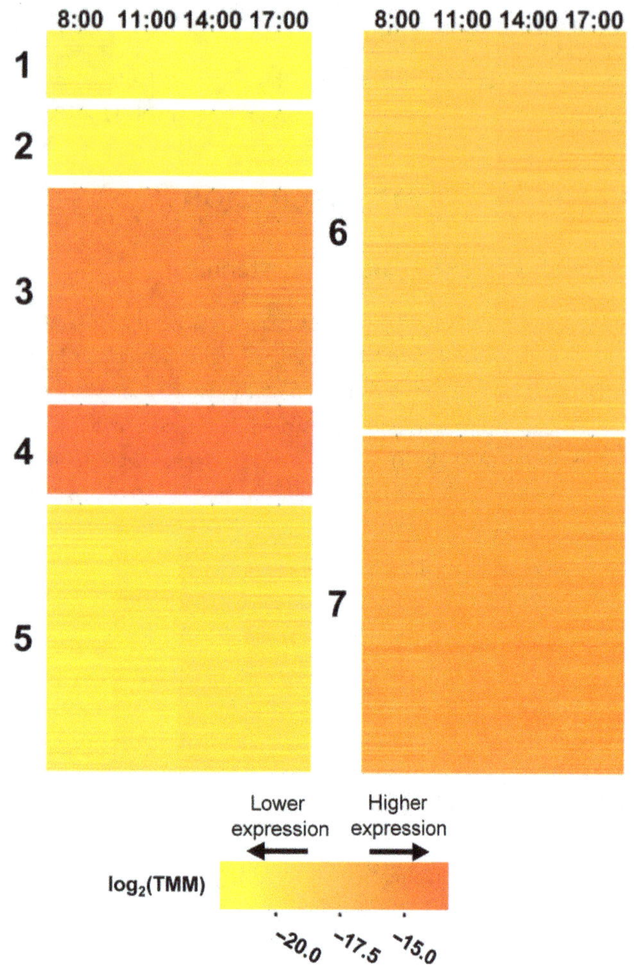

FIGURE 1 *k*-means clustering of diurnally differentially expressed genes in soybean. Normalized expression values [$\log_2$(TMM)] are shown (yellow = low expression, red = high expression) for four time points [8:00 (ZT2:20), 11:00 (ZT5:20), 14:00 (ZT8:20), 17:00 (ZT11:20)] measured during daylight hours

increased slightly from 8:00/ZT2:20 to 11:00/ZT5:20 and 11:00/ZT5:20 to 14:00/ZT8:20 and seemed to plateau at 14:00/ZT8:20. However, they differed in overall magnitude, with 5 < 6<7.

Gene lists for each cluster were tested for overrepresentation of GO terms for each distinct diurnal expression pattern (Supporting information Dataset S2). No specific GO terms showed significantly higher representation in clusters 1 and 2. mRNA 5′-UTR binding was the most overrepresented GO term in cluster 3, whereas Rubisco activator activity (i.e., Rubisco activase), was the most overrepresented GO term in cluster 4. Only one GO term, mitochondrial mRNA modification, was overrepresented in cluster 5. Protein import into the nucleus was the most highly overrepresented GO term in cluster 6, and cluster 7 showed the highest overrepresentation of the response to hormone stimulus GO term.

3.2 | CV analyses

Coefficient of variation was significantly different among time points for 3,231 genes, which was 8.96% of those included in the analysis (Supporting information Dataset S1), indicating stronger environmental impacts at some time points for those genes. For these genes, large variation in expression at one or two time points may have prevented a gene from being identified as DE, despite a trend in the means. This effect may have been exacerbated for genes with one or more very low-expression time points, due to the logarithmic transformation that was applied to TMM-normalized data, which was necessary to shape the data in a normal distribution as is appropriate for ANOVA and post-hoc tests.

The mean CV for each gene was calculated from the CVs of the four time points. Based on these mean CVs, 262 genes (0.7%) were extreme outliers for high mean CV (Supporting information Dataset S1), meaning that these genes had extremely high variation among sampling days relative to other genes. There were 964 genes (2.6%) considered as moderate outliers for high mean CV. As the lower outlier fence was below zero, no genes had a mean CV that fell below the lower outlier fence.

3.3 | Known circadian genes

We first examined the expression of key circadian-related genes in field-grown soybean leaves during daylight hours (Figure 2). TIMING OF CAB EXPRESSION1 (TOC1), LATE ELONGATION HYPOCOTYL (LHY), and CIRCADIAN CLOCK ASSOCIATED1 (CCA1) are considered as universal components of the circadian clock in plants. CCA1 and LHY are transcription factors with high expression at dawn that repress genes that are more highly expressed in the evening, such as *TOC1*. *GmTOC1*, *GmLHY/CCA1-LIKE 1* (*GmLCL1*), and *GmLCL2* are soybean circadian clock component genes with homology to *TOC1*, *LHY*, and *CCA1* genes in Arabidopsis, respectively. Thus, *GmLCL1* and *GmLCL2* also likely encode transcription factors that negatively regulate *GmTOC1* (Alabadí et al., 2001; Liu et al., 2009). *GmTOC1*, *GmLCL1*, and *GmLCL2* were DE in field-grown soybean leaves, and their transcription patterns were consistent with their homologs in other species (Figure 2). *GmTOC1* was assigned to cluster 2, with low transcript abundance in the morning that increased sixfold by 17:00/ZT11:20. Transcription of *GmLCL1* and *GmLCL2* declined over the course of the day, and both of these genes were assigned to cluster 4 (Figure 2). The transcription patterns of these genes match the patterns measured for chamber-grown soybean (Liu et al., 2009; Marcolino-Gomes et al., 2014).

Expression of several other circadian-regulated genes has been studied in chamber-grown soybean (Marcolino-Gomes et al., 2014). Therefore, we aimed to compare these previous findings to expression patterns from field-grown soybean. The expression profiles of soybean genes *EARLY FLOWERING 4* (*GmELF4*), *GmJumonji*, *LUX ARRHYTHMO* (*GmLUX*), and *PSEUDO-RESPONSE REGULATOR 3* (*GmPRR3*) from the controlled-environment study were similar to those found in this study (Figure 2). However, *GmPRR7* and *GmPRR9* expressions stabilized or continued to increase after 14:00/ZT8:20 in the field (Figure 2), which differed from a sharp decline after 8 h of light in the chamber experiment, despite similar day lengths in the field (13:23–13:47 h) and in the chamber studies (14 h) (Marcolino-Gomes et al., 2014). *GmGIGANTEA* (*GmGI*), showed a similar

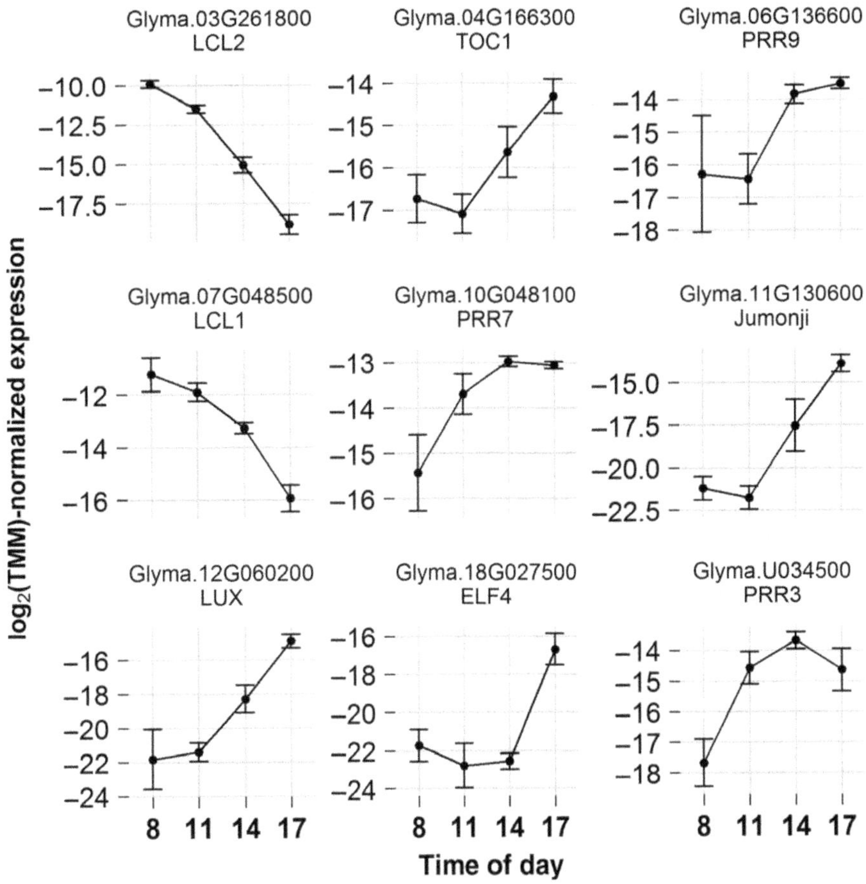

FIGURE 2 Diurnally differentially expressed circadian rhythm-related genes in field-grown soybean. Transcript abundance is shown as mean [$\log_2$(TMM)] ± *SD* for four time points [8:00 (ZT2:20), 11:00 (ZT5:20), 14:00 (ZT8:20), 17:00 (ZT11:20)] measured during daylight hours. *ELF4, EARLY FLOWERING 4; LCL1/2, LHY/CCA1-LIKE 1/2; LUX, LUX ARRHYTHMO; PRR3/7/9, PSEUDO-RESPONSE REGULATOR 3/7/9; TOC1, TIMING OF CAB EXPRESSION 1*

difference between field and chamber experiments (Supporting information Figure S3). Despite a trend in diurnal transcription, *GmGI* was not DE in this study, which is likely the result of high variability at 8:00/ZT2:20 and its non-normal distribution of residual error, which required analysis with the lower power, nonparametric test. CV was also significantly different among time points for *GmGI* (Supporting information Dataset S1). In rice, *GI* was sensitive to temperature only during part of the diurnal cycle (Nagano et al., 2012); this differential sensitivity to environmental conditions could also influence the variability observed among time points for *GmGI*. Stresses such as drought may alter circadian-regulated gene expression (Marcolino-Gomes et al., 2014), and soybeans often encounter drought in the field. However, the expression of these circadian-regulated genes (*GmPRR7*, *GmPRR9*, and *GmGI*) in the field did not match the patterns seen under growth chamber drought conditions. Thus, it is possible that diurnal gene regulation also depends on factors unique to field conditions that are not present in highly controlled environments. These results suggest that transcriptomic analyses conducted on soybeans grown in controlled environments may not fully represent the expression patterns of field-grown soybean.

3.4 | Photosynthesis: light absorption

Light absorption by chlorophyll is the first step in photosynthesis. GO terms for chlorophyll binding, chlorophyll catabolic processes, and chlorophyll biosynthetic processes were all significantly overrepresented in cluster 4, which included the most highly transcribed genes with diurnally decreasing transcription, so we further examined gene expression changes in the pathway for chlorophyll synthesis. Diurnal regulation of chlorophyll biosynthesis is light-dependent and likely mediated by phytochrome signaling pathways (Ilag, Kumar, & Söll, 1994; Reinbothe & Reinbothe, 1996). The first steps of the pathway are common to synthesis of all tetrapyrroles (chlorophyll, heme, and siroheme). Most of the DE genes in this portion of the pathway, including genes encoding porphobilinogen deaminase, uroporphyrinogen III synthase, and coproporphyrinogen III oxidase, were upregulated as the day progressed (Figure 3), which coincided with increases in total chlorophyll and chlorophylls *a* and *b* throughout the day (Figure 4a–c). However, the highly transcribed genes for glutamyl-tRNA reductase (GluTR), the rate-limiting enzyme for tetrapyrrole biosynthesis, showed a large decline in expression at 17:00/ZT11:20 compared to 8:00/ZT2:20 (Figure 3). Circadian studies of chlorophyll gene expression in Arabidopsis have shown a rapid increase in GluTR transcripts occurring within 3 h of illumination and declining during the second half of the day under constant light intensity (Matsumoto, Obayashi, Sasaki-Sekimoto, Ohta, & Takamiya, 2004), supporting other reports of light-induced expression of genes encoding GluTR (Tanaka et al., 1996). While we did not see the same initial increase in expression in this study, it is possible that GluTR transcript levels had already peaked at or around the first measurement of the day (8:00/ZT2:20), which was approximately 2 h after sunrise, despite light intensity peaking around 13:00.

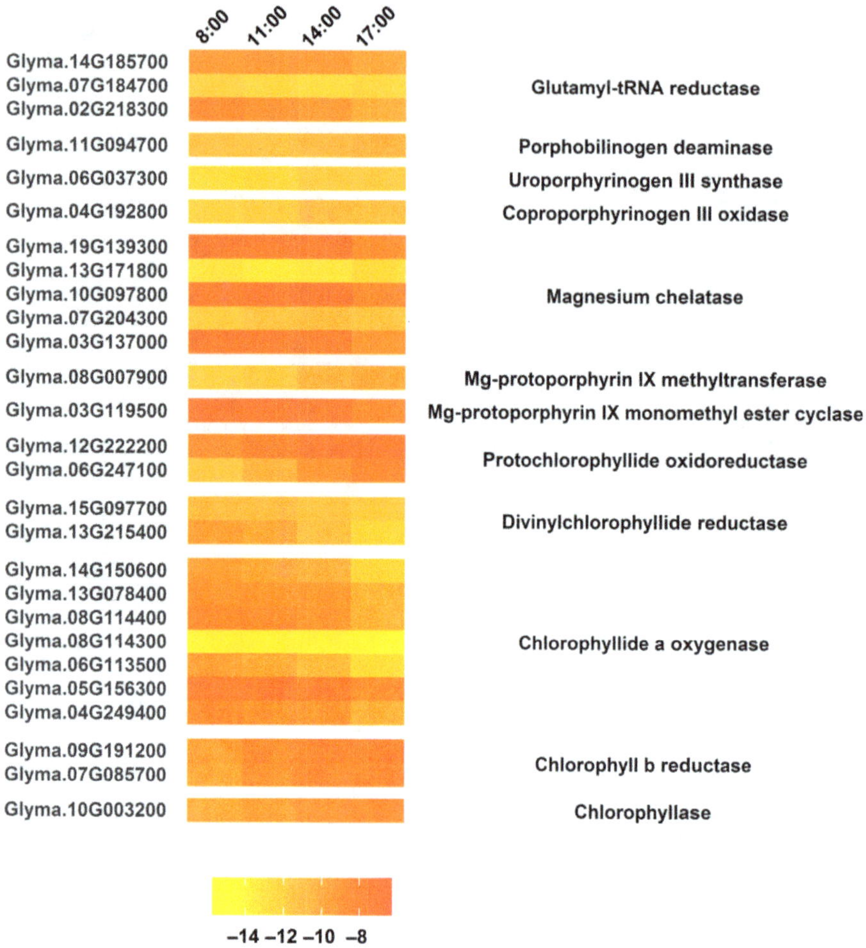

FIGURE 3 Diurnally differentially expressed genes within the chlorophyll biosynthetic pathway in field-grown soybean. Gene loci are listed on the left, and the corresponding enzymes are listed on the right. Expression values [log_2(TMM)] are shown (yellow = low expression, red = high expression) for four time points [8:00 (ZT2:20), 11:00 (ZT5:20), 14:00 (ZT8:20), 17:00 (ZT11:20)] measured during daylight hours

The insertion of Mg^{2+} into the protoporphyrin ring by magnesium chelatase (MgCh) is the first step in the chlorophyll-specific branch of chlorophyll synthesis. This enzyme contains three subunits which are encoded by several genes, the expression patterns of which varied during daylight hours (Figure 3). Gene expression varies among MgCh subunits in Arabidopsis, which may be due to involvement of different subunits in different roles in binding protoporphyrin and signaling of the chlorophyll synthesis pathway (Matsumoto et al., 2004) and roles of specific subunits in other processes, such as abscisic acid (ABA) signaling (Du et al., 2012). For field-grown soybean, expression of genes encoding successive enzymes in the chlorophyll branch [Mg-protoporphyrin IX methyltransferase (MgMT) and protochlorophyllide oxidoreductase (POR)] increased as the day progressed, and transcript abundance was highest at 17:00/ZT11:20 (Figure 3). Although the four time points in this study may not capture the true peaks of transcription for most genes, it is likely that peak transcript abundance for MgMT and POR occurred between 14:00/ZT8:20 and 17:00/ZT11:20, which was consistent with POR expression in chamber-grown Arabidopsis (Matsumoto et al., 2004). However, in chamber-grown Arabidopsis, MgMT transcription peaks at midday. This pattern of transcription is unlikely to have occurred in the field-grown plants, with solar midday at approximately 13:00 CDT, considering that the highest transcript abundance was measured at 17:00/ZT11:20, followed by 14:00/ZT8:20. Measured transcript abundance of the genes for Mg-protoporphyrin IX monomethyl ester cyclase and divinyl chlorophyllide reductase was highest around 11:00/ZT5:20 and declined later in the day (Figure 3) as was seen in Arabidopsis (Matsumoto et al., 2004).

The final steps of chlorophyll synthesis constitute the chlorophyll cycle (Tanaka & Tanaka, 2007), in which chlorophylls *a* and *b* are interconverted by three enzymes, chlorophyllide *a* oxygenase (CAO), chlorophyll *b* reductase (CBR), and 7-hydroxymethyl chlorophyll *a* reductase. This cycle serves to adjust the chlorophyll *a* to chlorophyll *b* ratio, which is thought to regulate light harvesting complex size and degradation (Tanaka, Kobayashi, & Masuda, 2011). CAO is putatively encoded by four genes in soybean and CBR by two genes in soybean [Phytozome (http://www.phytozome.net/soybean.php)], and the transcription of all of these decreased steadily from 8:00/ZT2:20 to 14:00/ZT8:20. As with GluTR, rapid increases in Arabidopsis CAO gene expression occur during the first 3 h of illumination (Matsumoto et al., 2004), so it is possible that CAO gene expression was highest at the 8:00/ZT2:20 measurement in this study. Transcription of the gene for chlorophyllase (Glyma.10G003200), which removes the phytol chain from chlorophyll *a* to form chlorophyllide *a* in the first step of chlorophyll catabolism, increased during daylight hours (Figure 3, Supporting information Dataset S1), but this enzyme may also be posttranslationally regulated (Harpaz-Saad et al., 2007).

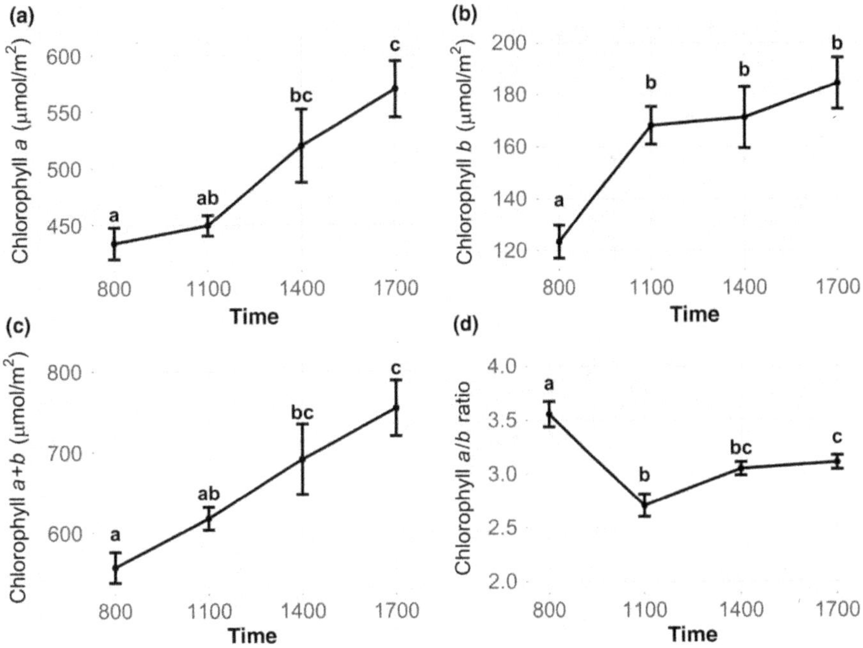

FIGURE 4 Diurnal chlorophyll contents and *a/b* ratios of field-grown soybean. (a) Chlorophyll *a*, (b) chlorophyll *b*, (c) total chlorophyll content, and (d) chlorophyll *a/b* ratios were measured at four time points [8:00 (ZT2:20), 11:00 (ZT5:20), 14:00 (ZT8:20), 17:00 (ZT11:20)] during daylight hours in leaves from field-grown soybean on DOY 228 in 2013. Values represent the means ± *SD* errors (n = 10). Different letters within panels represent significant differences ($p < 0.05$)

Since several genes in the chlorophyll biosynthesis pathway showed diurnal expression patterns (Figure 3) and chlorophyll content fluctuates on diurnal and circadian cycles in some plant species (Bukatsch & Rudolph, 1963), we measured diurnal chlorophyll content and *a/b* ratios in field-grown soybean, albeit not from the same plants used for RNA-seq analyses (Figure 4). Diurnal PPFD for the day on which chlorophyll was measured was similar to the days on which RNA samples were collected (Supporting information Figures S1, S4), with the exception of some possible cloud cover at 14:00/ZT8:20 on the chlorophyll sampling day. This lowered PPFD to approximately 750 µmol m^{-2} s^{-1} at 14:00/ZT8:20 on the chlorophyll sampling date, as compared to an average of approximately 1500 µmol m^{-2} s^{-1} at 14:00/ZT8:20 for the RNA sampling dates. Total chlorophyll content increased throughout the course of the day (Figure 4c). While both chlorophylls *a* and *b* increased from 8:00/ZT2:20 to 17:00/ZT11:20, the increase in total chlorophyll from 8:00/ZT2:20 to 11:00/ZT5:20 was largely due to an increase in chlorophyll *b* (Figure 4b), whereas total chlorophyll increased between later time points mostly due to an increase in chlorophyll *a* (Figure 4a). Thus, the chlorophyll *a/b* ratio changed significantly throughout the day, with a reduction occurring between 8:00/ZT2:20 and 11:00/ZT5:20 followed by a slight increase during the afternoon (Figure 4d). This pattern in soybean was similar to diurnal chlorophyll *a/b* ratios in wheat (*Triticum aestivum*; Busheva et al., 1991). While an earlier study reports no significant diurnal changes in total leaf chlorophyll content of soybean plants grown in greenhouses (Wickliff & Aronoff, 1962), a more recent study conducted in environmental growth chambers shows chlorophylls *a* and *b* in soybean leaves declining early during the light period and increasing after 4–8 h of illumination (Pan et al., 2015). Thus, by experiencing natural light conditions in the field, which are typically more intense and more varied than in growth chambers, soybean leaves in this study exhibited largely different patterns in diurnal chlorophyll content. These differing environmental conditions may also explain why gene expression was altered in our study compared to previous research in some chlorophyll synthesis-related genes, as the natural fluctuations in light intensity and/or quality in the field may regulate expression differently compared to the timed, typically flat light regimes used in controlled environments.

Diurnal changes in chlorophyll content and *a/b* ratios did not always align with the expression patterns of genes involved in chlorophyll biosynthesis. While this may partially be due to environmental conditions on the different sampling dates (Supporting information Figures S1, S4), it is also likely due to the posttranslational regulation of key enzymes. While total chlorophyll increased throughout the day (Figure 4c), GluTR transcripts decreased (Figure 3). However, previous research shows no correlation between GluTR protein abundance and chlorophyll synthesis (Nogaj, Srivastava, van Lis, & Beale, 2005). Instead, GluTR activity is negatively regulated by accumulation of pathway products, such as hemes, Mg-based rings, and divinyl protochlorophyllide *a* (Tanaka & Tanaka, 2007). In the present study, decreased CAO gene expression (Figure 3) was associated with a plateau in chlorophyll *b* content during the later three time points of daylight hours (Figure 4b), but evidence suggests that CAO's regulatory mechanisms are primarily posttranscriptional (Tanaka & Tanaka, 2011), so decreasing the transcription of CAO genes may not directly cause the shifts in chlorophyll *a/b* ratios of newly synthesized pigment. However, while posttranslational regulation may be a key for fine-tuning rates of chlorophyll-associated protein synthesis, the clear diurnal and circadian patterns associated with these genes indicate a role for diurnal regulation of transcription, even if the function is only to ensure sufficient protein levels for later fine-tuning via posttranslational control.

3.5 | Photosynthesis: energy transfer and electron transport

Chlorophyll *a/b* binding proteins have diurnal transcription fluctuations in several species that tend to peak around midday (Martino-Catt & Ort, 1992; Meyer, Thienel, & Piechulla, 1989). In field-grown soybean, most genes encoding chlorophyll binding proteins showed lowest transcript abundance at 8:00/ZT2:20, with expression increasing throughout the day or peaking in the afternoon (Figure 5). This increase could be related to the concurrently increasing gene expression of many components of the chlorophyll synthesis pathway, including MgMT and POR (Figure 3), and increased chlorophyll content in leaves throughout the day (Figure 4c).

Transcription of photosystem (PS)-related genes showed three distinct patterns. The most common pattern was low expression early and late in the day, with transcript abundance highest at either 11:00/ZT5:20 or 14:00/ZT8:20 (Figure 5). All of the genes with this pattern encoded proteins associated with PSII, including PsbD (or D2; Glyma.01G153500) and Psb28 (or PsbW; Glyma.10G040600, Glyma.13G127200). In wheat and Arabidopsis, *PsbD* shows similar circadian transcription patterns, with peak expression occurring during midday (Nakahira, Baba, Yoneda, Shiina, & Toyoshima, 1998; Noordally et al., 2013). While Psb28 is involved in PSII synthesis and assembly (Bečková et al., 2017; Dobakova, Sobotka, Tichy, & Komenda, 2008), it may also affect chlorophyll synthesis, which we found to change diurnally (Figure 4), and D1 synthesis (Dobakova et al., 2008; Mabbitt, Wilbanks, & Eaton-Rye, 2014). The next most common transcription pattern for PS-related genes was an increase in expression throughout the day (Figure 5), which included genes related to both PSI and PSII. The least common pattern was a decline in expression throughout the day (Figure 5), which was seen in the gene encoding PsaK (Glyma.09G087700), which may facilitate the interaction between light harvesting complex I and PSI (Jensen, Gilpin, Jurgen, & Scheller, 2000).

Expression of DEs related to other proteins and enzymes involved with electron transport and ATP synthase varied over the course of the day (Figure 5). Three of the four genes related to electron transport were involved with iron ion binding (Glyma.14G128500, Glyma.15G018000, Glyma.17G205000), but expression patterns differed for each gene.

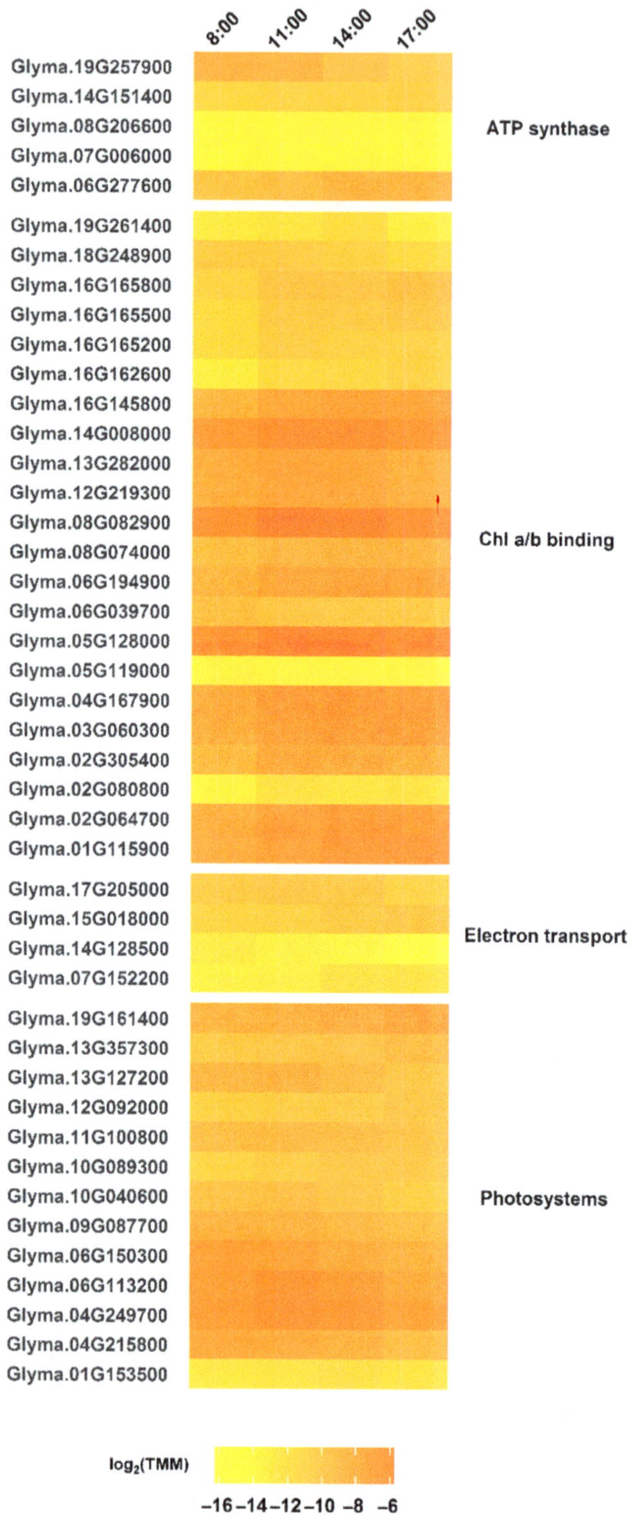

FIGURE 5 Diurnally differentially expressed genes related to photosynthetic energy transfer and electron transport in field-grown soybean. Gene loci are listed on the left, and the corresponding protein functions are listed on the right. Expression values [$\log_2$(TMM)] are shown (yellow = low expression, red = high expression) for four time points [8:00 (ZT2:20), 11:00 (ZT5:20), 14:00 (ZT8:20), 17:00 (ZT11:20)] measured during daylight hours

3.6 | Photosynthesis: C_3 cycle

Key steps in the C_3 cycle are catalyzed by the enzymes ribulose 1,5-bisphosphate carboxylase/oxygenase (Rubisco), fructose 1,6-bisphosphatase (FBPase), sedoheptulose 1,7-bisphosphatase (SBPase), and phosphoribulokinase (PRK). Rubisco proteins are expected to have a half-life of days, so diurnal transcriptional changes likely would not have much impact on protein abundance. Accordingly, none of the Rubisco small subunit-encoding genes were DE in our analysis.

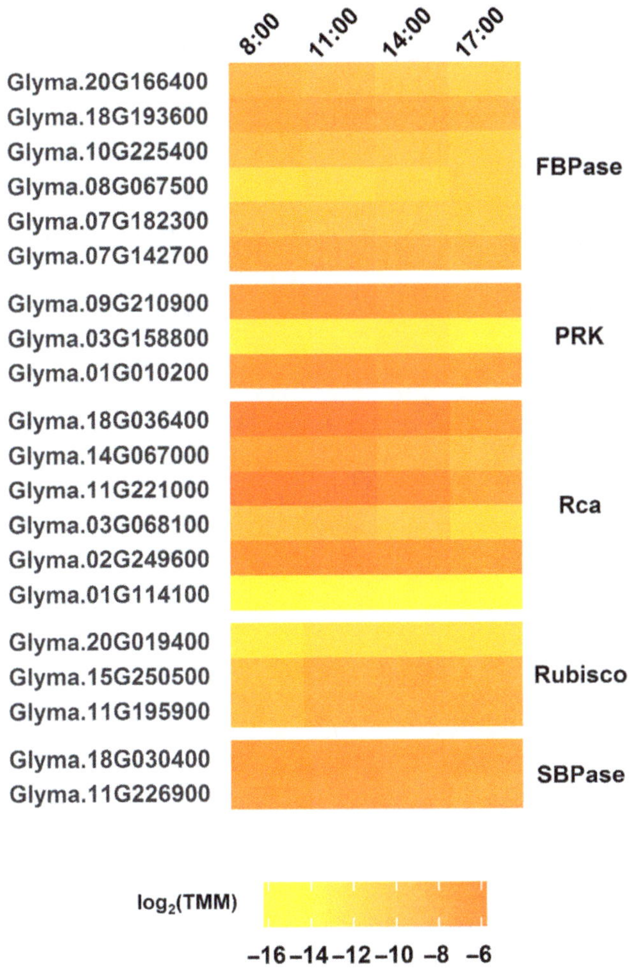

FIGURE 6 Diurnally differentially expressed genes related to key enzymes of the C_3 cycle in field-grown soybean. Gene loci are listed on the left, and the corresponding enzymes are listed on the right. Expression values [log_2(TMM)] are shown (yellow = low expression, red = high expression) for four time points [8:00 (ZT2:20), 11:00 (ZT5:20), 14:00 (ZT8:20), 17:00 (ZT11:20)] measured during daylight hours. FBPase, fructose 1,6-bisphosphatase; PRK, phosphoribulokinase; Rca, Rubisco activase; Rubisco, ribulose 1,5-bisphosphate carboxylase/oxygenase; SBPase, sedoheptulose 1,7-bisphosphatase

Because poly-A selection was performed on total RNA prior to RNA--seq library construction, transcription of the chloroplast-located Rubisco large subunit genes was not measured in our study. However, transcription of genes that encode Rubisco large subunit binding proteins (Glyma.11G195900, Glyma.15G250500, Glyma.20G019400) increased during the morning were highest at 14:00/ZT8:20 and then declined at 17:00/ZT11:20 (Figure 6). These binding proteins are involved in Rubisco holoenzyme assembly (Gutteridge & Gatenby, 1995), so their diurnal transcription could indicate a role for Rubisco complex assembly in diurnal adjustment of CO_2 fixation capacity, as incident light levels often drop below photosynthetic saturation by 17:00/ZT11:20. Rubisco activase (Rca) is required for Rubisco activity, and expression of diurnally affected Rca genes was generally highest at 8:00/ZT2:20 and 11:00/ZT5:20 but decreased strongly at later time points (Figure 6), which is similar to the circadian-regulated expression of Rca genes seen in Arabidopsis, apple (*Malus domestic*), rice, and tomato (*Lycopersicon esculentum*); (Liu, Taub, & McClung, 1996; Martino-Catt & Ort, 1992; Pilgrim & McClung, 1993; To, Suen, & Chen, 1999; Watillon, Kettmann, Boxus, & Burny, 1993). Similar to Rca gene expression, diurnal transcription of FBPase-, SBPase-, and PRK-encoding genes largely peaked in the morning and declined as the day progressed (Figure 6). The ferredoxin-thioredoxin system contributes to posttranslational regulation of these enzymes, and circadian rhythms have been observed in the expression of thioredoxin-type genes and proteins (Barajas-López et al., 2011). Thus, multiple levels of diurnal and/or circadian regulation are involved in the C_3 cycle. The decreased transcription of Rca genes and other C_3 cycle-related genes observed in this study, apparent as early as 14:00/ZT8:20, when diurnal photosynthesis typically

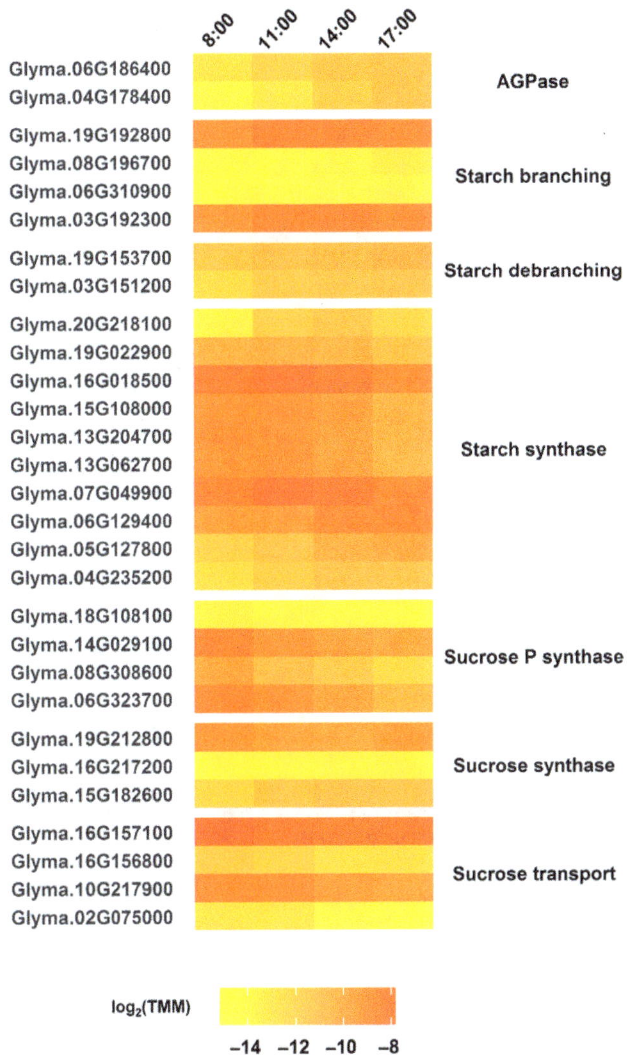

FIGURE 7 Diurnally differentially expressed genes related to starch and sucrose in field-grown soybean. Gene loci are listed on the left, and the corresponding enzymes or classes of enzymes are listed on the right. Expression values [log_2(TMM)] are shown (yellow = low expression, red = high expression) for four time points [8:00 (ZT2:20), 11:00 (ZT5:20), 14:00 (ZT8:20), 17:00 (ZT11:20)] measured during daylight hours. AGPase, ADP glucose pyrophosphorylase

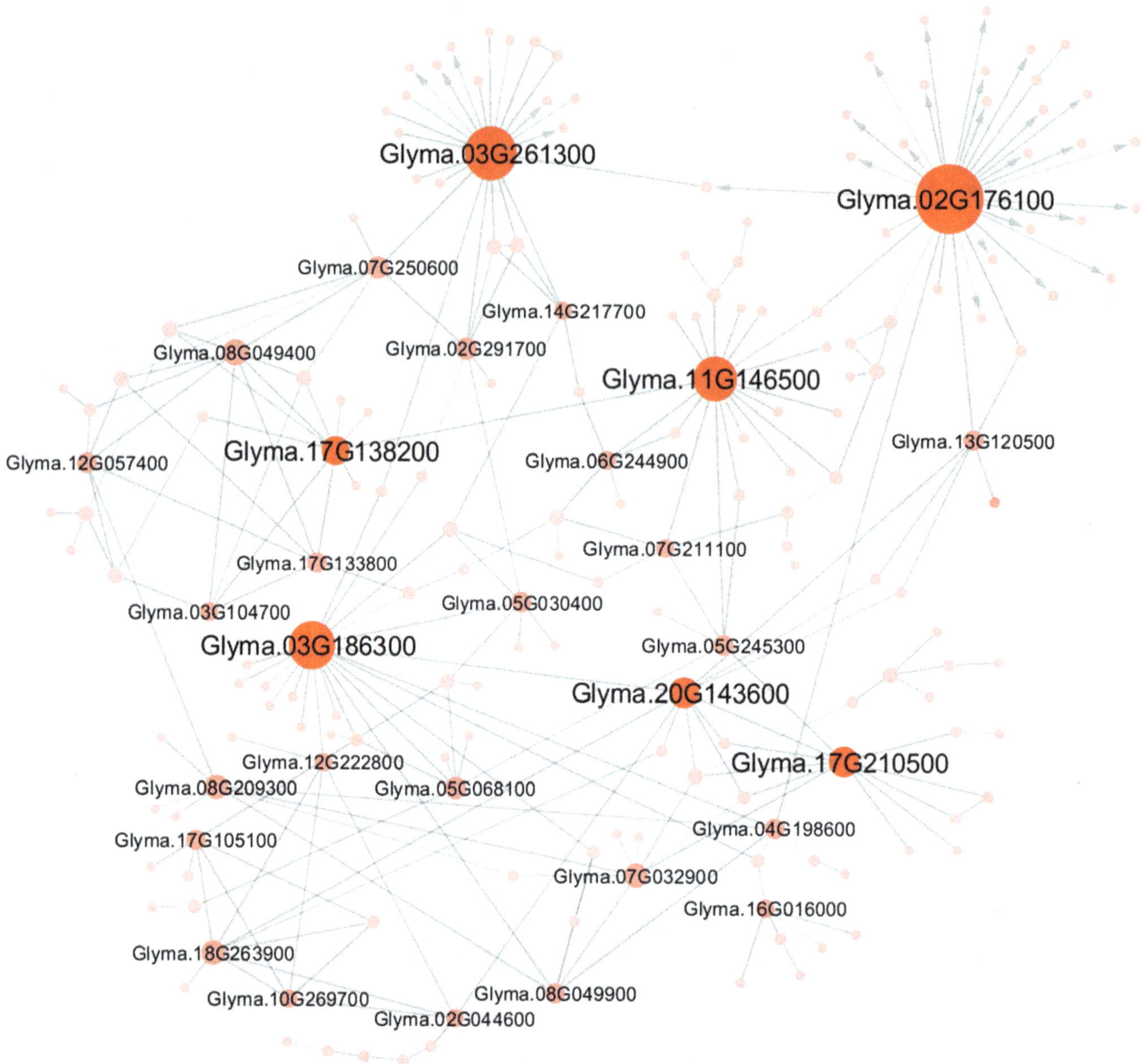

FIGURE 8 Gene regulatory network (GRN) analysis. Of the 1199 highly differentially expressed genes included in the analysis, 199 were assigned to the network (circles), which shows 267 significant "edges" (lines). Of these, 28 edges showed significant direction of causality (lines with arrowheads). Genes with ≥10 edges were considered "nodes" (large red circles). A very strict significance cutoff (fdr $p < 0.0001$) was used for determining edges and directionality

peaks for field-grown soybean (Rogers et al., 2004), might anticipate nightfall.

3.7 | Photosynthesis: Carbon metabolism

Leaf carbohydrate pools fluctuate diurnally. Starch is synthesized during the day when light can drive photosynthesis and is broken down during the night to fuel cellular respiration, and this has been measured previously in field-grown soybean (Rogers et al., 2004; Stitt & Zeeman, 2012). Starch degradation is strongly regulated by the circadian clock to ensure that starch stores last throughout the night (Graf & Smith, 2011), and several genes related to starch synthesis and breakdown were differentially regulated during the day in field-grown soybean in this study. Genes encoding ADP glucose pyrophosphorylase (AGPase), which catalyzes the entry step to starch synthesis, increased in expression as the day progressed (Figure 7). AGPase gene expression increases in the presence of sucrose (Sokolov, Dejardin, & Kleczkowski, 1998), the level of which increases in field-grown soybean during daylight hours (Rogers et al., 2004). However, other posttranscriptional modes of AGPase regulation lead to more fine-tuned responses of AGPase during fluctuating conditions (Geigenberger, 2011), and some of these processes, including the ferredoxin-thioredoxin system, have shown circadian rhythms (Barajas-López et al., 2011). A variety of starch synthase transcription patterns has been observed in Arabidopsis (Smith et al., 2004). Starch synthase genes in field-grown soybean also showed varied patterns of expression throughout the day, but expression was often highest at 8:00/ZT2:20 or 11:00/ZT5:20 and declined later in the day (Figure 7). Starch branching enzyme genes showed varied expression throughout the day, whereas expression of starch debranching-related genes increased as the day progressed (Figure 7).

Sucrose is exported from source leaves to be used for energy and biomass production in sink leaves, roots, and developing seeds. Although sucrose synthase gene expression varied during the day, expression of genes related to sucrose phosphate synthase and sucrose transport largely declined during daylight hours. This is the opposite pattern of sucrose accumulation resulting from photosynthesis over the course of the day. However, if these enzymes have a half-life of several hours, this could be indicative of the leaf anticipating the onset of darkness.

3.8 | Gene regulatory network (GRN)

To identify genes that are integral to diurnal cycles in soybean, a GRN was inferred using a very strict significance cutoff (fdr $p < 0.0001$) for both edges and directionality. Among the 1,199 highly DE genes included in the analysis, 199 were assigned to the network, and 267 significant relationships or "edges" were inferred among those (Figure 8, Supporting information Dataset S3). *A priori* identification of transcription factors and/or binding sites was not included in the analysis, so the inferred edges do not necessarily indicate direct regulation of transcription between nodes. The direction of causality was significantly determined for 28 of those relationships. Seven genes or "nodes" had 10 or more connections (Figure 8, Table 1), which are likely important regulators of diurnal functions in field-grown soybean. One of these nodes, *PSEUDO-RESPONSE REGULATOR 5* (*GmPRR5*; Table 1), is known to be directly involved with circadian rhythms (Bendix et al., 2015; Fujimori, Sato, Yamashino, & Mizuno, 2005; Nakamichi, Kita, Ito, Yamashino, & Mizuno, 2005) and interacts with CONSTANS to promote flowering in response to day length in Arabidopsis (Hayama et al., 2017). Two transcription factors, YABBY (GmYABBY16; (Zhao et al., 2017)) and the ethylene-responsive SHINE 2-RELATED (SHN2), were also identified as hub nodes (Table 1). YABBY plays an important role in plant development, including determining organ polarity, such as adaxial versus abaxial surfaces of leaves, and may be involved in flowering and seed development. In soybean, expression of *GmYABBY16* and other *GmYABBY* genes is affected by drought and salt stress and ABA application (Zhao et al., 2017), and YABBY likely regulates other transcription factors important to soybean development (Shamimuzzaman & Vodkin, 2013). SHN transcription factors are involved in regulating cutin, suberin, and wax protein levels in Arabidopsis (Licausi, Ohme-Takagi, & Perata, 2013; Shi et al., 2011), and the poplar (*Populus* spp.) SHN2 has been shown to regulate secondary cell wall formation in tobacco (*Nicotiana tabacum*) transformants (Liu et al., 2017).

TABLE 1 Hub nodes from gene regulatory network (GRN) inference analyses

Hub node genes	Protein annotation description[a]	Abbr.
Glyma.02G176100	Similarity (60%) with phosphoribosyl 1,2-cyclic phosphate phosphodiesterase and ribonuclease z.	PHNP
Glyma.03G186300	Phosphatidylinositol-4 -phosphate-5-kinase	PIP5K
Glyma.03G261300	Response regulator of two-component system (PSEUDO-RESPONSE REGULATOR 5)	PRR5
Glyma.11G146500	Aquaporin transporter	PIP2;4
Glyma.17G138200	YABBY protein	YABBY
Glyma.17G210500	Ethylene-responsive transcription factor	SHN2
Glyma.20G143600	Leucine-rich repeat N-terminal domain/carbohydrate binding protein of the endoplasmic reticulum	LRRNT_2

[a]Obtained from SoyBase and/or Phytozome12 (JGI).

Several nodes were identified as genes encoding proteins involved in cellular processes. Phosphatidylinositol-4-phosphate-5-kinase (PIP5K) was identified as a hub node (Table 1) and is involved in vesicle-mediated transport, cell adhesion, cell polarization, and cell migration (Irvine, 2003). In addition, water stress induces expression of specific PIP5Ks in Arabidopsis (Mikami, Katagiri, Iuchi, Yamaguchi-Shinozaki, & Shinozaki, 1998). Leucine-rich repeat N-terminal 2 (LRRNT_2) was also identified (Table 1) and is a carbohydrate-binding protein of the endoplasmic reticulum (Phytozome). The transcription of these genes was directly and negatively linked, although the direction of the interaction was not resolved. Similarly, the LRRNT_2 node was directly linked to SHN2 through a negative association, but again, the direction of repression was not evident. Another node, Glyma.02G176100, encoded a protein with 61% similarity to both a ribonuclease Z protein and phosphoribosyl 1,2-cyclic phosphate phosphodiesterase (Table 1).

The final hub node was a gene encoding an aquaporin, PLASMA MEMBRANE INTRINSIC PROTEIN2;4 (GmPIP2;4), that acts as a channel and regulates water flux through the plasma membrane (Table 1). Our RNA-seq analysis showed higher transcription of this gene at all time points compared to 8:00/ZT2:20, with the highest transcript abundance at 14:00/ZT8:20. Aquaporins show diurnal expression cycles in roots (Clarkson et al., 2000; Takase et al., 2011; Vandeleur et al., 2009) and may help regulate diurnal leaf water status in soybean leaves (Locke & Ort, 2015). As we reported previously, the diurnal transcription of *GmPIP2;4* in these leaves tracked atmospheric vapor pressure deficit, but its diurnal transcription pattern was inverse to the diurnal pattern of leaf hydraulic conductance (Locke & Ort, 2015). The GRN analysis indicated that *GmPIP2;4* expression influences or is influenced by many other genes, including a positive interaction with the YABBY node, although the direction of the relationship with *GmYABBY16* was unclear. As ABA treatment and salt stress upregulate but drought stress downregulates *GmYABBY16* expression (Zhao et al., 2017), this result is not surprising but deserves further analysis.

4 | CONCLUSION

This diurnal picture of the field-grown soybean transcriptome illustrates that genes in pathways either contributing to or depending on photosynthesis are often differentially transcribed over the course of the day. In addition, several genes act as hubs for regulating diurnal processes, either by directly regulating transcription of other genes or by integrating diurnal signals to control a key response (e.g., leaf water transport) with rapid and extensive downstream effects. Improving photosynthesis is becoming an increasingly important goal, as climate change and population growth test the limits of our agricultural production capacity. This data set can serve as a basis for further investigation of diurnally responsive pathways and eventual identification of genes and transcription factors that diurnally regulate photosynthesis and transpiration. If it is possible that diurnal variations in transcription of photosynthesis-regulating genes could be manipulated, this could greatly contribute to improving photosynthesis in light-limited conditions and water use efficiency at high midday vapor pressure deficit, and to better acclimate to changing climate. In addition, our comparisons with transcriptomic analyses conducted on soybean grown in controlled environments suggest that caution should be used when extrapolating gene expression patterns from controlled environments to field environments.

ACKNOWLEDGMENTS

We thank Nathan Couch, Courtney Leisner, Kannan Puthuval, Matthew Siebers, Kati Togliatti, and Adrian Zimbelman for assistance with field setup, maintenance, and tissue sampling, and Rachel Yelin for help with RNA isolation. We also thank Caitlin Moore for assistance obtaining meteorological data from the University of Illinois Energy Farm. This work was funded by the United States Department of Agriculture-Agricultural Research Service.

AUTHORS' CONTRIBUTIONS

AML and DRO designed the research; AML performed the research and analyzed the data; AML and RAS interpreted the data and wrote the paper with contributions from DRO.

REFERENCES

Alabadí, D., Oyama, T., Yanovsky, M. J., Harmon, F. G., Más, P., & Kay, S. A. (2001). Reciprocal regulation between *TOC1* and *LHY/CCA1* within the *Arabidopsis* circadian clock. *Science*, *293*, 880–883. https://doi.org/10.1126/science.1061320

Anders, S., Pyl, P. T., & Huber, W. (2015). HTSeq-a Python framework to work with high-throughput sequencing data. *Bioinformatics*, *31*, 166–169. https://doi.org/10.1093/bioinformatics/btu638

Barajas-López, J. D. D., Serrato, A. J., Cazalis, R., Meyer, Y., Chueca, A., Reichheld, J. P., & Sahrawy, M. (2011). Circadian regulation of chloroplastic *f* and *m* thioredoxins through control of the CCA1 transcription factor. *Journal of Experimental Botany*, *62*, 2039–2051. https://doi.org/10.1093/jxb/erq394

Bečková, M., Gardian, Z., Yu, J., Konik, P., Nixon, P. J., & Komenda, J. (2017). Association of Psb28 and Psb27 proteins with PSII-PSI supercomplexes upon exposure of *Synechocystis* sp. PCC 6803 to high light. *Molecular Plant*, *10*, 62–72.

Bendix, C., Marshall, C. M., & Harmon, F. G. (2015). Circadian clock genes universally control key agricultural traits. *Molecular Plant*, *8*, 1135–1152. https://doi.org/10.1016/j.molp.2015.03.003

Bernacchi, C. J., Leakey, A. D. B., Heady, L. E., Morgan, P. B., Dohleman, F. G., McGrath, J. M., … Ort, D. R. (2006). Hourly and seasonal variation in photosynthesis and stomatal conductance of soybean grown at future CO_2 and ozone concentrations for 3 years under fully open-air field conditions. *Plant, Cell and Environment*, *29*, 2077–2090. https://doi.org/10.1111/j.1365-3040.2006.01581.x

Bilgin, D. D., DeLucia, E. H., & Clough, S. J. (2009). A robust plant RNA isolation method suitable for Affymetrix GeneChip analysis and quantitative real-time RT-PCR. *Nature Protocols*, *4*, 333–340. https://doi.org/10.1038/nprot.2008.249

Blasing, O. E., Gibon, Y., Gunther, M., Hohne, M., Morcuende, R., Osuna, D., … Stitt, M. (2005). Sugars and circadian regulation make major contributions to the global regulation of diurnal gene expression in *Arabidopsis*. *Plant Cell*, *17*, 3257–3281. https://doi.org/10.1105/tpc.105.035261

Bolger, A. M., Lohse, M., & Usadel, B. (2014). Trimmomatic: A flexible trimmer for Illumina sequence data. *Bioinformatics*, *30*, 2114–2120. https://doi.org/10.1093/bioinformatics/btu170

Bukatsch, F., & Rudolph, E. (1963). New facts on the diurnal rhythm of chlorophyll content in growing and adult leaves. *Photochemistry and Photobiology*, *2*, 191–198. https://doi.org/10.1111/j.1751-1097.1963.tb08215.x

Busheva, M., Garab, G., Liker, E., Tóth, Z., Szèll, M., & Nagy, F. (1991). Diurnal fluctuations in the content and functional properties of the light harvesting chlorophyll *a/b* complex in thylakoid membranes: Correlation with the diurnal rhythm of the mRNA level. *Plant Physiology*, *95*, 997–1003. https://doi.org/10.1104/pp.95.4.997

Clarkson, D. T., Carvajal, M., Henzler, T., Waterhouse, R. N., Smyth, A. J., Cooke, D. T., & Steudle, E. (2000). Root hydraulic conductance: Diurnal aquaporin expression and the effects of nutrient stress. *Journal of Experimental Botany*, *51*, 61–70. https://doi.org/10.1093/jxb/51.342.61

Covington, M. F., Maloof, J. N., Straume, M., Kay, S. A., & Harmer, S. L. (2008). Global transcriptome analysis reveals circadian regulation of key pathways in plant growth and development. *Genome Biology*, *9*, R130. https://doi.org/10.1186/gb-2008-9-8-r130

Delhon, P., Gojon, A., Tillard, P., & Passama, L. (1995). Diurnal regulation of NO_3^- uptake in soybean plants. I. Changes in NO_3^- influx, efflux, and N utilization in the plant during the day/night cycle. *Journal of Experimental Botany*, *46*, 1585–1594. https://doi.org/10.1093/jxb/46.10.1585

Dobakova, M., Sobotka, R., Tichy, M., & Komenda, J. (2008). Psb28 protein is involved in the biogenesis of the photosystem II inner antenna CP47 (PsbB) in the cyanobacterium *Synechocystis* sp. PCC 6803. *Plant Physiology*, *149*, 1076–1086. https://doi.org/10.1104/pp.108.130039

Dodd, A. N., Salathia, N., Hall, A., Kévei, E., Tóth, R., Nagy, F., … Webb, A. A. R. (2005). Plant circadian clocks increase photosynthesis, growth, survival, and competitive advantage. *Science*, *309*, 630–633. https://doi.org/10.1126/science.1115581

Du, S.-Y., Zhang, X.-F., Lu, Z., Xin, Q., Wu, Z., Jiang, T., … Zhang, D.-P. (2012). Roles of the different components of magnesium chelatase in abscisic acid signal transduction. *Plant Molecular Biology*, *80*, 519–537. https://doi.org/10.1007/s11103-012-9965-3

Feltz, C. J., & Miller, G. E. (1996). An asymptotic test for the equality of coefficients of variation from *k* populations. *Statistics in Medicine*, *15*, 647–658. https://doi.org/10.1002/(ISSN)1097-0258

Fujimori, T., Sato, E., Yamashino, T., & Mizuno, T. (2005). PRR5 (PSEUDO-RESPONSE REGULATOR 5) plays antagonistic roles to CCA1 (CIRCADIAN CLOCK-ASSOCIATED 1) in *Arabidopsis thaliana*. *Bioscience, Biotechnology, and Biochemistry*, *69*, 426–430. https://doi.org/10.1271/bbb.69.426

Geigenberger, P. (2011). Regulation of starch biosynthesis in response to a fluctuating environment. *Plant Physiology*, *155*, 1566–1577. https://doi.org/10.1104/pp.110.170399

Geiger, D. R., & Servaites, J. C. (1994). Diurnal regulation of photosynthetic carbon metabolism in C_3 plants. *Annual Review of Plant Physiology and Plant Molecular Biology*, *45*, 235–256. https://doi.org/10.1146/annurev.pp.45.060194.001315

Graf, A., & Smith, A. M. (2011). Starch and the clock: The dark side of plant productivity. *Trends in Plant Science*, *16*, 169–175. https://doi.org/10.1016/j.tplants.2010.12.003

Grant, D., Nelson, R. T., Cannon, S. B., & Shoemaker, R. C. (2009). SoyBase, the USDA-ARS soybean genetics and genomics database. *Nucleic Acids Research*, *38*, 843–846.

Gutteridge, S., & Gatenby, A. A. (1995). Rubisco synthesis, assembly, mechanism, and regulation. *Plant Cell*, *7*, 809–819. https://doi.org/10.1105/tpc.7.7.809

Harpaz-Saad, S., Azoulay, T., Arazi, T., Ben-Yaakov, E., Mett, A., Shiboleth, Y. M., ... Eyal, Y. (2007). Chlorophyllase is a rate-limiting enzyme in chlorophyll catabolism and is posttranslationally regulated. *Plant Cell*, *19*, 1007–1022. https://doi.org/10.1105/tpc.107.050633

Hayama, R., Sarid-Krebs, L., Richter, R., Fernández, V., Jang, S., & Coupland, G. (2017). PSEUDO RESPONSE REGULATORs stabilize CONSTANS protein to promote flowering in response to day length. *EMBO Journal*, *36*, 904–918. https://doi.org/10.15252/embj.201693907

Ilag, L. L., Kumar, A. M., & Söll, D. (1994). Light regulation of chlorophyll biosynthesis at the level of 5-aminolevulinate formation in Arabidopsis. *Plant Cell*, *6*, 265–275. https://doi.org/10.1105/tpc.6.2.265

Irvine, R. F. (2003). Nuclear lipid signalling. *Nature Reviews Molecular Cell Biology*, *4*, 349–360. https://doi.org/10.1038/nrm1100

Jensen, E. P., Gilpin, M., Jurgen, K., & Scheller, H. V. (2000). The PSI-K subunit of photosystem I is involved in the interaction between light-harvesting complex I and the photosystem I reaction center core. *Journal of Biological Chemistry*, *275*, 24701–24708. https://doi.org/10.1074/jbc.M000550200

Kim, D., Pertea, G., Trapnell, C., Pimentel, H., Kelley, R., & Salzberg, S. L. (2013). TopHat2: Accurate alignment of transcriptomes in the presence of insertions, deletions and gene fusions. *Genome Biology*, *14*, R36. https://doi.org/10.1186/gb-2013-14-4-r36

Kinoshita, T., Doi, M., Suetsugu, N., Kagawa, T., Wada, M., & Shimazaki, K. (2001). phot1 and phot2 mediate blue light regulation of stomatal opening. *Nature*, *414*, 656–660. https://doi.org/10.1038/414656a

Ko, D. K., Rohozinski, D., Song, Q., Taylor, S. H., Juenger, T. E., Harmon, F. G., & Chen, Z. J. (2016). Temporal shift of circadian-mediated gene expression and carbon fixation contributes to biomass heterosis in maize hybrids. *PLoS Genetics*, *12*, 1–31.

Licausi, F., Ohme-Takagi, M., & Perata, P. (2013). APETALA2/Ethylene Responsive Factor (AP2/ERF) transcription factors: Mediators of stress responses and developmental programs. *New Phytologist*, *199*, 639–649. https://doi.org/10.1111/nph.12291

Lichtenthaler, H. (1987). Chlorophylls and carotenoids: Pigments of photosynthetic biomembranes. *Methods in Enzymology*, *148*, 350–382. https://doi.org/10.1016/0076-6879(87)48036-1

Liu, Z., Taub, C. C., & McClung, C. R. (1996). Identification of an *Arabidopsis thaliana* ribulose-1,5-bisphosphate carboxylase/oxygenase activase (RCA) minimal promoter regulated by light and the circadian clock. *Plant Physiology*, *112*, 43–51. https://doi.org/10.1104/pp.112.1.43

Liu, H., Wang, H., Gao, P., Xü, J., Xü, T., Wang, J., ... Fu, Y.-F. (2009). Analysis of clock gene homologs using unifoliolates as target organs in soybean (*Glycine max*). *Journal of Plant Physiology*, *166*, 278–289. https://doi.org/10.1016/j.jplph.2008.06.003

Liu, Y., Wei, M., Hou, C., Lu, T., Liu, L., Wei, H., ... Wei, Z. (2017). Functional characterization of *Populus PsnSHN2* in coordinated regulation of secondary wall components in tobacco. *Scientific Reports*, *7*, 1–11.

Locke, A. M., & Ort, D. R. (2015). Diurnal depression in leaf hydraulic conductance at ambient and elevated [CO_2] reveals anisohydric water management in field-grown soybean and possible involvement of aquaporins. *Environmental and Experimental Botany*, *116*, 39–46. https://doi.org/10.1016/j.envexpbot.2015.03.006

Mabbitt, P. D., Wilbanks, S. M., & Eaton-Rye, J. J. (2014). Structure and function of the hydrophilic photosystem II assembly proteins: Psb27, Psb28 and Ycf48. *Plant Physiology and Biochemistry*, *81*, 96–107. https://doi.org/10.1016/j.plaphy.2014.02.013

Marcolino-Gomes, J., Rodrigues, F. A., Fuganti-Pagliarini, R., Bendix, C., Nakayama, T. J., Celaya, B., ... Nepomuceno, A. (2014). Diurnal oscillations of soybean circadian clock and drought responsive genes. *PLoS ONE*, *9*, e86402. https://doi.org/10.1371/journal.pone.0086402

Martino-Catt, S., & Ort, D. R. (1992). Low temperature interrupts circadian regulation of transcriptional activity in chilling-sensitive plants. *Proceedings of the National Academy of Sciences of the United States of America*, *89*, 3731–3735. https://doi.org/10.1073/pnas.89.9.3731

Marwick, B, Krishnamoorthy, K (2018) cvequality: Tests for the equality of coefficients of variation among multiple groups. R software package version 0.1.3. Retrieved from https://github.com/benmarwick/cvequality on 20/08/2018

Matsumoto, F., Obayashi, T., Sasaki-Sekimoto, Y., Ohta, H., & Takamiya, K. (2004). Gene expression profiling of the tetrapyrrole metabolic pathway in Arabidopsis with a mini-array system. *Plant Physiology*, *135*, 2379–2391. https://doi.org/10.1104/pp.104.042408

McCarthy, D. J., Chen, Y., & Smyth, G. K. (2012). Differential expression analysis of multifactor RNA-Seq experiments with respect to biological variation. *Nucleic Acids Research*, *40*, 4288–4297. https://doi.org/10.1093/nar/gks042

Meyer, H., Thienel, U., & Piechulla, B. (1989). Molecular characterization of the diurnal/circadian expression of the chlorophyll a/b-binding proteins in leaves of tomato and other dicotyledonous and monocotyledonous plant species. *Planta*, *180*, 5–15. https://doi.org/10.1007/BF02411404

Mikami, K., Katagiri, T., Iuchi, S., Yamaguchi-Shinozaki, K., & Shinozaki, K. (1998). A gene encoding phosphatidylinositol-4-phosphate 5-kinase is induced by water stress and abscisic acid in *Arabidopsis thaliana*. *Plant Journal*, *15*, 563–568. https://doi.org/10.1046/j.1365-313X.1998.00227.x

Morales, A. M. A. P., O'Rourke, J. A., van de Mortel, M., Scheider, K. T., Bancroft, T. J., Borém, A., ... Graham, M. A. (2013). Transcriptome analyses and virus induced gene silencing identify genes in the *Rpp4*-mediated Asian soybean rust resistance pathway. *Functional Plant Biology*, *40*, 1029–1047. https://doi.org/10.1071/FP12296

Nagano, A. J., Sato, Y., Mihara, M., Anotnio, B. A., Motoyama, R., Itoh, H., ... Izawa, T. (2012). Deciphering and prediction of transcriptome dynamics under fluctuating field conditions. *Cell*, *151*, 1358–1369. https://doi.org/10.1016/j.cell.2012.10.048

Nakahira, Y., Baba, K., Yoneda, A., Shiina, T., & Toyoshima, Y. (1998). Circadian-regulated transcription of the *psbD* light-responsive promoter in wheat chloroplasts. *Plant Physiology*, *118*, 1079–1088. https://doi.org/10.1104/pp.118.3.1079

Nakamichi, N., Kita, M., Ito, S., Yamashino, T., & Mizuno, T. (2005). PSEUDO-RESPONSE REGULATORS, PRR9, PRR7 and PRR5, together play essential roles close to the circadian clock of *Arabidopsis thaliana*. *Plant and Cell Physiology*, *46*, 686–698. https://doi.org/10.1093/pcp/pci086

Nogaj, L. A., Srivastava, A., van Lis, R., & Beale, S. I. (2005). Cellular levels of glutamyl-tRNA reductase and glutamate-1-semialdehyde aminotransferase do not control chlorophyll synthesis in *Chlamydomonas reinhardtii*. *Plant Physiology*, *139*, 389–396. https://doi.org/10.1104/pp.105.067009

Noordally, Z. B., Ishii, K., Atkins, K. A., Wetherill, S. J., Kusakina, J., Walton, E. J., ... Dodd, A. N. (2013). Circadian control of chloroplast transcription by a nuclear-encoded timing signal. *Nature, 339*, 1316–1319.

Opgen-Rhein, R., & Strimmer, K. (2007). From correlation to causation networks: A simple approximate learning algorithm and its application to high-dimensional plant gene expression data. *BMC Systems Biology, 1*, 37. https://doi.org/10.1186/1752-0509-1-37

Pan, W. J., Wang, X., Deng, Y. R., Li, J. H., Chen, W., Chiang, J. Y., ... Zheng, L. (2015). Nondestructive and intuitive determination of circadian chlorophyll rhythms in soybean leaves using multispectral imaging. *Scientific Reports, 5*, 1–13.

Pilgrim, M. L., & McClung, C. R. (1993). Differential involvement of the circadian clock in the expression of genes required for ribulose-1,5-bisphosphate carboxylase/oxygenase synthesis, assembly, and activation in *Arabidopsis thaliana*. *Plant Physiology, 103*, 553–564. https://doi.org/10.1104/pp.103.2.553

Porra, R., Thompson, W., & Kriedemann, P. (1989). Determination of accurate extinctions coefficients and simultaneous equations for assaying chlorophylls *a* and *b* extracted with four different solvents: Verification of the concentration of chlorophyll standards by atomic absorption spectrosc. *Biochimica et Biophysica Acta, 975*, 384–394. https://doi.org/10.1016/S0005-2728(89)80347-0

Reinbothe, S., & Reinbothe, C. (1996). Regulation of chlorophyll biosynthesis in angiosperms. *Plant Physiology, 111*, 1–7. https://doi.org/10.1104/pp.111.1.1

Robinson, M. D., McCarthy, D. J., & Smyth, G. K. (2009). edgeR: A Bioconductor package for differential expression analysis of digital gene expression data. *Bioinformatics, 26*, 139–140.

Robinson, M., & Oshlack, A. (2010). A scaling normalization method for differential expression analysis of RNA-seq data. *Genome Biology, 11*, R25. https://doi.org/10.1186/gb-2010-11-3-r25

Rogers, A., Allen, D. J., Davey, P. A., Morgan, P. B., Ainsworth, E. A., Bernacchi, C. J., & Cornic, G. (2004). Leaf photosynthesis and carbohydrate dynamics of soybeans grown throughout their life-cycle under Free-Air Carbon dioxide Enrichment. *Plant, Cell and Environment, 27*, 449–458. https://doi.org/10.1111/j.1365-3040.2004.01163.x

Schaefer, J., Opgen-Rhein, R., & Strimmer, K. (2015). GeneNet: modeling and inferring gene networks.

Schmutz, J., Cannon, S. B., Schlueter, J., Ma, J., Mitros, T., Nelson, W., ... Jackson, S. A. (2010). Genome sequence of the palaeopolyploid soybean. *Nature, 463*, 178–183. https://doi.org/10.1038/nature08670

Shamimuzzaman, M., & Vodkin, L. (2013). Genome-wide identification of binding sites for NAC and YABBY transcription factors and co-regulated genes during soybean seedling development by ChIP-Seq and RNA-Seq. *BMC Genomics, 14*, 477. https://doi.org/10.1186/1471-2164-14-477

Shi, J. X., Malitsky, S., de Oliveira, S., Branigan, C., Franke, R. B., Schreiber, L., & Aharoni, A. (2011). SHINE transcription factors act redundantly to pattern the archetypal surface of Arabidopsis flower organs. *PLoS Genetics, 7*, e1001388. https://doi.org/10.1371/journal.pgen.1001388

Smith, S. M., Fulton, D. C., Chia, T., Thorneycroft, D., Chapple, A., Dunstan, H., ... Smith, A. M. (2004). Diurnal changes in the transcriptome encoding enzymes of starch metabolism provide evidence for both transcriptional and posttranscriptional regulation of starch metabolism in Arabidopsis leaves. *Plant Physiology, 136*, 2687–2699. https://doi.org/10.1104/pp.104.044347

Sokolov, L. N., Dejardin, A., & Kleczkowski, L. A. (1998). Sugars and light/dark exposure trigger differential regulation of ADP-glucose pyrophosphorylase genes in *Arabidopsis thaliana* (thale cress). *Biochemical Journal, 336*, 681–687. https://doi.org/10.1042/bj3360681

Stitt, M., & Zeeman, S. C. (2012). Starch turnover: Pathways, regulation and role in growth. *Current Opinion in Plant Biology, 15*, 282–292. https://doi.org/10.1016/j.pbi.2012.03.016

Takase, T., Ishikawa, H., Murakami, H., Kikuchi, J., Sato-Nara, K., & Suzuki, H. (2011). The circadian clock modulates water dynamics and aquaporin expression in Arabidopsis roots. *Plant and Cell Physiology, 52*, 373–383. https://doi.org/10.1093/pcp/pcq198

Tanaka, R., Kobayashi, K., & Masuda, T. (2011). Tetrapyrrole metabolism in *Arabidopsis thaliana*. *Arabidopsis Book, 9*, e0145. https://doi.org/10.1199/tab.0145

Tanaka, R., & Tanaka, A. (2007). Tetrapyrrole biosynthesis in higher plants. *Annual Review of Plant Biology, 58*, 321–346. https://doi.org/10.1146/annurev.arplant.57.032905.105448

Tanaka, R., & Tanaka, A. (2011). Chlorophyll cycle regulates the construction and destruction of the light-harvesting complexes. *Biochimica et Biophysica Acta (BBA) – Bioenergetics, 1807*, 968–976. https://doi.org/10.1016/j.bbabio.2011.01.002

Tanaka, R., Yoshida, K., Nakayashiki, T., Masuda, T., Tsuji, H., Inokuchi, H., & Tanaka, A. (1996). Differential expression of two *hemA* mRNAs encoding glutamyl-tRNA reductase proteins in greening cucumber seedlings. *Plant Physiology, 110*, 1223–1230. https://doi.org/10.1104/pp.110.4.1223

To, K., Suen, D., & Chen, S. G. (1999). Molecular characterization of ribulose-1,5-bisphosphate carboxylase/oxygenase activase in rice leaves. *Planta, 886*, 66–76. https://doi.org/10.1007/s004250050607

Upmeyer, D. J., & Koller, H. R. (1973). Diurnal trends in net photosynthetic rate and carbohydrate levels of soybean leaves. *Plant Physiology, 51*, 871–874. https://doi.org/10.1104/pp.51.5.871

Vandeleur, R. K., Mayo, G., Shelden, M. C., Gilliham, M., Kaiser, B. N., & Tyerman, S. D. (2009). The role of plasma membrane intrinsic protein aquaporins in water transport through roots: Diurnal and drought stress responses reveal different strategies between isohydric and anisohydric cultivars of grapevine. *Plant Physiology, 149*, 445–460. https://doi.org/10.1104/pp.108.128645

Watillon, B., Kettmann, R., Boxus, P., & Burny, A. (1993). Developmental and circadian pattern of rubisco activase mRNA accumulation in apple plants. *Plant Molecular Biology, 23*, 501–509. https://doi.org/10.1007/BF00019298

Wickliff, J. L., & Aronoff, S. (1962). Evidence for absence of diurnal variation of chlorophyll content in mature leaves of soybean. *Plant Physiology, 37*, 590–594. https://doi.org/10.1104/pp.37.5.590

Zeeman, S. C., Smith, S. M., & Smith, A. M. (2007). The diurnal metabolism of leaf starch. *Biochemical Journal, 401*, 13–28. https://doi.org/10.1042/BJ20061393

Zhao, S.-P., Lu, D., Yu, T.-F., Ji, Y.-J., Zheng, W.-J., Zhang, S.-X., ... Cui, X.-Y. (2017). Genome-wide analysis of the YABBY family in soybean and functional identification of *GmYABBY10* involvement in high salt and drought stresses. *Plant Physiology and Biochemistry, 119*, 132–146. https://doi.org/10.1016/j.plaphy.2017.08.026

6

Nutrient sink limitation constrains growth in two barley species with contrasting growth strategies

Angela C. Burnett[1] | Alistair Rogers[2] | Mark Rees[1] | Colin P. Osborne[1]

[1]Department of Animal and Plant Sciences, University of Sheffield, Sheffield, UK

[2]Environmental and Climate Sciences Department, Brookhaven National Laboratory, Upton, New York

Correspondence
Angela C. Burnett, Department of Animal and Plant Sciences, University of Sheffield, Sheffield, UK.
Email: aburnett@bnl.gov

Present address
Angela C. Burnett, Environmental and Climate Sciences Department, Brookhaven National Laboratory, Upton, New York 11973.

Funding information
United States Department of Energy, Grant/Award Number: DE-SC0012704; Society for Experimental Biology

Abstract

Mineral nutrients exert important limitations on plant growth. Growth is limited by the nutrient source when it is constrained by nutrient availability and uptake, which may simultaneously limit investment in photosynthetic proteins, leading to carbon source limitation. However, growth may also be limited by nutrient utilization in sink tissue. The relative importance of these processes is contested, with crop and vegetation models typically assuming source limitations of carbon and mineral nutrients (especially nitrogen). This study compared the importance of source and sink limitation on growth in a slower-growing wild perennial barley (*Hordeum bulbosum*) and a faster-growing domesticated annual barley (*Hordeum vulgare*), by applying a mineral nutrient treatment and measuring nitrogen uptake, growth, allocation, and carbon partitioning. We found that nitrogen uptake, growth, tillering, shoot allocation, and nitrogen storage were restricted by low nutrient treatments. Multiple lines of evidence suggest that low nutrient levels do not limit growth via carbon acquisition: (a) Carbohydrate storage does not increase at high nutrient levels. (b) Ratio of free amino acids to sucrose increases at high nutrient levels. (c) Shoot allocation increases at high nutrient levels. These data indicate that barley productivity is limited by the capacity for nutrient use in growth. Models must explicitly account for sink processes in order to properly simulate this mineral nutrient limitation of growth.

KEYWORDS
barley, crop, growth, nitrogen, nutrients, source-sink

1 | INTRODUCTION

The use of fertiliser in agriculture is an increasing cause of concern from both economic and environmental perspectives (Masclaux-Daubresse et al., 2010). In plants, after carbon, nitrogen is the element required in largest quantities (Sakakibara, Takei, & Hirose, 2006) and is therefore the primary mineral nutrient and a vital resource that affects growth, allocation and phenology (Stitt & Krapp, 1999). Crops are routinely treated with NPK fertiliser, containing the macronutrients nitrogen, phosphorus, and potassium. Fertiliser use has greatly improved crop yields, but the production and excess application of mineral nutrient fertiliser have simultaneously decreased agricultural nitrogen use efficiency and increased environmental damage (Erisman, Sutton, Galloway, Klimont, & Winiwarter, 2008). For instance, the manufacture of nitrogen fertiliser via the Haber-Bosch process and its denitrification in soils are major sources of anthropogenic greenhouse gas emissions in food production (Goucher, Bruce, Cameron, Lenny Koh, & Horton, 2017), whilst agricultural run-off has led to devastating eutrophication (fuelled by nitrogen and phosphorus) in areas such as the Gulf of Mexico (Rabalais, Turner, &

Wiseman, 2002). Appropriate fertiliser application will reduce the financial, environmental, and ecological costs of intensive farming, yet the crop requirement for nutrients must be met in order to maintain and increase yields. Nutrient use efficiency is therefore a major target for crop improvement, with nitrogen being the key element of interest (Perchlik & Tegeder, 2018; Tegeder & Masclaux-Daubresse, 2018). The timing of fertiliser application has been extensively researched from an agronomic perspective, and much is known at the molecular level about nitrate transport and signalling mechanisms (Kiba, Kudo, Kojima, & Sakakibara, 2011; Miller, Fan, Orsel, Smith, & Wells, 2007; Sakakibara et al., 2006). However, the whole-plant physiological behaviour linking the ecology of crops on one hand, and molecular physiology on the other, remains poorly characterised.

The relationships between sources and sinks are vital determinants of growth (Chang & Zhu, 2017; White, Rogers, Rees, & Osborne, 2016). Source strength may be defined as the capacity of the plant to take up a resource from the external environment—in this case, mineral nutrients from the soil—whilst sink strength is the internal capacity of the plant to utilize that resource in storage or growth. In each case, strength is the product of the size and activity of the source or sink tissue (Geiger & Shieh, 1993; White et al., 2016). For the primary mineral nutrient nitrogen, the source:sink ratio is regulated by interacting molecular mechanisms, yet our understanding of how, and the extent to which, nitrogen source and sink strengths limit growth at different times during ontogeny remains incomplete (White et al., 2016). Broadly speaking, plants are thought to transit from carbon sink to source limitation during development (Arp, 1991; Marschner, 1995), but the generality of this principle is unclear because differences in carbon source and sink limitation have been observed for a variety of crops, particularly at grain filling (Acreche & Slafer, 2009; Álvaro, Royo, García del Moral, & Villegas, 2008; Jaikumar, Snapp, Flore, & Loescher, 2014; Peterhansel & Offermann, 2012; Slewinski, 2012). Many studies focus on reproductive growth in order to examine source-sink relations with respect to yield, such that understanding of source-sink relations is especially deficient for the vegetative stage (Burnett, Rogers, Rees, & Osborne, 2016). The majority of studies focus on sources and sinks for carbon which are of great importance (Paul, Oszvald, Jesus, Rajulu, & Griffiths, 2017) yet nutrient sources and sinks are critical factors that interact with carbon sources and sinks (Ruiz-Vera, De Souza, Long, & Ort, 2017) and underpin plant growth in their own right (Sonnewald & Fernie, 2018; White et al., 2016).

Growth is controlled by a network of processes, the strengths of which vary with nutrient supply (White et al., 2016). Therefore, understanding how nutrient uptake and utilization limits growth is important for improving crop yield and sustainability (Burnett et al., 2016; White et al., 2016). In particular, the processes responsible for nutrient limitation of growth are contested. For example, low nitrogen uptake by the root could primarily constrain growth by one of two mechanisms: limiting the synthesis of photosynthetic proteins and thereby causing carbon source limitation; or directly limiting the synthesis of proteins and other compounds required for sink tissue expansion (Fatichi, Leuzinger, & Korner, 2014; Körner, 2015; Poorter, Anten, & Marcelis, 2013; Stitt & Schulze, 1994). Process-based vegetation and crop models often represent nitrogen limitation by focusing solely on root nitrogen uptake and nitrogen use in photosynthesis (Bao, Hoogenboom, McClendon, & Vellidis, 2017; Oleson et al., 2010; Wärlind, Smith, Hickler, & Arneth, 2014), without considering the direct limiting effects of nitrogen availability on tissue growth. However, there are several exceptions (Zaehle et al., 2014).

The natural diversity of plant growth rates can be used to investigate the factors that limit growth, and ecological research has advanced our understanding of how growth and nutrient use are adapted to different soil conditions. Growth rate is considered an important adaptation to variation in soil fertility, such that nutrient-poor environments are dominated by slower-growing plants, and nutrient-rich environments are dominated by faster-growing plants (Aerts & Chapin, 2000). This relationship is hypothesised to be a significant factor underlying global trait variation (Díaz et al., 2016; Wright et al., 2004). Life history strategy (i.e. annual or perennial) is also an important axis of growth rate variation among wild species (Garnier, 1992; Grime & Hunt, 1975), and the growth rate differences between annuals and perennials are well documented (Houghton, Thompson, & Rees, 2013). Annuals grow quickly in order to make a high investment in reproduction during their single year of life, whereas perennials grow more slowly to conserve resources for future years (Bennett, Roberts, & Wagstaff, 2012; Garnier, 1992; Iwasa, 2000). Therefore, annuals generally have a higher relative growth rate (RGR) than perennials, which is especially clear when congeneric species are compared (Garnier, 1992). Despite the dominance of slow growers in nutrient-poor environments (Aerts & Chapin, 2000), fast-growing species still grow faster than slow-growing ones in infertile soil (Campbell & Grime, 1992). Fast-growers have a high nutrient uptake capacity and a greater flexibility to alter their uptake capacity in response to nutrient availability, that is, a greater physiological plasticity (Aerts & Chapin, 2000; Garnier, Koch, Roy, & Mooney, 1989).

This study compares the responses of two barley species to a nutrient gradient, in order to investigate how nutrient availability and uptake limit vegetative growth. It aims to elucidate the relative importance of indirect limitation via carbon source strength, by measuring carbohydrate content and amino acid:sucrose ratio, and direct limitation of sink tissue expansion growth, by measuring relative growth rate (RGR) and tissue composition. The work compares an elite fast-growing domesticated annual spring malting barley (*Hordeum vulgare* cv. NFC Tipple) with a slower-growing wild perennial relative (*Hordeum bulbosum*). By working with species with different life history strategies, this approach uses pre-existing variation in growth rate to probe the nature of the annual crop system. Previous work on these species during the vegetative growth stage shows significant carbon sink limitation of growth in the annual, as evidenced by a lack of plasticity of photosynthesis and allocation, and a weak growth response to elevated CO_2 concentration (Burnett et al., 2016). In contrast, the perennial exhibits carbon source limitation, and is able to increase sink development and utilize the additional carbon available from photosynthesis under elevated CO_2 (Burnett et al., 2016). This study investigates the role of nutrient limitation in this system, testing the alternative hypotheses that: (a) nutrients (primarily nitrogen) limit photosynthesis, thereby causing carbon limitation of growth or (b) that nutrients directly limit expansion growth. It also tests the hypothesis (c) that vegetative growth in the fast-growing annual is more limited by its ability to take up soil nutrients (source limitation), while the slow-growing perennial is more limited by its ability to utilize these nutrients (sink limitation).

2 | MATERIALS AND METHODS

2.1 | Plant material, growth conditions, and nutrient treatment

Seeds of *Hordeum vulgare* cv. NFC Tipple from the UK HGCA (2014) recommended list and *Hordeum bulbosum* (Accessions GRA1031 and GRA947) from Turkey (von Bothmer, 1996) were obtained from Syngenta and IPK Gatersleben, respectively. Data were collected until 42 days after germination, in order to focus measurements on the period of maximal vegetative growth, since previous work had found that maximum RGR occurs approximately 28 days after germination during the vegetative growth stage in these species (Burnett et al., 2016). Seeds were first germinated on wet filter paper, then transplanted to 4-L pots filled with a 1:10 sand:vermiculite mix topped with an additional layer of sand to aid seedling root development. This mix was designed to provide a very low-nutrient, nitrogen-free substrate to which varying levels of nutrient solution could be added.

Plants were grown at the University of Sheffield in controlled environment plant growth chambers (BDR 16, Conviron, Isleham, UK), modified to scrub CO_2 using soda lime to achieve the current (2015) ambient atmospheric level of 400 μmol/mol CO_2. Plants were randomised between three chambers, with the following growth conditions: 12-hr photoperiod with day/night temperatures of 20/18°C, 65% relative humidity, 400 μmol/mol CO_2, and daytime light levels of 600 μmol photons m^{-2} s^{-1} to provide a daily light integral of 25.9 mol m^{-2} day^{-1}.

For the first week, plants were watered daily with Reverse Osmosis water. Thereafter, plants were watered three times per week with 250 ml Long Ashton nutrient solution, applied at different concentrations (nutrient treatments): "low nutrients" (1% of Long Ashton stock solution), "medium nutrients" (20% of stock), and "high nutrients" (100% of stock). One hundred percent Long Ashton solution contains 4.00 mM potassium nitrate, 3.92 mM calcium nitrate tetrahydrate, 1.73 mM sodium dihydrogen orthophosphate, 1.49 mM magnesium sulphate heptahydrate, 0.09 mM ethylenediaminetetraacetic acid ferric monosodium salt, 1.92 μM manganese sulphate tetrahydrate, 0.35 μM zinc sulphate, 0.19 μM copper sulphate pentahydrate, 9.65 μM boric acid, 0.11 μM sodium molybdate, and 19.27 μM sodium chloride, in Reverse Osmosis water. Long Ashton's solution provides mineral nutrients in proportions appropriate to the requirements of the plant, so the treatment altered the application of all mineral nutrients in accordance with these proportions. Plants did not display visible signs of mineral deficiency or toxicity (Supporting Information Figure S1).

2.2 | ^{15}N uptake

Nitrate labelled with the ^{15}N stable isotope ($^{15}NO_3^-$) was fed 24 hr before the final harvest at 42 days after germination. A ^{15}N-labelled potassium nitrate and calcium nitrate (10% labelled atom) were substituted for the same mass of unlabelled potassium nitrate and calcium nitrate in the Long Ashton nutrient solution, such that the plants received their usual dose of all nutrients.

A subsample of the ground, freeze-dried material obtained for metabolite measurements (see below) was analysed for ^{15}N enrichment using an ANCA GSL 20-20 Mass Spectrometer (Sercon PDZ Europa, Crewe, UK). Enrichment was calculated as $^{15}N/(^{14}N + ^{15}N)$ and adjusted for percentage labelled atom fed, and values from control samples which did not receive ^{15}N were subtracted from these data in order to give ^{15}N enrichment relative to the baseline level for each organ type and nutrient treatment (baseline ^{15}N of control samples did not differ between the two species).

2.3 | Growth and allocation

Shoot area was obtained by photographing plants twice per week starting 8 days after germination, using the method described by Burnett et al. (2016). For very young plants, only one photograph was required, whereas six photographs from evenly spaced angles were used for larger plants. Shoot area was calibrated with whole-plant dry mass using a batch of 29 additional plants, not used in the main study, which were also photographed twice weekly. These additional individuals were harvested and oven-dried 12, 19, 26, 33, and 40 days after germination in order to calibrate shoot area to leaf area and dry mass (r^2 = 0.88), and to analyse partitioning to different organs. Biomass data for plants in the main study were predicted using this calibration. Mass-based relative growth rate (RGR) was obtained for each individual in the main study by converting non-destructive projected area measurements to whole-plant dry mass using the calibration from the additional batch of destructively harvested plants and calculating the change in the natural logarithm of dry mass over time. Tillers were counted on all plants in the main study immediately prior to metabolite harvests.

Dry mass fractions of leaf, leaf sheath, and root were calculated for each oven-dried individual from the additional batch of plants and used to calculate leaf mass ratio (LMR; kg leaf dry mass per kg plant dry mass). Mean LMR values for each species and nitrogen concentration were then used to predict LMR for plants in the main study. Specific leaf area (SLA; m^2 leaf per kg leaf) was recorded for all plants in the main study by measuring the area and fresh mass of harvested leaves, and converting fresh mass into dry mass using the calibration obtained from oven-dried plants. Leaf area ratio (LAR; m^2 leaf per kg plant) for each individual in the main study was obtained by multiplying SLA by predicted LMR.

Data from the additional subset of 29 destructively harvested and oven-dried plants were used only for LMR (Table 1) and for calibrating RGR and LMR for plants in the main study. All other data were obtained from the plants in the main study.

2.4 | Metabolites

Metabolite harvests were carried out on plants from the main study. Harvests were carried out at 14, 28, and 42 days after germination; all these timepoints were within the vegetative growth stage. Plants were always watered 24 hr before harvest so that all samples would be equivalent with respect to watering schedule. For each nutrient level and species, one replicate from each of the three growth chambers was taken, giving a total of 18 plants at each timepoint. Due to the small size of plants harvested 14 days after germination, it was necessary to pool two individuals of the same species and nutrient treatment for each replicate, to obtain enough material for analysis. Plants were harvested within 1 hr before dawn, in order to capture pre-dawn metabolite levels, when the diurnal concentrations of carbohydrate are expected to be minimal, following previous work (Burnett et al., 2016). Plants were separated into leaf, leaf sheath, and root, flash-frozen in liquid nitrogen and stored at −80°C; samples were subsequently freeze-dried prior to analysis. For small plants, the entire plant was harvested; for larger plants, representative samples of each organ from both young and old tissue were harvested. Metabolite assays were carried out as described previously (Burnett et al., 2016): glucose, fructose, sucrose, low and high degree of polymerisation (LDP and HDP) fructans, and starch were quantified using continuous enzymatic substrate assays, whilst nitrate, amino acids, and proteins were quantified using the Griess reaction, fluorescamine assay, and bicinchoninic acid assay, respectively. Elemental CHN was analysed using a 2400 Series II CHN analyser (PerkinElmer, Waltham, MA, USA). More samples were available for CHN analysis than for metabolite assays, due to the small size of some plants. Metabolites and CHN were expressed per gram carbohydrate-corrected dry weight (CCDW), obtained by subtracting the total mass of non-structural carbohydrate (the sum of: glucose, fructose, sucrose, LDP and HDP fructan, and starch) from the dry mass of each sample. Metabolite data are available in Supporting Information Table 1.

TABLE 1 Root:shoot, leaf mass, sheath mass, and root mass ratios in annual and perennial barley at different nutrient treatment levels

	Species	Low nutrients	Medium nutrients	High nutrients
Root:shoot ratio	Annual	1.60 ± 0.2	0.80 ± 0.2	0.54 ± 0.1
	Perennial	1.54 ± 0.3	0.83 ± 0.2	0.59 ± 0.07
Leaf mass ratio	Annual	0.33 ± 0.03	0.42 ± 0.03	0.46 ± 0.01
	Perennial	0.27 ± 0.01	0.35 ± 0.01	0.44 ± 0.04
Sheath mass ratio	Annual	0.07 ± 0.004	0.17 ± 0.03	0.24 ± 0.02
	Perennial	0.14 ± 0.05	0.20 ± 0.07	0.19 ± 0.02
Root mass ratio	Annual	0.60 ± 0.03	0.41 ± 0.06	0.30 ± 0.01
	Perennial	0.59 ± 0.05	0.45 ± 0.05	0.37 ± 0.03

Allocation to roots increases at lower nutrient supplies ($p < 0.001$) and allocation to leaves and sheaths decreases at lower nutrient supplies ($p < 0.001$). These data are obtained from the subset of 29 additional plants harvested for biomass calibration described in the Materials and Methods section. For root:shoot ratio, root mass is divided by the sum of leaf and sheath mass for each individual and the results averaged. For leaf, sheath, and root mass ratios, the ratio is the dry mass of that organ divided by the dry mass of the whole plant. 18 annuals and 11 perennials were harvested at five timepoints. Data show mean ± *SE* (at low, medium, high nutrients, annual n = 6, 6, 5; perennial n = 4, 2, 5). For brevity, age effects have not been included in this summary, but they are discussed elsewhere in the manuscript.

2.5 | Statistical methods

All data were analysed in R (R Core Team, 2015) using Type II ANOVA. Natural logarithmic transformations were performed prior to analysis to satisfy the normality assumptions of ANOVA.

3 | RESULTS

Net nitrogen uptake (Figure 1) decreases significantly as nutrient treatment is lowered ($F_{2,10} = 54.6$, $p < 0.001$), but there is no significant species effect (either main effect or interaction, $p > 0.5$ in each case) on net nitrogen uptake (Figure 1). Relative growth rate (RGR; Figure 2) is consistently higher in annual than in perennial barley, suggesting that the annual invests more resources into growth than the perennial, even at low nutrient levels (significant effect of species: $F_{1,11} = 9.77$, $p < 0.01$). This shows that the nutrient use efficiency of growth is higher in the annual. RGR decreases significantly in annual and perennial barley when nutrient treatment is lowered

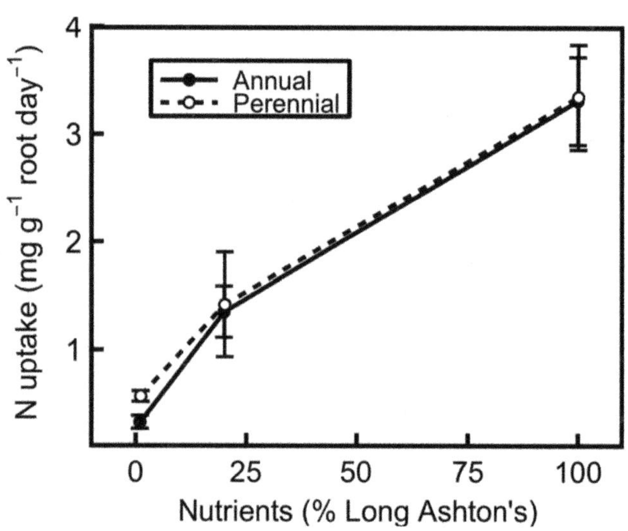

FIGURE 1 Net nitrogen uptake rate of annual (filled circles, solid line) and perennial barley (hollow circles, dashed line) decreases as nutrient treatment level is lowered ($p < 0.001$). Uptake was measured over 24 hr before harvesting plants 42 days after germination. Data show mean ± *SE* (n = 3)

($F_{2,11}$ = 7.98, $p < 0.01$), and this effect is especially strong when nutrients are decreased from the medium to low treatment level.

Biomass partitioning is also affected by nutrient treatment. In accordance with the change in RGR, tillering decreases substantially at low nutrient levels (Figure 3); this effect is greater in older plants (significant nutrient × age interaction for tillering: $F_{4,24}$ = 13.1, $p < 0.001$). However, whilst the nutrient effect is strong, there is neither a difference between species nor a nutrient × species interaction ($p > 0.6$ in each case). Allocation to roots (Table 1) increases at lower nutrient supplies (for root:shoot ratio, $F_{1,19}$ = 40.6, $p < 0.001$; for root mass ratio (RMR), $F_{1,19}$ = 38.9, $p < 0.001$), but does not differ between species. In both species, sheath mass ratio and leaf mass ratio (SMR and LMR; Table 1) decrease at low nutrient levels (for sheaths, $F_{1,19}$ = 25.9, $p < 0.001$; for leaves, $F_{1,19}$ = 30.1, $p < 0.001$). The annual allocates more biomass to roots earlier in development and more to sheaths later in development whilst the opposite trend is seen in the perennial (significant species × age interactions: for SMR, $F_{1,19}$ = 16.8, $p < 0.001$; for root:shoot ratio, $F_{1,19}$ = 10.6, $p < 0.01$).

Leaf area ratio (LAR) does not differ between species, although specific leaf area (SLA) is slightly higher in the perennial (Table 2; SLA: $F_{1,45}$ = 5.97, $p < 0.05$). LAR and SLA decrease with age in both species (LAR: $F_{1,45}$ = 34.6, $p < 0.001$; SLA: $F_{1,45}$ = 42.2, $p < 0.001$), and LAR decreases at lower nutrient treatment levels ($F_{1,45}$ = 56.6, $p < 0.001$), in addition to the decrease in LMR documented above (Tables 1 and 2). Thus at lower nutrient treatment levels, proportionately less biomass is allocated to leaves, and there is a lower leaf area per unit of plant biomass.

Annual plants have a greater absolute size than perennials (Table 3). Predicted biomass increases with age, especially in annuals (species × age interaction: $F_{1,415}$ = 25.1, $p < 0.001$) and especially at higher nutrient levels (nutrient × age interaction: $F_{2,415}$ = 48.3, $p < 0.001$). For brevity, only predicted biomass data collected before each destructive harvest are presented in Table 3, but the effects reported here are for all predicted biomass data (obtained twice weekly).

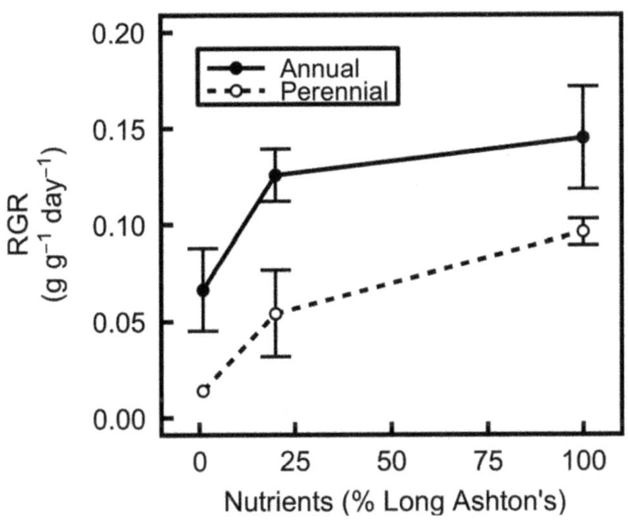

FIGURE 2 Relative growth rate (RGR, g g^{-1} day^{-1}) is higher in annual (solid line) than in perennial (dashed line) barley ($p < 0.01$). Data are from plants harvested 28 and 42 days after germination. Data show mean ± *SE* (for annuals, n = 5, 6, 4, and for perennials, n = 1, 3, 3 at low, medium, and high nutrients, respectively)

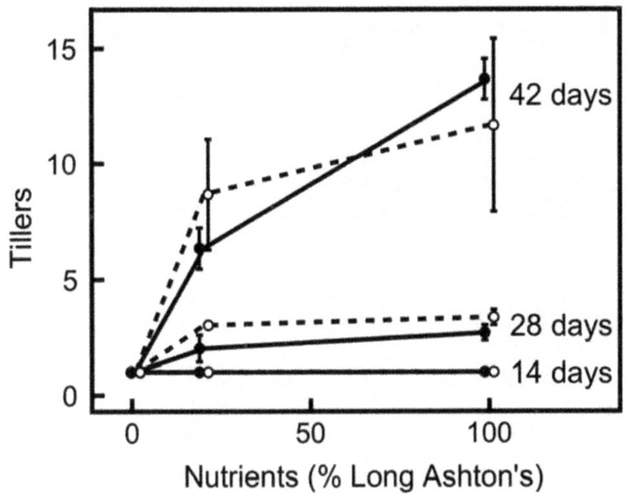

FIGURE 3 Tillering increases with nutrient treatment and this effect increases with age in both annual (filled circles, solid line) and perennial (hollow circles, dashed line) barley ($p < 0.001$). Data show mean ± *SE* (n = 3), for plants harvested 14, 28, and 42 days after germination; at all three ages, plants at low nutrient treatment have only one tiller. Points are offset with respect to x-axis position, to increase readability

TABLE 2 Leaf area ratio (LAR) and specific leaf area (SLA) in annual and perennial barley grown at different nutrient levels and harvested 14, 28, and 42 days after germination

	Species	Low nutrients	Medium nutrients	High nutrients
LAR at 14 days	Annual	11.3 ± 0.5	14.7 ± 1	20.0 ± 2
	Perennial	9.5 ± 0.8	16.1 ± 0.7	19.9 ± 2
LAR at 28 days	Annual	12.0 ± 0.2	13.7 ± 0.5	17.3 ± 1
	Perennial	12.3 ± 3	13.6 ± 1	16.1 ± 2
LAR at 42 days	Annual	9.8 ± 0.1	10.3 ± 0.6	12.3 ± 0.2
	Perennial	7.7 ± 0.4	10.6 ± 1	13.2 ± 0.3
SLA at 14 days	Annual	34.3 ± 1	35.0 ± 3	43.4 ± 4
	Perennial	35.1 ± 3	46.1 ± 2	45.3 ± 4
SLA at 28 days	Annual	36.5 ± 0.7	32.6 ± 1	37.6 ± 2
	Perennial	45.7 ± 12	38.9 ± 3	36.6 ± 4
SLA at 42 days	Annual	29.6 ± 0.3	24.6 ± 1	26.8 ± 0.4
	Perennial	28.6 ± 2	30.3 ± 3	30.1 ± 0.7

LAR decreases as nutrient treatment level is lowered ($p < 0.001$). SLA is higher in perennial than annual barley overall ($p < 0.05$). Both LAR and SLA are lower in older plants ($p < 0.001$). These data are for plants in the main study. LAR is expressed here as m^2 leaf/kg plant and SLA is expressed here as m^2 leaf/kg leaf. LAR is the product of SLA (measured on all plants in the main study at the time of metabolite harvest) and LMR (from the subset of plants harvested for biomass calibration; see Table 1). Data show mean ± *SE* (n = 3).

TABLE 3 Predicted biomass in annual and perennial barley grown at different nutrient levels, at 12, 22, and 40 days after germination

	Species	Low nutrients	Medium nutrients	High nutrients
12 days	Annual	0.034 ± 0.004	0.042 ± 0.01	0.052 ± 0.01
	Perennial	0.017 ± 0.003	0.062 ± 0.03	0.022 ± 0.003
22 days	Annual	0.042 ± 0.01	0.172 ± 0.02	0.228 ± 0.03
	Perennial	0.023 ± 0.01	0.056 ± 0.01	0.070 ± 0.01
40 days	Annual	0.149 ± 0.01	0.630 ± 0.09	1.367 ± 0.07
	Perennial	0.032 ± 0.01	0.236 ± 0.09	0.492 ± 0.2

Predicted biomass increases with age especially in annual barley, and especially at higher nutrient treatment levels ($p < 0.001$ for each of these interactions). These data are for plants in the main study. Biomass (g) is predicted from leaf area data, using the correlation obtained for the subset of additional plants harvested for biomass calibration. Predicted biomass is derived from photographs preceding destructive harvests (carried out at 14, 28, and 42 days after germination). Data show mean ± *SE*, with n decreasing over time due to destructive harvests (at 12 days, at low, medium, high nutrients, annual n = 17, 18, 18; perennial n = 7, 13, 15; at 22 days, at low, medium, high nutrients, annual n = 10, 10, 10; perennial n = 6, 6, 9; at 40 days, at low, medium, high nutrients, annual n = 5, 5, 5; perennial n = 3, 3, 4).

Overall, a decrease in nutrient treatment leads to a decrease in the concentration of nitrogen metabolites in annual and perennial barley (Figure 4), especially when nutrient treatment is decreased from medium to low. A comparatively smaller decrease is observed between high and medium nutrient levels, similar to the response of RGR to nutrient treatment (Figures 2 and 4). Nitrate and amino acids generally show a response to nutrient treatment level across the plant. In annual and perennial barley, with decreasing nutrient treatment, there is a significantly lower nitrate concentration in leaf ($F_{2,24}$ = 64.5, $p < 0.001$), leaf sheath ($F_{1,11}$ = 17.9, $p < 0.01$), and root ($F_{2,25}$ = 182.7, $p < 0.001$). Nitrate in the sheath tissue is significantly higher in the annual ($F_{1,11}$ = 45.9, $p < 0.001$). Amino acid concentration in leaf and root also decreases at lower nutrient treatment levels, and this effect is greater in older plants (nutrient treatment × age interaction in leaf: $F_{4,24}$ = 12.1, $p < 0.001$; in root: $F_{4,25}$ = 7.9, $p < 0.001$). Amino acid concentration in the sheath is not affected by nutrient treatment, but increases with age (Figure 4d,e; $F_{1,11}$ = 20.8, $p < 0.001$). In the root, amino acid content is higher in perennial than annual barley ($F_{1,24}$ = 26.1, $p < 0.001$). In contrast to nitrate and amino acids, protein does not show a whole-plant response to nutrient treatment level. Protein concentration shows an overall decrease from high to low nutrient treatment levels in the leaf (Figure 4; $F_{2,25}$ = 13.9, $p < 0.001$), but does not show a significant change in sheath or root.

TNC in the leaf is significantly higher in the annual ($F_{1,24}$ = 14.8, $p < 0.001$), as would be expected if the annual is more carbon sink limited than the perennial. Leaf TNC is also affected by nutrient treatment ($F_{2,24}$ = 6.1, $p < 0.01$): leaf TNC increases when nutrient treatment is lowered from medium to low in younger plants (Figure 5a), yet decreases at low nutrient levels in older plants (Figure 5b,c). However, contrary to expectations that carbon source limitation would increase at low nutrient levels, leaf TNC increases when nutrient treatment is lowered from high to medium, that is, leaf TNC accumulates more at medium nutrient treatment levels than at high levels. This indicates that carbon source limitation does not increase when nutrient treatment is decreased. Rather, carbon source limitation is greater at high nutrient treatment levels. Root TNC shows a significant nutrient treatment × age interaction ($F_{4,25}$ = 9.3, $p < 0.001$): the increase in TNC at lower nutrient treatment is most pronounced 14 days after germination (Figure 5f).

Elemental N content (Figure 6a–c) decreases at lower nutrient treatment level in leaf (nutrient treatment × age interaction: $F_{4,24}$ = 6.5, $p < 0.01$), sheath (nutrient treatment effect: $F_{1,11}$ = 50.8, $p < 0.001$), and root (nutrient treatment × age interaction: $F_{4,25}$ = 9.0, $p < 0.001$). In the leaf, elemental N content increases with age especially at medium nutrient levels; in the root, elemental N increases with age especially at low nutrient levels. Elemental N is also higher in the perennial in leaf tissue ($F_{1,24}$ = 10.8, $p < 0.01$), but there is no species × nutrient treatment interaction. Carbon:nitrogen (C:N) ratio (Figure 6d–f) increases with lowering nutrient treatment in leaf (nutrient treatment × age interaction: $F_{4,24}$ = 7.2, $p < 0.001$), sheath (nutrient treatment effect: $F_{1,11}$ = 34.5, $p < 0.001$), and root (nutrient treatment × age interaction: $F_{4,25}$ = 9.2, $p < 0.001$). In both leaf and root, C:N ratio decreases as plant age increases and this effect is strongest at low and medium nutrient treatment levels. Furthermore, C:N ratio in the leaf is higher in annual barley ($F_{1,24}$ = 10.7, $p < 0.01$), and C:N ratio in the sheath is higher in perennial barley ($F_{1,11}$ = 14.0, $p < 0.01$).

A high amino acid:sucrose ratio indicates that plants are carbon source limited. Perennial barley has a higher amino acid:sucrose ratio than the annual in leaf ($F_{1,22}$ = 30.8, $p < 0.001$) and root ($F_{1,25}$ = 43.6, $p < 0.001$; Figure 7), indicating greater carbon source limitation in perennial than annual barley. There was no significant effect of age on this ratio. Contrary to hypothesis (1), root amino acid:sucrose ratio increases at higher nutrient treatment levels ($F_{2,25}$ = 64.9, $p < 0.001$), especially between low and medium nutrient levels (Figure 7b), suggesting that carbon source limitation is increased rather than decreased at high nutrient levels, rather than carbon source limitation increasing at low nutrient levels.

4 | DISCUSSION

In this study, annual and perennial barley were grown along a nutrient gradient to examine the processes through which mineral nutrients limit growth (hypotheses 1, 2) and elucidate the relative contributions of nutrient source and sink strengths to growth in each species (3). The work focused on measurements of nitrogen uptake and concentration, since nitrogen is the primary mineral nutrient. There are multiple lines of evidence consistent with the hypothesis that nutrients directly limit expansion growth (2) rather than causing an indirect carbon limitation via photosynthesis (1). If low nutrient levels were causing carbon source limitation, via a lack of photosynthetic proteins, TNC would be expected to rise with increasing

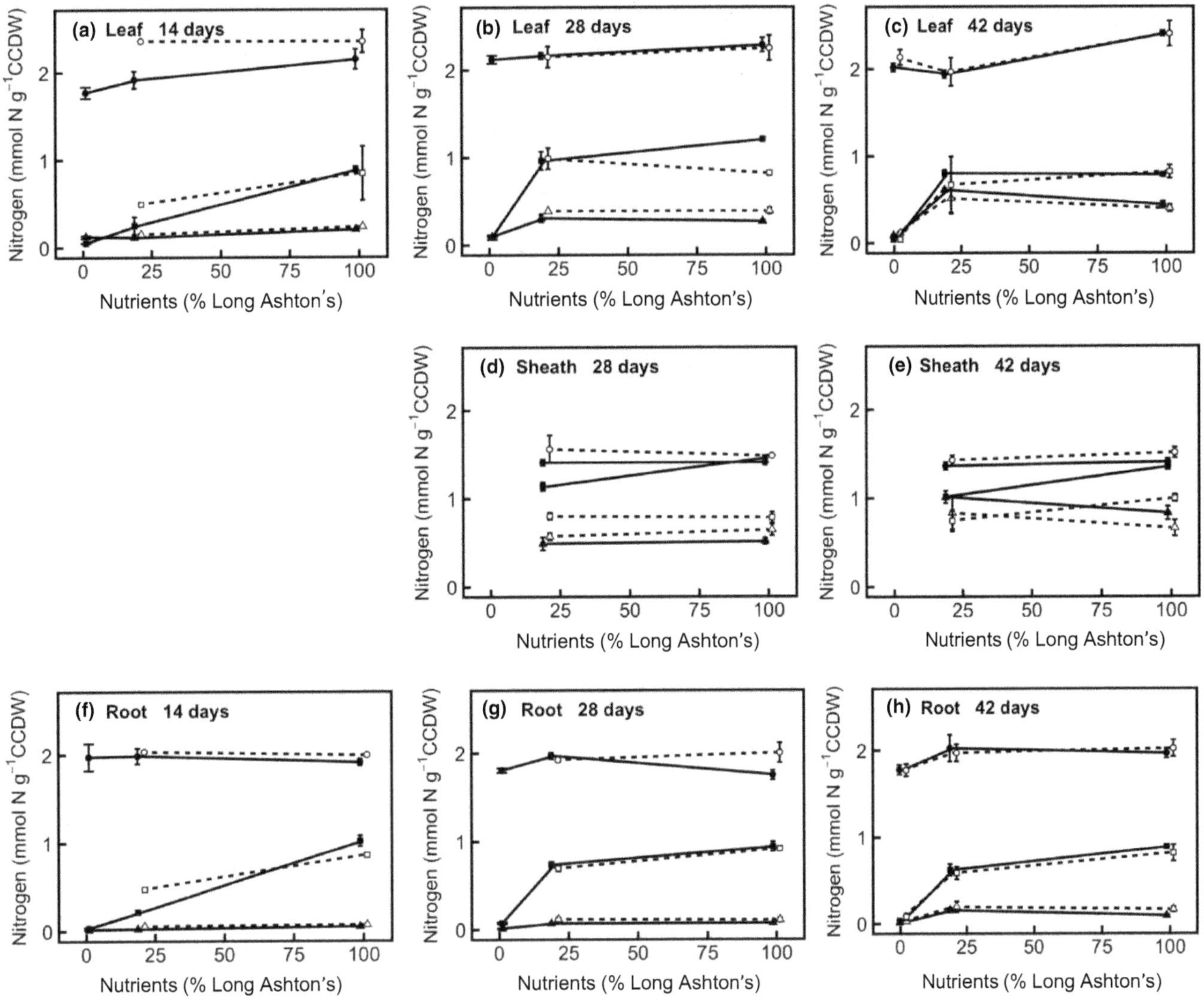

FIGURE 4 Partitioning of nitrogen resources in different organs of annual (filled symbols, solid line) and perennial (hollow symbols, dashed line) barley. Statistics are outlined in the main text. Panel shows allocation of nitrogen to protein (filled and hollow circles), free nitrate (filled and hollow squares), and free amino acids (filled and hollow triangles). All data are expressed in mmol N per gram carbohydrate-corrected dry weight (CCDW). CCDW is obtained by subtracting the mass of total non-structural carbohydrate (the sum of: glucose, fructose, sucrose, LDP and HDP fructan, and starch) from the dry mass of each sample. (a) Leaf 14 days after germination; (b) leaf 28 days; (c) leaf 42 days; (d) leaf sheath 28 days; (e) leaf sheath 42 days; (f) root 14 days; (g) root 28 days; (h), root 42 days. Leaf and leaf sheath tissues were pooled for the harvests at 14 days. Data show mean ± *SE* (n = 3). Points are offset with respect to *x*-axis position, to increase readability

nutrient level. However, this is not always the case (Figure 5), even though leaf protein shows a small increase in the higher nutrient treatment levels (Figure 4), indicating that carbon source limitation is not alleviated by nutrients. In addition, amino acid:sucrose ratio, another indicator of carbon source limitation, increases in the root at higher nutrient treatment levels, further indicating that carbon source limitation increases at high nutrient levels rather than decreasing (Figure 7b). Consistent with this hypothesis, carbon becomes an increasingly limiting resource at higher nutrient treatment levels, and the leaf and sheath mass ratios in both species increase to compensate for this effect (Table 1). Taken together, this evidence shows that the nutrient limitation on growth is mediated by a direct constraint on expansion growth (hypothesis (2)) rather than acting via carbon limitation (hypothesis (1)).

Regarding hypothesis (3), the investment of nitrogen into growth and storage shows large decreases when nutrient treatment is decreased from a medium to a low supply, but this response is smaller when nutrients are decreased from a high to a medium nutrient supply, especially in annual barley. These results indicate a significant nutrient source limitation at low nutrient levels, but a nutrient sink limitation at high nutrient levels (Figures 2, 4 and 6). The relatively small difference in nutrient investment between medium and high nutrient levels shows that plants are nutrient sink limited under medium nutrient treatment levels. Although the annual invests more resources in overall growth (Figure 2, Table 3), the perennial is better able to increase investment in growth under high nutrient levels, in relative terms, despite its smaller size (Figure 2). This result suggests that growth in the perennial is less limited by sink development than

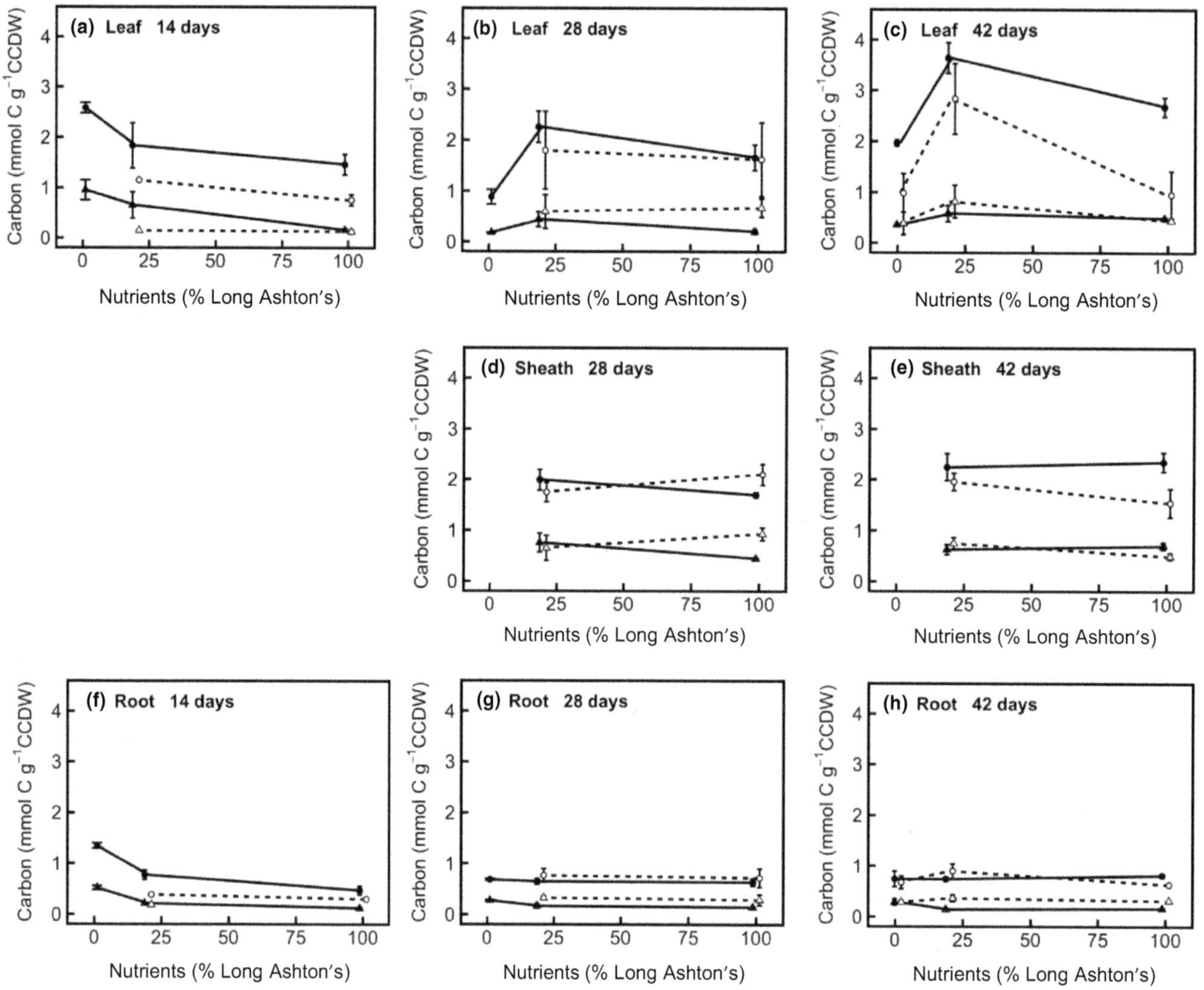

FIGURE 5 Partitioning of total non-structural carbohydrates (TNC) in different organs of annual (filled symbols, solid line) and perennial (hollow symbols, dashed line) barley. Statistics are outlined in the main text. Panel shows TNC (filled and hollow circles) and the proportion of TNC allocated to fructans (filled and hollow triangles). TNC is the sum of glucose, fructose, sucrose, and LDP and HDP fructans. All data are expressed in mmol C per gram carbohydrate-corrected dry weight (CCDW). CCDW is obtained by subtracting the mass of total non-structural carbohydrate (the sum of: glucose, fructose, sucrose, LDP and HDP fructan, and starch) from the dry mass of each sample. (a) Leaf 14 days after germination; (b) leaf 28 days; (c) leaf 42 days; (d) leaf sheath 28 days; (e) leaf sheath 42 days; (f) root 14 days; (g) root 28 days; (h) root 42 days. Leaf and leaf sheath tissues were pooled for the harvests at 14 days. Data show mean ± *SE* (n = 3). Points are offset with respect to *x*-axis position, to increase readability

growth in the annual, so improving nutrient sink strength will be important for improving the growth and yield of annual barley.

4.1 | Perennial barley shows a greater growth response to nutrients

Annual and perennial barley have the same net nitrogen uptake rate (Figure 1), and both show a substantial decrease in nitrogen uptake with decreasing nutrient treatment level (Figure 1), consistent with expectations. As the nutrient treatment decreases from medium to low nutrient supply, plants show large growth and storage responses, with both the amount of plant material (Figure 2) and the concentration of nitrogen (Figure 4) decreasing significantly.

Annual barley grows faster overall (Figure 2) and is larger than the perennial (Table 3) indicating higher nutrient use efficiency of growth. RGR is more strongly limited by nutrient supply in the perennial, suggesting that the lower nutrient efficiency of growth in the perennial leads to stronger nutrient limitation when nutrient supply is restricted. The perennial shows a relatively greater growth response to nutrient level than the annual (Figure 2) and has a higher leaf nitrogen concentration (Figure 6a). This suggests that, under nitrogen-limited conditions (i.e. the low nutrient treatment), the perennial preferentially allocates nitrogen to storage rather than growth, consistent with the hypothesis that the perennial will conserve mineral nutrients (Campbell & Grime, 1992). However, contrary to the hypothesis that growth in the perennial would be more

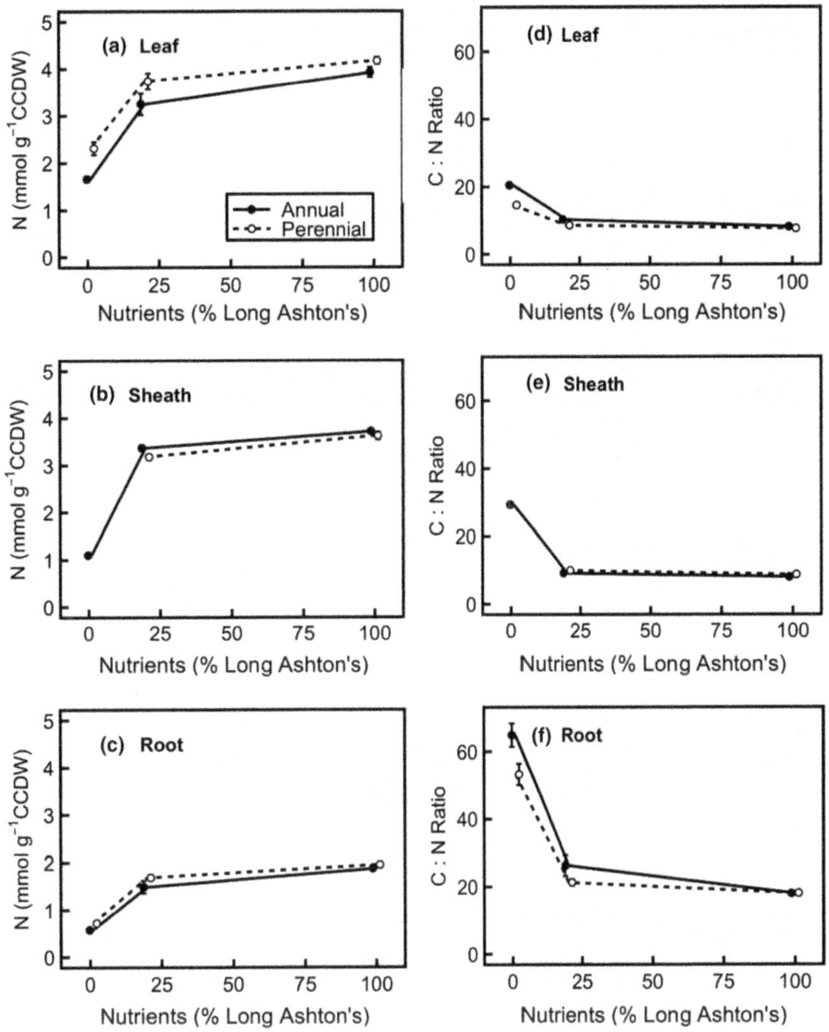

FIGURE 6 Total nitrogen concentration and carbon:nitrogen (C:N) ratio in annual (filled circles, solid line) and perennial (hollow circles, dashed line) barley. Statistics are outlined in the main text. Carbohydrate-corrected dry weight (CCDW) was used for these elemental N and C concentrations. CCDW is obtained by subtracting the mass of total non-structural carbohydrate (the sum of: glucose, fructose, sucrose, LDP and HDP fructan, and starch) from the dry mass of each sample. (a) Leaf nitrogen; (b) sheath nitrogen; (c) root nitrogen; (d) leaf C:N ratio; (e) sheath C:N ratio; (f) root C:N ratio. Data show mean across all ages (14, 28, 42 days after germination) due to lack of significant age effects ± *SE* (at low, medium, high nutrients, annual n = 9, 9, 9 in leaf and root and 1, 6, 6 for sheath; perennial n = 4, 7, 9 for leaf, 0, 5, 5 for sheath, and 5, 7, 9 for root). There are more samples for CHN analysis than for metabolite analysis due to the small size of some samples. Points are offset with respect to x-axis position, to increase readability

nutrient sink limited and growth in the annual more nutrient source limited, the perennial shows a greater response to increased nutrient treatment. This indicates that growth in the perennial is in fact less nutrient sink limited than growth in the annual.

Contrary to expectations, the perennial has a higher SLA (Table 2) and higher leaf nitrogen concentration (Figure 6) than the annual—traits that are generally associated with fast-growing species in the ecological literature (Reich et al., 2003; Wright et al., 2004). This is consistent with previous work on this species (Burnett et al., 2016), which showed carbon source limitation of growth in the perennial, since it is investing in carbon acquisition by the leaves in order to match carbon and nitrogen supply. Indeed, the higher SLA and leaf nitrogen concentration observed here for the perennial indicate that it may have the potential to be a rather fast-growing species despite its perennial life history strategy. Potential RGR has previously been correlated with nitrogen uptake capacity (Garnier et al., 1989) and here the net nitrogen uptake rates are very similar for annual and perennial barley, although the perennial never matches the RGR of the annual. RGR (Figure 2) is higher in the annual, but LAR (the product of SLA and LMR) does not differ between species (Table 2). Whilst SLA tends to be the major contributor to LAR and thus RGR in herbaceous species (Poorter & van der Werf, 1998), some studies have found that RGR correlates with LMR rather than SLA (references within Garnier, 1992), which is consistent with these data. Furthermore, since RGR is the product of net assimilation rate (NAR) and leaf area ratio (LAR; Poorter & Remkes, 1990), the higher RGR of annual compared to perennial barley seen here is likely due to higher photosynthetic assimilation, as seen in previous work with this species (Burnett et al., 2016). Two further points of contention are: whether RGR regulates resource uptake or whether resource uptake is regulated by RGR (Garnier et al., 1989; Rodgers & Barneix, 1988); and the extent to which uptake is regulated by demand (Taulemesse, Le Gouis, Gouache, Gibon, & Allard, 2015). Indeed, nitrate itself is an important regulator of nitrogen uptake (Masclaux-Daubresse et al., 2010). In addition to elucidating the relative contribution of nutrient source and sink to growth, a deeper understanding of the molecular drivers that underpin regulation of nutrient uptake will be an important component of improving crop nutrient use efficiency.

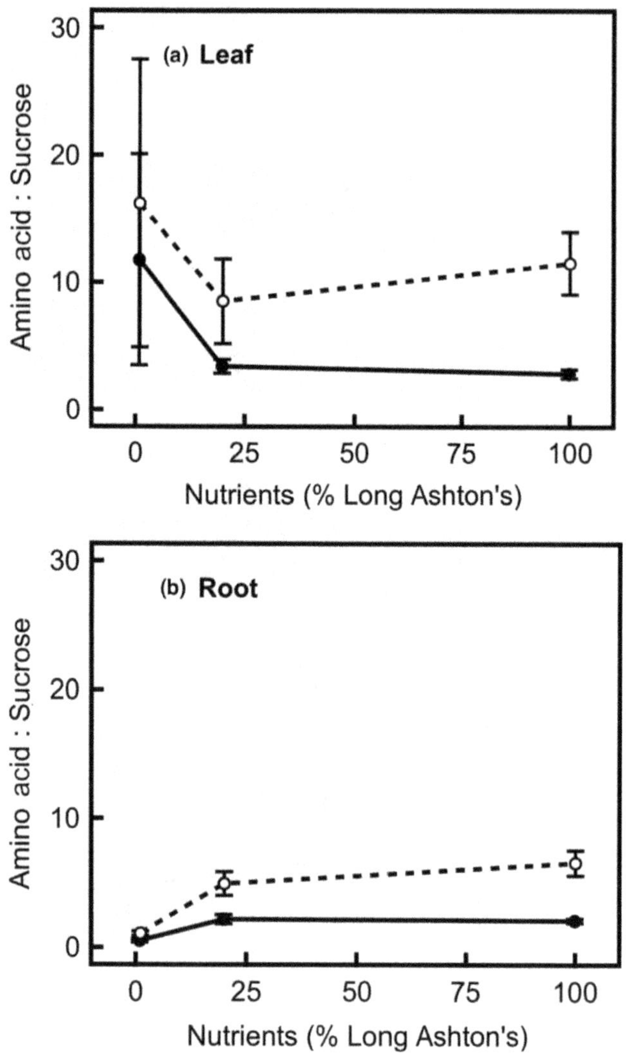

FIGURE 7 Amino acid:sucrose ratio is higher in perennial (hollow circles, dashed line) than in annual (filled circles, solid line) barley in both (a) leaf and (b) root ($p < 0.001$). In the root, amino acid:sucrose ratio increases with increasing nutrient treatment level ($p < 0.001$). Data show mean across all ages (14, 28, 42 days after germination) ± *SE* (in leaf at low, medium, high nutrients, annual $n = 6, 8, 9$, perennial $n = 9, 8, 9$; in root at low, medium, high nutrients, annual $n = 2, 7, 8$, perennial $n = 3, 6, 7$). Insufficient data were available for sheath

4.2 | Annual and perennial barley do not store excess nitrogen as protein

Nitrate is a labile store and therefore shows a particularly strong response to nutrient treatment (Figure 4); this metabolite also shows a strong response to nitrogen treatment in wheat (Devienne, Mary, & Lamaze, 1994) and *Arabidopsis* (Tschoep et al., 2009) and, like protein, constitutes a key store for nitrogen in herbaceous plants (Millard, 1988). Here, nitrogen is taken up from the soil as nitrate, which is readily accumulated in plant tissues without the energetic cost or carbon skeletons needed to reduce nitrate to amino acids. Accumulating nitrate therefore means that nitrogen storage can increase under high nutrient treatment levels, even if carbon is limiting. Such carbon limitation could arise quickly in young plants. Consistent with this idea, amino acid accumulation is greater in older plants (Figure 4). Compared to nitrate and amino acids, protein concentration remains relatively constant at all nutrient treatment levels, except in leaf tissue (Figure 4) where a small decrease occurs when nutrient treatment level decreases. Excess nitrogen is thus preferentially stored as nitrate (short term storage) or amino acids (intermediate storage), or invested in growth, rather than being used to elevate whole-plant protein concentration. The lack of an ontogenetic effect on leaf protein during vegetative growth is also notable (Figure 4): rather than regulating leaf protein concentration as leaves get older, barley plants create more leaf tissue and maintain the same protein concentrations; this contradicts the way in which many earth system models deal with nutrient limitation and is therefore an important avenue for further investigation.

4.3 | High nutrient levels increase carbon source limitation

Storage of nitrogen as nitrate and amino acids suggests that plants are carbon source limited as protein synthesis requires additional carbon. Both species show a decrease in N concentration and an increase in C:N ratio as the nutrient treatment level is lowered (Figure 6). This corresponds with a shift from excess nitrogen to excess carbon in the plants (Stitt & Krapp, 1999). However, this effect is observed when nutrient treatment is decreased from medium to low, but not between high and medium nutrient levels (Figure 6) despite a high nitrate availability and uptake rate, suggesting that plants are reaching their maximum capacity for nitrogen storage at high nutrient treatment levels, and are carbon source limited. Leaf TNC data (Figure 5) also show carbon source limitation at high nutrient levels.

The higher LMR and SMR in both species at higher nutrient treatment levels (Table 1) enables greater acquisition of carbon, which becomes an increasingly limiting resource at higher nutrient treatment levels; conversely, more biomass is allocated to roots in low nutrient environments (Aerts & Chapin, 2000). Not only does LMR increase with nutrient treatment in both species, but LAR also increases (Tables 1 and 2), as observed by Garnier et al. (1989), enabling greater photosynthetic carbon acquisition since there is a greater, thinner leaf area for light harvesting. This suggests an increase in carbon source limitation at high nutrient levels. Both species show an increase in tillering as nutrient treatment is increased (Figure 3), as observed in wheat by Taulemesse et al. (2015), and especially when plants are older, facilitating a rapid increase in allocation to shoots and thus enabling greater carbon acquisition.

In addition to these structural changes, and the differences in elemental carbon and nitrogen concentrations, root amino acid:sucrose ratio increases at higher nutrient treatment levels (Figure 7b). This indicates that carbon source limitation is increasing, since the available amino acids outsupply the corresponding supply of available carbon necessary to fuel growth (Isopp, Frehner, Long, & Nösberger, 2000; Paul & Driscoll, 1997; Stitt & Krapp, 1999). The increase is particularly pronounced between low and medium nutrient treatment levels. The higher amino acid:sucrose ratio in the perennial, and lower TNC concentration indicates that the perennial is more carbon source limited than the annual (Figures 5–7); this corroborates the evidence from previous work on these species (Burnett et al., 2016).

4.4 | Annual barley is more nutrient sink limited than perennial barley at moderate nutrient supply

In addition to the differences in the nutrient responses of each species, there are broad similarities revealed by the nutrient treatment. The treatment conditions are sufficiently strong that, at low nutrient treatment, growth in both annual and perennial barley is strongly nutrient source limited, as shown by high C:N ratios (Figure 6d–f), low nutrient uptake rates (Figure 1) and low growth rates compared to medium nutrient levels (Figure 2). At high nutrient levels, growth in the annual is more nutrient sink limited than the perennial, shown by its lower relative ability to increase growth (Figure 2). Growth at the medium nutrient supply appears to be more nutrient sink limited in annual than perennial barley, since increasing the nutrient level further has a much greater effect in the perennial.

In general, the higher RGR of annual plants arises from their large investment in leaf area and photosynthetic capacity: specific leaf area (SLA, mm^2 leaf per gram leaf), nitrogen content, and partitioning of that nitrogen to the photosynthetic machinery are higher in the leaves of annuals, enabling greater carbon acquisition due to a larger light-harvesting area per unit leaf mass and a greater nitrogen density associated with the photosynthetic machinery, thus facilitating high rates of carbon assimilation and faster growth (Garnier & Laurent, 1994; Grime et al., 1997; Pierce, Brusa, Vagge, & Cerabolini, 2013; Poorter & van der Werf, 1998). In contrast, perennials tend to have a lower SLA and nitrogen concentration, and invest more resources in the construction of robust, long-lived leaves. Plants with lower SLA may also invest proportionately less leaf nitrogen in the photosynthetic machinery (Hikosaka, Hanba, Hirose, & Terashima, 1998).

The nutrient sink limitation uncovered here for annual barley at medium nutrient treatment levels implies that barley crops are unable to invest excess nutrients into growth and storage during the vegetative stage. Although nitrogen is taken up and stored as nitrate, the subsequent reduction of nitrate to organic forms of nitrogen is lacking, coupled with a lack of proportional increases in expansion growth to create a nutrient sink. Both the efficiency of nutrient acquisition and the efficiency of nutrient utilisation are important for breeders (Santa-Maria, Moriconi, & Oliferuk, 2015), such that sink development in addition to source strength is vital for realising improved crop productivity (Burnett et al., 2016; White et al., 2016). Regarding the key mineral nutrient nitrogen, nitrogen transporters have been identified as a key target for improving the nitrogen source:sink balance (Tegeder & Masclaux-Daubresse, 2018), whilst nitrogen allocation patterns are important for nitrogen use efficiency and yield (Perchlik & Tegeder, 2018). Additional nutrient storage, in order to build up nutrient reserves for subsequent grain filling, would require larger nutrient sinks—such as increased capacity for expansion growth—to develop during the vegetative growth stage. These could allow farmers to reduce the dosage level of fertiliser application later in development and still increase crop yield (by increasing grain size and number rather than by increasing grain nitrogen concentration), which is of interest for breeders and farmers working with malting barley (Syngenta Crop Protection, 2011). The source: sink balance of the primary mineral nutrient nitrogen is important for crop improvement (Sonnewald & Fernie, 2018), and this element is of global ecological importance (Taylor & Menge, 2018).

5 | CONCLUSIONS

Concurrent measurements of growth and metabolite concentrations during barley development indicate that the nutrient limitation of vegetative growth is mediated via a direct effect of the nutrient supply on tissue expansion rather than an indirect limitation mediated via photosynthesis. The development of sinks for nutrient utilization becomes more limiting for growth at high nutrient supply. These results indicate the importance of sink development in mediating crop responses to nutrient supply, and highlight the importance of considering sink growth in crop models. In addition to an understanding of the contribution of source and sink strengths to growth, it will be important to investigate mechanisms linking supply and demand to effectively improve the mineral nutrient source:sink balance.

ACKNOWLEDGMENTS

This work was supported by a PhD studentship from the Society for Experimental Biology to ACB and by United States Department of Energy contract No. DE-SC0012704 to Brookhaven National Laboratory.

AUTHOR CONTRIBUTIONS

ACB, AR, MR, and CPO designed the research. ACB performed the research with practical assistance from AR. ACB and MR analysed the data. ACB and CPO wrote the manuscript with contributions from AR and MR.

REFERENCES

Acreche, M. M., & Slafer, G. A. (2009). Grain weight, radiation interception and use efficiency as affected by sink-strength in Mediterranean wheats released from 1940 to 2005. *Field Crops Research*, *110*, 98–105. https://doi.org/10.1016/j.fcr.2008.07.006

Aerts, R., & Chapin, F. S. (2000). The mineral nutrition of wild plants revisited: A re-evaluation of processes and patterns. *Advances in Ecological Research*, *30*, 1–67.

Álvaro, F., Royo, C., García del Moral, L. F., & Villegas, D. (2008). Grain filling and dry matter translocation responses to source–sink modifications in a historical series of durum wheat. *Crop Science*, *48*, 1523.

Arp, W. J. (1991). Effects of source-sink relations on photosynthetic acclimation to elevated CO_2. *Plant, Cell & Environment*, *14*, 869–875. https://doi.org/10.1111/j.1365-3040.1991.tb01450.x

Bao, Y., Hoogenboom, G., McClendon, R., & Vellidis, G. (2017). A comparison of the performance of the CSM-CERES-Maize and EPIC models using maize variety trial data. *Agricultural Systems*, *150*, 109–119. https://doi.org/10.1016/j.agsy.2016.10.006

Bennett, E., Roberts, J. A., & Wagstaff, C. (2012). Manipulating resource allocation in plants. *Journal of Experimental Botany*, *63*, 3391–400. https://doi.org/10.1093/jxb/err442

Burnett, A. C., Rogers, A., Rees, M., & Osborne, C. P. (2016). Carbon source–sink limitations differ between two species with contrasting growth strategies. *Plant, Cell & Environment*, *39*, 2460–2472. https://doi.org/10.1111/pce.12801

Campbell, B. D., & Grime, J. P. (1992). An experimental test of plant strategy theory. *Ecology*, *73*, 15–29. https://doi.org/10.2307/1938717

Chang, T., & Zhu, X. (2017). Source-sink interaction: A century old concept under the light of modern molecular systems biology. *Journal of Experimental Botany*, *68*, 4417–4431. https://doi.org/10.1093/jxb/erx002

Devienne, F., Mary, B., & Lamaze, T. (1994). Nitrate transport in intact wheat roots: II. Long-term effects of NO_3^- concentration in the nutrient solution on NO_3^- unidirectional fluxes and distribution within the tissues. *Journal of Experimental Botany*, *45*, 677–684. https://doi.org/10.1093/jxb/45.5.677

Díaz, S., Kattge, J., Cornelissen, J. H. C., Wright, I. J., Lavorel, S., Dray, S., … Gorné, L. D. (2016). The global spectrum of plant form and function. *Nature*, *529*, 167–183. https://doi.org/10.1038/nature16489

Erisman, J. W., Sutton, M. A., Galloway, J., Klimont, Z., & Winiwarter, W. (2008). How a century of ammonia synthesis changed the world. *Nature Geoscience*, *1*, 636–639. https://doi.org/10.1038/ngeo325

Fatichi, S., Leuzinger, S., & Korner, C. (2014). Moving beyond photosynthesis: From carbon source to sink-driven vegetation modeling. *New Phytologist*, *201*, 1068–95.

Garnier, E. (1992). Growth analysis of congeneric annual and perennial grass species. *Journal of Ecology*, *80*, 665–675. https://doi.org/10.2307/2260858

Garnier, E., Koch, G. W., Roy, J., & Mooney, H. A. (1989). Responses of wild plants to nitrate availability. *Oecologia*, *79*, 542–550. https://doi.org/10.1007/BF00378673

Garnier, E., & Laurent, G. (1994). Leaf anatomy, specific mass and water-content in congeneric annual and perennial grass species. *New Phytologist*, *128*, 725–736. https://doi.org/10.1111/j.1469-8137.1994.tb04036.x

Geiger, D. R., & Shieh, W. (1993). Sink strength: Learning to measure, measuring to learn. *Plant, Cell & Environment*, *16*, 1017–1018. https://doi.org/10.1111/j.1365-3040.1996.tb02048.x

Goucher, L., Bruce, R., Cameron, D. D., Lenny Koh, S. C., & Horton, P. (2017). The environmental impact of fertilizer embodied in a wheat-to-bread supply chain. *Nature Plants*, *3*, 17012. https://doi.org/10.1038/nplants.2017.12

Grime, J. P., & Hunt, R. (1975). Relative growth-rate: Its range and adaptive significance in a local flora. *Journal of Ecology*, *63*, 393–422. https://doi.org/10.2307/2258728

Grime, J. P., Thompson, K., Hunt, R., Hodgson, J. G. G., Cornelissen, J. H. C. H. C., Rorison, I. H. H., … Whitehouse, J. (1997). Integrated screening validates primary axes of specialisation in plants. *Oikos*, *79*, 259–281. https://doi.org/10.2307/3546011

HGCA (2014). HGCA Recommended List Spring barley 2014.

Hikosaka, K., Hanba, Y. T., Hirose, T., & Terashima, I. (1998). Photosynthetic nitrogen-use efficiency in leaves of woody and herbaceous species. *Functional Ecology*, *12*, 896–905. https://doi.org/10.1046/j.1365-2435.1998.00272.x

Houghton, J., Thompson, K., & Rees, M. (2013). Does seed mass drive the differences in relative growth rate between growth forms? *Proceedings of the Royal Society B: Biological Sciences*, *280*, 20130921. https://doi.org/10.1098/rspb.2013.0921

Isopp, H., Frehner, M., Long, S. P., & Nösberger, J. (2000). Sucrose-phosphate synthase responds differently to source-sink relations and to photosynthetic rates: *Lolium perenne* L. growing at elevated p(CO_2) in the field. *Plant, Cell and Environment*, *23*, 597–607. https://doi.org/10.1046/j.1365-3040.2000.00583.x

Iwasa, Y. (2000). Dynamic optimization of plant growth. *Evolutionary Ecology Research*, *2*, 437–455.

Jaikumar, N. S., Snapp, S. S., Flore, J. A., & Loescher, W. (2014). Photosynthetic responses in annual rye, perennial wheat, and perennial rye subjected to modest source:sink ratio changes. *Crop Science*, *54*, 274–283. https://doi.org/10.2135/cropsci2013.04.0280

Kiba, T., Kudo, T., Kojima, M., & Sakakibara, H. (2011). Hormonal control of nitrogen acquisition: Roles of auxin, abscisic acid, and cytokinin. *Journal of Experimental Botany*, *62*, 1399–1409. https://doi.org/10.1093/jxb/erq410

Körner, C. (2015). Paradigm shift in plant growth control. *Current Opinion in Plant Biology*, *25*, 107–114. https://doi.org/10.1016/j.pbi.2015.05.003

Marschner, H. (1995). *Mineral nutrition of higher plants*, 2nd ed. London, UK: Academic Press.

Masclaux-Daubresse, C., Daniel-Vedele, F., Dechorgnat, J., Chardon, F., Gaufichon, L., & Suzuki, A. (2010). Nitrogen uptake, assimilation and remobilization in plants: Challenges for sustainable and productive agriculture. *Annals of Botany*, *105*, 1141–1157. https://doi.org/10.1093/aob/mcq028

Millard, P. (1988). The accumulation and storage of nitrogen by herbaceous plants. *Plant, Cell & Environment*, *11*, 1–8. https://doi.org/10.1111/j.1365-3040.1988.tb01769.x

Miller, A. J., Fan, X., Orsel, M., Smith, S. J., & Wells, D. M. (2007). Nitrate transport and signalling. *Journal of Experimental Botany*, *58*, 2297–2306. https://doi.org/10.1093/jxb/erm066

Oleson, K. W., Lawrence, D. M., Bonan, G. B., Flanner, M. G., Kluzek, E., Lawrence, P. J., … Zeng, X. (2010). Technical description of version 4.0 of the Community Land Model (CLM) NCAR technical note NCAR/TN-478 + STR, NCAR, Boulder, Colorado.

Paul, M. J., & Driscoll, S. P. (1997). Sugar repression of photosynthesis: The role of carbohydrates in signalling nitrogen deficiency through source:sink imbalance. *Plant, Cell & Environment*, *20*, 110–116. https://doi.org/10.1046/j.1365-3040.1997.d01-17.x

Paul, M. J., Oszvald, M., Jesus, C., Rajulu, C., & Griffiths, C. A. (2017). Increasing crop yield and resilience with trehalose 6-phosphate: Targeting a feast-famine mechanism for better source-sink optimization. *Journal of Experimental Botany*, *68*, 4455–4462. https://doi.org/10.1093/jxb/erx083

Perchlik, M., & Tegeder, M. (2018). Leaf amino acid supply affects photosynthetic and plant nitrogen use efficiency under nitrogen stress. *Plant Physiology*, *178*, 174–188. https://doi.org/10.1104/pp.18.00597

Peterhansel, C., & Offermann, S. (2012). Re-engineering of carbon fixation in plants – challenges for plant biotechnology to improve yields in a high-CO_2 world. *Current Opinion in Biotechnology*, *23*, 204–8. https://doi.org/10.1016/j.copbio.2011.12.013

Pierce, S., Brusa, G., Vagge, I., & Cerabolini, B. E. L. (2013). Allocating CSR plant functional types: The use of leaf economics and size traits to classify woody and herbaceous vascular plants. *Functional Ecology*, *27*, 1002–1010. https://doi.org/10.1111/1365-2435.12095

Poorter, H., Anten, N., & Marcelis, L. F. M. (2013). Physiological mechanisms in plant growth models: Do we need a supra-cellular systems biology approach? *Plant, Cell & Environment*, *36*, 1673–1690. https://doi.org/10.1111/pce.12123

Poorter, H., & Remkes, C. (1990). Leaf area ratio and net assimilation rate of 24 wild species differing in relative growth rate. *Oecologia*, *83*, 553–559. https://doi.org/10.1007/BF00317209

Poorter, H., & van der Werf, A. (1998). Is inherent variation in RGR determined by LAR at low irradiance and by NAR at high irradiance? A review of herbaceous species. In H. Lambers, H. Poorter, & M. van Vuuren (Eds.), *Inherent variation in plant growth. Physiological mechanisms and ecological consequences* (pp. 309–336). Leiden, The Netherlands: Backhuys Publishers.

R Core Team (2015). R: A language and environment for statistical computing. Vienna, Austria: R Foundation for Statistical Computing. Retreived from https://www.R-project.org/

Rabalais, N. N., Turner, R. E., & Wiseman, W. J. (2002). Gulf of Mexico hypoxia, A.K.A. "The dead zone". *Annual Review of Ecology and Systematics*, *33*, 235–263. https://doi.org/10.1146/annurev.ecolsys.33.010802.150513

Reich, P. B., Wright, I. J., Cavender-Bares, J., Craine, J. M., Oleksyn, J., Westoby, M., & Walters, M. B. (2003). The evolution of plant functional variation: Traits, spectra, and strategies. *International Journal of Plant Sciences*, *164*, S143–S164. https://doi.org/10.1086/374368

Rodgers, C. O., & Barneix, A. J. (1988). Cultivar differences in the rate of nitrate uptake by intact wheat plants as related to growth rate. *Physiologia Plantarum*, *72*, 121–126. https://doi.org/10.1111/j.1399-3054.1988.tb06632.x

Ruiz-Vera, U. M., De Souza, A. P., Long, S. P., & Ort, D. R. (2017). The role of sink strength and nitrogen availability in the down-regulation of photosynthetic capacity in field-grown *Nicotiana tabacum* L. at elevated CO_2 concentration. *Frontiers in Plant Science*, *8*, 998. https://doi.org/10.3389/fpls.2017.00998

Sakakibara, H., Takei, K., & Hirose, N. (2006). Interactions between nitrogen and cytokinin in the regulation of metabolism and development. *Trends in Plant Science*, *11*, 440–448. https://doi.org/10.1016/j.tplants.2006.07.004

Santa-Maria, G. E., Moriconi, J. I., & Oliferuk, S. (2015). Internal efficiency of nutrient utilization: What is it and how to measure it during vegetative plant growth? *Journal of Experimental Botany*, *66*, 3011–3018. https://doi.org/10.1093/jxb/erv162

Slewinski, T. L. (2012). Non-structural carbohydrate partitioning in grass stems: A target to increase yield stability, stress tolerance, and biofuel production. *Journal of Experimental Botany*, *63*, 4647–70. https://doi.org/10.1093/jxb/ers124

Sonnewald, U., & Fernie, A. R. (2018). Next-generation strategies for understanding and influencing source-sink relations in crop plants. *Current Opinion in Plant Biology*, *43*, 63–70. https://doi.org/10.1016/j.pbi.2018.01.004

Stitt, M., & Krapp, A. (1999). The interaction between elevated carbon dioxide and nitrogen nutrition: The physiological and molecular background. *Plant, Cell & Environment*, *22*, 583–621. https://doi.org/10.1046/j.1365-3040.1999.00386.x

Stitt, M., & Schulze, D. (1994). Does Rubisco control the rate of photosynthesis and plant growth? An exercise in molecular ecophysiology. *Plant, Cell & Environment*, *17*, 465–487. https://doi.org/10.1111/j.1365-3040.1994.tb00144.x

Syngenta Crop Protection (2011). NFC Tipple.

Taulemesse, F., Le Gouis, J., Gouache, D., Gibon, Y., & Allard, V. (2015). Post-flowering nitrate uptake in wheat is controlled by N status at flowering, with a putative major role of root nitrate transporter NRT2.1. *PLoS ONE*, *10*, 1–23.

Taylor, B. N., & Menge, D. N. L. (2018). Light regulates tropical symbiotic nitrogen fixation more strongly than soil nitrogen. *Nature Plants*, *4*, 655–661. https://doi.org/10.1038/s41477-018-0231-9

Tegeder, M., & Masclaux-Daubresse, C. (2018). Source and sink mechanisms of nitrogen transport and use. *New Phytologist*, *217*, 35–53. https://doi.org/10.1111/nph.14876

Tschoep, H., Gibon, Y., Carillo, P., Armengaud, P., Szecowka, M., Nunes-Nesi, A., … Stitt, M. (2009). Adjustment of growth and central metabolism to a mild but sustained nitrogen-limitation in *Arabidopsis*. *Plant, Cell & Environment*, *32*, 300–318. https://doi.org/10.1111/j.1365-3040.2008.01921.x

von Bothmer, R. (1996). Distribution and habitat preferences in the genus *Hordeum* in Iran and Turkey. *Annalen des Naturhistorischen Museums Wien, Series B*, *98*(Suppl.), 107–116.

Wärlind, D., Smith, B., Hickler, T., & Arneth, A. (2014). Nitrogen feedbacks increase future terrestrial ecosystem carbon uptake in an individual-based dynamic vegetation model. *Biogeosciences*, *11*, 6131–6146. https://doi.org/10.5194/bg-11-6131-2014

White, A. C., Rogers, A., Rees, M., & Osborne, C. P. (2016). How can we make plants grow faster? A source–sink perspective on growth rate. *Journal of Experimental Botany*, *67*, 31–45. https://doi.org/10.1093/jxb/erv447

Wright, I. J., Reich, P. B., Westoby, M., Ackerly, D. D., Baruch, Z., Bongers, F., … Villar, R. (2004). The worldwide leaf economics spectrum. *Nature*, *428*, 821–827. https://doi.org/10.1038/nature02403

Zaehle, S., Medlyn, B. E., De Kauwe, M. G., Walker, A. P., Dietze, M. C., Hickler, T., … Norby, R. J. (2014). Evaluation of 11 terrestrial carbon–nitrogen cycle models against observations from two temperate Free-Air CO_2 Enrichment studies. *New Phytologist*, *202*, 803–822. https://doi.org/10.1111/nph.12697

7

Epigenomic plasticity of Arabidopsis msh1 mutants under prolonged cold stress

Sunil Kumar Kenchanmane Raju[#] | **Mon-Ray Shao**[#] | **Yashitola Wamboldt** | **Sally Mackenzie**

Department of Agronomy and Horticulture, University of Nebraska-Lincoln, Lincoln, Nebraska

Correspondence
Sally Mackenzie, Pennsylvania State University, University Park, PA.
Email: sam795@psu.edu

Present address

Sally Mackenzie, Departments of Biology and Plant Science, Pennsylvania State University, University Park, Pennsylvania.

Abstract

Dynamic transcriptional and epigenetic changes enable rapid adaptive benefit to environmental fluctuations. However, the underlying mechanisms and the extent to which this occurs are not well known. *MutS Homolog 1* (*MSH1*) mutants cause heritable developmental phenotypes accompanied by modulation of defense, phytohormone, stress-response, and circadian rhythm genes, as well as heritable changes in DNA methylation patterns. Consistent with gene expression changes, *msh1* mutants display enhanced tolerance for abiotic stress including drought and salt stress, while showing increased susceptibility to freezing temperatures. Despite changes in defense and biotic stress-response genes, msh1 mutants showed increasing susceptibility to the bacterial pathogen *Pseudomonas syringae*. Our results suggest that chronic cold and low light stress (10°C, 150 μmol m^{-2} s^{-1}) influences non-CG methylation to a greater degree in *msh1* mutants compared to wild-type Col-0. Furthermore, CHG changes are more closely pericentromeric, whereas CHH changes are generally more dispersed. This increased variation in non-CG methylation pattern does not significantly affect the *msh1*-derived enhanced growth behavior after mutants are crossed with isogenic wild type, reiterating the importance of CG methylation changes in *msh1*-derived enhanced vigor. These results indicate that *msh1*methylome is hyper-responsive to environmental stress in a manner distinct from the wild-type response, but CG methylation changes are potentially responsible for growth vigor changes in the crossed progeny.

KEYWORDS
abiotic stress, DNA methylation, phenotypic plasticity

1 | INTRODUCTION

Plants have developed mechanisms to overcome constantly changing environments. Species that are more adaptable to changing environments through phenotypic plasticity and selection of adaptable traits survive. These changes occur at different levels, from morphological and physiological changes to modulations in gene expression and chromatin behavior, allowing plants to cope with the challenges of nature. While a major source of this adaptive response can be attributed to genetic variation (Franks & Hoffmann, 2012), recent studies are pointing toward the potential role of chromatin

[#]Both authors contributed equally to the work.

modifications and epigenetics in plant responses to environmental changes (Bilichak & Kovalchuk, 2016). Environment-induced epigenetic modifications are generally transient, and the consistency of the environmental cue perceived by plants plays a role in inducing epigenetic changes and their inheritance (Uller, English, & Pen, 2015).

Cytosine DNA methylation is a heritable epigenetic modification involving the addition of a methyl (-CH3) group to the fifth carbon of the pyrimidine ring of cytosine nucleotides. This addition is catalyzed by DNA methyltransferases, commonly found in most eukaryotes (Cheng, 1995). In plants, DNA methylation can occur in three sequence contexts: the symmetric CG and CHG contexts, and the asymmetric CHH context, where H represents A, C, or T nucleotides (Law & Jacobsen, 2010). Methylation in these different contexts displays distinct genomic patterning within genes, repeat regions, and transposable elements. While CG methylation is largely concentrated within genes and transposable elements, CHG and CHH methylation contexts are usually associated with repeat regions and transposable elements (Cokus et al., 2008).

One role of DNA methylation is to silence transposable elements, which can become activated during stress conditions (Slotkin & Martienssen, 2007). In some cases, changes in DNA methylation have also been associated with stress-induced gene regulation, such as during phosphate starvation or *Pseudomonas syringae* infection (Dowen et al., 2012; Yong-Villalobos et al., 2015) and may provide the mechanistic basis for memory (Dowen et al., 2012; Kinoshita & Seki, 2014). Despite major progress in dissecting the genetic pathways responsible for establishment and maintenance of context-specific DNA methylation patterns (Stroud, Greenberg, Feng, Bernatavichute, & Jacobsen, 2013), functions of DNA methylation, particularly genic CG methylation, have remained mysterious (Zilberman, 2017).

MutS Homolog 1 (*MSH1*) is a plant-specific, nuclear-encoded gene that targets its protein to both plastids and mitochondria. Arabidopsis *msh1* mutants display a range of altered phenotypes that include variegation, dwarfing, delayed maturity transition, delayed flowering, and partial male sterility (Xu et al., 2011). The T-DNA *msh1* mutants display higher tolerance to heat, high light, and drought stress (Shedge, Davila, Arrieta-Montiel, Mohammed, & Mackenzie, 2010; Virdi et al., 2016; Xu et al., 2011), particularly in individuals showing stronger developmental phenotypes. *MSH1* phenotypes are conserved between monocots and eudicots. This conservation is evidenced in the RNAi suppression phenotypes, and the consistent observation that subsequent MSH1-RNAi transgene segregation gives rise to trans-generational *msh1* memory in sorghum, pearl millet, tomato, tobacco, and soybean (de la Rosa Santamaria et al., 2014; Kenchanmane Raju et al., 2018; Xu et al., 2011, 2012; Yang et al., 2015). These memory lines show attenuated *msh1* phenotype, even though *MSH1* transcripts are back to wild-type levels (Raju et al., 2018).

Disruption of *MSH1* causes genome-wide methylome repatterning in both CG and non-CG context (Virdi et al., 2015), along with large-scale changes in gene expression related to abiotic and biotic stress response, phytohormone pathways, circadian rhythm, defense response, and signaling (Shao, Kumar Kenchanmane Raju, Laurie, Sanchez, & Mackenzie, 2017). Arabidopsis *msh1* memory lines show a subset (ca 10%) of the gene expression changes of the T-DNA insertion mutant, with enrichment in circadian rhythm, ABA signaling, and light-response pathways, and with methylome repatterning predominantly in CG context (Sanchez et al., 2018).

In this study, we investigated the stress response behavior of plants following *msh1* developmental reprogramming. We show that *msh1* mutants display a differential response to abiotic and biotic stress, which could be partly explained by transcriptome changes. Epi-lines, deriving from crosses of *msh1* with wild type, showed increased seed yield and higher tolerance to salt, freezing, and mild heat stress. Under prolonged cold stress, *msh1* mutants showed increased variation in DNA methylation, particularly in non-CG context, and this increased CHG and CHH methylation pattern variation did not appear to influence the *msh1* crossing-derived vigor phenotype. Taken together, the data imply that developmental phenotypes in the *msh1* mutants are caused by large-scale gene expression changes associated with stress response, along with genome-wide methylome repatterning. Methylome changes in non-CG context were disproportionately affected by cold stress and were hyper-responsive to environmental changes, whereas changes in CG context appeared to be stable and to influence plant phenotype.

2 | MATERIALS AND METHODS

2.1 | Plant growth conditions and PCR genotyping

The genetic background used throughout the study was Arabidopsis Col-0 ecotype. For phenotypic measurements, seeds were sown into plastic pots containing Fafard germination mix with Turface MVP added. After 48–72 hr of cold stratification at 4°C in a dark chamber, pots were moved to growth chambers set at 22°C. The *msh1* T-DNA mutant was obtained from Arabidopsis Biological Resource Center (SAIL_877_F01, stock number CS877617) and genotyped as described previously (Shao et al., 2017). Mutants #12-29, #12-4, and #9 are three selections from the parental T-DNA line. Epi-lines were developed by crossing wild type with *msh1* mutants, some of which had been exposed to cold stress (S), and subsequently self-pollinating filial generations. PCR genotyping as previously described (Shao et al., 2017) was performed on the F_2 population, and only plants with wild-type *MSH1/MSH1* were evaluated and forwarded. Yield and stress tests were performed on bulked epi-F_3 populations or $F_{2:3}$ lines developed from progeny of individual F_2 plants. A Supporting information Figure S1 describes the stress treated *msh1* mutants and development of epi-lines.

2.2 | Abiotic and biotic stress treatments

All stress treatments were performed on wild-type Col-0, *msh1* mutants #9, #12-4 and #12-29, epiF_3 populations derived from crosses WT x *msh1*-N, WT x *msh1*-VD, WT x *msh1*-N(S), and WT x *msh1*-VD(S) that involved the two phenotypic classes of *msh1* mutants, normal phenotype (N), and variegated dwarf (VD), with and without exposure to stress (S).

Seeds for stress treatments were bleach sterilized and sown on half-strength MS medium containing 1.5% sucrose and 0.5% MES, pH 5.7, solidified with 4% agar in sterile plastic Petri plates. For 200 mM salt germination tests, 11.7 g of NaCl was added to the growth media before sterilization. After 48–72 hr of cold stratification in a dark room at 4°C, plates were moved to Percival growth chambers set at 22°C and 16/8 light/dark cycle. Germination was scored based on root length of more than 3 mm at 2 weeks after plates were moved to the growth chamber.

For freezing tolerance, 2-week-old seedlings were cold acclimatized for 1 week at 4°C in 12/12 hr light/dark photoperiod. Freezing tests were performed as previously described (Barnes, Benning, & Roston, 2016), with necessary modifications. Specifically, postfreezing plates were placed in a 4°C dark chamber for 24 hr before recovery in control growth conditions for 5–7 days. Survival was scored as plants having fully expanded green rosette leaf after recovery. The *sfr2-3* mutant (Moellering, Muthan, & Benning, 2010), used as negative control, was a kind gift from Dr. Rebecca Roston.

Two independent *MSH1* epi-lines for each phenotypic class of *msh1* mutant were developed, WT x *msh1*-N1 and WT x *msh1*-N2, created by crossing two -*msh1* mutants with a normal phenotype (N1, N2), and WT x *msh1*-VD1 and WT x *msh1*-VD2 developed from two *msh1* mutants with a variegated dwarf phenotype (VD1, VD2). Seed yield was measured as total seed weight at maturity. Floral stems of 6-week-old plants were tied to a wooden stake and the plant enclosed completely using Arabisifter (Lehle Seeds, SNS-03), making a pouch-like structure in the bottom to collect shattered seeds. All four epi-F_3s and wild type were grown in a completely randomized design in a growth chamber at 22°C or 32°C, 16/8-hr light/dark cycle. Seeds were carefully harvested from each population ($n > 18$ plants) at maturity. Seeds were dried in a 37°C chamber for 48–72 hr before recording seed weights.

Pseudomonas syringae pv. tomato DC3000 strains were grown for 24 hr at 30°C on King's B media (King, Ward, & Raney, 1954) with the appropriate antibiotics, and resuspended to an OD_{600} of 0.2 (2×10^8 cells ml^{-1}) in 10 mM $MgCl_2$. The resuspended culture was sprayed uniformly on upper and lower surfaces of fully expanded leaves of 4-week-old wild type, *msh1* mutant, and *msh1*-derived epilines using a jet-spray bottle. Treated plants were well-watered and kept in a dark room for 5 days, followed by five to 7 days in a growth chamber at 22°C and 16/8-hr light/dark cycle before scoring for survival.

2.3 | RNA extraction and sequencing analysis

Four-week-old plants grown in 22°C were transferred to a growth chamber set at 10°C, 150 μE m^{-2} s^{-1}, for 30 days. Tissue from four fully expanded rosette leaves was sampled before and after 10°C transfer with three replicates per group. For each sample, frozen tissue was ground and total RNA extracted using a standard TRIzol reagent protocol. RNA samples were then treated with DNaseI (Qiagen catalog #79254). Qiagen RNeasy Plant Mini Kit (Qiagen catalog #74904) was used to purify total RNA samples prior to RNA sequencing (RNAseq). Poly(A)-enriched RNAseq was performed by Beijing Genomics Institute (BGI), generating at least 59.6 M paired-end, 100-bp reads per sample. Reads were trimmed and aligned to the Arabidopsis TAIR10 reference genome sequence with annotation from Araport11 PreRelease3 using TopHat2 (Kim et al., 2013). The DESeq2 method (Love, Huber, & Anders, 2014) was used to identify differentially expressed genes (cutoff of FDR <0.05, |log2(fold change)| ≥1, and mean FPKM ≥1). Gene Ontology (GO) enrichment analysis was performed using the DAVID database (Huang, Sherman, & Lempicki, 2009). GO terms with *p*-value <0.05 after Benjamini-Hochberg (Benjamini & Hochberg, 1995) correction for multiple testing were considered statistically significant in each comparison.

For transposable element (TE) family expression analysis, reads were aligned using the STAR 2-pass method (Dobin et al., 2013), allowing up to 100 multi-mapped locations as per the recommendation of TEtranscripts (Jin, Tam, Paniagua, & Hammell, 2015). Quantification and testing for differential expression of TEs were performed using TEtranscripts with the developer-provided Arabidopsis TE family annotation.

2.4 | Cold stress methylome analysis

To obtain whole-genome bisulfite sequencing data for the cold stress experiment, plants were grown in a controlled growth chamber set to 10°C, 150 μE m^{-2} s^{-1} or 500 μE m^{-2} s^{-1}, and 12/12 day/night photoperiod for 21 days, beginning from germination, then moved to recovery at 22°C, 250 μE m^{-2} s^{-1} for 18 days before sampling. Control plants were grown continuously at 22°C, 250 μE m^{-2} s^{-1} from sowing, and sampled upon reaching a similar developmental stage as cold-stress recovered plants. Four fully expanded rosette leaves from each individual plant were harvested and DNA extracted as previously described (Li & Chory, 1998), with two replicates per group. Library generation and bisulfite-sequencing were performed by BGI on a Hiseq2000. Reads were aligned to the TAIR10 reference genome using Bismark (Krueger & Andrews, 2011) with default mismatch parameters. Due to the potential for artifacts, cytosines of CCC context were excluded from CHH analysis. Methylation conversion rates as determined by unmethylated chloroplast DNA ranged from 98.91% to 99.29% (Supporting information Dataset S1).

The R package *methylKit* 1.1.8 (Akalin et al., 2012) was used to call DMRs, based on 100 bp nonoverlapping windows, separately for CG, CHG, and CHH contexts. Only cytosine base positions with ≥3 reads were retained for analysis, and normalized methylation counts for each cytosine were used based on coverage. Windows with ≥5 cytosines (of the given context) were considered for analysis, to rule out low information regions. The principal component analysis was performed using the *PCASamples* function. Subsequent comparison between treatment (cold or control) and genotype (*msh1* T-DNA or wild type) combinations were performed by logistic regression with *methylKit*. DMRs for each context were identified based on a methylation difference of at least 10% absolute value and a *q*-value <0.05, then clustered using Ward's method (Ward, 1963). For CG context, genes overlapping DMRs within each cluster were identified and subjected to GO enrichment analysis using the DAVID database (Huang et al., 2009). For CHG and CHH contexts, TE's overlapping DMRs within each cluster were identified and tested for enrichment of TE families and superfamilies (annotated by TAIR10) using the hypergeometric test (FDR < 0.01).

3 | RESULTS

3.1 | The *msh1* mutant shows variable abiotic and biotic stress tolerance

Previous studies have shown that *msh1* mutants are more tolerant to drought, high light, and heat stress (Shedge et al., 2010; Virdi et al., 2016; Xu et al., 2011). We tested for other abiotic stress effects, focusing first on salt and freezing temperature. Seeds of *msh1* mutant and wild type were grown on plates with half-strength MS media and 200 mM NaCl. The 200 mM NaCl concentration is highly selective for germination tests in Col-0 (Wibowo et al., 2016). Germination was scored based on root length of greater than 3 mm at 2 weeks after sowing, assessed in three independent experiments. Only 32% percent of wild-type seeds germinated on 200 mM NaCl plates. Two selections from the *msh1* mutants, #9 and #12-29, showed significantly higher germination than wild type (*p*-value 1.25×10^{-10} and 1.28×10^{-4}, respectively), while *msh1*#12-4 did not show significant difference (*p*-value 0.148). These results suggest higher salinity tolerance in *msh1* mutants, with variation in mutant sub-populations (Figure 1a). This result is consistent with gene expression data from *msh1* mutants (Shao et al., 2017), which show differential expression for 493 (Supporting information Dataset S2) of the 1667 salt stress–responsive genes identified through comparative microarrays (Sham et al., 2015).

To examine whether or not *msh1* mutants also showed tolerance to freezing temperatures, 2-week-old seedlings of wild-type and *msh1* T-DNA mutants were cold acclimatized for a week before exposure to −2°C for 4 hr, followed by nucleation and −10°C for 12 hr. Survival was scored as the presence of green rosette leaves 1 week after recovery under normal growth conditions (22°C, 16/8 light/dark cycle). Surprisingly, the survival rate of *msh1* mutants was lower than that of wild type (Figure 1b), indicating that the mutants are not tolerant to all stresses. Under our experimental conditions, 34.5% wild type survived −10°C. The *msh1* mutants #9 and #12-29 showed significantly higher susceptibility to freezing temperatures (*p*-value 0.002 and 0.009, respectively), while *msh1* mutant #12-4 was not significantly different from wild type (*p*-value 0.42651). From a set of 590 differentially expressed genes correlated with acclimated and nonacclimated freezing tolerance (Hannah et al., 2006), only 64 were altered in expression in the *msh1* variegated dwarf mutant (Supporting information Dataset S3).

Because *msh1* mutants have increased tolerance to abiotic stresses such as drought, heat, high light, and salt, we tested whether or not they were likewise more resistant to biotic stress. We challenged *msh1* mutants with the gram-negative bacterial pathogen *Pseudomonas syringae* pv. *tomato* DC3000, which causes bacterial speck disease in tomato and is pathogenic to Arabidopsis. The *msh1* mutants showed susceptibility to the bacterial pathogen. While 87.5% of wild-type plants survived the stress, *msh1* mutant sub-populations #12-29 and #12-4 showed significantly higher susceptibility (Supporting information Figure S2a: *p*-value 0.006 and 0.081, respectively). Within one population, *msh1* #9, plants with variegation and dwarfing showed significantly higher susceptibility (*p*-value $4.35 \times 10^{-5}$and 7.15×10^{-7}, respectively) to *P. syringae* than *msh1* mutants with a mild phenotype (Supporting information Figure S2b: *p*-value 0.327). Thus, *msh1* mutants are susceptible to biotic stress despite markedly increased expression of biotic stress–responsive pathways in *msh1* mutants (Shao et al., 2017), and the biotic stress response appears related to the severity of the *msh1* phenotype.

To investigate the inheritance of these stress responses following crossing between *msh1* mutants (T-DNA) and wild type, seed

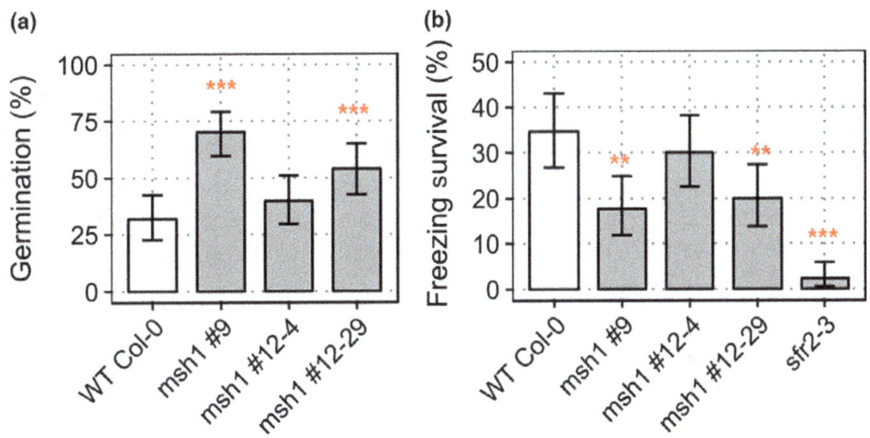

FIGURE 1 Abiotic stress tests in Arabidopsis *msh1* mutants. (a) Percent germination rate of wild type Col-0 and, *msh1* mutants #9, #12-4, and #12-29 at 200 mM NaCl-supplemented growth media, scored after 2 weeks postsowing [*n* = 100 plants each; error bars represent standard error of means (*SEM*)]. (b) Proportion of recovered (survived) plants 7 days postfreezing treatment at −10°C. *sfr2-3* was used as negative control for freezing tolerance (*n* = 130, error bars represent *SEM*). Significance at '***' 0.001 '**' 0.01 '*' 0.05 '.' 0.1

germination rate in 200 mM NaCl concentration and survival of seedlings at −10°C freezing temperatures were assayed in three independent experiments for *msh1*-derived epi-lines in the F_3 generation, with wild type as a control. Epi-lines were created by crossing wild-type Col-0 with *msh1* mutants as pollen donor, and self-pollinating filial generations after PCR genotyping for *MSH1/MSH1* in F_2s to obtain epi-F_3 bulks (Supporting information Figure S1). When seeds were germinated on plates with 200 mM NaCl, WT x *msh1*-N, and WT x *msh1*-VD showed significantly higher germination rate (*p*-value 7.15×10^{-14} and 3.20×10^{-12}, respectively) than wild type (Supporting information Figure S3a). This result was consistent with the parental *msh1* mutant, which showed a similar increase in salt tolerance (Figure 1a). However, epiF_3 population WT x *msh1*-VD also showed higher tolerance to freezing (Supporting information Figure S3b: *p*-value 0.015), where *msh1* mutant showed greater susceptibility. These observations are consistent with the expectation that *msh1* x wild-type crosses produce a different epigenetic state, thus resulting in distinctive phenotypes.

To evaluate the response of progeny from crossing under less severe, nonlethal stress, we subjected epi-F_3 plants to mild heat stress and measured total seed weight at harvest. For this experiment, four epi-lines and wild type were grown in growth chambers under control (22°C) or mild heat stress (32°C) throughout the plant life cycle. Epi-lines showed 9.7%–19.6% increase in seed yield compared to wild type in control conditions (Figure 2a). Three of the epi-lines also performed significantly better than wild type under mild heat stress, showing 9.5%–16.5% increase in yield (Figure 2a). The lower yield penalty under mild heat stress in the three epi-lines, coupled with the enhanced salt and cold tolerance, provides an indicator of greater yield stability and lower environmental effects on the *MSH1* growth-enhanced phenotype (Figure 2b). These results resemble the higher yield stability observed in soybean *MSH1* epi-lines grown across four different locations in Nebraska (Raju et al., 2018).

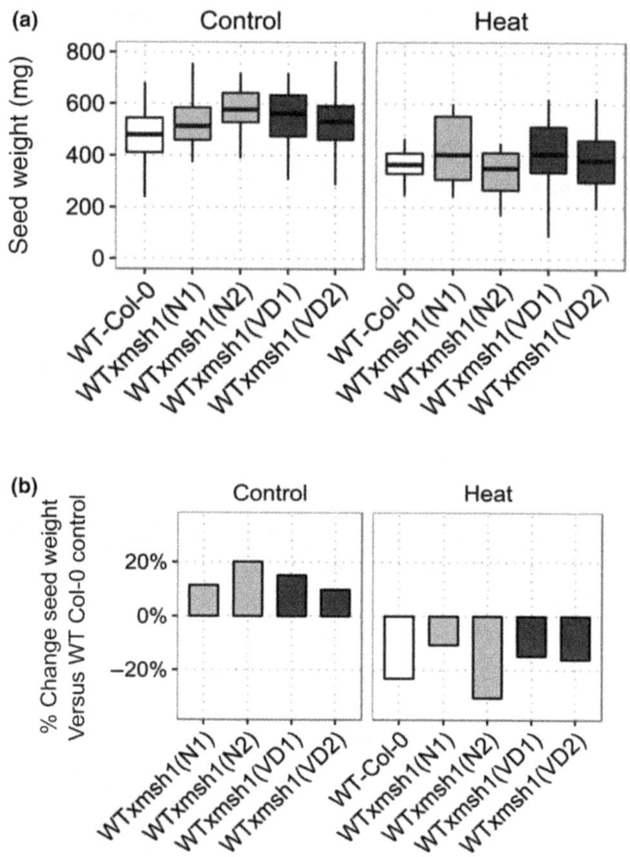

FIGURE 2 Total yield measurements for *msh*-derived epi-lines compared to wild-type Col-0. (a) Whisker plots showing differences in seed yield between wild type and *msh1* epi-lines under control (22°C) and mild heat-stress (32°C) growth conditions [Control (n = 36), heat stress (n = 18)]. (b) Percent change in seed weight for epi-lines under control and heat stress condition compared to seed weight of wild type under control growth conditions

3.2 | The methylome of *msh1* is hyper-responsive to cold stress with disproportionately higher CHH hypomethylation

Transcriptome studies of *msh1* showed clear enrichment of biotic and abiotic stress response genes, including response to cold. Despite changes in cold-responsive transcription factors (Shao et al., 2017), *msh1* mutants showed susceptibility to freezing temperatures. These observations led us to test whether *msh1* mutants would show differential methylome and transcriptome response to low-temperature stress.

To evaluate the extent of DNA methylation changes related to long-term cold stress, *msh1* mutants and wild-type plants were grown at 10°C for 18 days under 12/12 light/dark cycle, then allowed to recover at 22°C for 18 days before sampling for DNA extraction. Plants were allowed to recover prior to sampling for two reasons: Plant growth was slower under cold stress, complicating the collection of sufficient tissue for methylome sequencing. In addition, we wanted to avoid transient methylation changes present during plant exposure to cold treatment.

To facilitate comparison of each region between different genotype and treatment combinations, methylome analysis was performed using fixed 100-bp nonoverlapping windows. Principal component analysis plots from the first two principal components using the upper 0.9 quantile of variable windows showed CG methylation separating by genotype between wild type and *msh1*mutants, with or without stress (Figure 3a). These observations are consistent with studies of the *msh1* memory lines, where CG methylation is predominant in association with a memory phenotype (Sanchez et al., 2018). CHG methylation showed a similar pattern, although cold-stressed samples were discriminated from control samples in *msh1* mutants more than in wild type (Figure 3a). Notably, CHH methylation showed the greatest degree of discrimination for the cold stress treatment, predominantly in *msh1* mutants (Figure 3a). Together, these results indicate that cold stress influences DNA methylation in all methylation contexts, but there is evidence of interaction with the *msh1* background, amplifying the effect in CHH context.

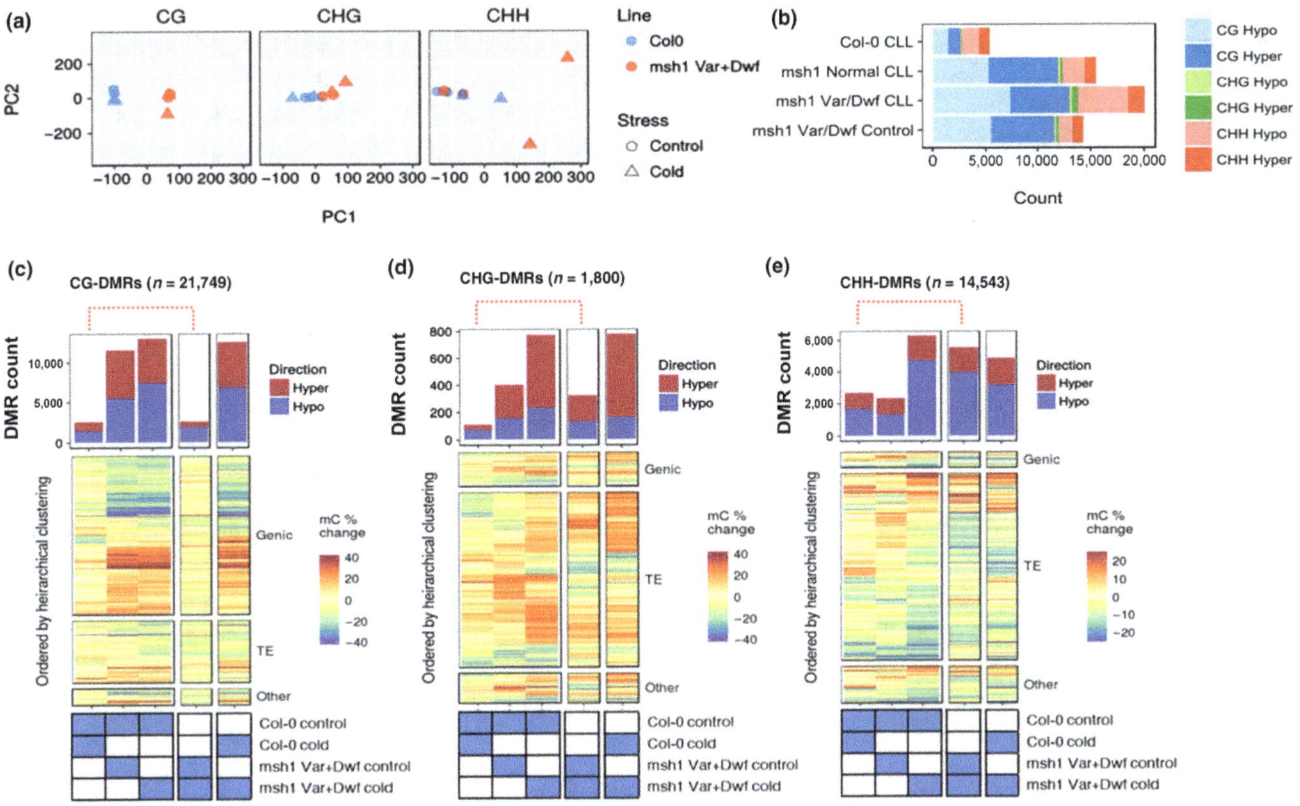

FIGURE 3 Methylome changes in *msh1* mutants and wild-type Col-0 from long-term cold stress. (a) Principal component analysis (PCA) plots for methylation levels within 100-bp windows separated for nucleotide context; CG, CHG, and CHH (H represents A, C, or T). (b) Graph of total DMR numbers in each comparison, showing hyper and hypomethylation in all three contexts. (c–e) DMR counts and hierarchical clustering of all pairwise comparisons, for CG (c), CHG (d), and CHH (e) contexts. Red dotted lines highlight *msh1* cold response relative to wild type under cold stress

3.3 | Genomewide distribution of DMRs in wild type and *msh1* mutants in response to cold stress

We investigated the number and genomic distribution of differentially methylated regions (DMRs). DMR calling was based on logistic regression over 100 bp nonoverlapping window, using a threshold of more than 10% absolute change in methylation level in each cytosine context. The resulting number of DMRs (Table 1, Figure 3b) confirmed trends observed by principal component analysis (Figure 3a). As expected, CG-DMRs mostly occurred over genes and were relatively few between cold and control treatments in wild type or *msh1* while comparing any *msh1*group to wild type (Supporting information Figure S4a). We found 11,579 CG-DMRs, 399 CHG-DMRs, and 2332 CHH-DMRs when comparing *msh1* to wild type under control conditions. Almost equal numbers of DMRs were hyper or hypomethylated in symmetric methylation context, while in CHH context, there were 30% more hypomethylated DMRs in *msh1* (Table 1, Figure 3b). We also detected 2,592 CG-DMRs, 109 CHG-DMRs, and 2658 CHH-DMRs induced by cold stress alone in the wild type. The magnitude of CG and CHG changes in the *msh1* mutant was four times higher than changes induced by cold stress alone in wild type, while the magnitude of CHH changes was similar. This implies that relative to CG and CHG methylation, cold stress disproportionately affects CHH methylation, consistent with previous reports of methylome behavior under low temperature (Dubin et al., 2015).

We examined whether *msh1* background affects methylation changes upon cold stress. We found 2626 CG-DMRs, 321 CHG-DMRs, and 5539 CHH-DMRs between stressed and unstressed *msh1* mutant (Table 1, Figure 3b). Thus, CHH methylation, primarily over transposable elements (Supporting information Figure S4a), showed the greatest effect of cold treatment within the *msh1*

TABLE 1 Number of DMRs in all three cytosine contexts across multiple comparisons

Comparison	Direction	CG	CHG	CHH
msh1-VD vs WT	Hyper	6,085	240	996
	Hypo	5,494	159	1,336
msh1-VD (S) vs *msh1-VD*	Hyper	700	186	1,574
	Hypo	1,926	135	3,965
WT(S) vs WT	Hyper	1,154	37	977
	Hypo	1,438	72	1,681
msh1-VD (S) vs WT	Hyper	5,626	538	1,548
	Hypo	7,400	232	4,723
msh1-VD (S) vs WT(S)	Hyper	5,707	611	1,675
	Hypo	6,829	167	3,176

background, consistent with separation seen in the PCA plot (Figure 3a). Although CHH methylation is affected by cold stress in wild type, CHH DMRs in cold-stressed *msh1* are twice as abundant as in similar wild type comparisons. Whereas CG DMR patterns were nearly identical for cold-stressed and control *msh1* mutants when compared to wild type, CHG and CHH DMRs showed a clear distinction in patterns, with several loci switching between hyper and hypomethylation (Figure 3c–e). These results indicate an interaction between the *msh1*effect and cold stress, such that non-CG methylation patterns are disproportionately affected.

3.4 | Non-CG methylome changes in association with transposable elements

To investigate the genomic distribution of non-CG changes in response to stress, we clustered non-CG DMRs and looked for enrichment of TE superfamilies in these clusters. Both CHG- and CHH-DMRs formed 4 clusters each (Supporting information Dataset S4). While all four clusters in CHG-DMRs showed enrichment for DNA/En-spm, LTR/COPIA, and LTR/Gypsy elements, clusters three (hyper) and one (hypo), which showed similar trends in all comparisons, were also enriched in LINE/L1 elements. In CHH-DMR clusters, DNA/MuDR elements were enriched in all clusters. Cluster one, which contained the most DMRs and hypomethylation in all three comparisons (wild-type stressed vs wild type, *msh1* vs wild type, and *msh1*-stressed vs wild type) showed enrichment for LINE/L1, LTR/COPIA, and LTR/Gypsy elements. Clusters three and four, which showed hypermethylation in *msh1*-stressed versus wild type, showed over-representation of DNA/Mariner and RC/Helitron elements.

Genomic distribution of DMRs matched with known behaviors within each cytosine context. CG-DMRs between *msh1* mutants and wild type were distributed evenly across the chromosome (Supporting information Figure S5b,c,e), while CG DMRs from cold stress were primarily limited to heterochromatin (Supporting information Figure S5a,d). CHG-DMRs and CHH-DMRs were mainly in heterochromatic regions for both comparisons. This finding is consistent with previous reports of cold stress methylome changes showing heterochromatin bias (Dubin et al., 2015). We examined expression changes in genes related to DNA methylation machinery. Interestingly, CHROMO METHYLTRANSFERASE 3 (CMT3) and DECREASE IN DNA METHYLATION 1 (DDM1) expression were down-regulated in cold-stressed *msh1* mutants compared to unstressed mutants and wild type (Supporting information Figure S6). As *cmt3* and *ddm1* mutants are known to increase heterochromatic TE de-repression, these observations appear consistent with CHH hypomethylation of heterochromatin in the interaction of *msh1* effect and low-temperature stress.

3.5 | Transcriptome response of Arabidopsis *msh1* mutants under chronic cold stress

We evaluated the effect of cold stress on the transcriptome of *msh1* mutants. Wild-type Col-0 and *msh1* plants were grown at 22°C for 4 weeks before leaf tissue was collected (control group), or grown at 10°C for an additional 30 days before sampling tissues for RNA extraction (cold-stressed group). Transcriptome analysis showed cold stress to be the largest contributor to transcriptional changes within the experimental groups, evident from the groups formed in PCA plotting with normalized log values of gene expression (Figure 4a). Although the magnitude of gene expression change was lower than transcriptome change in our earlier report (Shao et al., 2017), similar pathways were modulated in both *msh1* mutants with or without severe phenotype, including defense, jasmonic acid, abiotic stress response, photosynthesis, and oxidative stress (Supporting information Dataset S5). Technical differences, like differential developmental staging and changes in circadian phase (Hsu & Harmer, 2012), might explain the differences in the magnitude of transcriptome changes. Pathways affected in *msh1* appear to be induced by cold alone in wild type, suggesting that *msh1* mutants have stress response pathways activated in the absence of any environmental cues. Response to abiotic stress (cold, salt, light, and wounding) and biotic stress (response to chitin and jasmonic acid) is activated as a

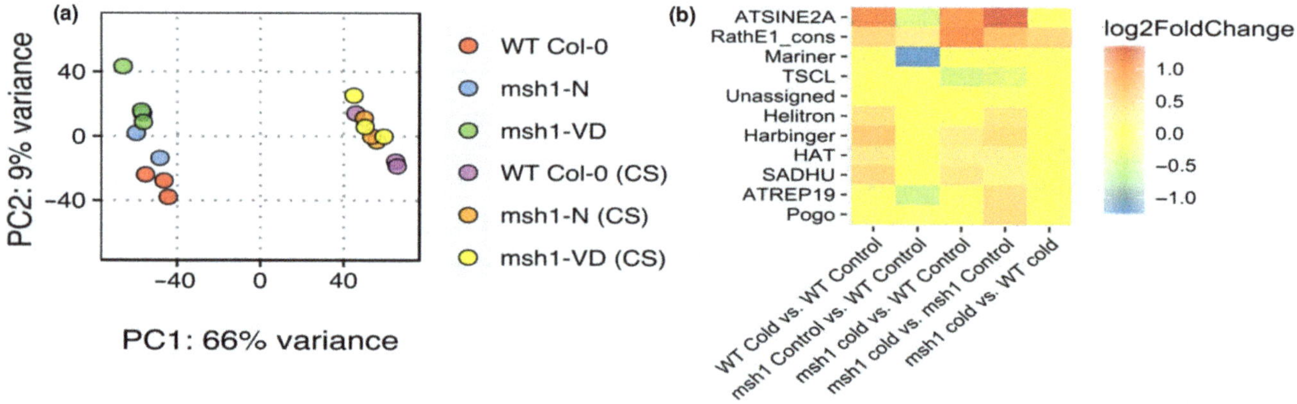

FIGURE 4 Transcriptome changes in *msh1* mutants before and after chronic cold stress. (a) PCA plot from normalized log values of gene expression from wild-type and *msh1* mutants under control or cold stress. (b) Heat map showing differential expression of transposable element (TE) super families from each corresponding comparison within cold stress experiment

cold stress response in wild type and is also activated in *msh1* (Supporting information Figure S7a). Defense response, jasmonic acid-mediated signaling, and photosynthesis-related genes were specifically enriched in *msh1* (Supporting information Figure S7a: Dataset S5). Taken together, these results suggest that unlike methylome, transcriptome changes do not show increased plasticity in *msh1* mutants under cold stress.

We also looked into differences in transposable element expression using TEtranscripts (Jin et al., 2015). Mariner, ATREP19, and SINE TE superfamilies were significantly down-regulated in *msh1* mutants compared to wild type, while Rath elements were up-regulated in all comparisons (Figure 4b). In contrast, SINE elements showed clear cold-stress induced up-regulation. Similarly, Helitron, Harbinger, HAT, and SADHU elements were up-regulated in wild type under cold stress (Figure 4b). At the family level, ATCOPIA28 and ATCOPIA31A showed clear stress-induced up-regulation, while VANDAL5A, ATREP3, ATCOPIA44, ATCOPIA 78, ATCOPIA 93, and ATMU1 showed down-regulation in *msh1* mutants (Supporting information Figure S7b).

3.6 | *MSH1*-induced CG methylation changes are associated with enhanced growth in progeny from *msh1* crossing

To evaluate the extent to which non-CG methylome divergences affect the *msh1* crossing-derived enhanced growth phenotype in Arabidopsis (Virdi et al., 2015), we investigated the epi-lines from *msh1* mutants with or without cold stress (see Methods). We assayed rosette diameter, days to flowering, and total seed weight from F_2 and selected $F_{2:3}$ lines (progeny from selected individual F_2 plants) under control growth conditions. The F_2 population WT x *msh1*-N(S) showed higher mean rosette diameter compared to wild type, measured at 6 weeks after sowing (Wilcox test, padj 0.004, Figure 5a). This population flowered an average of 2 days earlier (Wilcox test, padj 0.002, Supporting information Figure S8). Both WT x WT(S) and WT x *msh1*-N populations showed smaller rosette diameter compared to wild type (Figure 5a). Mean seed yield, measured in milligrams, for WT x *msh1*-VD(S) and WT x *msh1*-VD was significantly higher than wild-type Col-0 (Wilcox test, padj 0.015, Figure 5b), whereas no significant difference was found between WT x *msh1*-VD(S) and WT x *msh1*-VD (*t* test, *p*-value 0.80). These results show that for epiF_2s, WT x *msh1*-VD(S) and WT x *msh1*-VD showed 20% and 17.9% increase in yield compared to wild type, but stressing the *msh1* mutant prior to crossing does not have a significant effect on yield. We also noticed that while WT x *msh1*-VD and WT x *msh1*-VD(S) showed increases in seed yield, WT x *msh1*-N(S) showed higher rosette diameter compared to wild type, suggesting the possibility of selection for separate traits of vegetative biomass heterosis and seed yield heterosis in *msh1* derived epi-lines.

We evaluated total seed weight for $F_{2:3}$ lines following selection of the upper 20% for seed weight in each F_2 population under control conditions. Although the selection was performed on seed weight, $F_{2:3}$ epi-lines 3C2, derived from WT x *msh1*-VD(S), and 4C2, derived from WT x *msh1*-VD, showed significantly higher rosette diameter (Wilcox test, padj = 0.018, Figure 6a) compared to wild type. Both epi-lines also showed significantly higher seed weight compared to average wild type (*t* test, *p*-value = 0.0008 and 0.043 respectively, Figure 6b), reflecting a response to selection. Similar to F_2 results, there was no significant difference in seed weight between 3C2 and 4C2 (*t* test, *p*-value = 0.203), confirming that stress treatment of *msh1* mutants does not have an effect on *msh1*-derived growth enhancement in progeny from crosses.

We subsequently tested whether or not observed methylome changes had an effect on stress adaptation of derived epi-lines. Surprisingly, epi-lines derived from the cold-stressed *msh1* mutant as parent showed a different response to salt and freezing stress. Whereas WT x *msh1*-VD(S) and WT x *msh1*-N(S) epi-F_3 populations were significantly more tolerant to salt stress (*p*-value 9.1×10^{-4} and 1.8×10^{-3}, respectively), although lesser in magnitude to comparable populations from unstressed (Figure 7a), they were not significantly different from wild type in their response to freezing stress (*p*-value 0.867 and 0.904, respectively, Figure 7b). These

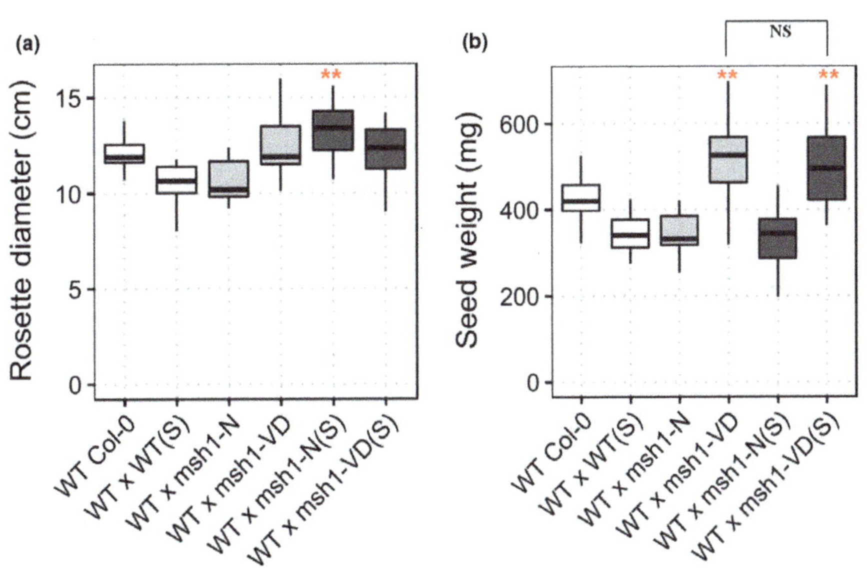

FIGURE 5 Variation in rosette diameter and seed weight in *MSH1* epi-F_2 population. (a) Whisker plot showing variation in rosette diameter measured at 6 weeks after sowing (*n* = 18). (b) Whisker plot showing variation in total seed weight measured carefully after bagging the plants with Arabisifter (Lehle Seeds, SNS-03). Epi-F_2 populations were developed from cold stressed (S) and unstressed *msh1* mutants with (VD) or without (N) phenotype. F_2 plants were selected after genotyping for *MSH1/MSH1* wild-type allele. Rosette diameter and seed weight were measured from the same set of F_2 plants. Significance at '***' 0.001 '**' 0.01 '*' 0.05 '.' 0.1

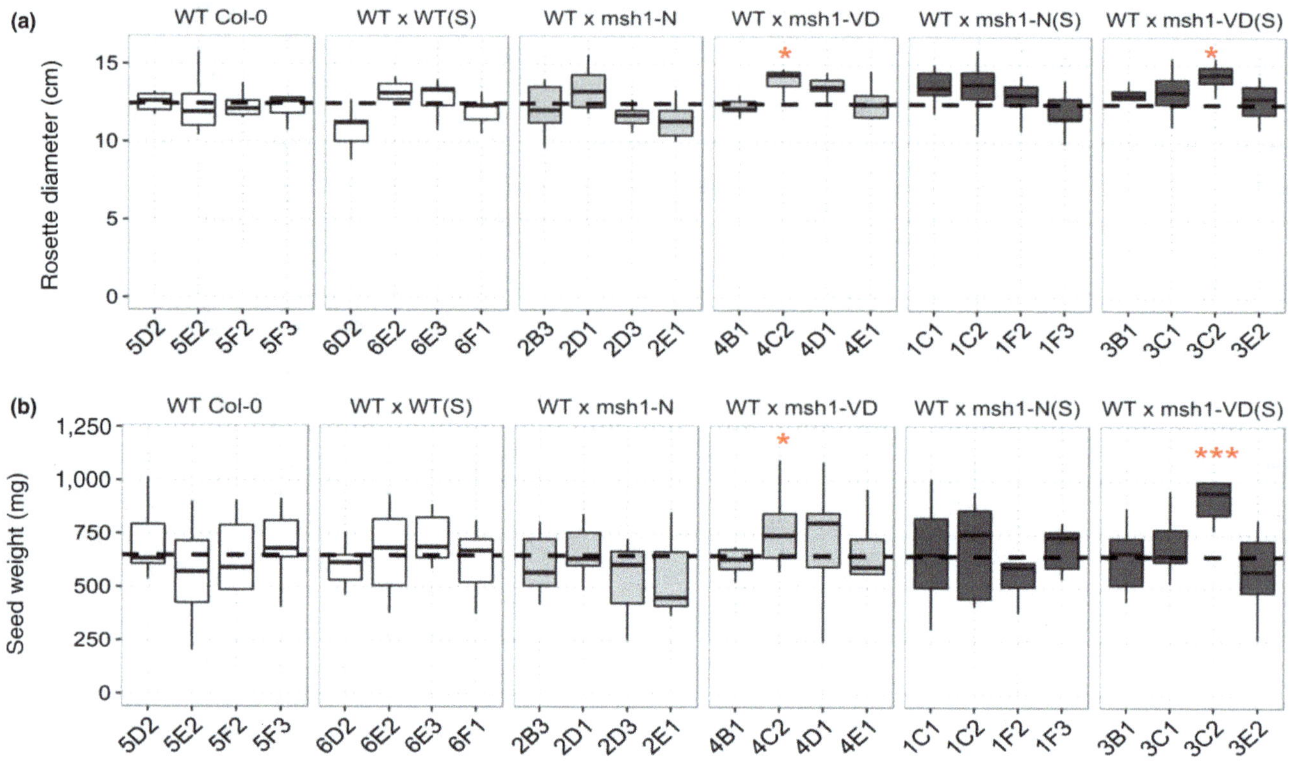

FIGURE 6 Rosette diameter and total seed weight in selected epi$F_{2:3}$. (a) Seed weight measurements from top 20% selection in each epi-F_2 population, including four sub-lines for wild type and WT x WT(S). Seed weight was measured in milligrams dried seeds collected from plants bagged with Arabisifter (Lehle Seeds, SNS-03). Black dotted line represents wild-type average (n = 9 plants each). (b) Rosette diameter measured from the same plants at 6 weeks postsowing. Significance at '***' 0.001 '**' 0.01 '*' 0.05 '.' 0.1

results suggest that an interaction exists between *msh1* effect and cold stress effects. Taken together, data indicate that stressing *msh1* mutants triggers a disproportionate increase in non-CG methylation, but these changes do not affect *msh1*-derived growth vigor and can negatively affect stress adaptation in *msh1*-derived epi-lines.

4 | DISCUSSION

Previous studies have shown *msh1* mutants to be more tolerant to high light, drought, and heat stress (Shedge et al., 2010; Virdi et al., 2016; Xu et al., 2011), consistent with enrichment for abiotic stress response genes (Shao et al., 2017). While we saw increased tolerance for salt stress, *msh1* mutants showed lower survival rate at freezing temperature and in response to the bacterial pathogen *P. syringae*. The seeming incongruity between activation of multiple stress pathways and susceptibility to freezing temperatures may be due to specific mechanisms underlying freezing tolerance in plants, which include plastid membrane remodeling (Moellering et al., 2010). Indeed, low frequency, localized plastid genome changes are reported in *msh1* mutants, along with a reduction in the number of plastids per cell and dramatically altered thylakoid membrane structure in a portion of the organelle population (Xu et al., 2011). Also, loss of *MSH1* might affect the functions of its putative protein interactors, such as the PsbP family protein PPD3 (Virdi et al., 2016), which could further impact the plastid. Alternatively, *msh1* mutants may be unable to mount an appropriate response to freezing stress due to desynchronization of the circadian clock (Shao et al., 2017), which influences freezing tolerance (Maibam et al., 2013). Freezing tolerance is impaired in *cca1-11/lhy-21* double mutants, and *gi-3* mutants are susceptible to freezing stress due to impaired sugar metabolism (Cao, Ye, & Jiang, 2005; Dong, Farré, & Thomashow, 2011). Also, CBF1 and CBF3 genes, which are positive regulators of cold acclimatization (Novillo, Medina, & Salinas, 2007), are down-regulated in *msh1* mutants (Shao et al., 2017).

A recent study has suggested that miR163 is a negative regulator of defense response to *P. syringae* in Arabidopsis (Chow & Ng, 2017). Interestingly, *msh1* mutants with variegation and dwarfing showed up-regulation of miR163, while *msh1* mutants with subtle mutant phenotype did not show any changes (Shao et al., 2017). The up-regulation of miR163 in *msh1* mutants with pronounced phenotype corresponds well with their susceptibility to the bacterial pathogen, while mutants with mild *msh1* phenotype show survival rates similar to wild type (Supporting information Figure S2b). Therefore, one possible explanation for observed stress responses is that organellar changes and/or modulation of key regulatory genes might affect particular stress response, while the vast majority of transcriptional changes may comprise a compensatory response that does not affect the phenotypic outcome.

Whole-genome bisulfite sequencing of *msh1* mutants previously revealed numerous changes in DNA methylation over both genic

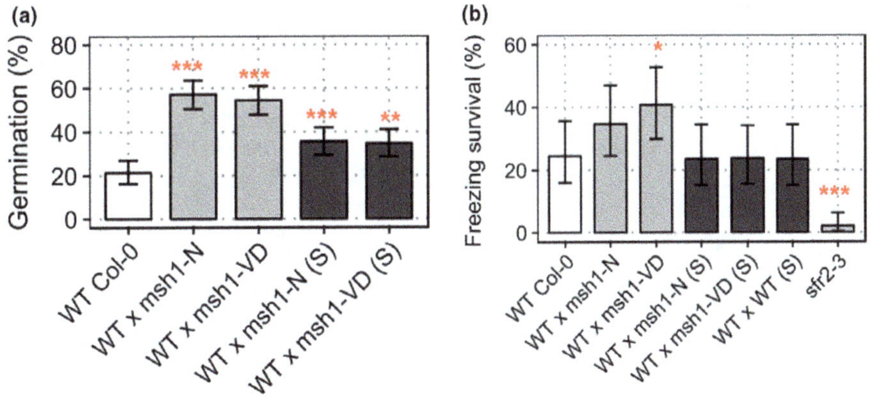

FIGURE 7 Abiotic stress tolerance in epi-F_3s derived from cold-stressed (S) and unstressed *msh1* mutants. (a) Percent germination of wild-type Col-0 and epi-F_3 bulks; WT x *msh1*-N, WT x *msh1*-VD, WT x *msh1*-N(S), WT x *msh1*-VD(S) at 200 mM NaCl-supplemented growth media. Each bar represents three replicates (n = 225), and error bars show SEM. (b) Percent survival of wild type and epi-F_3s, after 12 hr at −10°C. *sfr2-3* was used as negative control. Survival was scored after 1 week recovery from three replicates (n = 100), and error bars represent *SEM*

regions and transposable elements (Virdi et al., 2015), raising the possibility of epigenetic feedback as a response to *MSH1* loss, and heritable methylation changes at stress-responsive loci (Kinoshita & Seki, 2014). Enhanced tolerance to salt stress in epi-lines developed by crossing wild-type Col-0 with *msh1* mutants supports the heritability of methylation changes at stress–responsive loci. The derived epi-lines also showed tolerance to freezing stress, despite the parental *msh1* mutant showing susceptibility to freezing temperatures. It is possible that circadian regulation may resynchronize following the crossing of *msh1* mutants with wild type, which is known to influence freezing tolerance in plants (Maibam et al., 2013). One potential limitation of this study is that we did not sequence the methylomes of epi-lines derived from various *msh1* crosses, which could have shed more light on the regulation of genes responsible for differences in stress tolerance and growth behavior. In comparable soybean epi-F_4 lines, circadian genes *GI* and *PRR3/5/7* were up-regulated (Raju et al., 2018), suggesting that modulation of circadian regulators follows crosses with *msh1*. Derived epi-lines in soybean multi-location field trials display higher yield stability through reduced epitype-by-environment effect (Raju et al., 2018). In Arabidopsis, we likewise observed a lower yield penalty under mild heat stress in epi-lines compared to wild type (Figure 2b), implying higher buffering across environments. These observations invite more detailed investigation of the link between *msh1* derived epigenetic variation and decreased environment interaction in derived epi-lines.

Long-term cold stress disproportionately affects *msh1* mutants, which show an amplified CHH hypomethylation response primarily in the heterochromatic region. It is notable that epigenetic changes reported in *msh1* mutants under cold stress mainly involve non-CG methylation at TE sites, predominantly retroelements known to be affected by stress (Wessler, 1996). We also see down-regulation of *CMT3* and *DDM1* in cold-stressed *msh1* mutant compared to wild type and unstressed *msh1* (Supporting information Figure S6), which is implicated in heterochromatic TE derepression. The *msh1* mutants showed significant differences in expression of TE superfamilies. Differentially expressed TEs belonged to Rath elements, SINEs, and Mariner superfamilies known to contain shorter TEs, on average (Lewsey et al., 2016), that are usually methylated by the DRM1/2 pathway (Stroud et al., 2013, 2014). Mariner TE sequences are significantly underrepresented in exons and are often absent in GC-rich genic regions of the genome (Lockton & Gaut, 2009).

Transcriptome studies showed that stress was consistently the major contributor to gene expression changes in wild type and *msh1* mutants. This is expected as changes in CHG and CHH methylation in Arabidopsis are concentrated around the pericentromere, while CG changes are distributed throughout the genome, and non-CG methylation changes are unlikely to direct gene expression changes. A recent study involving multiple ecotypes in Arabidopsis has shown CHH hypomethylation from lower temperatures, with much of the temperature variation in CHH methylation due to components of the RdDM pathway (Dubin et al., 2015). Reports of chromatin changes and epigenetic features of stress memory in plants and observations that some epigenetic mutants have altered stress responses support the argument that these changes may have biological roles (Probst & Mittelsten Scheid, 2015). Interestingly, the increased variation in non-CG methylome divergence in *msh1* mutants does not seem to have any significant effect on the previously described *msh1*-derived enhanced growth phenotype (Virdi et al., 2015), emphasizing the importance of *MSH1*-induced CG methylation changes in this phenomenon. CG methylation changes are more stably transmitted to progeny than non-CG changes (Saze, Scheid, & Paszkowski, 2003). A recent study also showed that non-repetitive sequences and higher CG content predispose a region for the transgenerational stability of inherited epigenetic features (Catoni et al., 2017). Moreover, stress-induced epigenetic memory is conditionally heritable through the female germline (Wibowo et al., 2016). This excludes the possibility of inheritance of stress-induced methylation changes, particularly non-CG changes to the crossed progeny of this study since *msh1* mutants were used as the pollen donor.

Results from this study indicate that *msh1* methylomes are hyper-responsive to environmental stress in a manner distinct from the wild-type response, and identification of the *msh1* background as a modifier of cold-induced CHH hypomethylation provides an experimental system to further understand mechanisms that control temperature-responsive methylation changes and their inheritance behavior in crossed and selfed progeny. The experimental results indicate that CG methylation changes rather than non-CG in *msh1* mutants as an influence on growth behavior of epi-lines following crossing with wild type.

ACKNOWLEDGMENTS

We thank Bridget Bickner for technical assistance with phenotypic measurements and PCR analysis, Dr Rebecca Roston for help in establishing freezing tolerance tests, and Dr. John Laurie for valuable conversations early in the study. We also acknowledge funding to S.M. from the Bill and Melinda Gates Foundation (OPP1088661). SM and SKKR performed conceptualization; SKKR and SM performed experiment design; SKKR performed experiments, with YW in biotic stress tests; MRS and SKKR performed data analysis; SKKR performed writing—original draft: SM performed writing—review and editing; all authors read and approved the final manuscript.

REFERENCES

Akalin, A., Kormaksson, M., Li, S., Garrett-Bakelman, F. E., Figueroa, M. E., Melnick, A., & Mason, C. E. (2012). methylKit: A comprehensive R package for the analysis of genome-wide DNA methylation profiles. *Genome biology, 13*, R87. https://doi.org/10.1186/gb-2012-13-10-r87

Barnes, A. C., Benning, C., & Roston, R. (2016). Chloroplast membrane remodeling during freezing stress is accompanied by cytoplasmic acidification activating SENSITIVE TO FREEZING 2. *Plant physiology, 171*(3), 2140–2149. https://doi.org/10.1104/pp.16.00286

Benjamini, Y., & Hochberg, Y. (1995). Controlling the false discovery rate: A practical and powerful approach to multiple testing. *Journal of the royal statistical society. Series B (Methodological), 57*(1), 289–300.

Bilichak, A., & Kovalchuk, I. (2016). Transgenerational response to stress in plants and its application for breeding. *Journal of Experimental Botany, 67*, 2081–2092. https://doi.org/10.1093/jxb/erw066

Cao, S., Ye, M., & Jiang, S. (2005). Involvement of GIGANTEA gene in the regulation of the cold stress response in Arabidopsis. *Plant cell reports, 24*, 683–690. https://doi.org/10.1007/s00299-005-0061-x

Catoni, M., Griffiths, J., Becker, C., Zabet, N. R., Bayon, C., Dapp, M., … Paszkowski, J. (2017). DNA sequence properties that predict susceptibility to epiallelic switching. *The EMBO Journal, 36*(5), 617–628. https://doi.org/10.15252/embj.201695602

Cheng, X. (1995). Structure and function of DNA methyltransferases. *Annual Review of Biophysics and Biomolecular Structure, 24*(1), 293–318.

Chow, H. T., & Ng, D. W. (2017). Regulation of miR163 and its targets in defense against Pseudomonas syringae in Arabidopsis thaliana. *Scientific Reports (Nature Publisher Group), 7*, 46433. https://doi.org/10.1038/srep46433

Cokus, S. J., Feng, S., Zhang, X., Chen, Z., Merriman, B., Haudenschild, C. D., … Jacobsen, S. E. (2008). Shotgun bisulphite sequencing of the Arabidopsis genome reveals DNA methylation patterning. *Nature, 452*(7184), 215.

Dobin, A., Davis, C. A., Schlesinger, F., Drenkow, J., Zaleski, C., Jha, S., … Gingeras, T. R. (2013). STAR: Ultrafast universal RNA-seq aligner. *Bioinformatics, 29*, 15–21. https://doi.org/10.1093/bioinformatics/bts635

Dong, M. A., Farré, E. M., & Thomashow, M. F. (2011). Circadian clock-associated 1 and late elongated hypocotyl regulate expression of the C-repeat binding factor (CBF) pathway in Arabidopsis. *Proceedings of the National Academy of Sciences, 108*, 7241–7246. https://doi.org/10.1073/pnas.1103741108

Dowen, R. H., Pelizzola, M., Schmitz, R. J., Lister, R., Dowen, J. M., Nery, J. R., … Ecker, J. R. (2012). Widespread dynamic DNA methylation in response to biotic stress. *Proceedings of the National Academy of Sciences of the United States of America, 109*, E2183–E2191. https://doi.org/10.1073/pnas.1209329109

Dubin, M. J., Zhang, P., Meng, D., Remigereau, M.-S., Osborne, E. J., Casale, F. P., … Vilhjálmsson, B. (2015). DNA methylation in Arabidopsis has a genetic basis and shows evidence of local adaptation. *Elife, 4*, e05255.

Franks, S. J., & Hoffmann, A. A. (2012). Genetics of climate change adaptation. *Annual Review of Genetics, 46*, 185–208. https://doi.org/10.1146/annurev-genet-110711-155511

Hannah, M. A., Wiese, D., Freund, S., Fiehn, O., Heyer, A. G., & Hincha, D. K. (2006). Natural genetic variation of freezing tolerance in Arabidopsis. *Plant physiology, 142*, 98–112. https://doi.org/10.1104/pp.106.081141

Hsu, P. Y., & Harmer, S. L. (2012). Circadian phase has profound effects on differential expression analysis. *PLoS ONE, 7*, e49853. https://doi.org/10.1371/journal.pone.0049853

Huang, D. W., Sherman, B. T., & Lempicki, R. A. (2009). Systematic and integrative analysis of large gene lists using DAVID bioinformatics resources. *Nature protocols, 4*, 44–57. https://doi.org/10.1038/nprot.2008.211

Jin, Y., Tam, O. H., Paniagua, E., & Hammell, M. (2015). TEtranscripts: A package for including transposable elements in differential expression analysis of RNA-seq datasets. *Bioinformatics, 31*, 3593–3599. https://doi.org/10.1093/bioinformatics/btv422

Kim, D., Pertea, G., Trapnell, C., Pimentel, H., Kelley, R., & Salzberg, S. L. (2013). TopHat2: Accurate alignment of transcriptomes in the presence of insertions, deletions and gene fusions. *Genome Biology, 14*, R36. https://doi.org/10.1186/gb-2013-14-4-r36

King, E. O., Ward, M. K., & Raney, D. E. (1954). Two simple media for the demonstration of pyocyanin and fluorescin. *Journal of laboratory and clinical medicine, 44*, 301–307.

Kinoshita, T., & Seki, M. (2014). Epigenetic memory for stress response and adaptation in plants. *Plant and Cell Physiology, 55*, 1859–1863. https://doi.org/10.1093/pcp/pcu125

Krueger, F., & Andrews, S. R. (2011). Bismark: A flexible aligner and methylation caller for Bisulfite-Seq applications. *Bioinformatics, 27*, 1571–1572. https://doi.org/10.1093/bioinformatics/btr167

Law, J. A., & Jacobsen, S. E. (2010). Establishing, maintaining and modifying DNA methylation patterns in plants and animals. *Nature Reviews Genetics, 11*(3), 204.

Lewsey, M. G., Hardcastle, T. J., Melnyk, C. W., Molnar, A., Valli, A., Urich, M. A., … Ecker, J. R. (2016). Mobile small RNAs regulate genome-wide DNA methylation. *Proceedings of the National Academy of Sciences, 113*, E801–E810. https://doi.org/10.1073/pnas.1515072113

Li, J., & Chory, J. (1998). Preparation of DNA from Arabidopsis. In J. J. Sanchez-Serano & J. Salinas (Eds.), *Arabidopsis Protocols. Methods in Molecular Biology™ Book Series* (MIMB, vol *82*, pp. 55–60). New York, NY: Humana Press.

Lockton, S., & Gaut, B. S. (2009). The contribution of transposable elements to expressed coding sequence in Arabidopsis thaliana. *Journal of Molecular Evolution, 68*, 80–89. https://doi.org/10.1007/s00239-008-9190-5

Love, M. I., Huber, W., & Anders, S. (2014). Moderated estimation of fold change and dispersion for RNA-seq data with DESeq2. *Genome Biology, 15*, 550. https://doi.org/10.1186/s13059-014-0550-8

Maibam, P., Nawkar, G. M., Park, J. H., Sahi, V. P., Lee, S. Y., & Kang, C. H. (2013). The influence of light quality, circadian rhythm, and photoperiod on the CBF-mediated freezing tolerance. *International journal of molecular sciences*, *14*, 11527–11543. https://doi.org/10.3390/ijms140611527

Moellering, E. R., Muthan, B., & Benning, C. (2010). Freezing tolerance in plants requires lipid remodeling at the outer chloroplast membrane. *Science*, *330*, 226–228. https://doi.org/10.1126/science.1191803

Novillo, F., Medina, J., & Salinas, J. (2007). Arabidopsis CBF1 and CBF3 have a different function than CBF2 in cold acclimation and define different gene classes in the CBF regulon. *Proceedings of the National Academy of Sciences*, *104*, 21002–21007. https://doi.org/10.1073/pnas.0705639105

Probst, A. V., & Mittelsten Scheid, O. (2015). Stress-induced structural changes in plant chromatin. *Current Opinion in Plant Biology*, *27*, 8–16. https://doi.org/10.1016/j.pbi.2015.05.011

Raju, S. K. K., Shao, M. R., Sanchez, R., Xu, Y. Z., Sandhu, A., Graef, G., & Mackenzie, S. (2018). An epigenetic breeding system in soybean for increased yield and stability. *Plant Biotechnology Journal*, https://doi.org/10.1111/pbi.12919

de la Rosa Santamaria, R., Shao, M. R., Wang, G., Nino-Liu, D. O., Kundariya, H., Wamboldt, Y., … Mackenzie, S. A. (2014). MSH1-induced non-genetic variation provides a source of phenotypic diversity in Sorghum bicolor. *PLoS ONE*, *9*, e108407. https://doi.org/10.1371/journal.pone.0108407

Sanchez, R., Yang, X., Kundariya, H., Barreras, J.R., Wamboldt, Y., & Mackenzie, S. (2018). Enhancing resolution of natural methylome reprogramming behavior in plants. *bioRxiv*, 252106. https://doi.org/10.1101/252106

Saze, H., Scheid, O. M., & Paszkowski, J. (2003). Maintenance of CpG methylation is essential for epigenetic inheritance during plant gametogenesis. *Nature genetics*, *34*, 65–69. https://doi.org/10.1038/ng1138

Sham, A., Moustafa, K., Al-Ameri, S., Al-Azzawi, A., Iratni, R., & AbuQamar, S. (2015). Identification of Arabidopsis candidate genes in response to biotic and abiotic stresses using comparative microarrays. *PLoS ONE*, *10*, e0125666. https://doi.org/10.1371/journal.pone.0125666

Shao, M.-R., Kumar Kenchanmane Raju, S., Laurie, J.D., Sanchez, R., & Mackenzie, S.A. (2017). Stress-responsive pathways and small RNA changes distinguish variable developmental phenotypes caused by MSH1 loss. *BMC Plant Biology*, *17*, 47. https://doi.org/10.1186/s12870-017-0996-4

Shedge, V., Davila, J., Arrieta-Montiel, M. P., Mohammed, S., & Mackenzie, S. A. (2010). Extensive rearrangement of the Arabidopsis mitochondrial genome elicits cellular conditions for thermotolerance. *Plant physiology*, *152*, 1960–1970. https://doi.org/10.1104/pp.109.152827

Slotkin, R. K., & Martienssen, R. (2007). Transposable elements and the epigenetic regulation of the genome. *Nature Reviews Genetics*, *8*, 272–285. https://doi.org/10.1038/nrg2072

Stroud, H., Do, T., Du, J., Zhong, X., Feng, S., Johnson, L., … Jacobsen, S. E. (2014). Non-CG methylation patterns shape the epigenetic landscape in Arabidopsis. *Nature structural & molecular biology*, *21*, 64–72. https://doi.org/10.1038/nsmb.2735

Stroud, H., Greenberg, M. V., Feng, S., Bernatavichute, Y. V., & Jacobsen, S. E. (2013). Comprehensive analysis of silencing mutants reveals complex regulation of the Arabidopsis methylome. *Cell*, *152*, 352–364. https://doi.org/10.1016/j.cell.2012.10.054

Uller, T., English, S., & Pen, I. (2015). When is incomplete epigenetic resetting in germ cells favoured by natural selection? *Proceedings. Biological Sciences*, *282*, 20150682. In: *Proc. R. Soc. B* p. 20150682. The Royal Society. https://doi.org/10.1098/rspb.2015.0682

Virdi, K. S., Laurie, J. D., Xu, Y. Z., Yu, J., Shao, M. R., Sanchez, R., … Mackenzie, S. A. (2015). Arabidopsis MSH1 mutation alters the epigenome and produces heritable changes in plant growth. *Nature Communications*, *6*, 6386. https://doi.org/10.1038/ncomms7386

Virdi, K. S., Wamboldt, Y., Kundariya, H., Laurie, J. D., Keren, I., Kumar, K. S., … Elowsky, C. (2016). MSH1 is a plant organellar DNA binding and thylakoid protein under precise spatial regulation to alter development. *Molecular Plant*, *9*, 245–260. https://doi.org/10.1016/j.molp.2015.10.011

Ward, J. H. Jr (1963). Hierarchical grouping to optimize an objective function. *Journal of the American statistical association*, *58*, 236–244. https://doi.org/10.1080/01621459.1963.10500845

Wessler, S. R. (1996). Plant retrotransposons: Turned on by stress. *Current Biology*, *6*, 959–961. https://doi.org/10.1016/S0960-9822(02)00638-3

Wibowo, A., Becker, C., Marconi, G., Durr, J., Price, J., Hagmann, J., … Becker, J. (2016). Hyperosmotic stress memory in Arabidopsis is mediated by distinct epigenetically labile sites in the genome and is restricted in the male germline by DNA glycosylase activity. *Elife*, *5*, e13546.

Xu, Y. Z., Arrieta-Montiel, M. P., Virdi, K. S., de Paula, W. B., Widhalm, J. R., Basset, G. J., … Mackenzie, S. A. (2011). MutS HOMOLOG1 is a nucleoid protein that alters mitochondrial and plastid properties and plant response to high light. *Plant Cell*, *23*, 3428–3441. https://doi.org/10.1105/tpc.111.089136

Xu, Y. Z., Santamaria Rde, L., Virdi, K. S., Arrieta-Montiel, M. P., Razvi, F., Li, S., … Mackenzie, S. A. (2012). The chloroplast triggers developmental reprogramming when mutS HOMOLOG1 is suppressed in plants. *Plant Physiology*, *159*, 710–720. https://doi.org/10.1104/pp.112.196055

Yang, X., Kundariya, H., Xu, Y. Z., Sandhu, A., Yu, J., Hutton, S. F., … Mackenzie, S. A. (2015). MutS HOMOLOG1-derived epigenetic breeding potential in tomato. *Plant Physiology*, *168*, 222–232. https://doi.org/10.1104/pp.15.00075

Yong-Villalobos, L., González-Morales, S. I., Wrobel, K., Gutiérrez-Alanis, D., Cervantes-Peréz, S. A., Hayano-Kanashiro, C., … Herrera-Estrella, L. (2015). Methylome analysis reveals an important role for epigenetic changes in the regulation of the Arabidopsis response to phosphate starvation. *Proceedings of the National Academy of Sciences of the United States of America*, *112*, E7293–E7302. https://doi.org/10.1073/pnas.1522301112

Zilberman, D. (2017). An evolutionary case for functional gene body methylation in plants and animals. *Genome biology*, *18*, 87. https://doi.org/10.1186/s13059-017-1230-2

8

Consequences of the loss of catalytic triads in chloroplast CLPPR protease core complexes in vivo

Jui-Yun Rei Liao | Giulia Friso | Jitae Kim | Klaas J. van Wijk

Section of Plant Biology, School of Integrative Plant Sciences (SIPS), Cornell University, Ithaca, New York

Correspondence
Klaas J. van Wijk, Section of Plant Biology, School of Integrative Plant Sciences (SIPS), Cornell University, Ithaca, NY.
Email: kv35@cornell.edu

Funding information
This research was supported by a grant from the National Science Foundation (NSF), Division of Molecular and Cellular Biosciences (MCB) #1614629 to K.J.V.W., whereas J-Y.R.L was supported by a three-year fellowship from the Ministry of Education (MOE) of Taiwan

Abstract

The essential chloroplast CLP protease system consists of a tetradecameric proteolytic core with catalytic P (P1, 3–6) and non-catalytic R (R1–4) subunits, CLP chaperones and adaptors. The chloroplast CLP complex has a total of ten catalytic sites, but it is not known how many of these catalytic sites can be inactivated before plants lose viability. Here we show that CLPP3 and the catalytically inactive variant CLPP3S164A fully complement the developmental arrest of the *clpp3-1* null mutant, even under environmental stress. In contrast, whereas the inactive variant CLPP5S193A assembled into the CLP core, it cannot rescue the embryo lethal phenotype of the *clpp5-1* null mutant. This shows that CLPP3 makes a unique structural contribution but its catalytic site is dispensable, whereas the catalytic activity of CLPP5 is essential. Mass spectrometry of affinity-purified CLP cores of the complemented lines showed highly enriched CLP cores. Other chloroplast proteins were co-purified with the CLP cores and are candidate substrates. A strong overlap of co-purified proteins between the CLP core complexes with active and inactive subunits indicates that CLP cores with reduced number of catalytic sites do not over-accumulate substrates, suggesting that the bottle-neck for degradation is likely substrate recognition and unfolding by CLP adaptors and chaperones, upstream of the CLP core.

KEYWORDS
catalytic triad, chloroplast, CLP protease, CLPP3, CLPP5, substrates

1 | INTRODUCTION

ATP-dependent Clp proteases are present in bacteria, as well as mitochondria and plastids (organelles of bacterial origin), where they regulate accumulation levels of a broad range of substrates (Alexopoulos, Guarne, & Ortega, 2012; Liu, Ologbenla, & Houry, 2014; Nishimura, Kato, & Sakamoto, 2017; Nishimura & van Wijk, 2015; Sauer & Baker, 2011). The first step in the CLP degradation process requires the recognition of substrates by the CLP AAA+ chaperones, possibly aided by specific adaptors (also named recognins) that recognize and deliver specific substrates (Kuhlmann & Chien, 2017; Mahmoud & Chien, 2018). The ATP-dependent CLP chaperones then dock onto CLP protease core complexes consisting of two stacked heptameric rings, and unfold and direct substrates into the CLP protease complex (Olivares, Baker, & Sauer, 2018). The substrates are cleaved within the CLP protease complex and short (~7–10 amino acids) peptide fragments are released through lateral pores in the CLP protease (Alexopoulos et al., 2012; Liu et al., 2014; Peltier et al., 2004). To ensure optimal cellular levels of functional proteins and to remove unwanted proteins while avoiding uncontrolled degradation, substrate recognition and delivery by the adaptors and chaperones must be tightly regulated and depends also on the availability and exposure of degrons in the substrates.

The CLP proteolytic core *in Escherichia coli* consists of 14 identical CLPP subunits that belong to the family of serine-type proteases

(Wang, Hartling, & Flanagan, 1997). The photosynthetic bacterium *Synechococcus elongatus* contain three different CLPP proteins, and one CLPR proteins (Schelin, Lindmark, & Clarke, 2002), which is structurally similar to CLPP but lacks the catalytic residues for peptide bond hydrolysis. These cyanobacterial proteins assemble into an essential CLPR/P3 complex (identical heptameric rings with P3:R in a 3:4 ratio) and a non-essential CLPP1/P2 complex (Andersson et al., 2009; Stanne, Pojidaeva, Andersson, & Clarke, 2007). The apicoplast of *Plasmodium falsiparum* accumulates both a single CLPP and CLPR protein but they do appear to form separate homo-oligomers (El Bakkouri et al., 2013). Compared to these homologs, the chloroplast CLP system is far more diversified and complex (Nishimura & van Wijk, 2015; Olinares, Kim, Davis, & van Wijk, 2011; Yu & Houry, 2007).

The chloroplast CLP protease complex in Arabidopsis contains five different CLPP subunits namely plastid-encoded CLPP1 and the nuclear-encoded CLPP3-6, and four different non-catalytic CLPR subunits CLPR1-4. The five CLPP subunits with conserved catalytic sites accumulate in a 3:1:2:3:1 ratio for P1:P3:P4:P5:P6, whereas the non-catalytic CLPR subunits are each present in one copy. Three copies of CLPP1 and the four CLPR subunits form the heptameric R-ring, whereas CLPP3-6 form the heptameric P-ring. Therefore the chloroplast CLP complex has a total of ten catalytic sites, of which seven in the P-ring and only three in the R-ring. It is not known how many and which of these catalytic sites can be inactivated before plants lose viability. The chloroplast CLP core also associates with two plant specific proteins ClpT1 and ClpT2. These CLPT subunits likely function in CLPPR core formation, stabilization and activation (Kim et al., 2015; Sjogren & Clarke, 2011).

Arabidopsis CLPP and CLPR proteins are not very conserved to each other at the primary sequence level (24–48% identities between CLPPs; 28–38% identities between CLPRs). Severe phenotypes were found in Arabidopsis *clpp* and *clpr* mutants and in tobacco RNAi lines (Moreno et al., 2017) at various developmental stages (Kim et al., 2009, 2013; Koussevitzky et al., 2007; Rudella, Friso, Alonso, Ecker, & van Wijk, 2006; Zheng, MacDonald, Sutinen, Hurry, & Clarke, 2006). For example, null mutants in *CLPP3* (*clpp3-1*) are seedling lethal and are arrested in the cotyledon stage. However, the addition of sucrose breaks this developmental arrest and plants form pale-green leaves and eventually flower and produce seed (Kim et al., 2013). In contrast, null mutants in *CLPP5* (*clpp5-1*) are embryo lethal (Kim et al., 2009). These phenotypes showed that both CLP3 and CLP5 are required for plant development, and we speculated that the differential phenotype between *clpp3-1* and *clpp5-1* is in part due to the higher copy number of CLPP5 (three per complex) than CLPP3 (one per complex). Through site-directed mutagenesis and complementation of null mutants, this study tests if the catalytic contribution of CLPP3 and CLPP5 can be inactivated (while keeping the structural contribution) without functional consequences for growth and development.

So far, several (candidate) chloroplast CLP substrates have been identified based on their direct interaction with the CLPS1 adaptor, including GLUTR (Nishimura et al., 2013), which was subsequently confirmed as a CLP substrate (Apitz et al., 2016). Other candidate CLPS1 substrates are four enzymes in the chloroplast shikimate pathway (Nishimura et al., 2013). Furthermore, systematic screening of protein stability of the thylakoid copper transporter, PAA2/HMA8 (P-type ATPase of Arabidopsis2/Heavy-metal-associated8) in various plastid protease mutants identified the CLPPR core and CLPC1 as essential components to degrade PAA2 under copper replete conditions (Tapken, Kim, Nishimura, van Wijk, & Pilon, 2015). Additionally, DEOXYXYLULOSE 5-PHOSHATE (DXS) in the chloroplast isoprenoid biosynthesis pathway and PHYTOENE SYNTHASE (PSY) involved in carotenoid biosynthesis were reported as a CLP substrates (Pulido et al., 2016; Welsch et al., 2018). Other candidate substrates have been suggested based on comparative proteome analysis of a range of *clp* mutants in Arabidopsis, rice and tobacco, but here it is hard to distinguish between direct and indirect effects – reviewed in (Nishimura, Kato, & Sakamoto, 2016) and (Moreno et al., 2018; Wu et al., 2016).

An alternative approach to identify CLP substrates, termed substrate trapping, has been reported for various bacterial and fungal CLP protease systems (Feng et al., 2013; Flynn, Neher, Kim, Sauer, & Baker, 2003; Trentini et al., 2016). In this approach, the catalytic activity of the CLP core is inactivated through site-mutagenesis of the serine residue within the catalytic triad (Ser-His-Asp), resulting in accumulation of substrates within the CLP core central cavity and facilitating their identification by tandem mass spectrometry (MS/MS). This in vivo trapping strategy has been applied for the homo-tetrameric CLP complexes in the gram-negative bacterium *E. coli* (Flynn et al., 2003), the gram-positive *Staphylococcus aureus* (Feng et al., 2013) and *Bacillus subtilis* (Trentini et al., 2016), as well as the fungus *Podospora anserine* (Fischer, Langer, & Osiewacz, 2015), but not yet in plants. It is important to note that the inactivation of the CLP system in these species does not greatly affect viability. The subunit complexity and the essential nature of the chloroplast CLP core makes an in vivo CLP protease trapping approach in plants more challenging than in bacteria and fungi. In this study we used this approach for the *Arabidopsis* chloroplast CLP protease generating in vivo tagged CLP core complexes containing inactive serine to alanine variants in either CLPP3 or CLPP5. MS/MS of affinity-purified tagged CLP cores from the various transformants showed that we successfully obtained highly enriched STREPII-tagged CLP cores with all catalytically active CLPP3 and CLPP5 copies and with catalytically inactive CLPP3 or CLPP5 copies. Other chloroplast proteins were co-purified with the CLP cores and we evaluate and discuss their significance as candidate substrates in the context of bottlenecks in the chloroplast CLP adaptor-chaperone-protease system.

2 | METHODS

2.1 | Plant materials, cloning, transformation, genotyping, RT-PCR

The T-DNA insertion lines for *CLPP3* (At1g66670) and *CLPP5* (At1g02560) are SALK_065330 (*clpp3-1*) and SALK_007708 (*clpp5-1*), respectively, as described in (Kim et al., 2013). Heterozygous *clpp3-1* and *clpp5-1* were used for complementation. To generate transgene constructs, 3.5 kb of *CLPP3* and 3.7 kb of *CLPP5* genomic DNA including 1 kb upstream and 800 bp downstream from the coding region were cloned using gene-specific primers (primer set 1 – see Supporting information Table S1). A 24-nucleotide sequence, for the 8-amino acid strepII tag (WSHPQFEK), was introduced into the transgene before the stop codon by PCR amplification (primer set 2 – see Supporting information Table S1). To create the *CLPP3S164A-STREPII* and *CLPP5S193A-STREPII* constructs, the catalytic serine of CLPP3 and CLPP5 was substituted by alanine using PCR amplification using primer set 3. All PCR products were sub-cloned into pCR8/GW/TOPO®vector (Invitrogen) and binary vector pMDC123 using Gateway® LR Clonase® II enzyme mix (Invitrogen) and then introduced into the *Agrobacterium tumefaciens* GV3101 by electroporation. Agrobacterium-mediated transformation by the flower dip method was according to (Clough & Bent, 1998). Crude plant genomic DNA was extracted from grinding frozen leaf tissues in the gDNA extraction buffer (0.2 M Tris-HCl, pH 7.5, 0.25 M NaCl, 0.025 M EDTA, pH 8.0, 0.5% SDS) and precipitated by isopropanol. Genotyping was performed using *CLPP3/5*-specific (primer set 4), transgene-specific primers (primer set 5), and T-DNA specific primers (primer set 6) Total RNA was extracted from grinding frozen leaf tissues using RNeasy Plant Mini Kit (Qiagen). The first-strand of cDNA was synthesized using SuperScript® III First-Strand Synthesis System (Invitrogen) and then used for the synthesis of the second-strand cDNA. Primers are listed in Supporting information Table S1.

2.2 | Plant growth and phenotypic analysis

WT, *clpp3-1* null, and complemented CLPP3-STREPII and CLPP3S164A-STREPII T2 seeds were surface disinfected with 70% ethanol for 15 min and then rinsed with 95% ethanol for three times before sowing on the half Murashige and Skoog (1/2 MS) plates. These plates were put in a cold room (4°C, dark) for 3 days stratification and then transferred into the growth chamber with 16/8 hr light/dark at 70 μmol photons m^{-2} s^{-1} for 12 days. Plates were relocated and observed every day. For soil-grown phenotypic analysis, wt, complemented *CLPP3-STREPII* and *CLPP3S164A-STREPII* T3 seeds were sown on soil, put in a cold room for 3 days stratification, and then transferred in a growth chamber (16/8 hr light/dark at 120–150 μmol photons m^{-2} s^{-1}). These soil-grown plants were observed every 2 days. The plant height and the number of the rosette leaves were recorded for growth and development rates. Complemented CLPP5-STREPII and transgenic CLPP5S193A-STREPII were grown side by side on soil after 5 days stratification and then grown under 10/14 hr light/dark at 120–150 μmol photons m^{-2} s^{-1}. BASTA-spraying was performed every 2–3 days for the first 14 days. Surviving CLPP5-STREPII and CLPP5S193A-STREPII were genotyped and observed every week. At least ten developing siliques with similar size (length 1.8–2.0 cm) from heterozygous *clpp5-1* and heterozygous *clpp5-1* with *CLPP5S193A-STREPII* T2 plants were examined under the dissecting microscope. Seeds were counted according to their colors.

2.3 | SDS-PAGE, BN-PAGE, and immunoblot

Total leaf protein extraction under denaturing condition and non-denaturing condition was performed according to (Friso, Olinares, & van Wijk, 2011) and (Olinares et al., 2011), respectively. Isolation of stromal proteins was according to (Kim et al., 2015). For immunoblot analysis of transgene expression and the assembly state of CLPPRT complexes, 20 μg of total leaf soluble proteins or 10–50 μg of stromal proteins were separated on SDS-PAGE or Novex pre-cast Bis-Tris 4–16% gel (Invitrogen), transferred onto nitrocellulose or PVDF membranes, and stained with Ponceau S solution (0.3% Ponceau in 3% TCA). Immunoblot blotting using anti-STREPII, anti-CLPR2 and, anti-CLPP3 antiserum was according to (Kim et al., 2015).

2.4 | Affinity purification of STREPII-tagged CLP complexes and MS/MS analysis

At least 2 mg stromal proteins or 24 mg total leaf proteins extracted under the non-denaturing conditions was loaded on a self-packed StrepTactin column using the superflow high capacity resin (IBA) according to (Olinares et al., 2011), except that 5 mM biotin instead of 2.5 mM desthiobiotin was included in the elution step. To minimize endogenous biotin-conjugated proteins binding to the StrepTactin columns, avidin (0.2 mg per g leaf tissue) was included in the extraction buffer during total soluble protein extraction from CLPP5-STREPII and CLPP5S193A-STREPII lines. Amicon and Microcon spin concentrators (3 kDa) were used for concentration of the eluates before further analysis. Concentrated eluates were separated on 10.5–14% precast gels (Biorad), followed by the MS-compatible silver stain. Comparative proteome analysis using a LTQ-Orbitrap mass spectrometer, data processing, database searches, quantification of the relative protein abundance, as well as the selection of the best gene models followed the procedures described in Friso et al. (2011). We evaluated the samples for potential enrichment based on matched MS/MS adjusted spectra (adjSPC) normalized to the total number of adjSPC in each sample, resulting in NadjSPC. Annotations are from the Plant Proteome Data Base (http://ppdb.tc.cornell.edu/).

3 | RESULTS AND DISCUSSION

3.1 | Phenotypic analysis shows that the catalytic activity of CLPP3 is not needed for function

The catalytic triad Ser-His-Asp in CLPP proteases is widely conserved across bacterial CLPP proteins as well mitochondrial and plastid CLPP (but not in the catalytically inactive CLPR proteins) (Kim & Kim, 2008; Olinares et al., 2011; Peltier et al., 2004) (Supporting information Figure S1). Changing the serine residue into an alanine is sufficient to completely block proteolytic activity of CLPP proteases in all known cases, *e.g. E. coli* (Flynn, Levchenko, Sauer, & Baker, 2004), *Mycobacterium tuberculosis* (Raju et al., 2012), *Bacillus subtilis* (Trentini et al., 2016) and *Synechococcus elongates* (Andersson et al., 2009).

To determine if the contribution of CLPP3 to the core complex is through its unique structure or also by contributing a catalytic site to the core complex, we transformed heterozygous *clpp3-1* mutants with genomic *CLPP3* with a C-terminal STREPII tag (*CLPP3-STREPII*), or the same genomic construct but with a serine to alanine mutation in the catalytic site (*CLPP3S164A-STREPII*) rendering CLPP3 catalytically inactive. Following growth on selective medium and genotyping of the T1 and T2 populations, we obtained multiple homozygous *clpp3-1* lines expressing either *CLPP3-STREPII* or *CLPP3S164A-STREPII* (Figure 1a,b; Supporting information Figure S2A–C). No significant visible differences were observed among these complemented *clpp3-1* lines and wt, as they all germinated and developed without the need for sucrose, as well as displayed normal green rosettes and similar growth and development (Figure 1a,b). In contrast, *clpp3-1* nulls were arrested at the cotyledon stage (Figure 1a), as described previously (Kim et al., 2013).

The point mutation in CLPP3 was confirmed by DNA sequencing of RT-PCR products generated by *CLPP3* and *STREPII* specific primers (Figure 1c). Furthermore, MS/MS of affinity purified CLP complexes identified the specific point mutation S164A in CLPP3 (Figure 1d; Supporting information Figure S3), thus further confirming that the point mutation in CLPP3 was successfully introduced in these lines. Accumulation of the transgenic CLPP3-STREPII and CLPP3S164A-STREPII proteins in vivo was also confirmed by immunoblotting of the denaturing soluble proteome separated by SDS-PAGE and detected with anti-STREPII and anti-CLPP3 antisera (Figure 1e). To test if STREPII-tagged CLPP3 and CLPP3S164A proteins normally assembled into CLPRT complexes, soluble leaf proteomes were separated by native gel electrophoresis (BN-PAGE), followed by immunoblotting (Figure 1f). The similar size and migration pattern of CLPPRT complexes on BN-PAGE for wt and complemented lines indicated that (i) these STREPII-tagged transgenic proteins assembled into CLPPRT complexes in vivo, and (ii) the S-to-A change in CLPP3 and the STREPII tag did not interfere with the assembly state of CLPPRT complexes. Collectively, this shows that both catalytically active CLPP3-STREPII and catalytic inactive CLPP3S164A-STREPII successfully complement the developmental arrest of *clpp3-1* null mutants. It can thus be concluded that CLPP3 makes a unique structural contribution to the CLPPR complex but that its catalytic activity is dispensable for plant growth and development.

3.2 | The catalytic activity of CLPP5 is required for function; S-to-A change in CLPP5 prevents complementation of clpp5-1

Similar as for CLPP3, heterozygous *clpp5-1* was transformed with genomic *CLPP5* with a C-terminal STREPII tag (*CLPP5-STREPII*), or the same construct but with a serine to alanine mutation in the catalytic site (*CLPP5S193A-STREPII*) that rendered CLPP5 catalytically inactive. Expression of *CLPP5-STREPII* could fully complement the *clpp5-1* null mutant (Figure 2a – middle panel) but *CLPP5S193A-STREPII* could not. We also identified several lines expressing *CLPP5S193A-STREPII* in wt background or in the heterozygous *clpp5-1* background (Supporting information Figure S2A,D); these lines did not show any visible phenotype (Figure 2a). The S193A mutation was confirmed by sequencing of RT-PCR products, as well at the protein level by MS/MS-based identification of peptides covering this catalytic residue (Figure 2b,c; Supporting information Figure S4). Furthermore, similar as we determined for CLPP3, STREPII-tagged CLPP5 and CLPP5S193A accumulated at comparable levels (Figure 2d) and normally assembled into CLPPRT complexes (Figure 2e). The fact that *clpp5-1(Aa)/CLPP5S193A-STREPII* does not have a visible phenotype, indicates that accumulation of the catalytically inactive CLPP5 in presence of endogenous CLPP5 does not reduce the CLP protease capacity below the minimum threshold level required for chloroplast biogenesis, proteostasis, and function.

Despite extensive screening efforts, no homozygous *clpp5-1* lines expressing *CLPP5S193A-STREPII* were identified, in contrast to the many homozygous *clpp5-1* lines expressing *CLPP5-STREPII*. This suggests that complete loss of catalytic activity of CLPP5 results in embryo lethality. To further test this hypothesis, we analyzed the segregation pattern of the phenotypes of developing seeds in the siliques of wt, wt expressing *CLPP5S193A-STREPII*, heterozygous *clpp5-1*, and heterozygous *clpp5-1* expressing *CLPP5S193A-STREPII*. The first two lines produced only green seeds in developing siliques, while heterozygous *clpp5-1* made both green and white seeds in a 3:1 ratio (Figure 2f), similar as previously observed (Kim et al., 2013). Importantly, ~3:1 segregating ratios were also found in the progeny of heterozygous *clpp5-1* carrying *CLPP5S193A-STREPII* (Figure 2f,g). This confirms that the serine to alanine change of CLPP5 prevented complementation of *clpp5-1* null mutants, and thus that the complete loss of catalytic activity of CLPP5 which reduces the number of catalytic triads per complex from 10 to 7, results in embryo lethality.

3.3 | Affinity-purification of CLPP3 and CLPP5 STREPII-tagged complexes

To identify potential substrates using the CLP core trapping technique as explained in the INTRODUCTION, we carried out replicate affinity experiments for homozygous *clpp3-1* complemented with *CLPP3S164A-STREPII* and using *clpp3-1* complemented by *CLPP3-STREPII* as a control. In the case of CLPP5, we used heterozygous *clpp5-1* expressing *CLPP5S193A-STREPII*, with homozygous *clpp5-1* expressing *CLPP5-STREPII* as the control. The STREPII-tagged complexes were purified on streptactin columns. The affinity eluates were each run out on SDS-PAGE gels, and each gel lane was cut into gel slices, followed by digestion with trypsin and MS/MS analysis for protein identification.

First, we directly (*i.e.* eluates were run on the same SDS-PAGE gel and processed in parallel) compared a negative control affinity purification using soluble protein extracts of wt plants (*i.e.* these lack STREPII-tagged proteins) with CLPP3-STREPII and CLPP3S164A-STREPII purifications (Supporting information Table S2A). Biotin-containing proteins bind to streptactin columns and Arabidopsis contains

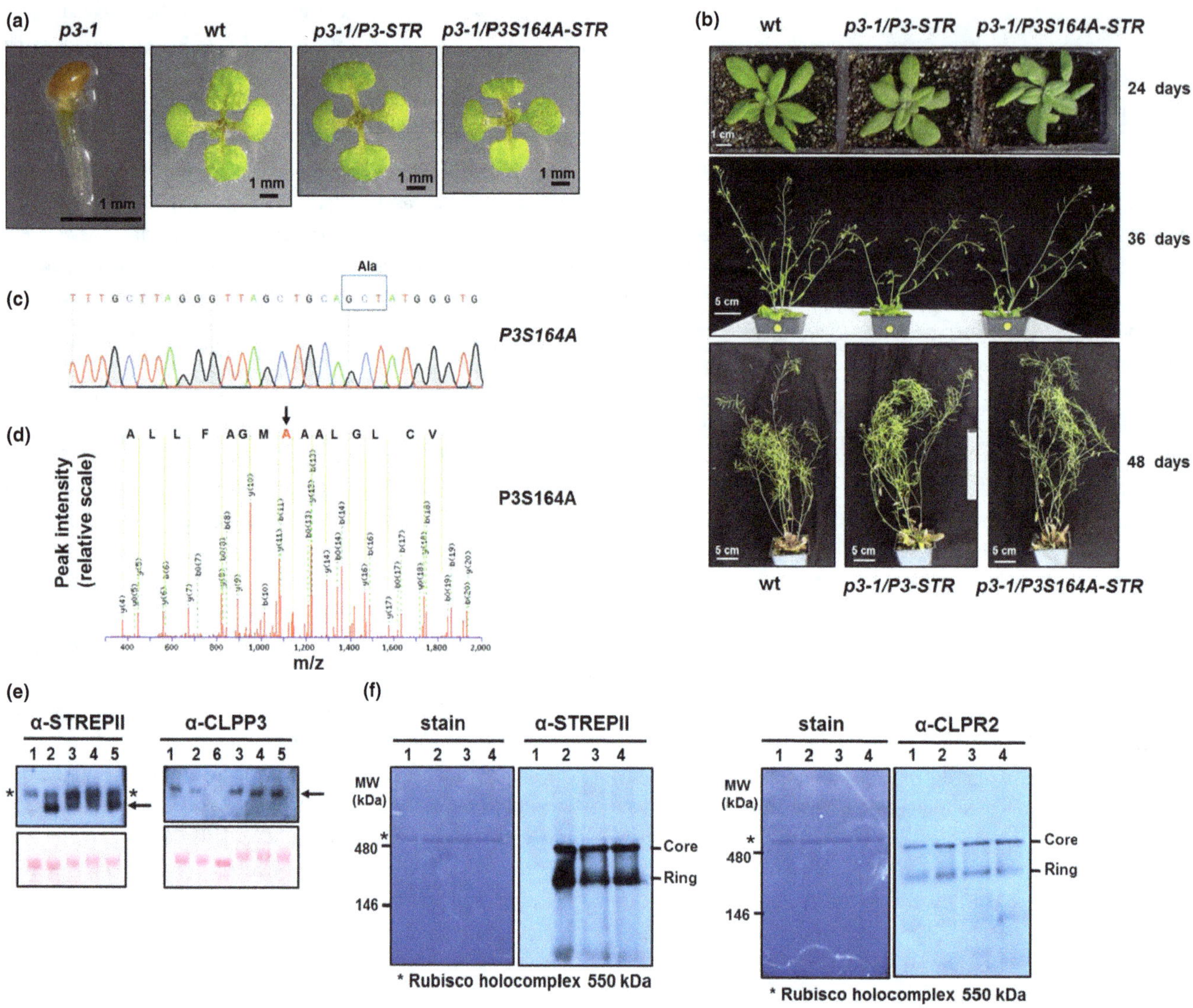

FIGURE 1 Characterization of *clpp3-1* complemented with *CLPP3-STREPII* and *CLPP3S164A-STREPII*. (a) Seedlings of wt, *clpp3-1* and *clpp3-1* complemented with either *CLPP3-STREPII* or *CLPP3S164A-STREPII* grown on half MS medium (no sucrose) for 12 days under 16/8 hr light/dark cycle at 70 μmol photons m^{-2} s^{-1}. Plants are homozygous for the STREPII transgenes. (b) Different developmental stages of soil-grown wt and *clpp3-1* complemented with either *CLPP3-STREPII* or *CLPP3S164A-STREPII*. Plants were grown on soil under 16/8 hr light/dark cycle at 120–150 μmol photons m^{-2} s^{-1}. Plants are homozygous for the STREPII transgenes. (c) Confirmation of the point mutation in the catalytic site of *CLPP3S164A-STREPII* by DNA sequencing of RT-PCR products (primers #6 and #9) amplified from transgenic plants. The codon responsible for the serine to alanine point mutation is boxed and is TCT in wt but GCT in the mutant. (d) Confirmation of the point mutation S164 to A164 in the catalytic site of CLPP3S164A-STREPII by MS/MS of the tryptic peptide (ADVSTVCLGLAAA$_{164}$MGAFLLASGSK) generated by tryptic digestion of affinity purified CLP complexes. The MS/MS spectrum is from a doubly charged precursor ion with *m/z* of 1155.5967 (2^+) with MASCOT ion score of 131 (0.38 ppm error) and supports the residue A164. The partial peptide sequence listed above the spectrum (ALLFAGMA$_{164}$AALGLCV) shown is based on y-ions explaining the reverse order of amino acids. A list of b- and y-ions is provided in Supporting information Figure S3A. An example of an MS/MS spectrum of the analogous CLPP3 wild-type peptide is provided in Supporting information Figure S3B. (e) SDS-PAGE gel and immunoblotting of total soluble protein from wt (lane 1), *clpp3-1* with CLPP3**-STREPII (lane 2), total soluble protein from *clpp3-1* with *CLPP3-STREPII* (lane 3) and *clpp3-1* with *CLPP3S164A-STREPII* (lanes 4 and 5), and stromal protein from *clpp3-1* (lane 6). Arrows indicate CLPP3-STREPII or CLPP3S164A-STREPII proteins. *Indicates the nonspecific reaction. The Ponceau red stain of the blot illustrates protein loading. **35-S driven cDNA of CLPP3-STREPII as described in (Kim et al., 2013); all other lines and samples in the current study are using genomic CLP DNA for transgene expression (see Methods). Anti-CLPP3 or anti-STREPII serum was used. (f) Native gels and immunoblotting of stromal proteomes of wt (lane 1), *clpp3-1* with CLPP3**-STREPII (lane 2), *clpp3-1* with CLPP3-STREPII ((lane 3), and *clpp3-1* with CLPP3S164A-STREPII (lane 4). The intact CLPPR core and individual rings are indicated. *Indicates the Rubisco complex. Anti-STREPII or anti-CLPR2 serum was used for visualization of the core CLP complex and R-ring. **35-S driven cDNA of CLPP3-strepII as described in (Kim et al., 2013)

three major endogenous biotin-binding complexes, namely cytosolic (ACC1) and plastidic acetyl-CoA carboxylase complexes (ACCase), mitochondrial 3-methylcrotonyl-CoA carboxylase complexes (MCC) (Nikolau, Ohlrogge, & Wurtele, 2003). Since the concentration of these complexes should be similar in wt and CLPP3-STREPII lines, they serve as internal controls. We detected high numbers of MS/

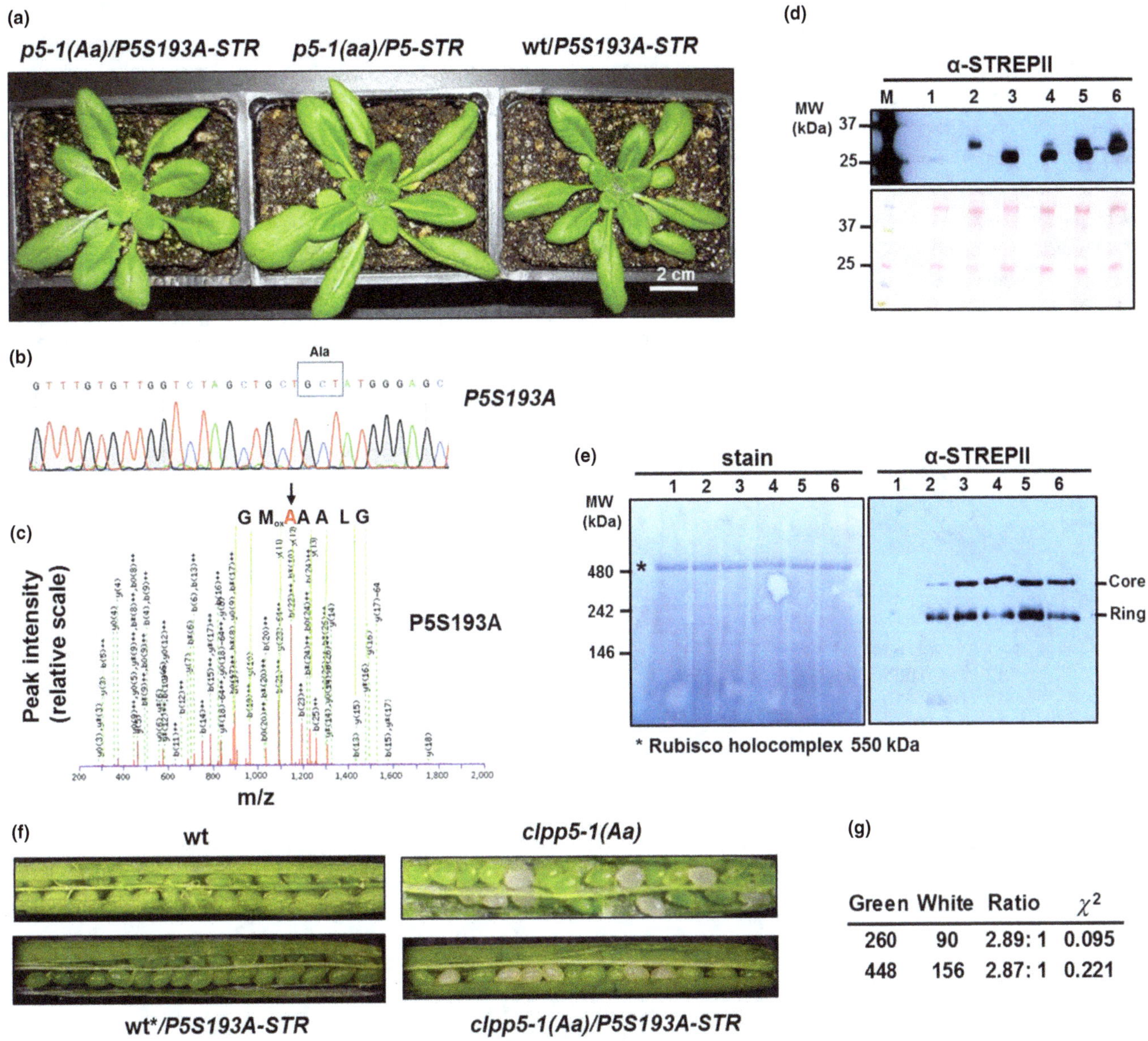

Green	White	Ratio	χ^2
260	90	2.89: 1	0.095
448	156	2.87: 1	0.221

FIGURE 2 Characterization of wt and *clpp5-1* lines expressing *CLPP5S193A-STREPII* or *CLPP5-STREPII* transgenes. (a) Phenotypic analysis of wt and heterozygous (Aa) *clpp5-1* expressing *CLPP5S193A-STREPII* or homozygous *clpp5-1* expressing *CLPP5-STREPII* grown on soil for 21 days under 10/14 hr light/dark cycle and then 11 days under 14/10 hr light/dark cycle all at 120–150 μmol photons m^{-2} s^{-1}. (b) Confirmation of the point mutation in the catalytic sites of *CLPP5S193A-STREPII* by DNA sequencing of RT-PCR products (primers #14 and #9) amplified from transgenic plants. The codon responsible for the Ser to Ala point mutation is boxed and is AGT in wt but GCT in the mutant. (c) Confirmation of the point mutation in the catalytic sites of CLPP5S193A-STREPII by MS/MS of the tryptic peptide (HIRPDVSTVCVGLAAA$_{193}$MGAFLLSAGTK) generated by tryptic digestion of affinity purified CLP complexes. The MS/MS spectrum is from a triply charged precursor ion with m/z of 914.8238 (3^+) with MASCOT ion score of 83 (−5 ppm error) and supports the residue A193. The partial peptide sequence listed above the spectrum (GMA$_{193}$AALGL) shown is based on y-ions explaining this reads in reverse order. A list of b- and y-ions is provided in Supporting information Figure S4A. An example of an MS/MS spectrum of the analogous CLPP5 wild-type peptide is provided in Supporting information Figure S4B. (d) SDS-PAGE and immunoblotting of total soluble protein from wt (lane 1), *clpp3-1* with *CLPP3-STREPII* (lane 2), *clpp5-1* with *CLPP5-STREPII* (lane 3), and *clpp5-1* heterozygous mutant with *CLPP5S193A-STREPII* (lanes 4–6). The Ponceau red stain of the blot is shown. Anti-STREPII serum was used. (e) Native gels and immunoblotting of total leaf protein from wt (lane 1), *clpp3-1* with *CLPP3-STREPII* (lane 2), *clpp5-1* with *CLPP5-STREPII* (lanes 3 and 5), the *clpp5-1* heterozygous mutant with *CLPP5S193A-STREPII* (lanes 4 and 6). The CLPPRT core and ring are indicated. Anti-STREPII serum was used. *Indicates the Rubisco complex. (f) Developing siliques of wt, heterozygous *clpp5-1 (Aa)*, heterozygous *clpp5-1 (Aa)* with *CLPP5S193A-STREPII*, and wt* with *CLPP5S193A-STREPII* obtained from a prior segregating progeny of heterozygous *clpp5-1* with *CLPP5S193A-STREPII*. The segregating white seeds are indicative of impaired chloroplast development due to the homozygous *clpp5-1* background.

(g) Segregation analysis of green and white seeds in developing siliques of two heterozygous *clpp5-1* mutants expressing *CLPP5S193A-STREPII*

MS spectra for these endogenous biotin binding proteins (ACC1, MCCA, MCCB, and BIOTIN CARBOXYLASE (BC) – part of the ACCase) (Figure 3a). Indeed, direct comparison between affinity eluates of wt and the two CLPP3-STREPII lines showed very similar levels of these endogenous biotin-containing complexes (average ratio 1.34) (Figure 3a). This provides an internal calibrant for evaluating proteins enriched in the affinity eluates. In contrast to these endogenous biotin binding complexes, CLP subunits were on average 25-fold higher in the CLPP3-STREPII lines than wt plants (Figure 3a), showing that the affinity purification of CLP complexes worked well.

A total of 16 successful independent affinity experiments for CLPP3 (four for each CLPP3-STREPII and CLPPP3S164A-STREPII) and CLPP5 (four for each CLPP5-STREPII and CLPP5S193A-STREPII) were carried out. We note that for all CLPP5 affinity purifications, we reduced the level of co-purifying endogenous biotin binding proteins by preincubation of total soluble leaf extract with avidin; yet, there was still sufficient endogenous biotin binders for internal calibration (Supporting information Figure S5). The identified proteins, their annotation and number of matched MS/MS spectra (adjSPC – see Methods for explanation) for each experiment are assembled in Table S2B. All chloroplast CLPPRT core subunits including CLPP (P1, P3-6), CLPR (R1-4), as well as CLPT1,2 subunits were identified. These subunits of the CLP complex represented on average 19% of all adjSPC matched to plastid proteins in the case of CLPP3-STREPII

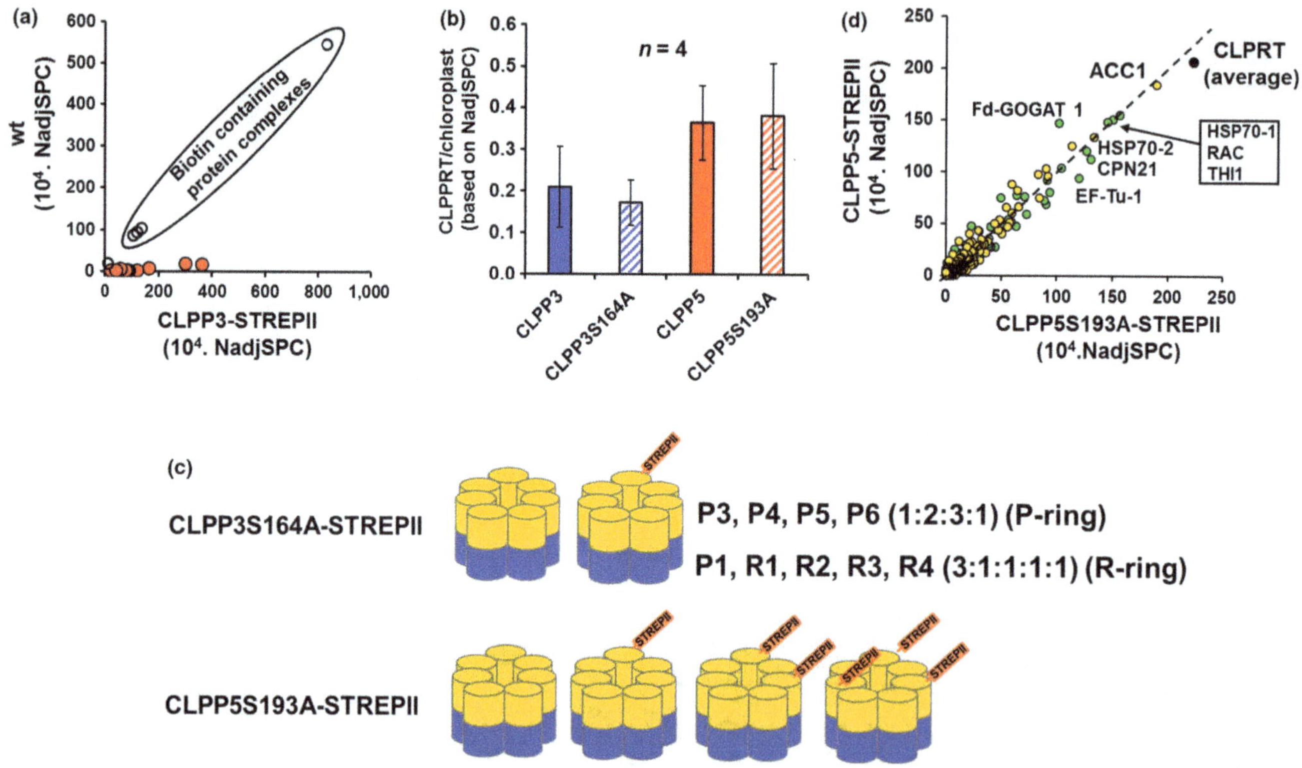

FIGURE 3 Affinity purification of STREPII-tagged CLP core complexes and MS/MS analysis. (a) Relative abundance of endogenous Arabidopsis biotin-binding ACC1, MCCA, BIOTIN CARBOXYLASE (BC) and MCCB, and subunits of the CLPPR complex in streptactin affinity eluates from wt (col-o) plants and CLPP3-STREPII transgenic lines. CLP subunits are marked in red-filled circles. In these experiments, no avidin was added prior to affinity purifications; addition of avidin does reduce the binding of endogenous biotin protein proteins (see Supporting information Figure S5). (b) The average yield of CLPPRT subunits for affinity-purified complexes containing CLPP3-STREPII, CLPP3S164A-STREPII, CLPP5-STREPII and CLPP5S193A-STREPII (n = 4 for each transgene). Yields were normalized to the total amount of detected plastid proteins, using NadjSPC as the basis for quantification. (c) Schematic representation of the wt and tagged CLPPR core complexes with P-rings and R-ring. Both ClpP3 and CLPP5 are part of the P-ring, with respectively one or three copies per ring in native complexes (Olinares et al., 2011). In the case of the STREPII-tagged CLPP3 and CLPP3S164A lines, each complex has one STREPII-tagged CLPP3 protein and no endogenous CLPP3. However in case of *CLPP5S193A-STREPII* lines, plants have both endogenous CLPP5 and CLPP5S193A-STREPII; affinity purified CLPP5 complexes from these lines have one, two or three CLPP5S193A-STREPII subunits per P-ring. We note that the transgenes are genomic constructs with their endogenous promoters. The positions of the STREPII tagged proteins within the P-ring are unknown. (d) Cross-correlation of proteins of average abundance (based on NadjSPC) identified in affinity eluates of *CLPP5-STREPII* and *CLPP5S193A-STREPII* lines. These proteins were observed two or more times across the eight preparations (4 of each STREPII line). Proteins localized in the chloroplasts are indicated in green and proteins located outside the chloroplast or without annotated subcellular locations are indicated in yellow; subcellular location was based on annotation of manually curated experimental evidence collected from information in the public domain and in-house data (from PPDB). The data point labeled 'CLPPRT' is the average of all 11 CLPP, CLPR and CLPT subunits. ACC1 (ACETYL-COA-CARBOXYLASE 1; AT1G36160) has biotin as the cofactor and has therefore high affinity to the streptactin column. HSP70-1 (AT4G24280); HSP70-2 (AT5G49910); RAC – RUBISCO ACTIVASE (AT2G39730); CPN21 (AT5G20720); EF-TU – ELONGATION FACTOR TU (AT4G20360); THI1-THIAMIN BIOSYNTHESIS 1 (AT5G54770)

and CLPP3S164A-STREPII, and on average 37% in the case of CLPP5-STREPII and CLPP5S193A-STREPII (Figure 3b). This shows that the serine to alanine mutation did not affect the yield of the CLPPR core complexes. We also identified low levels of CLPC1,2 chaperones in 14 out of 16 experiments without significant differences between the four STREPII tagged lines (Table S2B). We did not identify the adaptor CLPS1 and only once putative co-adaptor CLPF (with 3 MS/MS spectra in a CLPP5S193A experiment). The lack of consistent observation of these adaptors is consistent with their substrate delivery role and transient interaction to CLPC chaperones (Nishimura et al., 2013, 2015). Together, these affinity experiments indicate that (i) the loss of catalytic triads in CLPP3 or CLPP5 did not affect the yield of CLP core affinity purification, complex composition or interaction between the CLPC and the CLP core, and that (ii) the CLPP5-STREPII affinity purification was ~2-fold more efficient than the CLPP3-STREPII purification (Figure 3b), likely because there were on average more CLPP5-STREPII copies per CLP core complex as illustrated in Figure 3c.

3.4 | Trapping substrates in affinity-purified CLPP3S164A-STREPII and CLPP5S193A-STREPII complexes

We then evaluated the affinity experiments for other (non-CLP) chloroplast proteins enriched in these CLPP3- and CLPP5 STREPII-tagged complexes; these could be interactors to the CLP core or CLP substrates located within the chamber of the protease core. The theoretical basis for trying to identify substrates in CLP cores is based on the hypothesis that chloroplast CLP complexes that lack one or more of the 10 catalytic sites (three in the R ring and seven in the P ring) have slower catalytic rates and therefore the average time to degrade substrates is extended; consequently such substrates reside for a longer time in the CLP core complex. MS/MS of such affinity enriched complexes could allow recognition of such substrates as having increased abundance when compared to fully active affinity-enriched complexes. In the case of CLPP5, there are three endogenous copies per core (Olinares et al., 2011). Since we expressed the catalytically inactive CLPP5S193ASTREPII in the heterozygous *clpp5-1* mutant, there are both endogenous CLPP5 subunits and inactive tagged CLPP5 copies. We note that the CLPP3 and CLPP5 transgenes are generated from genomic sequences including the endogenous promoters, as to ensure more natural levels of expression. The affinity purified STREPII-tagged CLP cores must have either one, or possibly two or three inactive CLPP5 subunits (Figure 3c); in particular those complexes with 2 or 3 inactive CLPP5 subunits could have reduced CLP catalytic rates. However, since it is quite likely that the CLP protease is constantly involved in monitoring and degrading chloroplast proteins, it is possible that also fully active CLP core complexes contain substrates. Therefore, it is certainly possible that proteins observed in both unmodified CLPP3/5-STREPII and CLPP3/5-mutant-STREPII could be substrates. However, for such proteins there are no good criteria to distinguish between proteins identified in the eluates due to unspecific protein interactions with e.g. the outer surface of the CLP core complex, and proteins residing within the chamber of the proteolytic CLP core because they were selected for degradation. One of the key objectives of the in vivo trapping approach is to obtain a strong candidate list of substrates, such that these can then be further investigated in the context of selection and degradation by the CLP protease system.

We compared the eluates of the wt and mutant forms of CLPP3- and CLPP5-STREPII-tagged complexes based on the normalized number of matched MS/MS spectra (NadjSPC) (Supporting information Table S2B). We tested two types of thresholds, based on ideas from other protease trapping studies (Arends, Thomanek, Kuhlmann, Marcus, & Narberhaus, 2016; Arends et al., 2018; Fischer et al., 2015; Trentini et al., 2016) and our own experience with protein affinity experiments, *e.g.* (Kim et al., 2015; Nishimura et al., 2013) to identify possible candidate substrates: (i) only observed in eluates of mutant complexes and not in CLPP3/5-STREPII controls, or (ii) at least 2-fold enriched in eluates of the mutant complexes compared to CLPP3/5-STREPII controls. To reduce stochastic noise, we applied also either a minimum frequency of observation (2 or 3) across replicates or an abundance threshold of 1.10^{-3} based on NadjSPC (this corresponds to ~0.1% of protein mass).

In the case of CLPP3, we isolated chloroplasts and extracted soluble stromal proteomes and carried out two independent affinity purifications for each CLPP3-STREPII and CLPP3S164A-STREPII (trap) (Table S2B). 17 proteins were only found in the mutant and 12 proteins were >2× fold enriched in the mutant eluates; however because the number of matched MS/MS spectra were low and/or variable across replicates, we did not consider them strong candidate substrates. For comparison we note that the average ratio between CLP core subunits between CLPP3S164A-STREPII and CLPP3-STREPII was 0.7 and the ratio for stromal biotin carboxylase (BC) was 0.77. In a second set of CLPP3 experiments, we used soluble total leaf extracts and a total of four affinity purifications (two mutants and two controls). The relative average ratio of the yield CLP core subunits between CLPP3S164A-STREPII and CLPP3-STREPII was 1.0 and the ratio for endogenous biotin binder and internal control ACCase was 1.2. Applying these same thresholds (only in the mutant P3 or >2× enriched) we identified nine proteins that were more than 2-fold enriched; however, following closer inspection we do not suggest them to be strong candidates, based on the same arguments as for the stromal replicates (Supporting information Table S2B).

In the case of CLPP5, our dataset consists of four affinity experiments with each CLPP5S193A-STREPII and CLPP5-STREPII and using total soluble leaf extracts (Supporting information Table S2B). Figure 3d shows a cross-correlation of the average relative protein abundances (based on NadjSPC) across the replicate affinity purified CLPP5S193A-STREPII and CLPP5-STREPII complexes. This shows that the plastid and non-plastid proteins co-purified with the fully active and partially active complexes have generally similar relative abundances. Only one plastid protein was enriched, namely FRUCTOSE-BISPHOSPHATASE (FBPA; HCEF1; AT3G54050) (Livingston, Cruz, Kohzuma, Dhingra, & Kramer, 2010) with an average ratio of

2.2 (standard deviation 1.1). Interesting, the abundant stromal proteins THI1 and EF-TU were previously suggested to be CLP substrates (Moreno et al., 2018; Nishimura & van Wijk, 2015) and were observed with high abundance in the eluates of both CLPP5-STREPII and CLPP5S193A-STREPII (Figure 3d). However, they were also observed in the initial negative control using wt plants (*i.e.* they do not contain any STREPII-tagged CLP proteins) (Supporting information Table S2A), rendering their presence in the eluates of the CLPP5-STREPII and CLPP5S193A-STREPII less significant. Furthermore stromal CPN21, abundant in eluates of both CLPP5 constructs (mutant/control ratio 1.2 ± 1.2) (Figure 3d), was previously identified as a strong interactor in CLP affinity purification using CLPT1-STREPII and CLPT2-STREPII Arabidopsis lines (Kim et al., 2015).

3.5 | The bottle-neck for degradation is likely substrate recognition and unfolding by CLP adaptors and chaperones, upstream of the CLP core

We showed that CLPP3 (one copy per CLP core) makes an essential structural contribution, but that its catalytic site is dispensable for plant growth and development, whereas the catalytic activity of CLPP5 (3 copies per CLP core) is essential. Based on the extensive affinity experiments described in this study, we conclude that the CLP core catalytic activity in the CLPP3S164A-STREPII and CLPP5 S193A-STREPII complexes with reduced number of catalytic triads did not result in significant accumulation of substrates as compared to fully active CLP complexes. This suggests that the bottle-neck for degradation is likely substrate recognition and unfolding by CLP adaptors and chaperones, upstream of the CLP core. In vivo substrate trapping through partial inactivation of unfolding activity of the CLPC/D chaperones could provide an alternative strategy for identification of candidate substrates. This has been a successful approach for the CLPC homolog in the gram-positive bacterium *Staphylococcus aureus* where the CLP system is not essential for viability (Graham, Lei, & Lee, 2013).

ACKNOWLEDGEMENTS

The authors like to thank Nazmul Bhuiyan and other members of the van Wijk for help and support. No conflict of interest declared.

AUTHOR CONTRIBUTIONS

J.R.L. and K.J. V.W. designed the experiments, analyzed the data, and wrote the manuscript. J.R.L. and J.K. performed site-mutagenesis. J.R.L conducted plant and molecular experiments. J.R.L prepared protein samples; G.F. performed all MS analysis. K.J.v.W obtained funding and provided oversight of the project.

ORCID

Jui-Yun Rei Liao http://orcid.org/0000-0002-4995-6826
Giulia Friso http://orcid.org/0000-0002-7537-7887
Jitae Kim http://orcid.org/0000-0001-5402-7725
Klaas J. van Wijk http://orcid.org/0000-0001-9536-0487

REFERENCES

Alexopoulos, J. A., Guarne, A., & Ortega, J. (2012). ClpP: A structurally dynamic protease regulated by AAA+ proteins. *Journal of Structural Biology*, *179*, 202–210.

Andersson, F. I., Tryggvesson, A., Sharon, M., Diemand, A. V., Classen, M., Best, C., … Clarke, A. K. (2009). Structure and function of a novel type of ATP-dependent Clp protease. *Journal of Biological Chemistry*, *284*, 13519–13532.

Apitz, J., Nishimura, K., Schmied, J., Wolf, A., Hedtke, B., van Wijk, K. J., & Grimm, B. (2016). Posttranslational control of ALA synthesis includes GluTR degradation by Clp protease and stabilization by GluTR-binding protein. *Plant Physiology*, *170*, 2040–2051.

Arends, J., Griego, M., Thomanek, N., Lindemann, C., Kutscher, B., Meyer, H. E., & Narberhaus, F. (2018). An Integrated proteomic approach uncovers novel substrates and functions of the lon protease in *Escherichia coli*. *Proteomics*, *18*, e1800080.

Arends, J., Thomanek, N., Kuhlmann, K., Marcus, K., & Narberhaus, F. (2016). In vivo trapping of FtsH substrates by label-free quantitative proteomics. *Proteomics*, *16*, 3161–3172.

Clough, S. J., & Bent, A. F. (1998). Floral dip: A simplified method for Agrobacterium-mediated transformation of *Arabidopsis thaliana*. *Plant Journal*, *16*, 735–743.

El Bakkouri, M., Rathore, S., Calmettes, C., Wernimont, A. K., Liu, K., Sinha, D., … Houry, W. A. (2013). Structural insights into the inactive subunit of the apicoplast-localized caseinolytic protease complex of *Plasmodium falciparum*. *Journal of Biological Chemistry*, *288*, 1022–1031.

Feng, J., Michalik, S., Varming, A. N., Andersen, J. H., Albrecht, D., Jelsbak, L., … Frees, D. (2013). Trapping and proteomic identification of cellular substrates of the ClpP protease in *Staphylococcus aureus*. *Journal of Proteome Research*, *12*, 547–558.

Fischer, F., Langer, J. D., & Osiewacz, H. D. (2015). Identification of potential mitochondrial CLPXP protease interactors and substrates suggests its central role in energy metabolism. *Scientific Reports*, *5*, 18375.

Flynn, J. M., Levchenko, I., Sauer, R. T., & Baker, T. A. (2004). Modulating substrate choice: The SspB adaptor delivers a regulator of the extracytoplasmic-stress response to the AAA+ protease ClpXP for degradation. *Genes & Development*, *18*, 2292–2301.

Flynn, J. M., Neher, S. B., Kim, Y. I., Sauer, R. T., & Baker, T. A. (2003). Proteomic discovery of cellular substrates of the ClpXP protease reveals five classes of ClpX-recognition signals. *Molecular Cell*, *11*, 671–683.

Friso, G., Olinares, P. D. B., & van Wijk, K. J. (2011) The workflow for quantitative proteome analysis of chloroplast development and differentiation, chloroplast mutants, and protein interactions by spectral counting. In R. P. Jarvis (Eds.), *Chloroplast research in Arabidopsis* (pp. 265–282). New York: Humana Press.

Graham, J. W., Lei, M. G., & Lee, C. Y. (2013). Trapping and identification of cellular substrates of the *Staphylococcus aureus* ClpC chaperone. *Journal of Bacteriology*, *195*, 4506–4516.

Kim, D. Y., & Kim, K. K. (2008). The structural basis for the activation and peptide recognition of bacterial ClpP. *Journal of Molecular Biology*, *379*, 760–771.

Kim, J., Kimber, M. S., Nishimura, K., Friso, G., Schultz, L., Ponnala, L., & van Wijk, K. J. (2015). Structures, functions, and interactions of ClpT1 and ClpT2 in the Clp protease system of Arabidopsis chloroplasts. *Plant Cell*, *27*, 1477–1496.

Kim, J., Olinares, P. D., Oh, S. H., Ghisaura, S., Poliakov, A., Ponnala, L., & van Wijk, K. J. (2013). Modified Clp protease complex in the ClpP3 null mutant and consequences for chloroplast development and function in Arabidopsis. *Plant Physiology*, *162*, 157–179.

Kim, J., Rudella, A., Ramirez Rodriguez, V., Zybailov, B., Olinares, P. D., & van Wijk, K. J. (2009). Subunits of the plastid ClpPR protease complex have differential contributions to embryogenesis, plastid biogenesis, and plant development in Arabidopsis. *Plant Cell*, *21*, 1669–1692.

Koussevitzky, S., Stanne, T. M., Peto, C. A., Giap, T., Sjogren, L. L., Zhao, Y., … Chory, J. (2007). An *Arabidopsis thaliana* virescent mutant reveals a role for ClpR1 in plastid development. *Plant Molecular Biology*, *63*, 85–96.

Kuhlmann, N. J., & Chien, P. (2017). Selective adaptor dependent protein degradation in bacteria. *Current Opinion in Microbiology*, *36*, 118–127.

Liu, K., Ologbenla, A., & Houry, W. A. (2014). Dynamics of the ClpP serine protease: A model for self-compartmentalized proteases. *Critical Reviews in Biochemistry and Molecular Biology*, *49*, 400–412.

Livingston, A. K., Cruz, J. A., Kohzuma, K., Dhingra, A., & Kramer, D. M. (2010). An Arabidopsis mutant with high cyclic electron flow around photosystem I (hcef) involving the NADPH dehydrogenase complex. *Plant Cell*, *22*, 221–233.

Mahmoud, S. A., & Chien, P. (2018). Regulated proteolysis in bacteria. *Annual Review of Biochemistry*, *87*, 677–696.

Moreno, J. C., Martinez-Jaime, S., Schwartzmann, J., Karcher, D., Tillich, M., Graf, A., & Bock, R. (2018). Temporal proteomics of inducible RNAi lines of clp protease subunits identifies putative protease substrates. *Plant Physiology*, *176*, 1485–1508.

Moreno, J. C., Tiller, N., Diez, M., Karcher, D., Tillich, M., Schottler, M. A., & Bock, R. (2017). Generation and characterization of a collection of knock-down lines for the chloroplast Clp protease complex in tobacco. *Journal of Experimental Botany*, *68*, 2199–2218.

Nikolau, B. J., Ohlrogge, J. B., & Wurtele, E. S. (2003). Plant biotin-containing carboxylases. *Archives of Biochemistry and Biophysics*, *414*, 211–222.

Nishimura, K., Apitz, J., Friso, G., Kim, J., Ponnala, L., Grimm, B., & van Wijk, K. J. (2015). Discovery of a unique Clp component, ClpF, in chloroplasts: A proposed binary ClpF-ClpS1 adaptor complex functions in substrate recognition and delivery. *Plant Cell*, *27*, 2677–2691.

Nishimura, K., Asakura, Y., Friso, G., Kim, J., Oh, S. H., Rutschow, H., … van Wijk, K. J. (2013). ClpS1 is a conserved substrate selector for the chloroplast Clp protease system in Arabidopsis. *Plant Cell*, *25*, 2276–2301.

Nishimura, K., Kato, Y., & Sakamoto, W. (2016). Chloroplast proteases: Updates on proteolysis within and across suborganellar compartments. *Plant Physiology*, *171*, 2280–2293.

Nishimura, K., Kato, Y., & Sakamoto, W. (2017). Essentials of proteolytic machineries in chloroplasts. *Molecular Plant*, *10*, 4–19.

Nishimura, K., & van Wijk, K. J. (2015). Organization, function and substrates of the essential Clp protease system in plastids. *Biochimica et Biophysica Acta*, *1847*, 915–930.

Olinares, P. D., Kim, J., Davis, J. I., & van Wijk, K. J. (2011). Subunit stoichiometry, evolution, and functional implications of an asymmetric plant plastid ClpP/R protease complex in Arabidopsis. *Plant Cell*, *23*, 2348–2361.

Olivares, A. O., Baker, T. A., & Sauer, R. T. (2018). Mechanical protein unfolding and degradation. *Annual Review of Physiology*, *80*, 413–429.

Peltier, J. B., Ripoll, D. R., Friso, G., Rudella, A., Cai, Y., Ytterberg, J., … Van Wijk, K. J. (2004). Clp protease complexes from photosynthetic and non-photosynthetic plastids and mitochondria of plants, their predicted three-dimensional structures, and functional implications. *Journal of Biological Chemistry*, *279*, 4768–4781.

Pulido, P., Llamas, E., Llorente, B., Ventura, S., Wright, L. P., & Rodriguez-Concepcion, M. (2016). Specific Hsp100 chaperones determine the fate of the first enzyme of the plastidial isoprenoid pathway for either refolding or degradation by the stromal clp protease in Arabidopsis. *PLoS Genetics*, *12*, e1005824.

Raju, R. M., Unnikrishnan, M., Rubin, D. H., Krishnamoorthy, V., Kandror, O., Akopian, T. N., … Rubin, E. J. (2012). Mycobacterium tuberculosis ClpP1 and ClpP2 function together in protein degradation and are required for viability in vitro and during infection. *PLoS Pathogens*, *8*, e1002511.

Rudella, A., Friso, G., Alonso, J. M., Ecker, J. R., & van Wijk, K. J. (2006). Downregulation of ClpR2 Leads to reduced accumulation of the ClpPRS protease complex and defects in chloroplast biogenesis in Arabidopsis. *Plant Cell*, *18*, 1704–1721.

Sauer, R. T., & Baker, T. A. (2011). AAA+ proteases: ATP-fueled machines of protein destruction. *Annual Review of Biochemistry*, *80*, 587–612.

Schelin, J., Lindmark, F., & Clarke, A. K. (2002). The clpP multigene family for the ATP-dependent Clp protease in the cyanobacterium Synechococcus. *Microbiology*, *148*, 2255–2265.

Sjogren, L. L., & Clarke, A. K. (2011). Assembly of the chloroplast ATP-dependent Clp protease in Arabidopsis is regulated by the ClpT accessory proteins. *Plant Cell*, *23*, 322–332.

Stanne, T. M., Pojidaeva, E., Andersson, F. I., & Clarke, A. K. (2007). Distinctive types of ATP-dependent Clp proteases in cyanobacteria. *Journal of Biological Chemistry*, *282*, 14394–14402.

Tapken, W., Kim, J., Nishimura, K., van Wijk, K. J., & Pilon, M. (2015). The Clp protease system is required for copper ion-dependent turnover of the PAA2/HMA8 copper transporter in chloroplasts. *New Phytologist*, *205*, 511–517.

Trentini, D. B., Suskiewicz, M. J., Heuck, A., Kurzbauer, R., Deszcz, L., Mechtler, K., & Clausen, T. (2016). Arginine phosphorylation marks proteins for degradation by a Clp protease. *Nature*, *539*, 48–53.

Wang, J., Hartling, J. A., & Flanagan, J. M. (1997). The structure of ClpP at 2.3 A resolution suggests a model for ATP-dependent proteolysis. *Cell*, *91*, 447–456.

Welsch, R., Zhou, X., Yuan, H., Alvarez, D., Sun, T., Schlossarek, D., … Li, L. (2018). Clp protease and OR directly control the proteostasis of phytoene synthase, the crucial enzyme for carotenoid biosynthesis in Arabidopsis. *Molecular Plant*, *11*, 149–162.

Wu, Y., Mirzaei, M., Pascovici, D., Chick, J. M., Atwell, B. J., & Haynes, P. A. (2016). Quantitative proteomic analysis of two different rice varieties reveals that drought tolerance is correlated with reduced abundance of photosynthetic machinery and increased abundance of ClpD1 protease. *Journal of Proteomics*, *143*, 73–82.

Yu, A. Y., & Houry, W. A. (2007). ClpP: A distinctive family of cylindrical energy-dependent serine proteases. *FEBS Letters*, *581*, 3749–3757.

Zheng, B., MacDonald, T. M., Sutinen, S., Hurry, V., & Clarke, A. K. (2006). A nuclear-encoded ClpP subunit of the chloroplast ATP-dependent Clp protease is essential for early development in *Arabidopsis thaliana*. *Planta*, *224*, 1103–1115.

9

Upregulation of *DWARF27* is associated with increased strigolactone levels under sulfur deficiency in rice

Masato Shindo[1] | **Koichiro Shimomura**[1] | **Shinjiro Yamaguchi**[2] | **Mikihisa Umehara**[1]

[1]Graduate School of Life Sciences, Toyo University, Gunma, Japan

[2]Graduate School of Life Sciences, Tohoku University, Sendai, Japan

Correspondence
Mikihisa Umehara, Graduate School of Life Sciences, Toyo University, Gunma, Japan.
Email: umehara@toyo.jp

Funding information
This work was in part supported by grants from Toyo University (the Inoue Enryo Memorial Foundation for Promoting Sciences) to M.S. and from the Japan Society for the Promotion of Science (KAKENHI Grant Nos. 26450144 and 17K07650 to M.U.).

Abstract

Plants produce strigolactones (SLs) in roots in response to nitrogen or phosphate deficiency. To evaluate SL levels under other mineral deficiencies in rice, we cultivated rice seedlings in hydroponic media without nitrogen, phosphorus, potassium, sulfur, calcium, magnesium, and iron. Tiller bud outgrowth was stimulated under calcium deficiency because of low SL levels. SL levels increased under sulfur deficiency, in addition to phosphate, and nitrogen deficiencies. To explore which genes are key regulators of SL production under sulfur deficiency, we analyzed the expression of SL-related genes in sulfur-sufficient and sulfur-deficient conditions. An SL biosynthesis gene, *DWARF27* (*D27*), was strongly expressed under sulfur deficiency, and its expression was decreased by sulfur supply. The levels of *D10*, *D17*, and *OsMAX1* transcripts did not differ between sulfur-sufficient and sulfur-deficient conditions. These results suggest that the increased SL levels under sulfur deficiency are due to a high expression of *D27*. A combination of nitrogen, phosphorus, and sulfur deficiencies had no additive synergistic effect on SL production. Under combined phosphorus and sulfur deficiency, the expression levels of most SL biosynthesis genes were elevated. The number of tiller buds in the *d27* mutant was higher than in the wild type, but lower than in other *d* mutants. Under sulfur deficiency, the chlorophyll content of *d27* was lower than those of other *d* mutants. These results indicate that *D27* plays an important role in adaptation to sulfur deficiency in rice.

KEYWORDS
D27, leaf senescence, macronutrient deficiency, *Oryza sativa*, shoot branching, strigolactones

1 | INTRODUCTION

Strigolactones (SLs) are a class of terpenoid lactones that were originally discovered as stimulants of seed germination in root parasitic plants such as *Striga* spp., *Orobanche* spp., and *Phelipanche* spp. (for reviews, see Xie, Yoneyama, & Yoneyama, 2010; Yoneyama, Awad, Xie, Yoneyama, & Takeuchi, 2010). Later, SLs were found to induce hyphal branching of arbuscular mycorrhizal (AM) fungi, which support the acquisition of P and N in soil by the host plants (Akiyama, Matsuzaki, & Hayashi, 2005). Thus, SLs are thought to act as communication signals to recognize host plants for parasitism and symbiosis in the rhizosphere. In addition to the roles in the rhizosphere, SLs inhibit shoot branching (Gomez-Roldan et al., 2008; Umehara et al., 2008). Because SL-related mutants show pleiotropic phenotypes including enhanced shoot branching, SLs have recently been recognized as plant hormones that control plant growth at various developmental stages. In vascular plants, SLs stimulate stem thickening, leaf senescence, root hair elongation, and primary root growth and suppress adventitious root formation (Agusti et al., 2011; Kapulnik et al., 2011; Rasmussen et al., 2012; Ruyter-Spira et al., 2011; Ueda & Kusaba, 2015; Yamada et al., 2014). In the moss *Physcomitrella patens*, SLs suppress chloronema branching and colony expansion (Proust et al., 2011).

Carlactone (CL), a biosynthetic precursor of SLs, is synthesized from β-carotene through consecutive reactions catalyzed by the β-carotene isomerase DWARF27 (D27) and carotenoid cleavage dioxygenases 7 and 8 (CCD7 and CCD8) (Alder et al., 2012; Seto et al., 2014). In Arabidopsis, CL is converted to carlactonoic acid via oxidation by a cytochrome P450 encoded by *MORE AXILLARY GROWTH1* (*MAX1*) (Abe et al., 2014). In rice, *D17/HTD1* encodes CCD7, and *D10* encodes CCD8 (Arite et al., 2007; Zou et al., 2006), and there are five *MAX1* homologs: *Os01g0700900* (*Os900*), *Os01g0701400* (*Os1400*), *Os01g0701500* (*Os1500*), *Os02g0221900* (*Os1900*), and *Os06g0565100* (*Os5100*) (Nelson, Schuler, Paquette, Werck-Reichhart, & Bak, 2004) (Figure 1). Os900 catalyzes the oxidation of CL into the parent SL 4-deoxyorobanchol (4DO), and Os1400 catalyzes the hydroxylation of 4DO to orobanchol (Zhang et al., 2014). Canonical SLs have tricyclic lactone and methylbutenolide moieties connected by an enol ether bond. This bond and the methylbutenolide structure are essential for shoot branching inhibition (Boyer et al., 2012; Umehara et al., 2015). Configuration of the methylbutenolide moiety affects the biological activity of SLs in rice and Arabidopsis but not in pea (Boyer et al., 2012; Umehara et al., 2015). In SL signaling, D14 is a probable SL receptor and belongs to the α/β-fold hydrolase superfamily, which also includes the gibberellin receptor GID1 (Arite et al., 2009) (Figure 1). D14 is transported through the phloem to tiller buds in rice (Kameoka et al., 2016). D3 is a leucine-rich-repeat F-box protein that acts as a recognition subunit in a SKP1–CUL1–F-box-protein complex and binds target proteins for proteasomal degradation (Ishikawa et al., 2005) (Figure 1).

Several types of SLs have been identified; these are mainly produced in roots in response to P deficiency in various plant species including red clover, tomato, sorghum, *Lotus japonicus*, alfalfa, and rice (Lopez-Raez et al., 2008; Sugimoto & Ueyama, 2008; Umehara et al., 2008; Yoneyama, Xie, et al., 2007; Yoneyama, Yoneyama, Takeuchi, & Sekimoto, 2007; Yoneyama et al., 2012). In rice, the levels of *D27*, *D17*, *D17*, and *Os900* mRNAs are highly elevated in roots under P deficiency and contribute to the high levels of SLs, indicating that most SL biosynthesis genes are upregulated in response to P deficiency (Umehara, Hanada, Magome, Takeda-Kamiya, & Yamaguchi, 2010). Tiller bud outgrowth is not inhibited under P-sufficient conditions because SL levels are low in roots. Under P deficiency, SL levels in roots are highly elevated and SLs are probably transported to tiller buds to inhibit the outgrowth. However, shoot branching is not suppressed under P deficiency in SL mutants of rice and Arabidopsis (Kohlen et al., 2011; Umehara et al., 2010). In some plant species such as lettuce, marigold, sorghum, and rice, SL levels in roots increase under both N deficiency and P deficiency (Sun et al., 2014; Yoneyama, Xie, et al., 2007; Yoneyama, Yoneyama, et al., 2007), and N and P

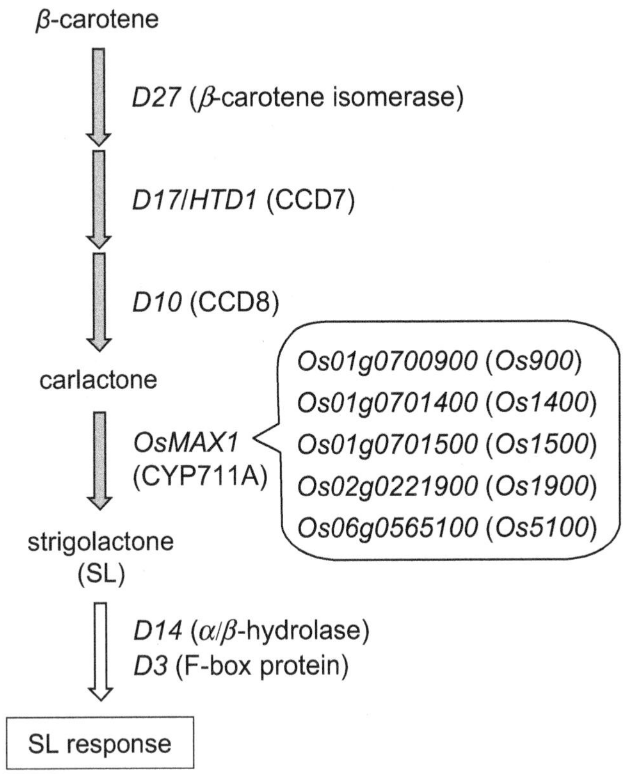

FIGURE 1 SL pathway in rice. Gray and white arrows indicate the SL biosynthesis and signaling, respectively

fertilization suppresses SL production and exudation in roots (Umehara et al., 2010; Yoneyama, Xie, Kisugi, Nomura, & Yoneyama, 2013; Yoneyama et al., 2012). Thus, SLs are thought to be key regulators of efficient nutrient allocation and adaption to nutrient-deficient environments (Umehara, 2011).

Little is known about how SL levels change in roots under deficiencies of macronutrients other than P and N. To address this question, here we examined the effects of deficiency in N, P, K, S, Ca, Mg, or Fe on SL production in rice. Here, we show that tiller buds of rice seedlings outgrow under Ca deficiency because SL levels decrease and that SL levels increase under S deficiency as well as N or P deficiency. We also show that *D27* is strongly expressed under S deficiency, and its expression is reduced by S supply, whereas the expression levels of *D10*, *D17*, and *OsMAX1* are not affected by S supply. These results suggest that SL production under S deficiency is due to a marked increase in the expression of *D27*. Under S deficiency, tiller bud outgrowth in a *d27* mutant was strongly suppressed compared with that in other *d* mutants, and chlorophyll content was lower than that in other *d* mutants, indicating that *D27* expression plays an important role in survival and adaptation to S deficiency.

2 | MATERIALS AND METHODS

2.1 | Plant materials and growth conditions

We used the rice (*Oryza sativa* L.) cultivar Shiokari as the WT and tillering dwarf mutants, *d10*, *d14*, and *d17* in the Shiokari background

(Ishikawa et al., 2005). Seeds were provided by Dr. Junko Kyozuka (Tohoku University) and were propagated in a glass room at the Research Center for Life and Environmental Sciences, Toyo University. Seedlings were grown hydroponically as described previously (Umehara et al., 2008). Surface-sterilized seeds were incubated in sterile water at 27°C in the dark for 1 day. Germinated seeds were put in hydroponic culture media solidified with 0.6% agar and cultured under fluorescent white light (130–180 μmol m^{-2} s^{-1}; 16 hr light at 25°C, 8 hr dark at 23°C) for 7 days. The seedlings were then transferred to glass vials containing hydroponic culture medium (Kamachi, Yamaya, Mae, & Ojima, 1991). To evaluate the effect of nutrient deficiency on SL production, we prepared hydroponic culture media as shown in Table S1. All media contained 1 mM 2-(*N*-morpholino) ethanesulfonic acid (MES) buffer adjusted to pH 5.7. Only tillers growing over 2 mm were counted (Umehara et al., 2008). Relative chlorophyll contents were measured as SPAD values using a leaf chlorophyll meter, SPAD-502Plus (Konica Minolta Inc., Japan) (Ata-Ul-Karim et al., 2016).

2.2 | SL analysis

To measure the amounts of 4DO released from roots, hydroponic media were extracted with ethyl acetate twice after adding d_1-labeled 4DO as an internal standard. The ethyl acetate phase was evaporated to dryness under nitrogen gas. The extracts were dissolved in ethyl acetate: *n*-hexane (15:85) and loaded onto 1-ml Sep-Pak Silica cartridges (Waters, MA, USA), washed with ethyl acetate: *n*-hexane (15:85) and then eluted with ethyl acetate: *n*-hexane (35:65).

To measure 4DO levels in roots, roots (ca. 650 mg) were homogenized in 10 ml acetone containing d_1-labeled 4DO; the homogenates were filtered with Bond Elute reservoirs (Agilent, CA, USA) and evaporated to dryness under nitrogen gas. The extracts were dissolved in water adjusted to pH 2–3 with 1 N HCl and extracted with ethyl acetate twice. The ethyl acetate phase was evaporated to dryness under nitrogen gas. The extracts were dissolved in 10% acetone, loaded onto Oasis HLB 3-ml cartridges (Waters), washed with 10% acetone, and eluted with 60% acetone. The eluates were dissolved in ethyl acetate: *n*-hexane (15:85), loaded onto Sep-Pak Silica 1-ml cartridges (Waters), washed with ethyl acetate: *n*-hexane (15:85), and eluted with ethyl acetate: *n*-hexane (35:65).

Purified 4DO-containing fractions were dissolved in 50% acetonitrile and subjected to liquid chromatography–tandem mass spectrometry (LC-MS/MS) analysis using a system consisting of a quadrupole tandem mass spectrometer (3200 QTRAP; Sciex, MA, USA) and a high-performance liquid chromatograph (Prominence, Shimadzu, Kyoto, Japan) equipped with a reverse-phase column (Acquity UPLC BEH-C18, 2.1 × 50 mm, 1.7 μm, Waters). Data were analyzed in Analyst 1.5.1 and Multi Quant 2.0.2 (Sciex, MA, USA).

2.3 | Gene expression analysis

Total RNA was extracted from roots using an RNeasy Maxi kit (Qiagen, Hilden, Germany) and concentrated using an RNeasy Mini kit (Qiagen). A 3-μg aliquot of total RNA was used for cDNA synthesis with a ReverTra Ace qPCR RT Master Mix with gDNA Remover (Toyobo, Osaka, Japan). qRT-PCR was performed on a StepOnePlus RT-PCR system (Thermo Fisher Scientific, MA, USA) using a Thunderbird Probe qPCR Mix (Toyobo, Osaka, Japan), specific primers, and Taq-Man probes listed in Table S2. Ubiquitin expression was used as internal control. Total RNA extraction, cDNA synthesis, and qRT-PCR were conducted according to manufacturers' instructions.

2.4 | Statistical analysis

Statistical analysis was performed in SPSS 23 (IBM SPSS Inc., Armonk, NY, USA) using Student's *t* test for pairwise comparisons, ANOVA and Tukey's honestly significant difference test (HSD) for multiple comparisons.

3 | RESULTS

3.1 | Effect of macronutrient deficiency in hydroponic culture medium on tiller bud outgrowth and SL production in rice

To evaluate the effect of macronutrient deficiency on tiller bud outgrowth in rice, we cultivated WT seedlings in hydroponic media without N, P, K, S, Ca, Mg, or Fe for 1 week and counted the number of outgrowing tiller buds (>2 mm). Interestingly, the second leaf tillers grew in the absence of Ca but not of other nutrients (Figure S1A). Tiller bud outgrowth in the absence of Ca was inhibited by 1 μM GR24 that is a SL synthetic analog (Figure S1B). LC-MS/MS

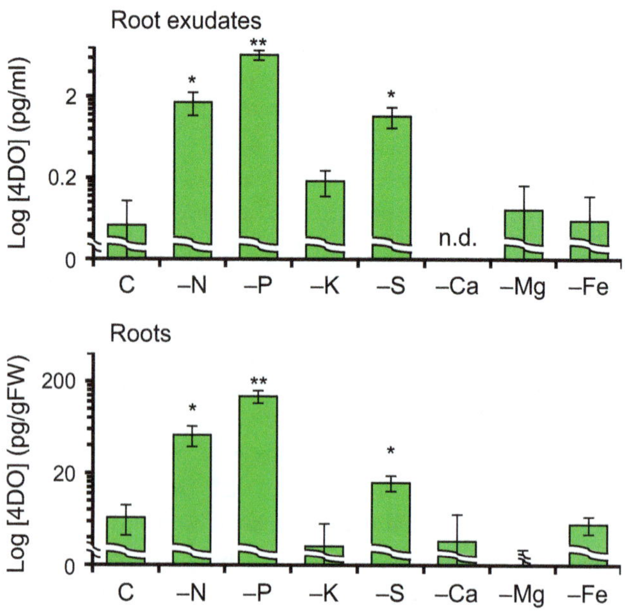

FIGURE 2 4DO levels in root exudates and roots under deficiency of N, P, K, S, Ca, Mg, or Fe. C, control. Samples were collected on day 7 after transfer to hydroponic media, and 4DO levels were analyzed using LC-MS/MS. n.d., not detected. Data are means ± *SE* (n = 4). *p < .05, **p < .01 vs. control (Student's *t* test)

analysis demonstrated that the 4DO level was approximately 0.05 pg/ml in root exudates and 6.6 pg/g fresh weight (FW) in roots of nutrient-sufficient control (Figure 2). Under Ca deficiency (–Ca), 4DO was not detected in root exudates and its content in roots (3.7 pg/g FW) was lower than that of control (Figure 2). The tiller bud outgrowth in –Ca plants was probably due to the reduction of 4DO levels in roots. Reduced SL levels under –Ca were reported in root exudates of red clover by liquid chromatography–tandem mass spectrometry (LC-MS/MS) analysis and seed germination assay of root parasitic plants (Yoneyama, Xie, et al., 2007; Yoneyama, Yoneyama, et al., 2007).

Under –P conditions, the 4DO levels were highest among all experimental conditions tested (~6.5 pg/ml in root exudates and 137 pg/g FW in roots; Figure 2). Under –N conditions, the 4DO levels were also higher than in control (~1.7 pg/ml in root exudates and 51.5 pg/g FW in roots). Under –S conditions, the 4DO levels increased to ~1.1 pg/ml in root exudates and 15.7 pg/g FW in roots (Figure 2). To further investigate the effects of S deficiency on SL production, we grew rice seedlings for 1 week in hydroponic media with different S concentrations (potassium sulfate: 0, 10, 110, 210, and 310 μM as the S source). 4DO levels were higher at 0 and 10 μM than at 110, 210, and 310 μM potassium sulfate (Figure S2).

3.2 | Time course analysis of SL levels and expression of SL-related genes

To investigate which genes regulate SL production under S deficiency, we analyzed the expression levels of SL-related genes using quantitative real-time polymerase chain reaction (qRT-PCR). We grew WT seedlings in hydroponic medium without S for 1 week,

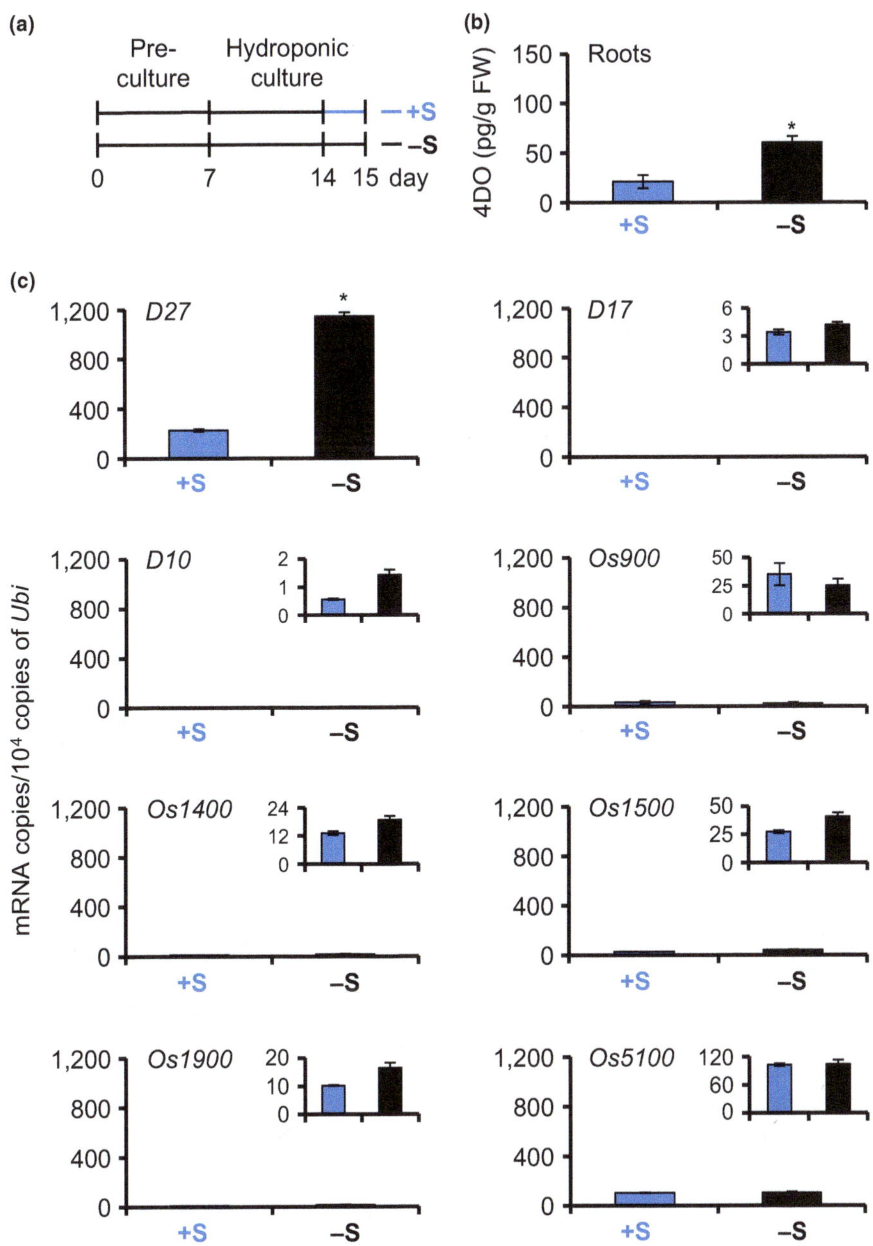

FIGURE 3 SL levels and expression of SL biosynthesis genes in WT rice seedlings under +S or –S conditions. (a) Schematic diagram showing the experimental conditions. Black and gray bars indicate +S and –S conditions, respectively. Seedlings were transferred to fresh +S or –S medium on day 14 of culture. 4DO levels and gene expression were analyzed on day 15. (b) 4DO levels in roots. Data are means ± *SE* ($n = 4$). (c) Transcript levels of SL biosynthesis genes in roots. Data are means ± *SE* ($n = 3$). *$p < .05$ (Student's *t* test)

transferred them to fresh +S or −S medium, and analyzed 4DO levels and gene expression levels at 0, 1, and 2 days after transfer (Figures S3A). 4DO levels and expression of SL biosynthetic genes at 1 day were picked up in Figure 3. Although 4DO levels remained high under −S conditions, they were significantly reduced by S supply on days 1 and 2 after transfer (Figures 3b and S3B). *D27* was strongly expressed under −S conditions, and its expression was significantly reduced by S supply (Figures 3c and S3C). The expression pattern of *D27* resembled the changes in 4DO levels (Figure S3B,C). mRNA levels of *D10*, *D17*, and the five *MAX1* homologs did not differ significantly between +S and −S conditions (Figures 3c and S3C). On the other hand, the expression levels of the SL signaling genes *D3* and *D14* strongly increased under +S conditions in comparison with −S (Figure S3C).

3.3 | SL production under N, P, and S deficiencies

To investigate the effects of combinations of N, P, and S deficiencies on 4DO levels, we grew WT seedlings under nutrient-sufficient conditions (control), single macronutrient deficiency, or combined deficiencies. 4DO levels increased significantly in all the treatments in comparison with control (Figure 4). The highest 4DO levels were observed in both root exudates and roots under P deficiency, but this increase was reduced in combinations of P deficiency with deficiency in N and/or S (Figure 4).

To evaluate how the expression of SL biosynthesis genes changes in response to a combination of nutrient deficiencies, we analyzed the 4DO levels and expression of SL biosynthesis genes under combined P and S deficiency (−PS). We grew WT seedlings in −PS medium for 1 week and then transferred them to fresh +PS or −PS medium (Figure 5a). 4DO levels and gene expression were analyzed 1 day after the transfer. 4DO levels were slightly higher in the roots of seedlings grown in −PS medium than in those of seedlings grown in −S medium and were significantly decreased by PS supply (Figures 3b, 4, and 5b). *D27* expression was lower under −PS than under −S conditions, whereas the expression of other SL biosynthesis genes such as *D17*, *D10*, and *OsMAX1s* was higher (Figures 3c and 5c). On the other hand, the expression levels of the SL signaling genes *D3* and *D14* strongly increased under +PS conditions in comparison with −PS (Figure S4).

3.4 | Effect of S deficiency on tiller bud outgrowth and leaf senescence

Shoot branching is suppressed and chlorosis is accelerated under S deficiency (Dobermann & Fairhurst, 2001). To evaluate whether shoot branching inhibition and chlorosis are caused by SL produced under S deficiency, we grew WT, the SL biosynthesis mutant *d10* and *d27*, and the SL signaling mutant *d14* in hydroponic culture medium with or without S for 22 days after preculture (Figure S5A). In WT, the number of outgrowing tillers over 2 mm gradually increased from 14 to 29 days under +S, but no tiller outgrowth was observed under −S (Figures 6a and S5B). In *d* mutants, the number of outgrowing tillers also gradually increased from 14 to 29 days under +S and was significantly reduced under −S (Figures 6a and S5B). The tiller bud outgrowth of *d27* was slower than those of other *d* mutants under +S, but achieved to similar levels at day 29 (Figure 6a). Under −S, the tiller number of *d27* was higher than that of WT but lower than those of other *d* mutants (Figure 6a).

Next, we pre-cultured WT, *d10*, *d14*, and *d27* for 1 week, then grew them in hydroponic culture medium under +S for 1 week and then under −S or +S for 15 days (Figure S6A) and measured SPAD value in the third leaves (the SPAD value indicates total chlorophyll content, and low SPAD values indicate leaf senescence) (Ata-Ul-Karim et al., 2016). The SPAD value tended to gradually decrease with time and was significantly lower under −S than under +S from day 20 to day 29 in WT, and on day 29 in *d14* and *d27*, but did not differ significantly between +S and −S in *d10* (Figures 6b and S6B). The SPAD values were mostly similar from day 14 to day 29 between WT and *d* mutants under +S, but the SPAD values of WT were lower than *d* mutants on days 26 and 29 (Figures 6b and S6B). Interestingly, the SPAD values of *d27* were similar to WT and other *d* mutants under +S, but lower than that in other *d* mutants under −S (Figure 6b). To check a possibility that *d27* may produce slight levels of 4DO, we confirmed 4DO levels in *d27* using LC-MS/MS. We could not detect 4DO in *d27* under either −S, −P, or nutrient-sufficient conditions (Figure S7). The reduction in the number of outgrowing tiller buds and SPAD value in *d27* under −S was not observed under −P (Figure S8).

4 | DISCUSSION

In this study, we found that 4DO levels increased under −S in addition to −P, and −N conditions in WT rice seedlings (Figure 2). Among SL

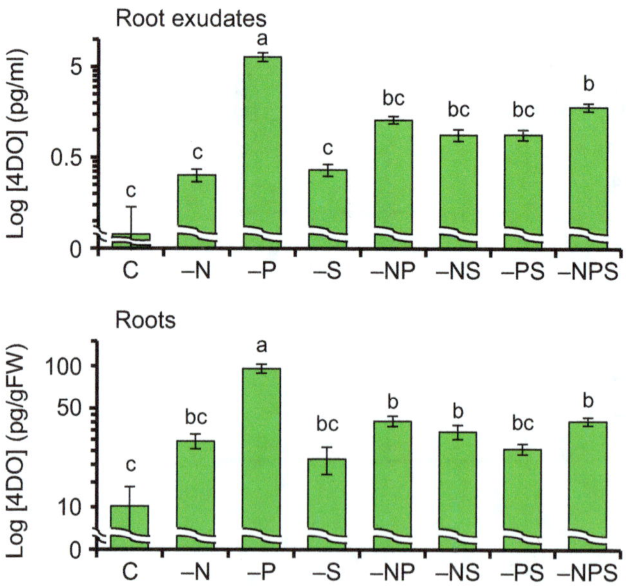

FIGURE 4 Effects of combined N, P, and S deficiencies on 4DO levels. The samples were collected on the day 7 after transfer to hydroponic culture media, and 4DO levels were analyzed using LC-MS/MS. Data are means ± *SE* ($n = 4$). *$p < .05$, **$p < .01$ (Student's *t* test)

biosynthesis genes, only *D27* (not *D10*, *D17*, or *OsMAX1s*) was highly expressed and thus contributed to SL production under −S (Figure 3). Thus, SL production under S deficiency is due to a marked increase in the expression of *D27*. In this time, we cannot exclude a possibility that activity of SL metabolism is low under −S because we still do not know the SL metabolic pathway. On the other hand, gene expression of *D3* and *D14*, which are the SL signaling genes, is rather upregulated under +S indicate that rice seedlings may increase the SL sensitivity because SL levels are low in S sufficient condition (Figures S3 and S4). When the *d27* mutant was grown under −S, it had fewer outgrowing tiller buds than other *d* mutants, and its chlorophyll content was lower than those of other *d* mutants (Figure 6). These results suggest that *D27* is involved in −S response and may maintain plant growth and development in −S conditions.

D27 was originally characterized in rice as a gene encoding an iron-containing protein that localizes in chloroplasts (Lin et al., 2009). *d27* mutants fail to produce SL, have increased tiller number and reduced plant height, and their phenotypes are rescued by exogenously applied SL, indicating that D27 is required for SL biosynthesis (Lin et al., 2009). In vitro experiments demonstrated that the D27 protein is an isomerase that converts all-*trans*-β-carotene to 9-*cis*-β-carotene in the first step of SL biosynthesis (Alder et al., 2012). In rice roots, high expression of most SL biosynthesis genes (*D27*, *D17*, *D10*, and *OsMAX1s*) increases SL levels under −P (Umehara et al., 2010). The expression levels of *D10* and *D17* increase under −N, but that of *D27* is not affected (Sun et al., 2014). In contrast, only *D27* is highly expressed under −S, and its level is 10 times that under −P (Umehara et al., 2010) (Figure 3). Under −PS, expression levels of *D27* decreased but that of the other SL biosynthetic genes increased compared with −S, resulting in increasing the flow of SL biosynthesis (Figures 3 and 5). Thus, 4DO content of PS deficiencies might be higher than that of S deficiency. D27 in Arabidopsis, AtD27, is located in plastids and acts upstream of MAX1 in the SL biosynthetic pathway (Waters, Brewer, Bussell, Smith, & Beveridge, 2012). Microarray data

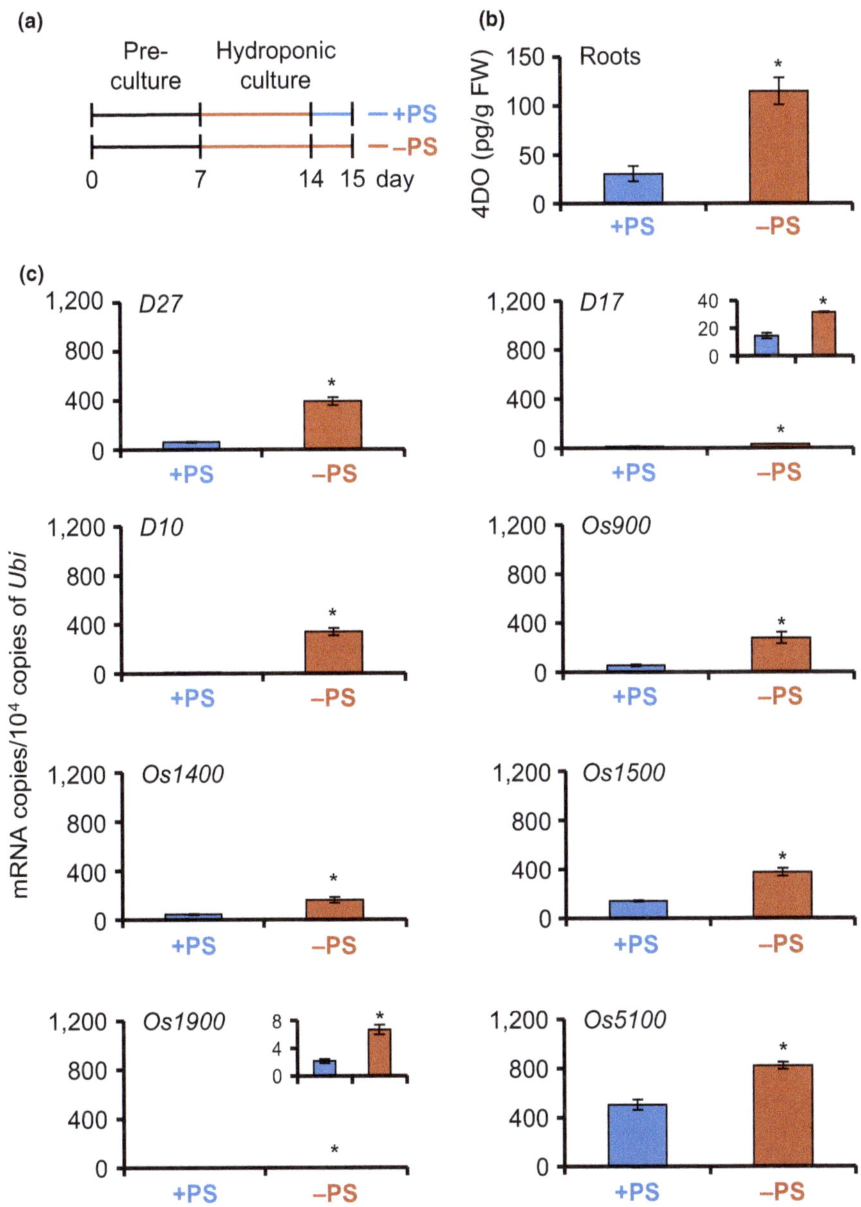

FIGURE 5 SL levels and expression of SL biosynthesis genes in WT rice seedlings under +PS or −PS conditions. (a) Schematic diagram showing the experimental conditions. Black and gray bars indicate +PS and −PS conditions, respectively. Seedlings were transferred to new +PS or −PS media on the day 14 of culture. 4DO levels and gene expression were analyzed on day 15. (b) 4DO levels in roots. Data are means ± *SE* (n = 4). (c) Transcript levels of SL biosynthesis genes in roots. Data are means ± *SE* (n = 3). *p < .05 (Student's t test)

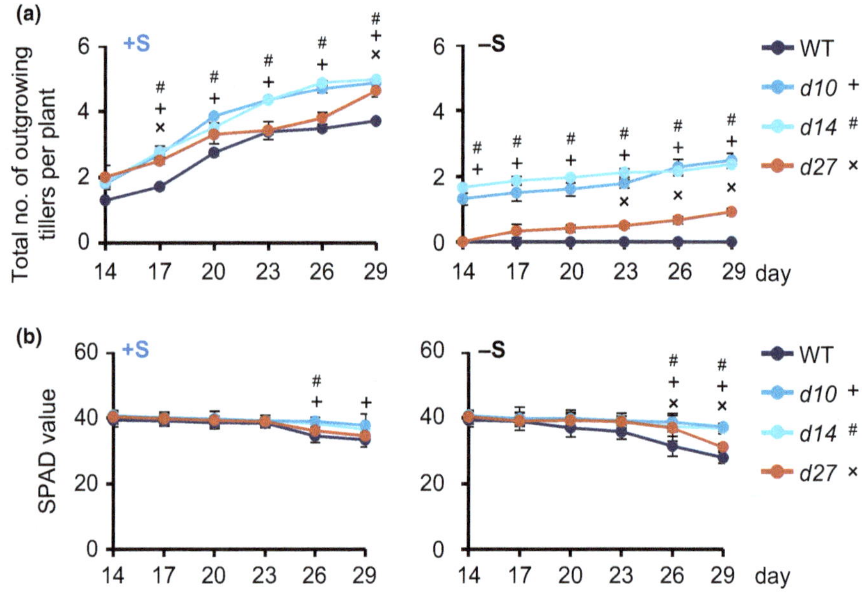

FIGURE 6 Phenotypical analysis of WT and SL mutants under +S and −S. (a) Total number of outgrowing tillers. Outgrowing tillers over 2 mm were counted in 21-day-old seedlings. Data are means ± *SE* (*n* = 3). (b) Comparison of the SPAD values of WT and the *d27*, *d10*, and *d14* mutants. Data are means ± *SE* (*n* = 3). Significant differences between WT and *d* mutants are shown (ANOVA, $p < .05$)

demonstrate that the expression level of *AtD27* in Arabidopsis roots increases under low S condition, but the expression of *MAX1* and *MAX3* do not (Iyer-Pascuzzi et al., 2011; Maruyama-Nakashita, Nakamura, Tohge, Saito, & Takahashi, 2006). These data support our results that *D27* expression is responsive to −S.

Sulfur is a macro-element required for plant growth because it is a component of biologically important compounds such as the amino acids Cys and Met, antioxidant tripeptide glutathione, *S*-adenosyl methionine, sulfolipids, and many secondary metabolites (Amtmann & Armengaud, 2009). In rice, S deficiency reduces chlorophyll content in young leaves, the number of tillers, and plant height (Dobermann & Fairhurst, 2001). These symptoms are similar to the phenotypes of SL biosynthesis- or signaling-deficient mutants. In WT, tiller bud outgrowth was completely suppressed under −S, whereas tiller buds of *d* mutants still grew (Figure 6a). The SPAD values of *d10* and *d14* were higher than that of WT under −S (Figure 6b). Thus, SLs regulate shoot branching and leaf senescence in response to S deficiency. However, number of tiller buds and SPAD value of *d* mutants were smaller under −S than under +S, indicating that SLs only partially mediate the regulation of shoot branching and leaf senescence, and effect of nutrient deficiency are involved directly.

In SL biosynthesis, the transcription factors NSP1 and NSP2 are indispensable in *Medicago* and rice (Liu et al., 2011). *D27* expression is upregulated by NSP1 and NSP2 under −P and in rhizobium symbiotic signaling (Liu et al., 2011; van Zeijl et al., 2015). However, the expression levels of *NSP1* and *NSP2* did not increase in the absence of N or P (Liu et al., 2011). NSP1 and NSP2 were originally found as the GRAS family transcription factors regulating nodule formation (Kalo et al., 2005; Smit et al., 2005). Later, they were also found to be components of the Myc signaling pathway of AM fungi, in addition to the Nod signaling pathway (Delaux, Becard, & Combier, 2013; Lauressergues et al., 2012). In rice, only *D27* transcript increase in AM colonized roots compared with non-AM colonized roots among SL biosynthetic genes (Kobae et al., 2018). Secreted SLs activate hyphal branching of AM fungi and trigger their symbiotic interactions with the host plants (Akiyama et al., 2005). AM fungi can supply S as well as N and P to the host (Allen & Shachar-Hill, 2009; Govindarajulu et al., 2005; Smith & Read, 2008). Therefore, expression of rice *D27* is involved in −S response, and *D27* might play an important role in effective S absorption via AM fungi. To understand the interactions between SL production under −S condition and symbiosis with AM fungi, further research would be required in the future.

ACKNOWLEDGMENTS

This work was in part supported by grants from Toyo University (the Inoue Enryo Memorial Foundation for Promoting Sciences to M.S.) and from the Japan Society for the Promotion of Science (KAKENHI Grant Nos. 26450144 and 17K07650 to M.U.). We thank Dr. Junko Kyozuka (Tohoku University) for providing rice seeds, Dr. Kohki Akiyama (Osaka Prefecture University) for providing d_1-labeled 4DO, Mr. Atsushi Hanada (Sciex) for his technical supports of LC-MS/MS, and Drs. Shosaku Kashiwada, Yuichi Iwasaki, and Uma Maheswari Rajagopalan (Toyo University) for constructive comments. This study was in part supported by the Research Center for Life and Environmental Sciences, Toyo University.

CONFLICT OF INTERESTS

None declared.

AUTHOR CONTRIBUTIONS

M.S. and M.U. performed the experiments and analyzed the data. K.S. and S.Y. contributed to the experimental design. M.U. directed the research and designed the experiments. M.S. and M.U. wrote the manuscript.

1981; Kozaki & Sasaki, 1999; Sasaki, Kozaki, & Hatano, 1997), which effects both lipid compositions (Maatta et al., 2012) and seed oil production in developing green oilseeds (Goffman, Alonso, Schwender, Shachar-Hill, & Ohlrogge, 2005; Li et al., 2006). In wild-type Arabidopsis, the amount of light is one of the major factors that lead to the variation of seed oil content reported in the literature of ~24%–43% seed oil by weight and ~3–9 μg oil per seed (Li et al., 2006). The amount of light used to grow HFA accumulating transgenic Arabidopsis plants reported in the literature also varies widely (if even reported at all). A few examples are as follows: just the photoperiod length reported such as continuous light (Dauk et al., 2007), or glasshouse supplemented with lamps to 16 hr day (van Erp et al., 2011); or both photoperiod and amount of light with a range of quantities such as, 16 hr and 150 μmol photons m^{-2} s^{-1} (Lu et al., 2006), or 24 hr and ~100–200 μmole photons m^{-2} s^{-1} (van Erp et al., 2011).

Considering the effect that both novel fatty acid production and light have on seed oil synthesis, we questioned if part of the variation in the reported successes of HFA engineering could be due to differences in light conditions between experiments and between laboratories. To move a proof-of-concept engineering strategy tested in Arabidopsis to an actual field crop will require that the engineering strategy be robust at any growth condition and not be dependent on select laboratory conditions such as very low light or 24-hr light conditions. Therefore, we investigated the effect of 16 versus 24-hr photoperiods, and three different 16-hr photoperiod light intensities on the seed oil content of three different transgenic Arabidopsis lines accumulating HFA, and the nontransgenic background line. Our results demonstrate that the interpretation of the success of multigene stacking for "increase in HFA content" and "recovery of reduced oil content" are both dependent on light conditions and the type of oil content measurement utilized to draw conclusions. These results will be valuable for designing experiments to evaluate future oilseed engineering strategies in Arabidopsis prior to oilseed crop engineering.

2 | METHODS

2.1 | Plant germination

The seeds were sterilized using an aqueous solution of 10% bleach, 27% ethanol, and 0.1% SDS, followed by five washes of water and final suspension in 0.1% Agar solution, and applied to germination plates (1x MS salts, 0.05% MES free acid, 1% sucrose, and 0.8% Agar, pH 5.7). The plates were incubated at 4°C for 3 days then placed under the low light condition (see below) until all lines germinated and produced two true leaves (approximately 7–10 days). The seedlings were then transferred to soil and placed in the proper light conditions for each light treatment below.

2.2 | Growth conditions for light treatments:

All plants were grown in Percival growth chambers under white fluorescent light; the chambers did not have humidity control. The 16-hr versus 24-hr photoperiod experiment was grown in a model AR-22L chamber which contains two independent growth compartments each with an interior volume of 15.7 ft^3. Each compartment was set to 200 μmol photons m^{-2} s^{-1} light with 23/20°C for the 16/8-hr day/night cycle and a constant 23°C for the continuous light treatment. The high and low light intensity experiment plants were grown in equivalent E41HO chambers each with an interior volume of 37.6 ft^3. The light intensities were 112 and 364 μmol photons m^{-2} s^{-1} in the low and high light treatments, respectively. Each chamber had the same day/night cycle of 16/8 hr and 23/20°C. All light measurements were made with a LI-COR® LI-250A light meter with quantum sensor LI-190 in the middle of the chamber at pot level prior to adding the plants to the chamber. In each growth chamber, the light intensity 10 inches above the pot height (average plant height) was approximately 50–60 μmol photons m^{-2} s^{-1} higher than at pot level. For each light treatment, 10–17 individual plants of each plant line were grown together randomized across the growth chamber to minimize the effects of position within the chamber. All plants were watered three times a week and fertilized once a week with Peter's NPK 20-20-20 fertilizer (0.957 g/l).

2.3 | Determination of total seed oil content

Total seed lipid content was determined by direct conversion to fatty acid methyl esters and quantification by gas chromatography with flame ionization detection (GC-FID) (Li et al., 2006). In brief, dry seeds were weighed (2–3 mg) or counted (n = 30) to determine the amount of FAME by dry weight or per seed. FAMEs were produced in 1 ml 5% sulfuric acid in methanol at 85°C for 1.5 hr together with 40 or 20 μg 17:0 TAG in 0.2 ml toluene as an internal standard. FAMEs were extracted by adding 0.2–0.5 ml hexane and 1.5 ml 0.88% potassium chloride. The hexane phase was analyzed by GC-FID separated on a RESTEK Rtx®-65 column (30 m, 0.25 mm ID, df = 0.25 μm). GC method parameters were as follows: carrier gas, He at constant linear velocity 30cm/s; temperature profile, 190°C for 2 min, increased 10°C/min to 270°C and held for 2 min.

2.4 | Statistical analysis

Calculations for oil quantity from GC data were made using Microsoft Excel and exported to GraphPad Prism 6.01 for statistical analysis and graphical representation. For each experiment, 10–17 individual plants per plant line represent the biological replicates and the data are presented as the average and *SEM*. For statistical analysis, ordinary two-way ANOVA test was conducted using uncorrected Fisher's LSD at 95% confidence interval. The propagation of error through ratio calculations $A\Delta a/B\Delta b = C\Delta c$ was done by the two formulas, where the ratio is $C = A/B$, and the standard deviation is $\Delta c = C \times \sqrt{((a \div A)^2 + (b \div B)^2)}$. Statistical relevance is represented directly onto the graph using different letters to denote significant differences. Upper-case letters are used to compare the same lines between different growth conditions while lower-case letters are used to compare differences between plant lines within an individual growth condition. If letters are different between two bars, then there is significant difference (p-value <0.05) between the two bars.

3 | RESULTS

3.1 | Effect of photoperiod length on novel fatty acid engineering of seed oil

To determine the effect of light conditions on both the oil amount and oil fatty acid composition of transgenic seeds producing novel fatty acids, we analyzed the seed lipid content in four different previously produced Arabidopsis lines (Table 1) grown under various light regimens. The novel fatty acids are hydroxylated fatty acids produced through the seed-specific expression of the castor fatty acid hydroxylase (*RcFAH12*). RcFAH12 hydroxylates oleic acid (18:1) to ricinoleic acid (18:1-OH) (Moreau & Stumpf, 1981; Vandeloo, Broun, Turner, & Somerville, 1995). In this study, all transgenic lines are in the *fae1* mutant background (Kunst, Taylor, & Underhill, 1992) which eliminates elongation of oleic acid to eicosenoic acid (20:1), accumulating approximately double the amount of 18:1 than in wild-type Col-0 seeds. Therefore, due to the higher amount of the 18:1 substrate that could be used by RcFAH12 to produce HFA, the *fae1* line was originally chosen as a background for HFA production in Arabidopsis seeds as a model for novel fatty acid engineering in transgenic plants (Lu et al., 2006). Therefore, in this study, the *fae1* line is our nontransgenic control. Previously, the amount of seed oil in Col-0 and *fae1* was demonstrated to be similar (Bates et al., 2014; Kunst et al., 1992). Line CL37 expresses the *RcFAH12* alone in *fae1* (Lu et al., 2006). The DGAT line coexpresses *RcDGAT2* with *RcFAH12* in *fae1* (Burgal et al., 2008), and the PDAT line coexpresses *RcPDAT1a* with *RcFAH12* in *fae1* (van Erp et al., 2011). The lines used in this study are summarized in Table 1. Previously, the CL37 line was demonstrated to have up to a 50% reduction in total seed oil, which was partially recovered by the coexpression of *RcDGAT2* or *RcPDAT1a*, as determined from 24-hr fluorescent light at ~100–170 μmol photons m^{-2} s^{-1} (Bates et al., 2014).

We initially set out to determine how growth of Arabidopsis under 24-hr continuous light may affect our interpretation of both percent HFA within the seed oil and the recovery of the reduced oil phenotype of CL37 by coexpression of *RcDGAT2* or *RcPDAT1a*, in the DGAT and PDAT lines, respectively. Therefore, we grew all four plant lines together in two separate equivalent Percival AR22L growth chambers both set for 200 μmol photons m^{-2} s^{-1} fluorescent white light at pot height in the center of the chamber. One chamber was set to 16-hr photoperiod (8-hr dark), and the other to 24 hr of continuous light. 10–17 individual plants from each plant line were randomized across the growth chamber to minimize the effects of position within the chamber. The oil content of mature seeds was determined both as a percentage of seed weight (Figure 1a) and as μg per seed (Figure 1b). In general, the continuous light treatment produced more oil by both measurements (Figure 1a,b). However, the relative amount of oil in each transgenic line compared to the nontransgenic control was not consistent between light treatments or between measurement types (Figure 1c,d). When grown under the 16/8-hr day/night cycle and measured as a percentage of seed weight, the oil content of CL37 was 80% of *fae1* and the DGAT and PDAT lines were partially recovered to 90% of *fae1* (Figure 1a,c). However, under continuous light, the CL37 oil content by weight was 90% of *fae1*, and the DGAT and PDAT lines were essentially the same as *fae1* (Figure 1a,c). Based on these results, the effect of HFA production on lowering seed oil content appears to minimal when Arabidopsis is grown at 200 μmol photons m^{-2} s^{-1} under 24-hr light and measured as a percent of seed weight.

When the oil content was measured per seed, the differences between each transgenic and the wild-type control were significantly greater than when measured as a percent of seed weight. CL37 oil content per seed was 68% of *fae1* for 16-hr photoperiod samples, which was recovered to approximately 88%–90% in the DGAT and PDAT lines. For the 24-hr light samples, CL37 was 73% of *fae1*, which was only partially recovered to 89 and 81% in the DGAT and PDAT lines, respectively (Figure 1b,d). Therefore, the 16-hr light growth conditions and measurements of oil content per seed demonstrate both the biggest reduction in oil content caused by expression of the hydroxylase in CL37 compared to wild type, and the largest recovery in oil content (increase from 68% to 90% of wild type) when the hydroxylase is coexpressed with HFA selective acyltransferases from castor.

In addition to differences between oil amounts between 16-hr and 24-hr light, there were also significant differences in the amounts of the novel HFA in the transgenic lines between light treatments. Figure 2 demonstrates the seed fatty acid composition (weight percent of total fatty acids). For all transgenic lines, there was a higher percentage of HFA in the seed oil when grown under 16/8-hr day/night cycles than for the 24-hr continuous light. The lower percent HFA at 24-hr light seeds was compensated for predominantly by higher levels of 18:1. While just percent fatty acid composition is a common way to report transgenic changes in seed fatty acid content, it is a measurement that is independent of the total amount of fatty acids in the seeds. Considering that light effects the quantity of seed oil, we determined the total amount of HFA as both a percentage of seed weight (Figure 3a) and as μg HFA per seed (Figure 3b). The apparent reduction in percent HFA composition when grown at 24-hr light in Figure 2 was not actually a reduction in total HFA amount. The total amount of HFA in all transgenic lines increased in 24-hr light conditions as compared to the day/night cycle (with the acceptation of the DGAT line measured only as a percentage of seed weight, Figure 3a). Therefore,

TABLE 1 Summary of Arabidopsis genotypes utilized

Name	Genotype	Reference
fae1	*fae1*	Kunst et al. (1992)
CL37	*fae1/RcFAH12*	Lu et al. (2006)
DGAT	*fae1/RcFAH12/RcDGAT2*	Burgal et al. (2008)
PDAT	*fae1/RcFAH12/RcPDAT1a*	van Erp et al. (2011)

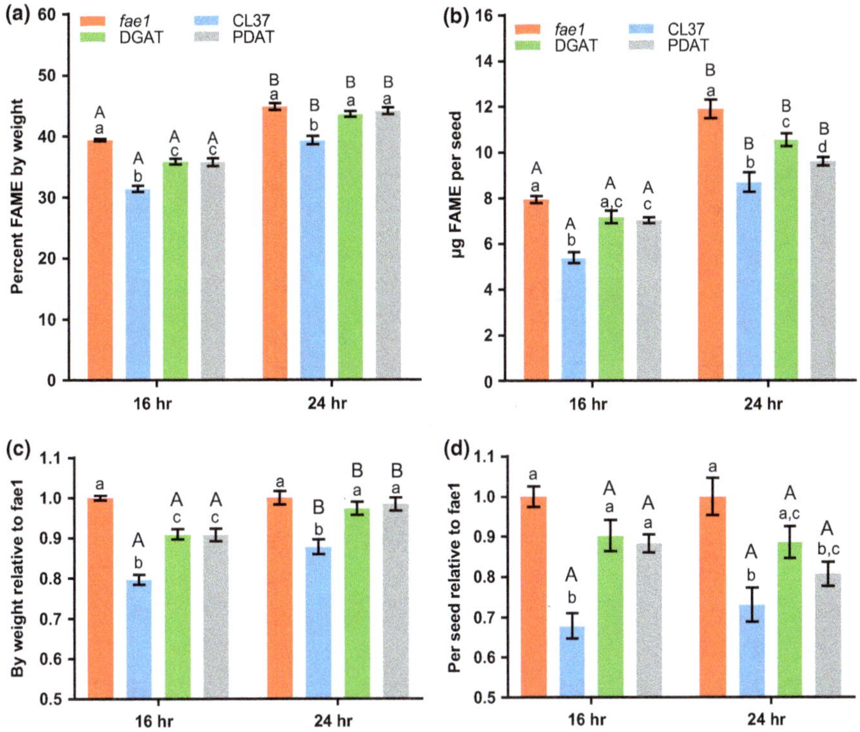

FIGURE 1 Seed oil content of day/night cycle versus continuous light. Mature seed oil content from plants grown under a 16-hr photoperiod and 8-dark day/night cycle versus continuous 24-hr photoperiod. Light intensity of 200 μmol photons m^{-2} s^{-1} for both light treatments. (a) Total seed lipid content as a percentage of seed dry weight. (b) Total lipid content per seed. Data represent average and *SEM* of 10–17 biological replicates for each measurement. (c) and (d) represent the relative lipid content of the transgenic with respect to *fae1* in a and b, respectively. If letters above two bars are different, then there is significant difference between the two bars. Upper-case letters compare the same line between light treatments; lower-case letters compare each line within a light treatment

considering that 24-hr light increases the total amount of all fatty acids including HFA in seeds, the reduction in the percent HFA composition (Figure 2) was due to a larger increase in accumulation of common endogenous fatty acids over that of the novel HFA within the transgenic lines.

In summary, 24-hr light produces more total fatty acids than a day–night cycle, but novel fatty acid production does not keep up with the increase in production of endogenous fatty acids, which effects the fatty acid percent composition. In addition, the reduced oil phenotypes of various transgenic lines are artificially minimized when grown under 24 hr and measured solely as oil content by weight.

3.2 | Effect of light intensity on novel fatty acid engineering of seed oil

To determine the effect of changing light intensity on novel fatty acid engineering, plants were grown in equivalent Percival E41HO chambers under a 16/8-hr day/night cycle at low and high light intensities (112 and 364 μmol photons m^{-2} s^{-1} white light, respectively) as compared to the intermediate light intensity (200 μmol photons m^{-2} s^{-1}) in the day/night versus continuous light experiment (Figure 1–3). All plant lines were grown together randomized across each growth chamber. The plants of the high light treatment grew and matured 2–3 weeks faster than the low light treatment. The seed oil quantity from each plant line was measured as a percentage of seed weight (Figure 4a) and as μg per seed (Figure 4b). The differences between low and high light intensity within a 16/8-hr day/night cycle were less clear-cut than the day/night versus continuous light experiments and were dependent on individual line, type of oil content measurement (Figure 4a vs. 4b), and comparison method utilized Figure 4a,b vs. 4c,d).

As expected from previous light intensity experiments with Col-0 (Li et al., 2006), the *fae1* line produced significantly more seed oil by weight in the high light treatment as compared to the low light treatment (Figure 4a). The same was also observed for the PDAT line; however, there was no statistical difference between the low and high light treatments for both the CL37 and DGAT lines when measured as a percentage of seed weight. When oil content was measured as μg per seed (Figure 4b), the high light treatment produced significantly more oil than the low light treatment for all four plant lines.

Comparing the amount of oil between the lines within a light treatment led to similar conclusions for both the oil measurement types (Figure 4a,b). At low light, the PDAT line significantly recovered the reduced oil phenotype of CL37 such that it was not statistically different than *fae1*. However, the DGAT line under low light did not recover the reduced oil phenotype of CL37 and was still significantly less than *fae1*. Under high light conditions, the amount of oil in both the PDAT and DGAT lines was significantly higher than

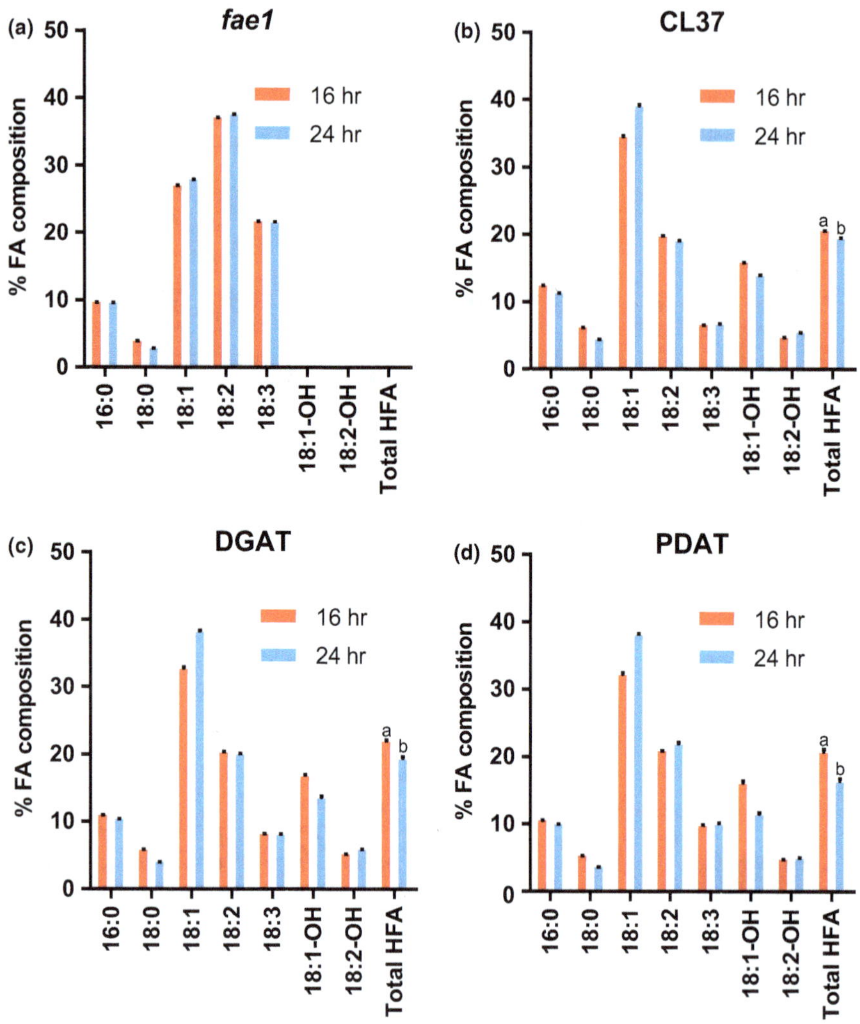

FIGURE 2 Seed fatty acid composition for day/night cycle versus continuous light. Fatty acid composition of mature seeds is percent by weight. (a) *fae1*. (b) CL37. (c) DGAT line. (d) PDAT line. Data represent average and *SEM* of 10–17 biological replicates for each measurement. The significance for total HFA content is represented by different letters above each bar. Fatty acid nomenclature: # carbons:# double bonds, -OH indicates presence of a hydroxyl

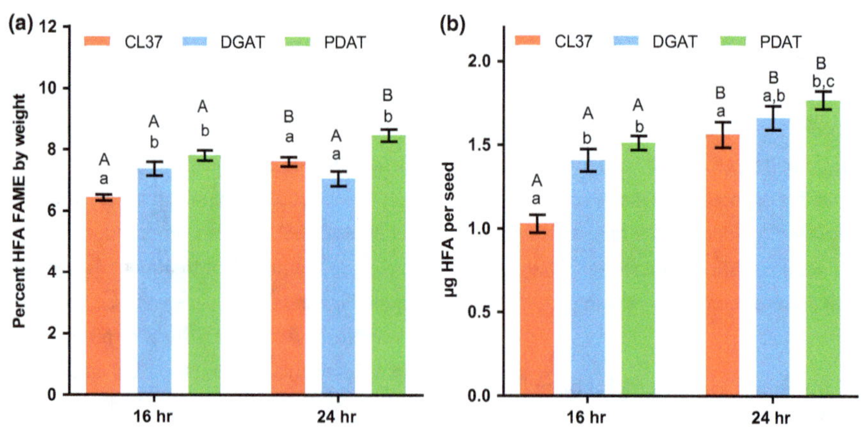

FIGURE 3 Total seed HFA accumulation for day/night cycle versus continuous light. Mature seed HFA content from plants grown under a 16-hr photoperiod and 8-dark day/night cycle versus continuous 24-hr photoperiod. Light intensity of 200 µmol photons m^{-2} s^{-1} for both light treatments. (a) Total HFA content as a percentage of seed dry weight. (b) Total HFA content per seed. Data represent average and *SEM* of 10–17 biological replicates for each measurement. If letters above two bars are different, then there is significant difference between the two bars. Upper-case letters compare the same line between light treatments; lower-case letters compare each line within a light treatment

CL37, but both were still significantly less than *fae1*. Interestingly, when the amount of oil in each transgenic line was analyzed relative to the *fae1* control within a light treatment and the subsequent transgenic/*fae1* oil content ratios compared between light treatments (Figure 4c,d), there was no statistical difference for the effect of light intensity on the relative amount of oil between the transgenics and the *fae1* control. The difference in significance of oil content between individual lines within a light treatment when comparing

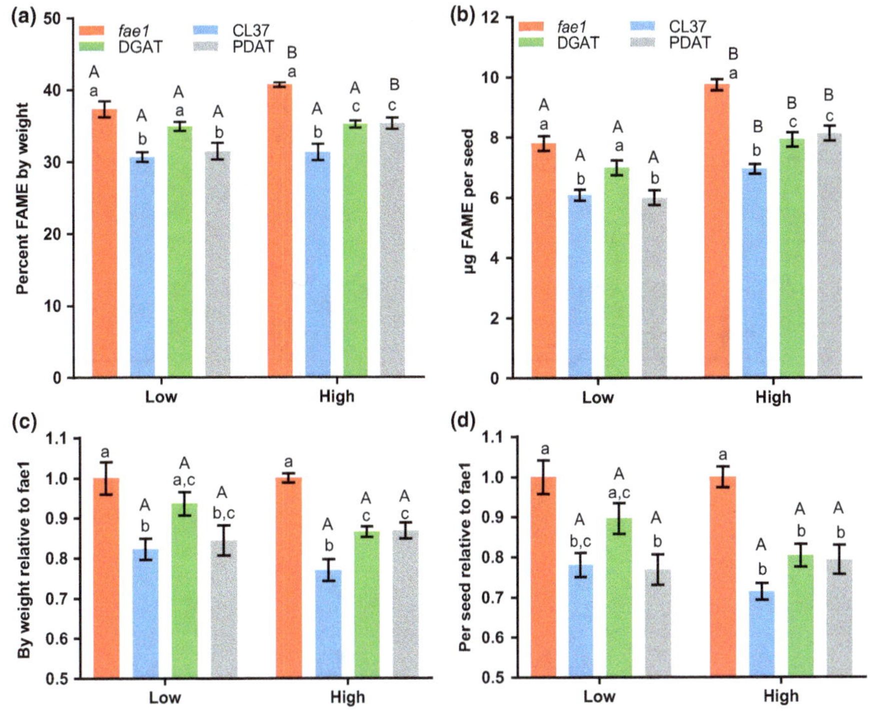

FIGURE 4 Seed oil content for day/night cycle grown at low and high light intensities. The 16/8-hr day/night cycle light intensities in μmol photons m^{-2} s^{-1} are as follows: Low, 112; and high, 364. (a) Total seed lipid content as a percentage of seed dry weight. (b) Total lipid content per seed. Data represent average and *SEM* of 10–17 biological replicates for each measurement. (c) and (d) represent the relative lipid content of the transgenic with respect to *fae1* in a and b, respectively. If letters above two bars are different, then there is significant difference between the two bars. Upper-case letters compare the same line between light treatments; lower-case letters compare each line within a light treatment

the analysis in Figure 4a versus Figure 4c and Figure 4b versus Figure 4d was due to the propagation of error during the ratio calculation which raised the ANOVA *p*-values out of the significant range (e.g. <0.05).

The effect of increasing light intensity within a 16/8-hr day/night cycle on fatty acid percent composition of the transgenic lines had a consistent trend with the least HFA in the low light treatment and the most HFA in the high light treatment in each of the transgenic lines (Figure 5). When total HFA abundance was considered as a percentage of seed weight, there was no significant difference between the light treatments for CL37, but there was a significant increase for both the PDAT and DGAT lines (Figure 6a). When HFA accumulation was measured as μg HFA per seed, all three transgenic lines showed a significant increase in HFA accumulation at high light over the low light treatment (Figure 6b).

In summary, the high light intensity within a 16/8-hr day/night cycle increased total fatty acid content as μg/seed in all lines as compared to low light, but the results were line specific when fatty acid accumulation was measured as a percentage of seed weight (Figure 4). Unlike the apparent decrease in percent HFA composition of seed oil in transgenic plants when the amount of light was increased by growing under 24-hr continuous light (Figure 2), the increase in light intensity from low to high within the day/night cycle led to an increase in both the percent HFA composition (Figure 5) and in the total accumulation of HFA (Figure 6).

4 | DISCUSSION

It has been well characterized that wild-type Arabidopsis seed oil content is strongly dependent on light conditions (Li et al., 2006). Our results comparing a 16/8-hr day/night cycle versus continuous light and different light intensities within a day/night cycle confirm these previous results with our nontransgenic control *fae1*. In addition, we provide additional results which indicate that various transgenic lines affected in fatty acid and oil biosynthesis do not respond to fluctuating light conditions the same as the nontransgenic control does. We also demonstrate the conclusions drawn about relative oil content between transgenic lines and the nontransgenic control, and between the different transgenic lines, are greatly dependent on the type of oil content measurement and analysis method used.

The transgenic lines utilized in these experiments have been engineered to accumulate novel HFA in Arabidopsis seed oil. However, the production of HFA has been previously characterized to induce inhibition of acetyl-CoA carboxylase activity (ACCase, the initial step of fatty acid synthesis) and thus reduce total oil accumulation in CL37 (Bates et al., 2014). The DGAT and PDAT lines coexpressing the castor hydroxylase *RcFAH12* with either *RcDGAT2* or *RcPDAT1a* more effectively incorporate HFA into TAG have increased ACCase activity and at least partially recover the reduced oil phenotype (Bates et al., 2014). The differential accumulation of seed oil between the control and each transgenic line under various light treatments (Figures 1 and 4) is likely due to the differential strength of the HFA-induced inhibition of ACCase between each transgenic line. The relative strength of ACCase inhibition likely limits the ability each transgenic line to upregulate of fatty acid synthesis in response to more light, as compared to the uninhibited nontransgenic control. For example, 24-hr light or the high light day/night cycle treatment both measured as μg fatty acids per seed (Figures 1b,c and 4b,d) indicated the biggest differences in relative oil content between *fae1* and the DGAT and PDAT lines, even though the total oil content of those lines was higher than all other light

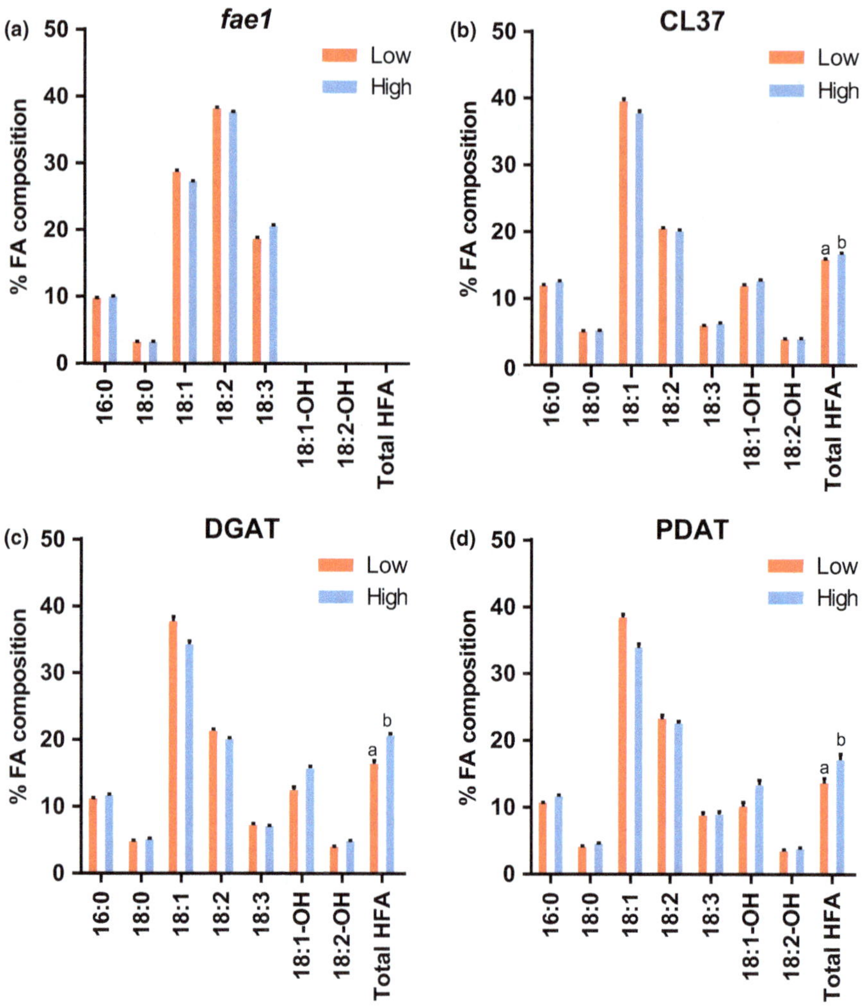

FIGURE 5 Seed fatty acid composition for day/night cycle grown at low and high light intensities. Fatty acid composition of mature seeds is percent by weight. (a) *fae1*. (b) CL37. (c) DGAT line. (d) PDAT line. Data represent average and *SEM* of 10–17 biological replicates for each measurement. The significance for total HFA content is represented by different letters above each bar. Fatty acid nomenclature: # carbons:# double bonds, -OH indicates presence of a hydroxyl

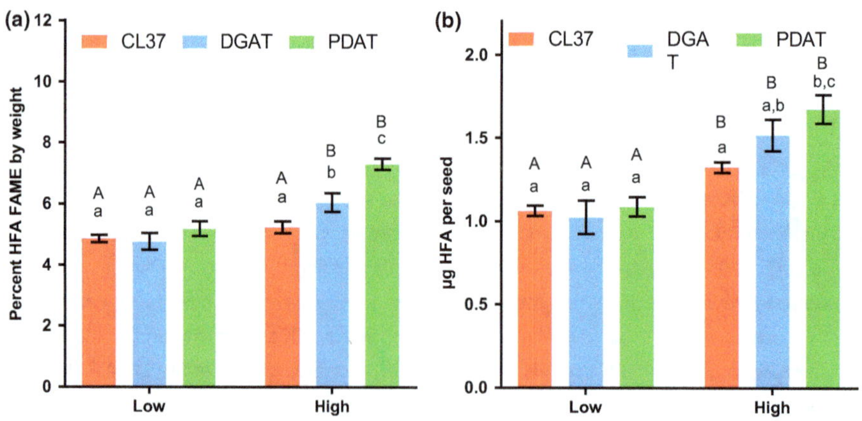

FIGURE 6 Total seed HFA accumulation for day/night cycle grown at low and high light intensities. The 16/8-hr day/night cycle light intensities in μmol photons m^{-2} s^{-1} are as follows: Low, 112; and high, 364. (a) Total HFA content as a percentage of seed dry weight. (b) Total HFA content per seed. Data represent average and *SEM* of 10–17 biological replicates for each measurement. If letters above two bars are different, then there is significant difference between the two bars. Upper-case letters compare the same line between light treatments; lower-case letters compare each line within a light treatment

treatments (Figures 1b and 4b). The likely reason for the low oil content of the transgenics relative to the nontransgenic control was that similar to wild type (Li et al., 2006) the *fae1* line has the ability to greatly increase fatty acid synthesis in response to additional light, but in each transgenic line the novel fatty acid-induced inhibition of fatty acids (Bates et al., 2014) limits this response to varying degrees (Figures 1b and 4b). Considering that reduced seed oil content has been indicated in many different instances of novel fatty acid composition engineering (Knutzon et al., 1999; Larson et al., 2002; Li et al., 2012; Lunn et al., 2018; Mansour et al., 2014; Shrestha et al., 2016), it is likely that other novel fatty acid engineering strategies will demonstrate similar seed oil results based on light conditions.

When interpreting conclusions on the effect of novel fatty acid engineering on total seed oil accumulation, we found that both the light conditions and the type of measurement utilized can influence the conclusions drawn. The maximal amount of seed fatty acids for all lines by both measurement types was with the 24-hr light treatment at 200 μmol photons m^{-2} s^{-1}, but the differences between the lines were dependent on measurement. Seed fatty acid content as a percentage of seed weight minimized the differences between the lines such that the there was no statistical difference between the *fae1* control and both the PDAT and DGAT lines under 24-hr light (Figure 1a). However, when fatty acid content was measured as μg per seed, each line was statistically different from each other (Figure 1b).

For plants grown under a day–night cycle, there was a general trend of a higher average oil content with higher light intensities, but the significance of the differences between light treatments was also dependent on the measurement type and the individual lines. When total fatty acids were measured as percent of seed weight, only the *fae1* and PDAT line had significant increases in the high light treatment over the low light treatment (Figure 4a), with PDAT having the largest increase in oil content from 31.5% ± 1.2 to 35.3% ± 0.8. However, the PDAT line oil content was still less than *fae1* in the same experiment which increased from 37.3% ± 1.1 to 40.7% ± 0.4. This suggests that the HFA-induced inhibition of fatty acid synthesis is least in the PDAT line over the other two transgenic lines, but fatty acid synthesis is still inhibited as compared to the nontransgenic control. When total fatty acids were measured as μg per seed, all lines grown under high light had a significant increase in oil content over the low light-treated plants (Figure 4b). Together these results suggest that light also affects seed weight by components other than lipids (e.g., protein, starch, and fiber), and that when focusing on lipid content alone measurements based off seed weight may not be as informative to the actual changes in lipid content as measurements on a per seed basis.

When considering the effects of just low light on the differences in oil content between the lines, low light minimized the difference between the nontransgenic control and the transgenic lines. For example, the PDAT line was not statistically different from *fae1* by either oil content measurements under low light (Figure 4a,b), but PDAT oil content was statistically less than *fae1* when measured as a percent of seed weight when plants were grown at a day/night cycle at 200 μmol photons m^{-2} s^{-1} (Figure 1a,c), and statistically less for both oil measurements for the high light treatment (Figure 4). These results suggest that suboptimal growth conditions (such as a growth chamber with old dim lights) may produce misleading results as to the ability of different plant lines to accumulate seed oils.

When seed oil content of the transgenic lines was analyzed as a ratio of the nontransgenic control, the differences between the light treatments and within each treatment by both oil measurements became less significant (Figure 1c,d and 4c,d). Only the day/night cycle versus 24-hr light experiment when measured as fatty acid content as a percentage of seed weight was significantly different between the two treatments for all transgenic lines when analyzed as a ratio of the control (Figure 1c). All other comparisons of light treatments as a ratio of transgenics to control were not significant (Figure 1d and 2c,d). A simple conclusion would be that even though the total oil amount is changing based on light, the relative amounts of oil between lines do not change. However, this does not fit with the direct oil abundance measurements. For example, it is clear that the pattern of oil accumulation among the four lines (and the significant differences between each line) is different between the light treatments in each experiment (Figure 1a,b and 4a,b). However, the differences between the lines were less significant after the ratio calculation because the error is propagated through the division calculation. Thus, this result implies larger numbers of replicates are likely needed for seed oil content measurements when the analysis method involves mathematical manipulation of the data and propagation of the error.

An interesting result from these experiments was the difference that extra light had on accumulation of HFA between the day/night versus continuous light experiment (Figures 2 and 3) and the light intensity experiment (Figures 5 and 6). Extra light appeared to reduce the HFA content as a percentage of total fatty acids when it was provided as 24-hr light (Figure 2), whereas extra light provided as a higher light intensity during a day/night cycle increased the percent HFA composition (Figure 5). However, both treatments increased the absolute levels of HFA in seeds for each line (Figures 3 and 6), with the exception of CL37 HFA measured just as a percent of seed weight for the low versus high light treatment (Figure 6a). The effect of light conditions on percent fatty acid composition (Figures 2 and 5) is likely due to the differences in rate of fatty acid synthesis versus fatty acid modification. Previously it was demonstrated that high rates of fatty acid synthesis proceed predominantly in the light (Browse et al., 1981), while the slower fatty acid desaturation can also continue in the dark leading to enhanced levels of desaturated fatty acids at the end of the night period versus at the end of the photoperiod (Maatta et al., 2012). Fatty acid hydroxylation is a variant of fatty acid desaturation (Vandeloo et al., 1995) and is likely slow relative to fatty acid synthesis as well. Therefore, in the continuous light plants, fatty acid hydroxylation likely cannot keep up with fatty acid synthesis to generate the same proportion of HFA as the during the day/night cycle. Likewise, the higher percent HFA with increasing light intensity with 16/8 day/night cycles is likely due to the enhanced production of the 18:1 substrate during the day, and during the subsequent night period the hydroxylase catches up and produce more HFA. When the HFA content between the three transgenic lines was compared just within the low light treatment, there were no statistical differences in HFA content between the CL37, DGAT, and PDAT lines. Therefore, suboptimal light conditions can minimize the effect of gene stacking engineering strategies to increase total levels of the desired product.

In conclusion, we demonstrate that the relative "success" of fatty acid engineering strategies for amount of novel fatty acids, their proportion within the oil, and their effect on total oil content relative to the oil content of controls and other engineered lines is highly dependent on light conditions, the type of oil content measurement, and data analysis methods (e.g., as exact values or as a ratio relative to the control). Therefore, our recommendations to the Arabidopsis seed oil engineering community are as follows: 1) to consider measuring seed oil and novel fatty acid quantity as part of the analysis of fatty acid engineering experiments rather than just fatty acid percent composition. 2) To utilize a day/night cycle growth condition instead of continuous light for studies quantifying seed oil content. Even though continuous light helps the plants to grow faster, it can lead to misleading results on both fatty acid composition and oil content as compared to nontransgenic controls and other transgenic lines. 3) To measure both oil content as a percentage of seed weight and oil content per seed because each has value and can provide alternative conclusions about the changes in oil content due to engineering. Finally, 4) to include detailed reporting of Arabidopsis growth conditions for each experiment in publications for the critical analysis of the success of oilseed engineering strategies between publications within the literature.

ACKNOWLEDGMENTS

The seed lines were provided by John Browse, Washington State University. This work was supported by the National Science Foundation (MCB #1613923) and The University of Southern Mississippi.

AUTHOR CONTRIBUTIONS

Both NK and PB planned experiments. NK conducted experiments. Both NK and PB analyzed data and wrote the manuscript.

REFERENCES

Adhikari, N., Bates, P. D., & Browse, J. (2016). WRINKLED1 Rescues feedback inhibition of fatty acid synthesis in hydroxylase-expressing seeds of Arabidopsis. *Plant Physiology*, *171*, 179–191. https://doi.org/10.1104/pp.15.01906

Aznar-Moreno, J. A., & Durrett, T. P. (2017). Review: Metabolic engineering of unusual lipids in the synthetic biology era. *Plant Science*, *263*, 126–131. https://doi.org/10.1016/j.plantsci.2017.07.007

Badami, R. C., & Patil, K. B. (1980). Structure and occurrence of unusual fatty acids in minor seed oils. *Progress in Lipid Research*, *19*, 119–153. https://doi.org/10.1016/0163-7827(80)90002-8

Bates, P. D. (2016). Understanding the control of acyl flux through the lipid metabolic network of plant oil biosynthesis. *Biochimica et Biophysica Acta*, *1861*, 1214–1225. https://doi.org/10.1016/j.bbalip.2016.03.021

Bates, P. D., & Browse, J. (2011). The pathway of triacylglycerol synthesis through phosphatidylcholine in Arabidopsis produces a bottleneck for the accumulation of unusual fatty acids in transgenic seeds. *The Plant Journal*, *68*, 387–399. https://doi.org/10.1111/j.1365-313X.2011.04693.x

Bates, P. D., Johnson, S. R., Cao, X., Li, J., Nam, J.-W., Jaworski, J. G., … Browse, J. (2014). Fatty acid synthesis is inhibited by inefficient utilization of unusual fatty acids for glycerolipid assembly. *Proceedings of the National Academy of Sciences of the United States of America*, *111*, 1204–1209. https://doi.org/10.1073/pnas.1318511111

Bayon, S., Chen, G., Weselake, R. J., & Browse, J. (2015). A small phospholipase A2-α from castor catalyzes the removal of hydroxy fatty acids from phosphatidylcholine in transgenic Arabidopsis seeds. *Plant Physiology*, *167*, 1259–1270. https://doi.org/10.1104/pp.114.253641

Broun, P., Boddupalli, S., & Somerville, C. (1998). A bifunctional oleate 12-hydroxylase: Desaturase from *Lesquerella fendleri*. *The Plant Journal*, *13*, 201–210. https://doi.org/10.1046/j.1365-313X.1998.00023.x

Broun, P., & Somerville, C. (1997). Accumulation of ricinoleic, lesquerolic, and densipolic acids in seeds of transgenic Arabidopsis plants that express a fatty acyl hydroxylase cDNA from castor bean. *Plant Physiology*, *113*, 933–942. https://doi.org/10.1104/pp.113.3.933

Browse, J., Roughan, P. G., & Slack, C. R. (1981). Light control of fatty-acid synthesis and diurnal fluctuations of fatty-acid composition in leaves. *Biochemical Journal*, *196*, 347–354. https://doi.org/10.1042/bj1960347

Burgal, J., Shockey, J., Lu, C. F., Dyer, J., Larson, T., Graham, I., & Browse, J. (2008). Metabolic engineering of hydroxy fatty acid production in plants: RcDGAT2 drives dramatic increases in ricinoleate levels in seed oil. *Plant Biotechnology Journal*, *6*, 819–831. https://doi.org/10.1111/j.1467-7652.2008.00361.x

Cahoon, E. B., Shockey, J. M., Dietrich, C. R., Gidda, S. K., Mullen, R. T., & Dyer, J. M. (2007). Engineering oilseeds for sustainable production of industrial and nutritional feedstocks: Solving bottlenecks in fatty acid flux. *Current Opinion in Plant Biology*, *10*, 236–244. https://doi.org/10.1016/j.pbi.2007.04.005

Carlsson, A. S., Yilmaz, J. L., Green, A. G., Stymne, S., & Hofvander, P. (2011). Replacing fossil oil with fresh oil - with what and for what? *European Journal of Lipid Science and Technology*, *113*, 812–831. https://doi.org/10.1002/ejlt.201100032

Dauk, M., Lam, P., Kunst, L., & Smith, M. A. (2007). A FAD2 homologue from Lesquerella lindheimeri has predominantly fatty acid hydroxylase activity. *Plant Science*, *173*, 43–49. https://doi.org/10.1016/j.plantsci.2007.03.015

Durrett, T. P., Benning, C., & Ohlrogge, J. (2008). Plant triacylglycerols as feedstocks for the production of biofuels. *The Plant Journal*, *54*, 593–607. https://doi.org/10.1111/j.1365-313X.2008.03442.x

Dyer, J. M., Stymne, S., Green, A. G., & Carlsson, A. S. (2008). High-value oils from plants. *The Plant Journal*, *54*, 640–655. https://doi.org/10.1111/j.1365-313X.2008.03430.x

Goffman, F. D., Alonso, A. P., Schwender, J., Shachar-Hill, Y., & Ohlrogge, J. B. (2005). Light enables a very high efficiency of carbon storage in developing embryos of rapeseed. *Plant Physiology*, *138*, 2269–2279. https://doi.org/10.1104/pp.105.063628

Gunstone, F. D., Harwood, J. L., & Dijkstra, A. J. (2007). *The lipid handbook with CD-ROM*. Boca Raton, FL: CRC Press.

Haslam, R. P., Ruiz-Lopez, N., Eastmond, P., Moloney, M., Sayanova, O., & Napier, J. A. (2013). The modification of plant oil composition via metabolic engineering—better nutrition by design. *Plant Biotechnology Journal*, *11*, 157–168. https://doi.org/10.1111/pbi.12012

Iskandarov, U., Kim, H., & Cahoon, E. (2014). Camelina: An emerging oilseed platform for advanced biofuels and bio-based materials. In M. C. McCann, M. S. Buckeridge, & N. C. Carpita (Eds.), *Plants and bioenergy* (Vol. 4, pp. 131–140). New York, NY: Springer. https://doi.org/10.1007/978-1-4614-9329-7

Knutzon, D. S., Hayes, T. R., Wyrick, A., Xiong, H., Maelor Davies, H., & Voelker, T. A. (1999). Lysophosphatidic acid acyltransferase from coconut endosperm mediates the insertion of laurate at the sn-2 position of triacylglycerols in lauric rapeseed oil and can increase total laurate levels. *Plant Physiology*, *120*, 739–746. https://doi.org/10.1104/pp.120.3.739

Koornneef, M., & Meinke, D. (2010). The development of Arabidopsis as a model plant. *The Plant Journal*, *61*, 909–921. https://doi.org/10.1111/j.1365-313X.2009.04086.x

Kozaki, A., & Sasaki, Y. (1999). Light-dependent changes in redox status of the plastidic acetyl-CoA carboxylase and its regulatory component. *Biochemical Journal*, *339*, 541–546. https://doi.org/10.1042/bj3390541

Kunst, L., Taylor, D. C., & Underhill, E. W. (1992). Fatty-acid elongation in developing seeds of *Arabidopsis-thaliana*. *Plant Physiology and Biochemistry*, *30*, 425–434.

Larson, T. R., Edgell, T., Byrne, J., Dehesh, K., & Graham, I. A. (2002). Acyl CoA profiles of transgenic plants that accumulate medium-chain fatty acids indicate inefficient storage lipid synthesis in developing oilseeds. *The Plant Journal*, *32*, 519–527. https://doi.org/10.1046/j.1365-313X.2002.01440.x

Lee, K.-R., Chen, G., & Kim, H. (2015). Current progress towards the metabolic engineering of plant seed oil for hydroxy fatty acids production. *Plant Cell Reports*, *34*, 603–615. https://doi.org/10.1007/s00299-015-1736-6

Li, Y. H., Beisson, F., Pollard, M., & Ohlrogge, J. (2006). Oil content of Arabidopsis seeds: The influence of seed anatomy, light and plant-to-plant variation. *Phytochemistry*, *67*, 904–915. https://doi.org/10.1016/j.phytochem.2006.02.015

Li, R., Yu, K., Wu, Y., Tateno, M., Hatanaka, T., & Hildebrand, D. F. (2012). Vernonia DGATs can complement the disrupted oil and protein metabolism in epoxygenase-expressing soybean seeds. *Metabolic Engineering*, *14*, 29–38. https://doi.org/10.1016/j.ymben.2011.11.004

Lu, C. F., Fulda, M., Wallis, J. G., & Browse, J. (2006). A high-throughput screen for genes from castor that boost hydroxy fatty acid accumulation in seed oils of transgenic Arabidopsis. *The Plant Journal*, *45*, 847–856. https://doi.org/10.1111/j.1365-313X.2005.02636.x

Lu, C. F., Napier, J. A., Clemente, T. E., & Cahoon, E. B. (2011). New frontiers in oilseed biotechnology: Meeting the global demand for vegetable oils for food, feed, biofuel, and industrial applications. *Current Opinion in Biotechnology*, *22*, 252–259. https://doi.org/10.1016/j.copbio.2010.11.006

Lunn, D., Wallis, J. G., & Browse, J. (2018). Overexpression of Seipin1 increases oil in hydroxy fatty acid-accumulating seeds. *Plant and Cell Physiology*, *59*, 205–214. https://doi.org/10.1093/pcp/pcx177

Maatta, S., Scheu, B., Roth, M. R., Tamura, P., Li, M., Williams, T. D., … Welti, R. (2012). Levels of *Arabidopsis thaliana* leaf phosphatidic acids, phosphatidylserines, and most trienoate-containing polar lipid molecular species increase during the dark period of the diurnal cycle. *Frontiers in Plant Science*, *3*, 49.

Mansour, M., Shrestha, P., Belide, S., Petrie, J., Nichols, P., & Singh, S. (2014). Characterization of oilseed lipids from "DHA-producing *Camelina sativa*": A new transformed land plant containing long-chain omega-3 oils. *Nutrients*, *6*, 776–789. https://doi.org/10.3390/nu6020776

Moire, L., Rezzonico, E., Goepfert, S., & Poirier, Y. (2004). Impact of unusual fatty acid synthesis on futile cycling through beta-oxidation and on gene expression in transgenic plants(1[w]). *Plant Physiology*, *134*, 432–442. https://doi.org/10.1104/pp.103.032938

Moreau, R. A., & Stumpf, P. K. (1981). Recent studies of the enzymic-synthesis of ricinoleic acid by developing castor beans. *Plant Physiology*, *67*, 672–676. https://doi.org/10.1104/pp.67.4.672

Napier, J. A. (2007). The production of unusual fatty acids in transgenic plants. *Annual Review of Plant Biology*, *58*, 295–319. https://doi.org/10.1146/annurev.arplant.58.032806.103811

Napier, J. A., Haslam, R. P., Beaudoin, F., & Cahoon, E. B. (2014). Understanding and manipulating plant lipid composition: Metabolic engineering leads the way. *Current Opinion in Plant Biology*, *19*, 68–75. https://doi.org/10.1016/j.pbi.2014.04.001

Provart, N. J., Alonso, J., Assmann, S. M., Bergmann, D., Brady, S. M., Brkljacic, J., … McCourt, P. (2016). 50 years of Arabidopsis research: Highlights and future directions. *New Phytologist*, *209*, 921–944. https://doi.org/10.1111/nph.13687

Rivero, L., Scholl, R., Holomuzki, N., Crist, D., Grotewold, E., & Brkljacic, J. (2014). Handling Arabidopsis plants: Growth, preservation of seeds, transformation, and genetic crosses. In J. J. Sanchez-Serrano, & J. Salinas (Eds.), *Arabidopsis protocols* (vol 1062, pp. 3–25). New York, NY: Humana Press. https://doi.org/10.1007/978-1-62703-580-4

Ruiz-Lopez, N., Usher, S., Sayanova, O. V., Napier, J. A., & Haslam, R. P. (2015). Modifying the lipid content and composition of plant seeds: Engineering the production of LC-PUFA. *Applied Microbiology and Biotechnology*, *99*, 143–154. https://doi.org/10.1007/s00253-014-6217-2

Sasaki, Y., Kozaki, A., & Hatano, M. (1997). Link between light and fatty acid synthesis: Thioredoxin-linked reductive activation of plastidic acetyl-CoA carboxylase. *Proceedings of the National Academy of Sciences of the United States of America*, *94*, 11096–11101. https://doi.org/10.1073/pnas.94.20.11096

Shrestha, P., Callahan, D. L., Singh, S. P., Petrie, J. R., & Zhou, X.-R. (2016). Reduced triacylglycerol mobilization during seed germination and early seedling growth in Arabidopsis containing nutritionally important polyunsaturated fatty acids. *Frontiers in Plant Science*, *7*, 1402.

Singh, S. P., Zhou, X. R., Liu, Q., Stymne, S., & Green, A. G. (2005). Metabolic engineering of new fatty acids in plants. *Current Opinion in Plant Biology*, *8*, 197–203. https://doi.org/10.1016/j.pbi.2005.01.012

van Erp, H., Bates, P. D., Burgal, J., Shockey, J., & Browse, J. (2011). Castor phospholipid: Diacylglycerol acyltransferase facilitates efficient metabolism of hydroxy fatty acids in transgenic Arabidopsis. *Plant Physiology*, *155*, 683–693. https://doi.org/10.1104/pp.110.167239

van Erp, H., Shockey, J., Zhang, M., Adhikari, N. D., & Browse, J. (2015). Reducing isozyme competition increases target fatty acid accumulation in seed triacylglycerols of transgenic Arabidopsis. *Plant Physiology*, *168*, 36–46. https://doi.org/10.1104/pp.114.254110

Vandeloo, F. J., Broun, P., Turner, S., & Somerville, C. (1995). An oleate 12-hydroxylase from *Ricinus communis* L. is a fatty acyl desaturase homolog. *Proceedings of the National Academy of Sciences of the United States of America*, *92*, 6743–6747. https://doi.org/10.1073/pnas.92.15.6743

Vanhercke, T., Wood, C. C., Stymne, S., Singh, S. P., & Green, A. G. (2013). Metabolic engineering of plant oils and waxes for use as industrial feedstocks. *Plant Biotechnology Journal*, *11*, 197–210. https://doi.org/10.1111/pbi.12023

11

A role for two-component signaling elements in the Arabidopsis growth recovery response to ethylene

Brad M. Binder[1] | Hyo Jung Kim[2,3] | Dennis E. Mathews[4] | Claire E. Hutchison[5] | Joseph J. Kieber[5] | G. Eric Schaller[2]

[1]Department of Biochemistry and Cellular & Molecular Biology, University of Tennessee, Knoxville, Tennessee

[2]Department of Biological Sciences, Dartmouth College, Hanover, New Hampshire

[3]Center for Plant Aging Research, Institute for Basic Science (IBS), Daegu, Korea

[4]Department of Molecular, Cellular, and Biomedical Sciences, University of New Hampshire, Durham, New Hampshire

[5]Department of Biology, University of North Carolina, Chapel Hill, North Carolina

Correspondence
Brad M. Binder, Department of Biochemistry and Cellular & Molecular Biology, University of Tennessee, Knoxville, TN.
Email: bbinder@utk.edu
and
G. Eric Schaller, Department of Biological Sciences, Dartmouth College, Hanover, NH.
Email: george.e.schaller@dartmouth.edu

Present address
Claire E. Hutchison, William Harvey Research Institute, Queen Mary University of London, Charterhouse Square, London, EC1M 6BQ, UK.

Funding information
NSF, Grant/Award Number: #IOS-1456487, #MCB-1716279, #IOS-1022053; the Human Frontier Science Program, Grant/Award Number: #LT000757/2009-L; Institute for Basic Science, Grant/Award Number: # IBS-R013-D1

Abstract

Previous studies indicate that the ability of Arabidopsis seedlings to recover normal growth following an ethylene treatment involves histidine kinase activity of the ethylene receptors. As histidine kinases can function as inputs for a two-component signaling system, we examined loss-of-function mutants involving two-component signaling elements. We find that mutants of phosphotransfer proteins and type-B response regulators exhibit a defect in their ethylene growth recovery response similar to that found with the loss-of-function ethylene receptor mutant *etr1-7*. The ability of two-component signaling elements to regulate the growth recovery response to ethylene functions independently from their well-characterized role in cytokinin signaling, based on the analysis of cytokinin receptor mutants as well as following chemical inhibition of cytokinin biosynthesis. Histidine kinase activity of the receptor ETR1 also facilitates growth recovery in the ethylene hypersensitive response, which is characterized by a transient decrease in growth rate when seedlings are treated continuously with a low dose of ethylene; however, this response was found to operate independently of the type-B response regulators. These results indicate that histidine kinase activity of the ethylene receptor ETR1 performs two independent functions: (a) regulating the growth recovery to ethylene through a two-component signaling system involving phosphotransfer proteins and type-B response regulators and (b) regulating the hypersensitive response to ethylene in a type-B response regulator independent manner.

KEYWORDS
Arabidopsis, ethylene, histidine kinase, phosphotransfer protein, receptor, response regulator

1 | INTRODUCTION

Two-component signaling systems involve histidine kinases, response regulators, and sometimes histidine-containing phosphotransfer proteins (Mizuno, 2005; Schaller, Shiu, & Armitage, 2011). Two-component signaling systems are found in bacteria, archaea, fungi, slime molds, and plants (Mizuno, 2005; Schaller et al., 2011). The two-component signaling system is so named because, in its simplest form, it incorporates a receptor histidine kinase and a response regulator (Gao & Stock, 2009; Stock, Robinson, & Goudreau, 2000). The receptor histidine kinase autophosphorylates on a conserved histidine residue in response to an environmental stimulus, and the phosphate is then transferred to a conserved aspartic acid residue within the receiver domain of a response regulator. Response regulators frequently serve as transcription factors, with phosphorylation modulating their ability to control gene expression. Plants make use of a permutation of the two-component system known as the multistep phosphorelay (Schaller, Kieber, & Shiu, 2008). As found in plants, the multistep phosphorelay typically incorporates three components: (a) a hybrid receptor kinase that contains both histidine kinase and receiver domains in one protein, (b) a histidine-containing phosphotransfer protein, and (c) a separate response regulator. In Arabidopsis, these are referred to as ARABIDOPSIS HISTIDINE KINASEs (AHKs), ARABIDOPSIS HISTIDINE-CONTAINING PHOSPHOTRANSMITTERs (AHPs), and ARABIDOPSIS RESPONSE REGULATORs (ARRs). The two-component signaling system has an established role in mediating cytokinin signal transduction in plants (Gruhn & Heyl, 2013; Hwang, Sheen, & Müller, 2012; Kieber & Schaller, 2014; To & Kieber, 2008; Werner & Schmülling, 2009), but a potential role in mediating ethylene signal transduction is still unclear despite the fact that several ethylene receptors have histidine kinase activity (Gamble, Coonfield, & Schaller, 1998; Moussatche & Klee, 2004).

The ethylene receptor family of plants is related to histidine kinases, containing sensor domains near their N-termini and histidine kinase-like domains in the C-terminal halves. In Arabidopsis, the ethylene receptor family consists of five members that divide into two subfamilies based on phylogenetic analysis and some shared structural features. Subfamily 1 is composed of ETHYLENE RESPONSE1 (ETR1) and ETHYLENE RESPONSE SENSOR1 (ERS1), and subfamily 2 is composed of ETR2, ERS2, and ETHYLENE INSENSITIVE4 (EIN4) (Chang & Stadler, 2001; Chen, Etheridge, & Schaller, 2005; Schaller & Kieber, 2002). The subfamily-1 receptors have canonical histidine kinase domains and exhibit histidine kinase activity based on in vitro analysis (Gamble et al., 1998; Moussatche & Klee, 2004), whereas the subfamily-2 receptors contain diverged histidine kinase-like domains and exhibit serine/threonine kinase activity based on in vitro analysis (Chen et al., 2009; Moussatche & Klee, 2004). Three of the five receptors are hybrid receptors (ETR1, ETR2, and EIN4) that contain receiver domains with all the conserved residues required for functioning in a multistep phosphorelay. In spite of these conserved features, no substantive role has been identified for the multistep phosphorelay in ethylene signaling (Binder, O'Malley, et al., 2004; Hall et al., 2012; Hass et al., 2004; Mason et al., 2005; Qu & Schaller, 2004). Instead, genetic analysis indicates that the primary elements functioning downstream of the ethylene receptors are the Raf-like kinase CONSTITUTIVE TRIPLE RESPONSE1 (CTR1), the transmembrane protein EIN2, and the EIN3 family of transcription factors (Alonso, Hirayama, Roman, Nourizadeh, & Ecker, 1999; Chao et al., 1997; Huang, Li, Hutchison, Laskey, & Kieber, 2003; Ju et al., 2012; Kieber, Rothenberg, Roman, Feldman, & Ecker, 1993; Qiao et al., 2012; Solano, Stepanova, Chao, & Ecker, 1998). However, it has recently been shown that ETR1 and ETR2 also signal independently of CTR1, supporting the possibility for other, non-canonical signaling pathways (Bakshi et al., 2018).

If not involved in the primary signaling pathway, what role does the histidine kinase activity of the ethylene receptors play in signal transduction? Several studies suggest that the enzymatic activity of ETR1 modulates ethylene signal output (Binder, O'Malley, et al., 2004; Hall et al., 2012; Qu & Schaller, 2004; Street et al., 2015). This might seem to imply that a multistep phosphorelay operates downstream of the receptors; however, the histidine kinase and receiver domains of the receptors physically interact with CTR1 and EIN2 and autophosphorylation may also modulate interactions with the established ethylene signaling pathway (Binder, Chang, & Schaller, 2012; Bisson & Groth, 2010; Clark, Larsen, Wang, & Chang, 1998; Gao et al., 2003). It is thus necessary to specifically evaluate the role of two-component signaling elements in ethylene-mediated responses to resolve their level of contribution.

One situation in which histidine kinase activity of the ethylene receptors appears to play a modulating role is in the ability of seedlings to recover normal growth following cessation of ethylene treatment (Binder, O'Malley, et al., 2004). Treatment of wild-type etiolated seedlings with ethylene inhibits their growth rate; upon removal of ethylene, the seedlings return to their basal growth rate within two hours. Loss-of-function receptor mutants for the family members with receiver domains (ETR1, ETR2, and EIN4) all result in a slower recovery to normal growth rate following removal of ethylene, the slow growth recovery phenotype being particularly strong in the *etr1-7* mutant. In addition, the double mutant *etr1-7;ers1-2* is delayed in its ability to recover normal growth rate, and this can be rescued by introducing a wild-type version of ETR1 but not by a kinase-inactive version of ETR1 (Binder, O'Malley, et al., 2004).

These results suggest that histidine kinase activity and, by extension, the multistep phosphorelay may mediate the ability of seedlings to recover normal growth following cessation of ethylene treatment. We tested this hypothesis by performing a kinetic analysis of the ethylene growth response and recovery in mutants of Arabidopsis two-component signaling elements. Results from this study indicate that histidine kinase activity of the ethylene receptor ETR1 performs two independent functions. First, it regulates the growth recovery to ethylene through a two-component signaling system involving AHPs and type-B ARRs. Second, it also plays a role in regulating growth recovery during a "hypersensitive" response to ethylene, which is characterized by a transient decrease in growth rate when seedlings are treated with a continuous very low dose of ethylene, but does so in a type-B ARR-independent manner. The potential mechanisms underlying these differences in histidine kinase-mediated regulation are discussed.

2 | MATERIALS AND METHODS

2.1 | Plant materials

Loss-of-function mutants analyzed were the *etr1-7* ethylene receptor mutant (Hua & Meyerowitz, 1998), *AHK* double mutants constructed from published T-DNA insertion lines (Argueso et al., 2012), *AHP* mutants *ahp5-2*, *ahp2 ahp3*, and *ahp2 ahp3 ahp5-2* (Hutchison et al., 2006), type-B *ARR* mutants *arr2-4*, *arr10-2 arr12-1*, and *arr2-2 arr10-2 arr12-1* (Mason et al., 2005), and are all of the Arabidopsis ecotype Col-0. Construction of the transgenic *etr1-7 ers1-2* lines containing wild-type *ETR1* (*gETR1*) or kinase-inactivated ETR1 (*getr1-HGG*) was as described (Wang, Hall, O'Malley, & Bleecker, 2003).

2.2 | Kinetic analysis of hypocotyl growth rate

Measurements were performed using two-day-old etiolated Arabidopsis seedlings as described previously (Binder, O'Malley, et al., 2004). To examine the growth response and recovery kinetics to ethylene, 10 μl/L ethylene was introduced 1 hr after measurements were initiated and then removed 2 hr later. To examine the hypersensitive growth response to ethylene, a concentration of 8.7 nl/L ethylene was used. All data represent the mean of at least three seedlings for the recovery analysis and of at least four seedlings for the hypersensitive response analysis. The Kolmogorov–Smirnov test, performed with JMP 10.0, was used to assess the significance of differences in growth curves. For kinetic recovery analysis, the 3-5 hr time points were compared for significance; for the hypersensitive response, the 1.5-3.5 hr time points were compared for significance.

2.3 | Hypocotyl growth response to cytokinin

Dose–response for the effect of the cytokinin *trans*-zeatin on hypocotyl elongation of 4-d-old dark-grown seedlings was performed as described (Argyros et al., 2008). Seedlings were grown on vertical plates on MS media with 1% (w/v) sucrose, along with the indicated concentrations of cytokinin, scanned, and lengths measured using ImageJ software. All data represent the mean of at least nine seedlings.

2.4 | Accession numbers

Sequence data from this article can be found in the EMBL/GenBank data libraries under accession number(s): AHK2 (AT5G35750), AHK3 (AT1G27320), AHK4 (AT2G01830), AHP2 (AT3G29350), AHP3 (AT5G39340), AHP5 (AT1G03430), ARR2 (AT4G16110), ARR10 (AT4G31920), ARR12 (AT2G25180), ERS1 (AT2G40940), ETR1 (AT1G66340).

3 | RESULTS

3.1 | Two-component signaling elements regulate the growth recovery to ethylene

We examined short-term changes in seedling growth response and recovery to ethylene using two-day-old etiolated seedlings. Under these growth conditions, wild-type seedlings exhibit a growth rate of approximately 0.4 mm/hr in the absence of ethylene (Figure 1). Treatment with 10 μl/L exogenous ethylene results in a rapid decrease in the growth rate, initiated within 15 min of the treatment and reaching a new steady-state growth rate approximately 75 min after ethylene addition. As has been observed previously (Binder, Mortimore, Stepanova, Ecker, & Bleecker, 2004; Binder, O'Malley, et al., 2004), the growth response has two kinetic phases: a rapid inhibition response lasting for approximately 15 min, followed by a slower inhibition response that lasts for approximately 60 min. Removal of ethylene results in a recovery of seedling growth rate, with the initial growth rate being attained approximately 90 min after ethylene removal.

Loss-of-function mutations in the ethylene receptors *ETR1*, *ETR2*, or *EIN4* affect these growth recovery kinetics to ethylene (Binder, O'Malley, et al., 2004). As shown in Figure 1, the loss-of-function *etr1-7* mutant exhibits a normal growth response to ethylene, but is significantly delayed in its recovery kinetics compared to wild type ($p < 0.0001$), taking approximately 30 min longer than wild type to attain its initial growth rate. We examined loss-of-function mutants involving the histidine kinase-linked cytokinin receptors AHK2, AHK3, and AHK4 to determine whether the slow growth recovery phenotype was unique to ethylene receptor mutants (Figure 1). Because there is functional redundancy in the cytokinin receptor family, we examined the double mutants *ahk2 ahk3*, *ahk2 ahk4*, and *ahk3 ahk4*, all of which have demonstrated effects on shoot and root growth (Argueso et al., 2012; Higuchi et al., 2004; Nishimura et al., 2004; Riefler, Novak, Strnad, & Schmülling, 2006). All three *ahk* double mutant combinations exhibited ethylene growth response and recovery kinetics similar to those observed in wild type (Figure 1), indicating that ethylene receptors play a role in mediating these growth responses that does not require activity of the cytokinin receptors.

Arabidopsis contains five genes encoding phosphotransfer proteins (*AHP1*, *AHP2*, *AHP3*, *AHP4*, and *AHP5*) that are predicted to contain the conserved histidine for phosphorylation (Schaller et al., 2008). Single loss-of-function mutants do not display significant effects on cytokinin signaling, but higher order mutants involving *ahp1*, *ahp2*, *ahp3*, and *ahp5* result in decreased cytokinin sensitivity (Hutchison et al., 2006). We analyzed mutants involving *ahp2*, *ahp3*, and *ahp5* for their ethylene growth response and recovery kinetics (Figure 1). All the mutants exhibited an ethylene response similar to wild type, but the *ahp5*, *ahp2 ahp3*, and *ahp2 ahp3 ahp5* mutants exhibited a significantly slower growth recovery phenotype ($p < 0.0001$).

Arabidopsis contains 11 type-B response regulators (type-B ARRs), which serve as transcription factors to mediate the final step in the multistep phosphorelay (Schaller et al., 2008). *ARR1*, *ARR2*, *ARR10*, and *ARR12* are broadly expressed and based on genetic analysis contribute the most to cytokinin signal transduction (Argyros et al., 2008; Ishida, Yamashino, Yokoyama, & Mizuno, 2008; Mason, Li, Mathews, Kieber, & Schaller, 2004; Mason et al., 2005). *ARR2* has been implicated in ethylene signaling (Hass et al., 2004), and we therefore focused our analysis on *arr2* mutants (Figure 1). As with the other mutants examined, all the type-B *arr* mutants exhibited an ethylene response similar to wild type, but the *arr2*, *arr10 arr12*, and *arr2 arr10 arr12* mutants exhibited a significantly slower growth recovery phenotype ($p < 0.0001$).

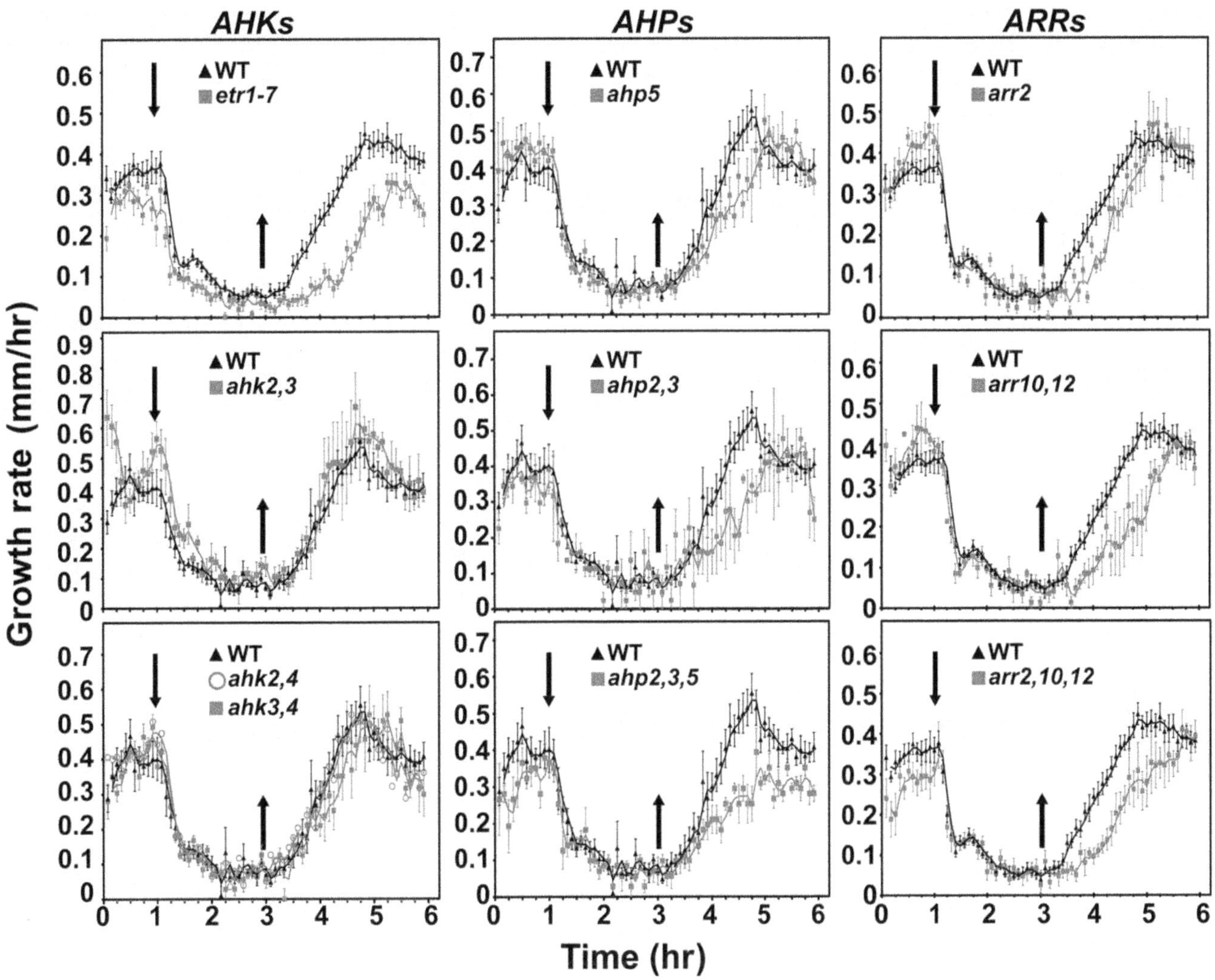

FIGURE 1 Growth kinetics of two-day-old etiolated Arabidopsis hypocotyls containing mutations in two-component signaling elements. Short-term growth kinetic analysis in response to 10 μl/L ethylene was performed on mutants involving receptors (column 1), AHPs (column 2), and type-B ARRs (column 3) and compared to the wild type. Ethylene was introduced one hour after measurements were initiated (down arrow) and then removed two hours later (up arrow). For receptor mutants, we examined the *etr1-7* mutant as well as *ahk* double mutants. For *ahp* mutants, we examined *ahp5-2*, *ahp2 ahp3*, and *ahp2 ahp3 ahp5-2*. For type-B *arr* mutants, we examined *arr2-4*, *arr10-2 arr12-1*, and *arr2-2 arr10-2 arr12-1*. Error bars indicate *SE* ($n \geq 3$)

3.2 | Two-component signaling elements regulate ethylene growth recovery independently from their role in cytokinin signaling

We performed several controls to confirm that the two-component signaling elements regulate the growth recovery to ethylene independently from their role in cytokinin signaling. First, as described earlier, double mutants of the cytokinin receptors do not exhibit the same ethylene recovery phenotype as mutants involving the downstream AHPs and ARRs, which is not what would be predicted if cytokinin played a role in the ethylene response (Figure 1). We note that we could not test the *ahk* triple mutant because it is infertile and needs to be propagated in a segregating population, and the homozygous triple mutant cannot be differentiated phenotypically from other genotypes in the etiolated seedlings used for the kinetic analysis. To confirm the lack of correlation between the effects of the mutants on the cytokinin and ethylene responses, we specifically examined how mutations in the cytokinin receptors (*AHKs*), *AHPs*, and type-B *ARRs* affect the hypocotyl growth response to cytokinin (Figure 2), doing so because our kinetic results are based on a hypocotyl growth response to ethylene. We observed that the *ahk* double mutants *ahk2 ahk4* and *ahk3 ahk4* were less responsive to cytokinin than any of the *ahp* and type-B *arr* mutant combinations we tested, even though the *ahp* and *arr* mutants exhibit a stronger effect on ethylene growth recovery kinetics.

As an alternative approach to assess a potential role for cytokinin in the ethylene growth response, we performed kinetic analysis in the presence of the cytokinin biosynthesis inhibitor lovastatin (Crowell & Salaz, 1992). Lovastatin is an inhibitor of the cytosolic pathway for isoprenoid biosynthesis, one of the two pathways that generate the isopentenyl groups used in the biosynthesis of cytokinins. Treatment of four-day-old dark-grown seedlings with 1 μM lovastatin inhibited hypocotyl growth, reducing hypocotyl length from 9.87 mm (±0.15 *SE*) for the untreated control to 7.10 mm (±0.07 *SE*) for the lovastatin-treated seedlings (i.e., a 28% reduction in growth). When examined by kinetic analysis, the control seedlings exhibiting a growth rate of 0.33 mm/hr (± 0.01 *SE*) compared to 0.24 mm/hr (± 0.01 *SE*) for the lovastatin-treated seedlings (i.e., a 23% reduction in growth rate). This lovastatin-induced reduction in hypocotyl growth is consistent with that previously observed (Nagata, Suzuki, Yoshida, & Muranaka, 2002), is indicative of the efficacy of the inhibitor, and also still allowed sufficient growth to perform kinetic analysis (Figure 3). We found that the ethylene growth response and recovery kinetics are not affected by lovastatin, consistent with these growth responses being independent of cytokinin (Figure 3). Thus, the effects we observe for the mutants and lovastatin on the ethylene response do not correlate with their effects on cytokinin signaling responses.

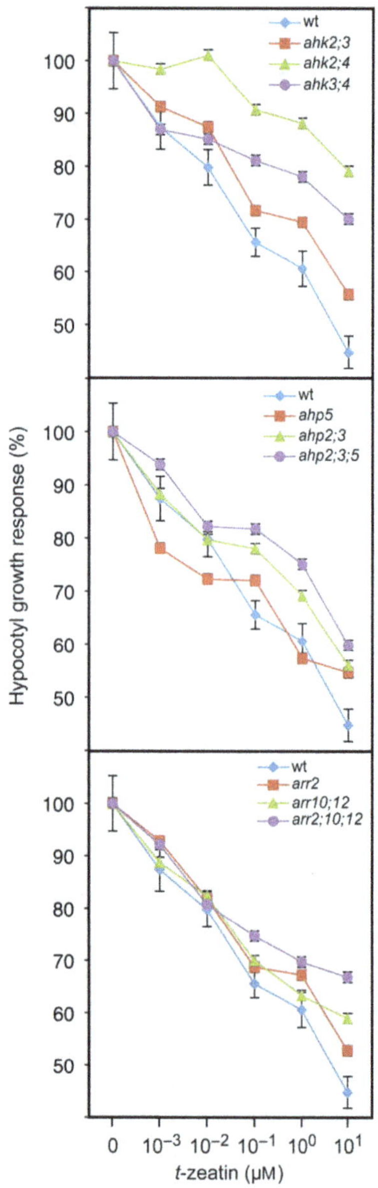

FIGURE 2 Hypocotyl growth response of two-component mutants to cytokinin. Dose–response for the effect of the cytokinin *t*-zeatin on hypocotyl elongation of four-day-old etiolated seedlings, performed as described (Argyros et al., 2008). The hypocotyl length of each line in the presence of cytokinin is expressed as a percentage of the control. Error bars indicate *SE* ($n \geq 9$); error bars not shown if smaller than symbol

3.3 | Histidine kinase activity of ETR1 regulates the hypersensitive response to ethylene in a type-B ARR-independent manner

The first phase for ethylene growth inhibition, characterized by a rapid decrease in seedling growth, can be differentiated from the second slower response phase based on its hypersensitivity to ethylene (Binder, Mortimore, et al., 2004). As shown in Figure 4, wild-type seedlings treated with 8.7 nl/L ethylene (an approximately 1000-fold lower dose than that used in Figure 1) exhibit a rapid but transient decrease in their growth rate, recovering to their pretreatment growth rate in approximately 2 hr, even though the seedlings are still in the presence of ethylene. We examined the receptor mutant *etr1-7* to determine whether ETR1 plays a role in the hypersensitive ethylene response. The *etr1-7* mutant responded similarly to wild type, but its reduced growth rate was significantly prolonged compared to wild type ($p < 0.0001$), requiring approximately 1 hr longer to return to the initial growth rate (Figure 4). To determine whether type-B ARR transcription factors might play a role in the hypersensitive response, we examined the ethylene response of *arr2 arr10 arr12* but observed similar growth kinetics to wild type (Figure 4). To determine whether the histidine kinase activity of ETR1 plays a role in the hypersensitive response, we made use of the subfamily-1 loss-of-function mutant *etr1-7 ers1-2* transformed with

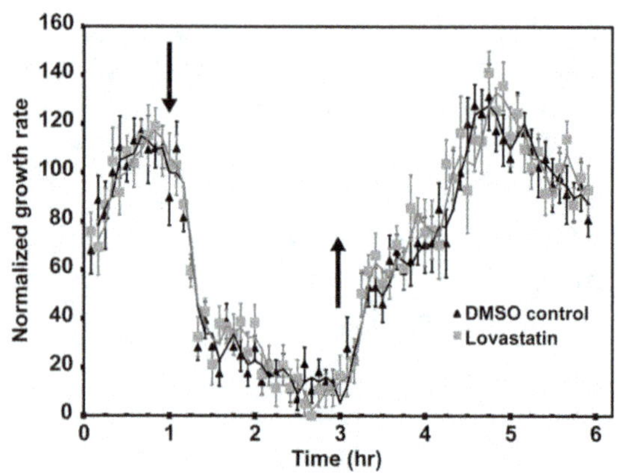

FIGURE 3 Lovastatin does not affect growth kinetics of two-day-old etiolated Arabidopsis hypocotyls. Short-term growth kinetic analysis in response to 10 μl/L ethylene was performed on wild-type seedlings grown in the absence (DMSO control) or the presence of 1 μM lovastatin. Data were normalized to growth rate in air prior to treatment with ethylene to facilitate the comparison. Unnormalized growth rates in air were 0.33 ± 0.01 mm/hr for the control and 0.24 ± 0.01 mm/hr for the lovastatin-treated seedlings. Error bars indicate *SE* ($n \geq 3$)

wild-type *ETR1* (*gETR1*) or a kinase-inactive *ETR1* (*getr1-HGG*) (Binder, O'Malley, et al., 2004). The *gETR1* line exhibited a hypersensitive response similar to wild type, but the *getr1-HGG* line exhibited a significantly prolonged hypersensitive response ($p < 0.0001$), suggesting that histidine kinase activity plays a role in the response.

4 | DISCUSSION

These results indicate that histidine kinase activity of the ethylene receptor ETR1 performs two independent functions: (a) regulating the growth recovery to ethylene through a two-component signaling system involving AHPs and type-B ARRs and (b) regulating the hypersensitive response to ethylene in a type-B ARR-independent manner. Several points can be made about the role of the two-component system in ethylene signaling. First, loss-of-function mutants in ethylene receptors *ETR1*, *ETR2*, and *EIN4* (Binder, O'Malley, et al., 2004), the *AHPs*, and the type-B *ARRs*, although having no effect on the initial ethylene response, all affect ethylene growth recovery kinetics in a similar manner, consistent with their functioning in the same regulatory pathway (Figure 5). The 45-min delay in recovery time observed in some of the mutants is substantive when one considers that growth recovery usually occurs in less than 2 hr, and indicates that the two-component signaling pathway plays a significant role in mediating the plant's responsiveness to changes in its phytohormone environment. Second, there is functional overlap within members of each two-component family for mediating the ethylene recovery response, similar to what has been found in the cytokinin signaling pathway (Schaller et al., 2008; To & Kieber, 2008; Werner & Schmülling, 2009), a finding consistent with physical interactions being detected for multiple AHPs with ETR1 and the type-B ARRs (Dortay et al., 2008; Scharein, Voet-van-Vormizeele, Harter, & Groth, 2008; Urao, Miyata, Yamaguchi-Shinozaki, & Shinozaki, 2000). Third, given the role of type-B ARRs as transcription factors, the effect of the two-component signaling pathway is likely to be on transcriptional output from ethylene signaling, potentially regulating the expression of genes involved in the growth response (Hass et al., 2004). Fourth, the AHPs and type-B ARRs that function in ethylene signaling also play roles in cytokinin signaling (Kieber & Schaller, 2014; To & Kieber, 2008; Werner & Schmülling, 2009), raising the possibility of cross talk between these phytohormone signaling pathways as well as the question as to whether and how specificity is obtained when utilizing the same signaling elements.

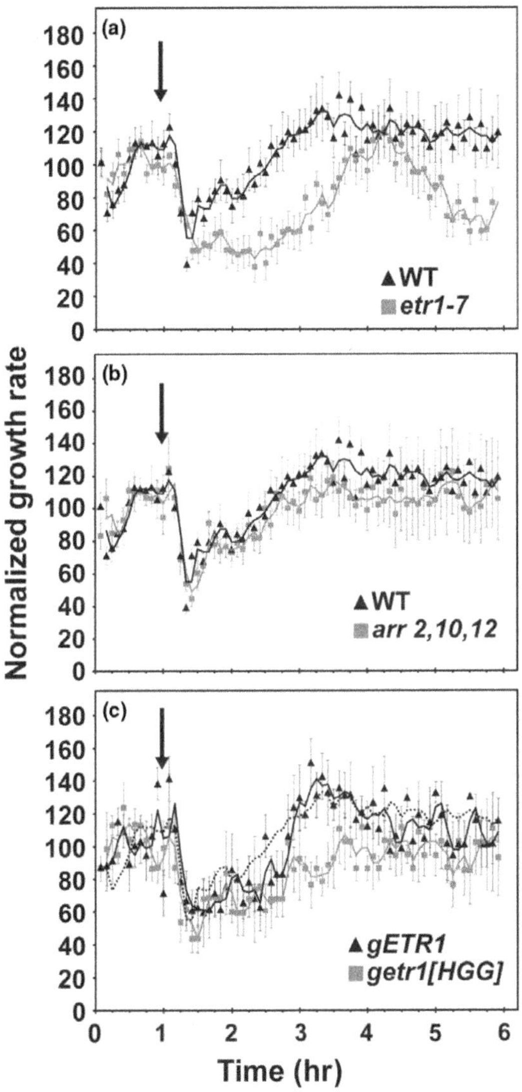

FIGURE 4 Kinetic analysis of the hypersensitive response to ethylene. Growth kinetic analysis was performed with 8.7 nl/L ethylene introduced 1 hr after measurements were initiated (down arrow). The hypersensitive response of *etr1-7* (a), *arr2-2 arr10-2 arr12-1* (b), and transgenic *etr1-7 ers1-2* lines containing wild-type *ETR1* (*gETR1*), or kinase-inactive ETR1 (*getr1-HGG*) (c) are all compared to wild type. The dotted line in panel C shows data for wild type for comparison. Data were normalized to growth rate in air prior to treatment with ethylene to facilitate comparisons. Error bars indicate *SE* ($n \geq 4$)

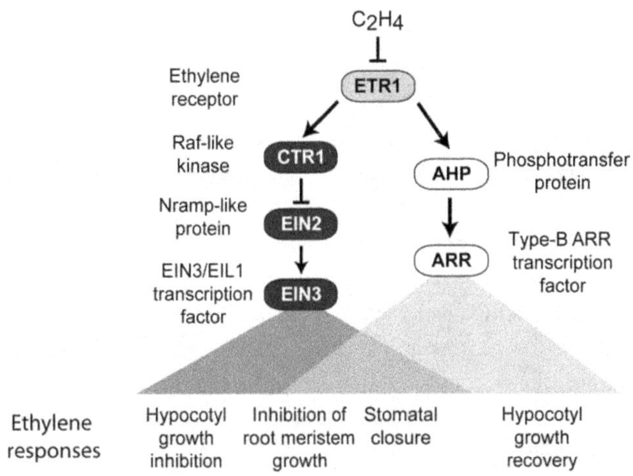

FIGURE 5 Model for ethylene signal transduction. Signal output from the ethylene receptors diverges to regulate the CTR1/EIN2/EIN3 and AHP/ARR pathways. ETR1 is shown here as a representative ethylene receptor that has histidine kinase activity. Ethylene responses may be dependent on an individual pathway or have varying contributions from both pathways

Our results complement prior studies that have implicated two-component signaling elements in the regulation of ethylene responses, including effects on hypocotyl growth, stomatal closure, and root meristem development (Figure 5). The type-B response regulator ARR2 was previously implicated in ethylene responses based on an *arr2* mutant exhibiting a hyposensitive ethylene response for the inhibition of hypocotyl elongation, and the ARR2 protein being phosphorylated in an ETR1-dependent manner and binding to a promoter element for the ethylene-induced gene ERF1 (Hass et al., 2004). In contrast to this study by Hass et al. (2004), we do not observe an effect of *arr2* mutants in the hypocotyl growth response assay, either as a long-term effect (Mason et al., 2005) or when assessing the short-term kinetic response as described here in our current study; however, our results do support a role for ARR2 in the hypocotyl based on the ethylene growth recovery kinetics of *arr2*. Differences observed between the two studies on *arr2* mutant phenotypes may relate to growth conditions used for studying the effects of ethylene on hypocotyl growth. In addition to potentially mediating ethylene effects on hypocotyl growth, ARR2 as well as three AHPs (AHP1, AHP2, and AHP3) have been implicated in mediating the stimulatory effects of ethylene on stomatal closure (Desikan et al., 2006; Mira-Rodado et al., 2012). Furthermore, the type-B response regulator ARR1, but not ARR2, ARR10, or ARR12, contributes to ethylene-mediated inhibition of cell proliferation at the primary root meristem; kinase-inactive versions of ETR1 similarly reduce this ethylene response, consistent with it being mediated through a phosphorelay initiated at the ethylene receptors (Street et al., 2015).

Our study supports the existence of two independent ethylene signaling pathways initiated at the receptors: the well-documented CTR1/EIN2/EIN3-dependent pathway as well as an AHP/ARR-dependent pathway (Figure 5). The degree to which these pathways contribute to various ethylene responses varies depends on the response assayed. For example, both pathways contribute to a similar extent in ethylene-mediated regulation of stomatal closure, but the CTR1/EIN2/EIN3 pathway is predominant in the regulation of cell proliferation at the root meristem (Desikan et al., 2006; Mira-Rodado et al., 2012; Street et al., 2015). Although genetic analysis reveals a role for the AHP/ARR pathway in mediating the growth recovery response to ethylene, whether the CTR1/EIN2/EIN3 pathway also plays a role cannot be readily assessed because this pathway mediates the growth inhibition response (i.e., a growth recovery response cannot be assessed unless there is an initial growth inhibition response) (Binder, Mortimore, et al., 2004). As noted above, both pathways contribute to varying extents in the control of different ethylene responses. A potential mechanism that would facilitate such control is combinatorial regulation between the EIN3/EIL and type-B ARR transcription factors, EIN3 and ARR having each individually been found to coregulate gene expression in conjunction with other transcription factors (Feng et al., 2017; Liu et al., 2017; Marin-de la Rosa et al., 2015).

Through analysis of the hypersensitive response to ethylene, our results also indicate that the histidine kinase activity of ETR1 may play a transcriptionally independent role in ethylene signaling. Previous genetic analysis demonstrated that the hypersensitive response of Arabidopsis to ethylene is independent of the transcription factors EIN3 and EIL1 (Binder, Mortimore, et al., 2004), our results here demonstrating that this response is also independent of the type-B ARR transcription factors ARR2, ARR10, and ARR12. Nevertheless, histidine kinase activity of ETR1 appears to play a role in the hypersensitive response, because loss of ETR1 or transformation of Arabidopsis with a kinase-inactivated ETR1 results in a prolonged response. Because the ethylene receptors are components within larger signaling complexes (Chen et al., 2010), phosphorylation could potentially modulate the interactions and activity of proteins physically associated with ETR1 such as CTR1, EIN2, and/or other ethylene receptors (Clark et al., 1998; Gao et al., 2003, 2008; Grefen et al., 2008). However, the output in such a case would not be through an established transcriptionally based pathway but potentially through a phosphorylation/dephosphorylation cascade, which often mediates rapid eukaryotic signaling responses.

ACKNOWLEDGMENTS

We thank Tobias Zutz for technical assistance in the kinetic analysis, and Brad Taylor for assistance in the statistical analysis. This work was supported by the NSF grants #IOS-1456487 to GES, #MCB-1716279 to BMB, and #IOS-1022053 to JJK, DEM, and GES, and the Human Frontier Science Program grant #LT000757/2009-L and Institute for Basic Science # IBS-R013-D1 to HJK.

AUTHOR CONTRIBUTIONS

BMB, JJK, and GES conceived and supervised the study. BMB and HJK designed and performed experiments. DEM and CEH generated mutant lines used in the study. BMB, HJK, and GES analyzed data. GES wrote the manuscript with contributions from all other authors.

REFERENCES

Alonso, J. M., Hirayama, T., Roman, G., Nourizadeh, S., & Ecker, J. R. (1999). EIN2, a bifunctional transducer of ethylene and stress responses in Arabidopsis. *Science*, *284*, 2148–2152. https://doi.org/10.1126/science.284.5423.2148

Argueso, C. T., Ferreira, F. J., Epple, P., To, J. P., Hutchison, C. E., Schaller, G. E., ... Kieber, J. J. (2012). Two-component elements mediate interactions between cytokinin and salicylic acid in plant immunity. *PLoS Genetics*, *8*, e1002448. https://doi.org/10.1371/journal.pgen.1002448

Argyros, R. D., Mathews, D. E., Chiang, Y.-H., Palmer, C. M., Thibault, D. M., Etheridge, N., ... Schaller, G. E. (2008). Type B response regulators of Arabidopsis play key roles in cytokinin signaling and plant development. *Plant Cell*, *20*, 2102–2116. https://doi.org/10.1105/tpc.108.059584

Bakshi, A., Piya, S., Fernandez, J. C., Chervin, C., Hewezi, T., & Binder, B. M. (2018). Ethylene receptors signal via a noncanonical pathway to regulate abscisic acid responses. *Plant Physiology*, *176*, 910–929. https://doi.org/10.1104/pp.17.01321

Binder, B. M., Chang, C., & Schaller, G. E. (2012). Perception of ethylene by plants: Ethylene receptors. In M. T. McManus (Ed.), *Annual plant reviews: The plant hormone ethylene*, Vol 44 (pp. 117–145). Hoboken, NJ: Wiley-Blackwell. https://doi.org/10.1002/9781118223086.ch5

Binder, B. M., Mortimore, L. A., Stepanova, A. N., Ecker, J. R., & Bleecker, A. B. (2004). Short-term growth responses to ethylene in Arabidopsis seedlings are EIN3/EIL1 independent. *Plant Physiology*, *136*, 2921–2927. https://doi.org/10.1104/pp.104.050393

Binder, B. M., O'Malley, R. C., Wang, W., Moore, J. M., Parks, B. M., Spalding, E. P., & Bleecker, A. B. (2004). Arabidopsis seedling growth response and recovery to ethylene. A kinetic analysis. *Plant Physiology*, *136*, 2913–2920. https://doi.org/10.1104/pp.104.050369

Bisson, M. M., & Groth, G. (2010). New insight in ethylene signaling: Autokinase activity of ETR1 modulates the interaction of receptors and EIN2. *Molecular Plant*, *3*, 882–889. https://doi.org/10.1093/mp/ssq036

Chang, C., & Stadler, R. (2001). Ethylene hormone receptor action in Arabidopsis. *BioEssays*, *23*, 619–627. https://doi.org/10.1002/(ISSN)1521-1878

Chao, Q., Rothenberg, M., Solano, R., Roman, G., Terzaghi, W., & Ecker, J. R. (1997). Activation of the ethylene gas response pathway in Arabidopsis by the nuclear protein ETHYLENE-INSENSITIVE3 and related proteins. *Cell*, *89*, 1133–1144. https://doi.org/10.1016/S0092-8674(00)80300-1

Chen, Y. F., Etheridge, N., & Schaller, G. E. (2005). Ethylene signal transduction. *Annals of Botany*, *95*, 901–915. https://doi.org/10.1093/aob/mci100

Chen, Y. F., Gao, Z., Kerris, R. J. 3rd, Wang, W., Binder, B. M., & Schaller, G. E. (2010). Ethylene receptors function as components of high-molecular-mass protein complexes in Arabidopsis. *PLoS ONE*, *5*, e8640. https://doi.org/10.1371/journal.pone.0008640

Chen, T., Liu, J., Lei, G., Liu, Y. F., Li, Z. G., Tao, J. J., … Zhang, J. S. (2009). Effects of tobacco ethylene receptor mutations on receptor kinase activity, plant growth and stress responses. *Plant and Cell Physiology*, *50*, 1636–1650. https://doi.org/10.1093/pcp/pcp107

Clark, K. L., Larsen, P. B., Wang, X., & Chang, C. (1998). Association of the Arabidopsis CTR1 Raf-like kinase with the ETR1 and ERS1 ethylene receptors. *Proceedings of the National Academy of Sciences of the United States of America*, *95*, 5401–5406. https://doi.org/10.1073/pnas.95.9.5401

Crowell, D. N., & Salaz, M. S. (1992). Inhibition of growth of cultured tobacco cells at low concentrations of lovastatin is reversed by cytokinin. *Plant Physiology*, *100*, 2090–2095. https://doi.org/10.1104/pp.100.4.2090

Desikan, R., Last, K., Harrett-Williams, R., Tagliavia, C., Harter, K., Hooley, R., … Neill, S. J. (2006). Ethylene-induced stomatal closure in Arabidopsis occurs via AtrbohF-mediated hydrogen peroxide synthesis. *The Plant Journal*, *47*, 907–916. https://doi.org/10.1111/j.1365-313X.2006.02842.x

Dortay, H., Gruhn, N., Pfeifer, A., Schwerdtner, M., Schmülling, T., & Heyl, A. (2008). Toward an interaction map of the two-component signaling pathway of *Arabidopsis thaliana*. *Journal of Proteome Research*, *7*, 3649–3660. https://doi.org/10.1021/pr0703831

Feng, Y., Xu, P., Li, B., Li, P., Wen, X., An, F., … Guo, H. (2017). Ethylene promotes root hair growth through coordinated EIN3/EIL1 and RHD6/RSL1 activity in Arabidopsis. *Proceedings of the National Academy of Sciences*, *114*, 13834–13839. https://doi.org/10.1073/pnas.1711723115

Gamble, R. L., Coonfield, M. L., & Schaller, G. E. (1998). Histidine kinase activity of the ETR1 ethylene receptor from Arabidopsis. *Proceedings of the National Academy of Sciences of the United States of America*, *95*, 7825–7829. https://doi.org/10.1073/pnas.95.13.7825

Gao, Z., Chen, Y. F., Randlett, M. D., Zhao, X. C., Findell, J. L., Kieber, J. J., & Schaller, G. E. (2003). Localization of the Raf-like kinase CTR1 to the endoplasmic reticulum of Arabidopsis through participation in ethylene receptor signaling complexes. *Journal of Biological Chemistry*, *278*, 34725–34732. https://doi.org/10.1074/jbc.M305548200

Gao, R., & Stock, A. M. (2009). Biological insights from structures of two-component proteins. *Annual Review of Microbiology*, *63*, 133–154. https://doi.org/10.1146/annurev.micro.091208.073214

Gao, Z., Wen, C.-K., Binder, B. M., Chen, Y.-F., Chang, J., Chiang, Y.-H., … Schaller, G. E. (2008). Heteromeric interactions among ethylene receptors mediate signaling in Arabidopsis. *Journal of Biological Chemistry*, *283*, 23081–23810.

Grefen, C., Städele, K., Růžička, K., Obrdlik, P., Harter, K., & Horák, J. (2008). Subcellular localization and in vivo interaction of the *Arabidopsis thaliana* ethylene receptor family members. *Molecular Plant*, *1*, 308–320. https://doi.org/10.1093/mp/ssm015

Gruhn, N., & Heyl, A. (2013). Updates on the model and the evolution of cytokinin signaling. *Current Opinion in Plant Biology*, *16*, 569–574. https://doi.org/10.1016/j.pbi.2013.09.001

Hall, B. P., Shakeel, S. N., Amir, M., Haq, N. U., Qu, X., & Schaller, G. E. (2012). Histidine kinase activity of the ethylene receptor ETR1 facilitates the ethylene response in Arabidopsis. *Plant Physiology*, *159*, 682–695. https://doi.org/10.1104/pp.112.196790

Hass, C., Lohrmann, J., Albrecht, V., Sweere, U., Hummel, F., Yoo, S. D., … Harter, K. (2004). The response regulator 2 mediates ethylene signalling and hormone signal integration in Arabidopsis. *EMBO Journal*, *23*, 3290–3302. https://doi.org/10.1038/sj.emboj.7600337

Higuchi, M., Pischke, M. S., Mahonen, A. P., Miyawaki, K., Hashimoto, Y., Seki, M., … Kakimoto, T. (2004). In planta functions of the Arabidopsis cytokinin receptor family. *Proceedings of the National Academy of Sciences*, *101*, 8821–8826. https://doi.org/10.1073/pnas.0402887101

Hua, J., & Meyerowitz, E. M. (1998). Ethylene responses are negatively regulated by a receptor gene family in *Arabidopsis thaliana*. *Cell*, *94*, 261–271. https://doi.org/10.1016/S0092-8674(00)81425-7

Huang, Y., Li, H., Hutchison, C. E., Laskey, J., & Kieber, J. J. (2003). Biochemical and functional analysis of CTR1, a protein kinase that negatively regulates ethylene signaling in Arabidopsis. *The Plant Journal*, *33*, 221–233. https://doi.org/10.1046/j.1365-313X.2003.01620.x

Hutchison, C. E., Li, J., Argueso, C., Gonzalez, M., Lee, E., Lewis, M. W., … Kieber, J. J. (2006). The Arabidopsis histidine phosphotransfer proteins are redundant positive regulators of cytokinin signaling. *Plant Cell*, *18*, 3073–3087. https://doi.org/10.1105/tpc.106.045674

Hwang, I., Sheen, J., & Müller, B. (2012). Cytokinin signaling networks. *Annual Review of Plant Biology*, *63*, 353–380. https://doi.org/10.1146/annurev-arplant-042811-105503

Ishida, K., Yamashino, T., Yokoyama, A., & Mizuno, T. (2008). Three type-B response regulators, ARR1, ARR10, and ARR12, play essential but redundant roles in cytokinin signal transduction throughout the life cycle of *Arabidopsis thaliana*. *Plant and Cell Physiology*, *49*, 47–57. https://doi.org/10.1093/pcp/pcm165

Ju, C., Yoon, G. M., Shemansky, J. M., Lin, D. Y., Ying, Z. I., Chang, J., … Chang, C. (2012). CTR1 phosphorylates the central regulator EIN2 to control ethylene hormone signaling from the ER membrane to the nucleus in Arabidopsis. *Proceedings of the National Academy of Sciences*, *109*, 19486–19491. https://doi.org/10.1073/pnas.1214848109

Kieber, J. J., Rothenberg, M., Roman, G., Feldman, K. A., & Ecker, J. R. (1993). *CTR1*, a negative regulator of the ethylene response pathway in Arabidopsis, encodes a member of the Raf family of protein kinases. *Cell*, *72*, 427–441. https://doi.org/10.1016/0092-8674(93)90119-B

Kieber, J. J., & Schaller, G. E. (2014). Cytokinins. *The Arabidopsis Book*, *12*, e0168. https://doi.org/10.1199/tab.0168

Liu, X., Liu, R., Li, Y., Shen, X., Zhong, S., & Shi, H. (2017). EIN3 and PIF3 Form an Interdependent Module That Represses Chloroplast Development in Buried Seedlings. *Plant Cell*, *29*, 3051–3067. https://doi.org/10.1105/tpc.17.00508

Marin-de la Rosa, N., Pfeiffer, A., Hill, K., Locascio, A., Bhalerao, R. P., Miskolczi, P., … Alabadi, D. (2015). Genome wide binding site analysis reveals transcriptional coactivation of cytokinin-responsive genes by DELLA proteins. *PLoS Genetics*, *11*, e1005337. https://doi.org/10.1371/journal.pgen.1005337

Mason, M. G., Li, J., Mathews, D. E., Kieber, J. J., & Schaller, G. E. (2004). Type-B response regulators display overlapping expression patterns in Arabidopsis. *Plant Physiology, 135*, 927–937. https://doi.org/10.1104/pp.103.038109

Mason, M. G., Mathews, D. E., Argyros, D. A., Maxwell, B. B., Kieber, J. J., Alonso, J. M., ... Schaller, G. E. (2005). Multiple type-B response regulators mediate cytokinin signal transduction in Arabidopsis. *Plant Cell, 17*, 3007–3018. https://doi.org/10.1105/tpc.105.035451

Mira-Rodado, V., Veerabagu, M., Witthoft, J., Teply, J., Harter, K., & Desikan, R. (2012). Identification of two-component system elements downstream of AHK5 in the stomatal closure response of Arabidopsis thaliana. *Plant Signaling & Behavior, 7*, 1467–1476. https://doi.org/10.4161/psb.21898

Mizuno, T. (2005). Two-component phosphorelay signal transduction systems in plants: From hormone responses to circadian rhythms. *Bioscience, Biotechnology, and Biochemistry, 69*, 2263–2276. https://doi.org/10.1271/bbb.69.2263

Moussatche, P., & Klee, H. J. (2004). Autophosphorylation activity of the Arabidopsis ethylene receptor multigene family. *Journal of Biological Chemistry, 279*, 48734–48741. https://doi.org/10.1074/jbc.M403100200

Nagata, N., Suzuki, M., Yoshida, S., & Muranaka, T. (2002). Mevalonic acid partially restores chloroplast and etioplast development in Arabidopsis lacking the non-mevalonate pathway. *Planta, 216*, 345–350. https://doi.org/10.1007/s00425-002-0871-9

Nishimura, C., Ohashi, Y., Sato, S., Kato, T., Tabata, S., & Ueguchi, C. (2004). Histidine kinase homologs that act as cytokinin receptors possess overlapping functions in the regulation of shoot and root growth in Arabidopsis. *Plant Cell, 16*, 1365–1377. https://doi.org/10.1105/tpc.021477

Qiao, H., Shen, Z., Huang, S. S., Schmitz, R. J., Urich, M. A., Briggs, S. P., & Ecker, J. R. (2012). Processing and subcellular trafficking of ER-tethered EIN2 control response to ethylene gas. *Science, 338*, 390–393. https://doi.org/10.1126/science.1225974

Qu, X., & Schaller, G. E. (2004). Requirement of the histidine kinase domain for signal transduction by the ethylene receptor ETR1. *Plant Physiology, 136*, 2961–2970. https://doi.org/10.1104/pp.104.047126

Riefler, M., Novak, O., Strnad, M., & Schmülling, T. (2006). Arabidopsis cytokinin receptor mutants reveal functions in shoot growth, leaf senescence, seed size, germination, root development, and cytokinin metabolism. *Plant Cell, 18*, 40–54. https://doi.org/10.1105/tpc.105.037796

Schaller, G. E., & Kieber, J. J. (2002). Ethylene. In C. Somerville, & E. Meyerowitz (Eds.), *The Arabidopsis Book*, Vol. *1* (p. e0071). Rockville, MD: American Society of Plant Biologists.

Schaller, G. E., Kieber, J. J., & Shiu, S.-H. (2008). Two-component signaling elements and histidyl-aspartyl phosphorelays. *The Arabidopsis Book*(C. Somerville, E. Meyerowitz, editors), *6*, e0112. https://doi.org/10.1199/tab.0112

Schaller, G. E., Shiu, S. H., & Armitage, J. P. (2011). Two-component systems and their co-option for eukaryotic signal transduction. *Current Biology, 21*, R320–R330. https://doi.org/10.1016/j.cub.2011.02.045

Scharein, B., Voet-van-Vormizeele, J., Harter, K., & Groth, G. (2008). Ethylene signaling: Identification of a putative ETR1-AHP1 phosphorelay complex by fluorescence spectroscopy. *Analytical Biochemistry, 377*, 72–76. https://doi.org/10.1016/j.ab.2008.03.015

Solano, R., Stepanova, A., Chao, Q., & Ecker, J. R. (1998). Nuclear events in ethylene signaling: A transcriptional cascade mediated by ETHYLENE-INSENSITIVE3 and ETHYLENE-RESPONSE-FACTOR1. *Genes & Development, 12*, 3703–3714. https://doi.org/10.1101/gad.12.23.3703

Stock, A. M., Robinson, V. L., & Goudreau, P. N. (2000). Two-component signal transduction. *Annual Review of Biochemistry, 69*, 183–215. https://doi.org/10.1146/annurev.biochem.69.1.183

Street, I. H., Aman, S., Zubo, Y., Ramzan, A., Wang, X., Shakeel, S. N., ... Schaller, G. E. (2015). Ethylene inhibits cell proliferation of the Arabidopsis root meristem. *Plant Physiology, 169*, 338–350. https://doi.org/10.1104/pp.15.00415

To, J. P. C., & Kieber, J. J. (2008). Cytokinin signaling: Two-components and more. *Trends in Plant Science, 13*, 85–92. https://doi.org/10.1016/j.tplants.2007.11.005

Urao, T., Miyata, S., Yamaguchi-Shinozaki, K., & Shinozaki, K. (2000). Possible His to Asp phosphorelay signaling in an Arabidopsis two-component system. *FEBS Letters, 478*, 227–232. https://doi.org/10.1016/S0014-5793(00)01860-3

Wang, W., Hall, A. E., O'Malley, R., & Bleecker, A. B. (2003). Canonical histidine kinase activity of the transmitter domain of the ETR1 ethylene receptor from Arabidopsis is not required for signal transmission. *Proceedings of the National Academy of Sciences, 100*, 352–357. https://doi.org/10.1073/pnas.0237085100

Werner, T., & Schmülling, T. (2009). Cytokinin action in plant development. *Current Opinion in Plant Biology, 12*, 527–538. https://doi.org/10.1016/j.pbi.2009.07.002

12

Lysyl oxidase-like protein secreted from an acidophilic red alga, *Cyanidium caldarium*

Tatsuya Tomo[1,2] | Akinori Okumura[3] | Takehiro Suzuki[4] | Mirai Okuhara[1] | Ruriko Katayama[1] | Noboru Isayama[1] | Ryo Nagao[5] | Masako Iwai[6] | Naoshi Dohmae[4] | Isao Enami[1]

[1]Department of Biology, Tokyo University of Science, Tokyo, Japan

[2]Graduate School of Science, Tokyo University of Science, Tokyo, Japan

[3]Department of Integrated Sciences in Physics and Biology, College of Humanities and Sciences, Nihon University, Tokyo, Japan

[4]Biomolecular Characterization Unit, RIKEN Center for Sustainable Resource Science, Saitama, Japan

[5]Research Institute for Interdisciplinary Science, Okayama University, Okayama, Japan

[6]School of Life Science and Technology, Tokyo Institute of Technology, Yokohama, Japan

Correspondence
Tatsuya Tomo, Department of Biology, Tokyo University of Science, Tokyo, Japan.
Email: tomo@rs.tus.ac.jp

Funding information
Grant-in-Aids for Scientific Research, Grant/Award Number: No. 26220801, 17K07453, 18H05177; JST PRESTO

Abstract

Cyanidium caldarium is a primitive acidophilic red alga which grown optimally at pH 1–3. When the alga was cultured at pH 6, which is the upper limit of acidity for its survival, most of the algal cells became large cells with four endospores which did not split into daughter cells. This suggests that the alga survives in the endospore state at pH 6 to protect against nutrient uptake deficiency due to low pH gradient across the cell membranes. The alga was also found to secrete an extracellular protein specifically at pH 6. The protein was identified to be lysyl oxidase-like protein, which had been reported to be widely distributed in the animal kingdom but not yet found in the plant kingdom. In the plant kingdom, only two primitive acidophilic algae, *C. caldarium* and *Cyanidioschyzon merolae*, possess a gene encoding this protein.

KEYWORDS

cell division, *Cyanidium*, endospore, eukaryote, lysyl oxidase

1 | INTRODUCTION

The transition from prokaryotic cells into eukaryotic cells is important in biological evolution. The order Cyanidiales that consist of three genera *Cyanidioschyzon*, *Cyanidium* and *Galdieria* are the only photosynthetic organisms which have existed for hundreds of millions of years in acidic hot springs (Reeb & Bhattacharya, 2010). Since the common characteristics of these species are shared with the prokaryotic cyanobacteria and with the eukaryotic Rhodophyta (red algae), this makes them good candidates for covering the evolutionary gap between the prokaryotic and eukaryotic cell (Seckbach, 1987). These species show progressive morphological and biochemical steps from primitive cyanobacterial features of *Cyanidioschyzon* to *Cyanidium* and to further advanced *Galdieria* (Seckbach, 1995). The first alga, *Cyanidioschyzon merolae*, resembles the cyanobacteria in its mode of division by binary fission and simple eukaryotic cellular features, and it has the smallest of nuclear and chloroplast DNA in eukaryotic algae (Matsuzaki et al., 2004). The second member, *Cyanidium caldarium*, indicates an intermediate status in cyanidiaceaen line. The third alga, *Galdieria sulphuraria* which shows advanced features, has more developed organelles, a multilobed chloroplasts, and storage glucans closer to the lower *Rhodophyta* (Seckbach, 1995).

Cyanidium caldarium strain RK-1 used in this study is a primitive unicellular red alga which can grow at pH values as low as 0 (Allen, 1959), and the pH optimum for growth is between 1.0 and 3.0 (Allen, 1959; Doemel & Brock, 1971). The alga is an obligate autotroph and has a simple cellular structure, with each cell having one nucleus, one mitochondrion, and a large single chloroplast but no vacuoles (Enami, Nagashima, & Fukuda, 1975). The algal cells multiply by internal divisions forming four endospores within a mother cell (Ford, 1984). It has a thick cell wall which is largely (50%–55%) proteinaceous, unlike most of the red algae (Bailey & Staehelin, 1968). Analysis of plastid gene clusters indicated that *Cyanidium* and *Cyanidioschyzon* were closely related and that they formed a sister group distinct from other rhodophytes (Ohta et al., 1997).

Despite the extremely acidic environment, the intracellular pH of *Cyanidium* cells has been indicated to be neutral (Enami & Kura-Hotta, 1984; Kura-Hotta & Enami, 1981, 1984). The intracellular pH of *Cyanidium* cells was directly measured as a function of external pH by ^{31}P-NMR (Enami, Akutsu, & Kyogoku, 1986). Cells incubated under aerobic conditions in the dark or anaerobic conditions in the light were found to maintain their intracellular pH within a narrow range (pH 6.8–7.0) even when the external pH was changed from pH 1.2 to 8.4 (Enami et al., 1986). In contrast, the intracellular pH of cells incubated under anaerobic conditions in the dark was acidified as the external pH decreased (Enami et al., 1986). These indicate that when ATP production by either respiration or photosynthesis is suppressed, passive transport of protons into the cells from acidic external medium results in an intracellular acidification. If ATP production is functional, cells are able to pump out protons and maintain their intracellular pH at a constant, physiological range. On the basis of the results described above, it was concluded that the active H^+ efflux is functioning to maintain the intracellular pH constant against passive H^+ influx according to the steep pH gradient across the cell membrane. The gene encoding the plasma membrane H^+-ATPase which is functioning in this active H^+ efflux was cloned and sequenced (Ohta et al., 1997)l.

The enormous proton gradient across the cell membrane can be used for uptake of nutrients by effective H^+-symport (Gross, 2000). As the pH gradient decreases, the uptake of nutrients should be suppressed. Since the intracellular pH in *Cyanidium* cells is maintained at a constant within pH 6.8–7.0 regardless of external pH (Enami et al., 1986), the pH gradient across the cell membrane can be changed by the external pH. Thus, this study started from examination of the upper limit of acidity for *Cyanidium* survival and showed that it was pH 6. Furthermore, when the alga was cultured at the upper limit of acidity, the algal cells were found to survive under the stage of endospores and to secrete lysyl oxidase-like protein (*Cyanidium* LOXL) as an extracellular enzyme.

Lysyl oxidases (LOX) and a number of closely related lysyl oxidase-like proteins (LOXL) are widely distributed in the animal kingdom but have not been yet found in the plant kingdom (Csiszar, 2001; Kagan & Li, 2003; Lucero & Kagan, 2006). These proteins play a critical role in the formation and repair of the extracellular matrix by oxidizing lysine residues in elastin and collagen, thereby initiating the formation of covalent crosslinkages which stabilize these fibrous proteins (Kagan & Li, 2003). These proteins have been also revealed to exhibit a number of biological functions such as developmental regulation, tumor suppression, cell mobility, and cellular senescence, suggesting that these play exceptionally important roles beyond the induction of crosslinkage formation in collagen and elastin (Csiszar, 2001; Kagan & Li, 2003). Recently, Grau-Bové, Ruiz-Trillo, and Rodriguez-Pascual (2015) have surveyed a wide selection of genomes in order to infer the evolutionary history of LOX(L) and found LOX(L) genomes not only in animals, but also in bacteria and archaea.

2 | RESULTS

2.1 | Upper limit of acidity for *Cyanidium* survival

Cyanidium caldarium was cultured at a wide pH range (pH 1–7) to determine the upper limit of acidity for its survival. Figure 1a shows photographs of algal cells cultured at various pHs for 10 days. The cells grew optimally at pH 2 and grew well at a wide pH range of pH 1–5. In contrast, when the alga was cultured at pH 6, its original blue-green color was kept but its color did not deepen, suggesting no multiplication of algal cells. Furthermore, the algal cells cultured at pH 7 showed complete chlorosis and died. This indicates that the upper limit of acidity is pH 6. To confirm this result, we measured induction of chlorophyll fluorescence from photosystem II (PS II) in intact cells cultured for 1–9 days at pH 2.5, 6, and 7. The Fv ($Fm - F_0$)/Fm ratio, which is an estimate of the photochemical efficiency of PS II (Clarke, Hurry, Gustafsson, & Oquist, 1993; Herbert, Samson, Fork, & Laudenbach, 1992; Kvíderová, 2012), showed a constant value of about 0.6 in the cells cultured at both pH 2.5 and pH 6 even after 9 days (Figure 1b). This result indicates that the algal cells cultured at pH 6 retain the same photosynthetic activity as those cultured at pH 2.5. On the other hand, the Fv/Fm ratio of the cells at pH 7 declined to 0.185 after 2 days (Figure 1b), indicating that the alga lost the photosynthetic activity. Because the pH gradient across the cell membrane is nearly zero at external pH 7, the uptake of nutrients due to the pH gradient is completely suppressed, which may result in the inactivation of photosynthesis.

2.2 | Suppression of cell division of *Cyanidium* cells at pH 6

When *Cyanidium* cells were cultured at the upper limit of acidity (pH 6), microscopic observation showed that the number of large cells with four endospores increased significantly. As shown in Figure 2, most of the cells were small spheres of approximately 2–3 μm in diameter when cultured at pH 2.5 (a), whereas most cells became larger with four endospores of approximately 6–8 μm in diameter after 15 days of the culture at pH 6 (b).

The number of cells and ratio of cells with four endospores were compared between algal cells cultured at pH 2.5 and those cultured

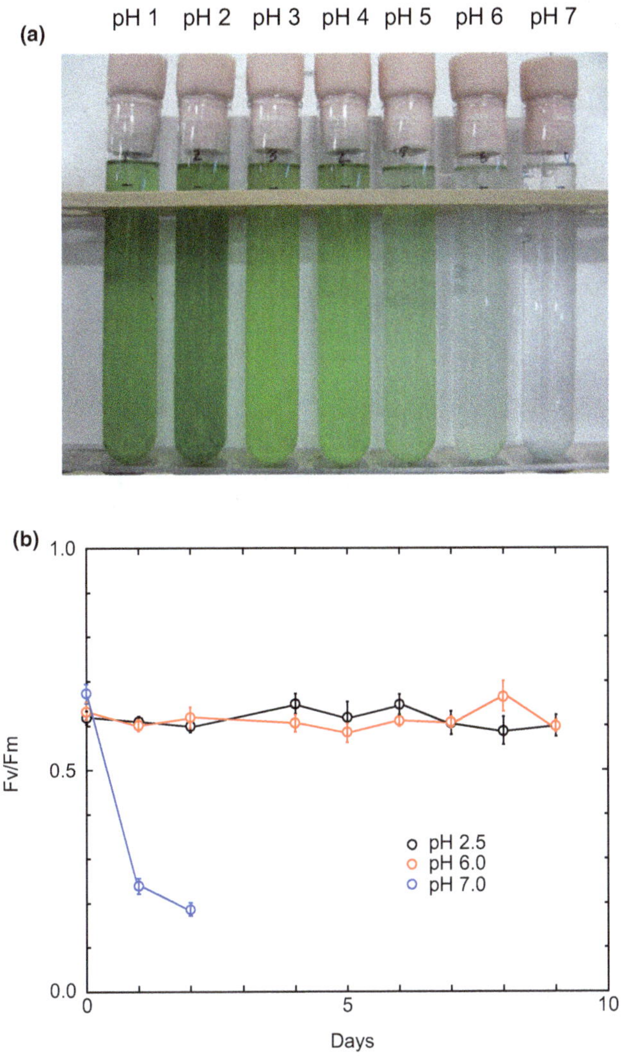

FIGURE 1 Photographs of *Cyanidium caldarium* which was cultured at various pHs (1–7) for 10 days (a), and Fv/Fm of intact cells of the alga cultured at pH 2.5 (black circles), at pH 6 (red circles), and at pH 7 (blue circles) for 0–9 days (b). Each value was obtained by repeating the experiment three times independently. Vertical bars show standard deviation

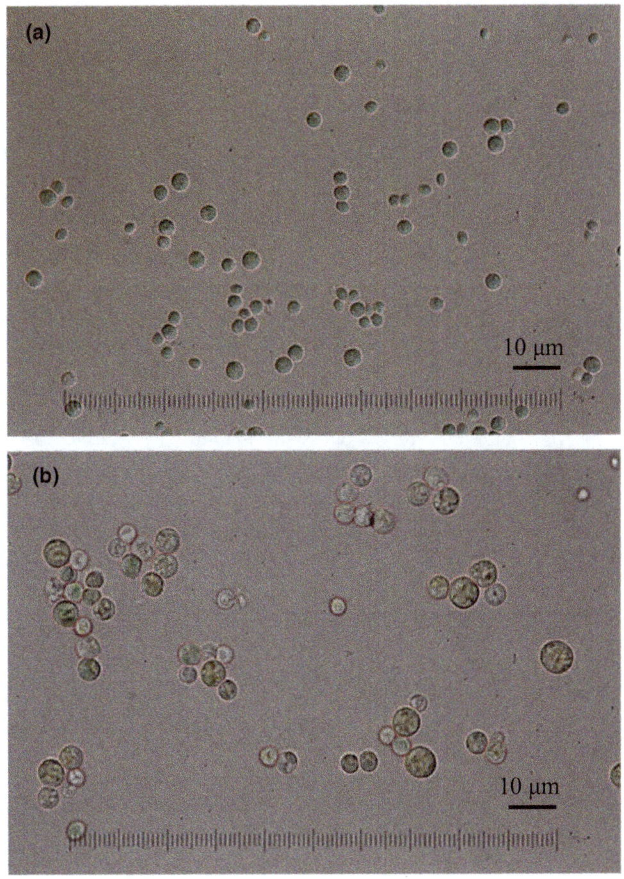

FIGURE 2 Micrographs of *Cyanidium caldarium* cultured at pH 2.5 (a) and at pH 6 (b). *Cyanidium* cells grown at pH 2.5 were subcultured in the medium at pH 6 and then cultured for 15 days

at pH 6. Figure 3a shows the growth curves assayed by counting the cell number. When the alga was cultured at pH 2.5, the cell number logarithmically increased after lag periods of 4 days, while no increase in the number of cells was observed in the alga cultured at pH 6 even after 12 days. In Figure 3b, the ratio of cells with four endospores was compared between algal cells cultured at pH 2.5 and those cultured at pH 6. The ratio of cells with four endospores in algal cells cultured at pH 2.5 showed a constant value of 25%–30% through 1–12 days, whereas those in algal cells cultured at pH 6 significantly increased and reached approximately 80% after more than 7 days. These results indicate that *Cyanidium* cells cultured at pH 6 are able to form four endospores but not to split into daughter cells. This suggests that the alga survives in the endospores state at pH 6 and suppresses the increase in the number of cells to protect against nutrient uptake deficiently due to low pH gradient across the cell membrane.

To confirm this suggestion, the algal cells cultured at pH 6 for 11 days were again cultured at pH 2.5. As shown in Figure 4a, the number of cells sharply increased after a long lag time of 10 days from changing to pH 2.5. The ratio of cells with four endospores gradually decreased and reached to about 30% after 25 days from changing to pH 2.5 (Figure 4b). This indicates that four endospores formed at pH 6 split into daughter cells at pH 2.5 in which sufficient nutrients are present due to the steep pH gradient across the cell membrane.

2.3 | Secretion of an extracellular enzyme, lysyl oxidase-like protein, from *Cyanidium* cells cultured at pH 6

When *Cyanidium* cells were cultured at the upper limit of acidity (pH 6), we found that the culture medium foamed up remarkably, suggesting that some extracellular proteins are secreted from algal cells. To confirm this, peptide components of the centrifuged supernatants after precipitation of algal cells were analyzed by SDS-PAGE. As shown in Figure 5a, one clear protein band appeared at an apparent molecular weight of 34 kDa after cultivation for more than 3 days.

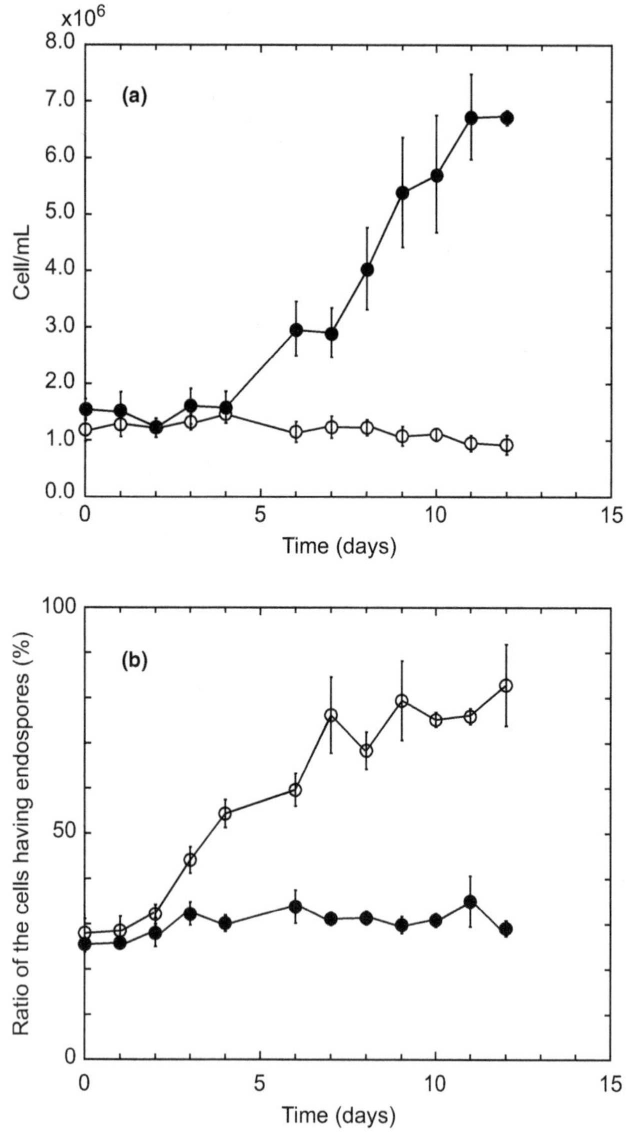

FIGURE 3 Growth curves (a) and ratio (%) of the cells with four endospores (b) of *Cyanidium caldarium* which was cultured at pH 2.5 (closed circles) and pH 6 (open circles). Each value was obtained by counting at least 500 cells and by repeating third times independently. Vertical bars show standard deviation

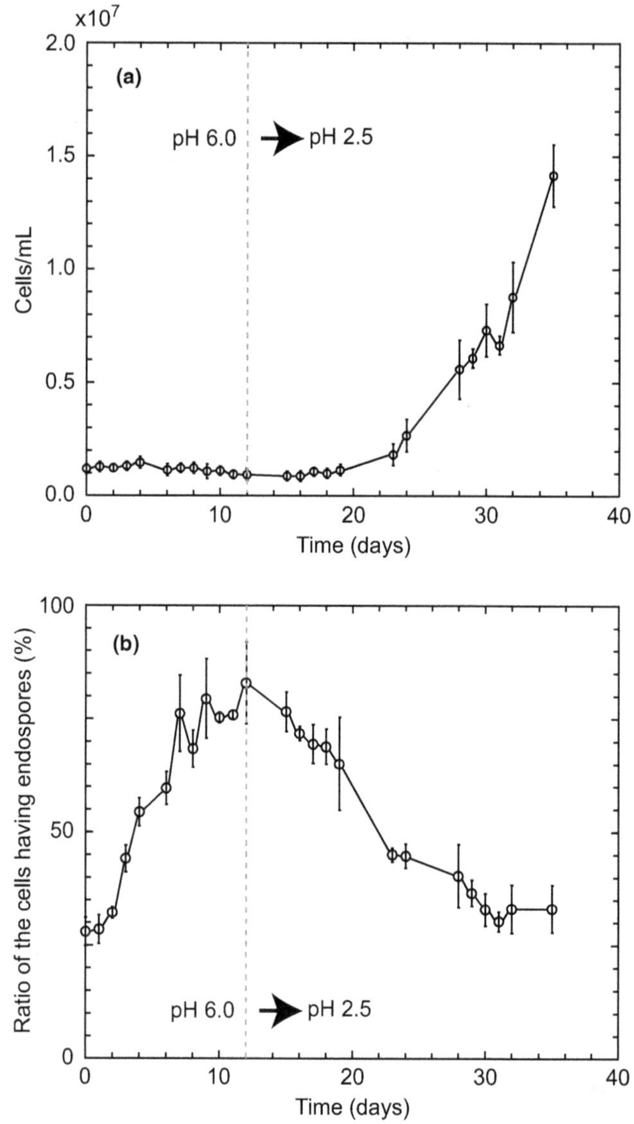

FIGURE 4 Growth curves (a) and ratio (%) of cells with four endospores (b) of *Cyanidium caldarium* which was cultured at pH 2.5 after the cultivation at pH 6 for 11 days. *Cyanidium* cells were cultured at pH 6 for 11 days and then centrifuged to remove medium at pH 6. Centrifuged precipitates were suspended in medium at pH 2.5 and then cultured. Each value was obtained by counting at least 500 cells and by repeating the experiment three times independently. Vertical bars show standard deviation

The 34 kDa extracellular protein was secreted from algal cells cultured at pH 6 but was not secreted from those cultured at pH 1–5 and pH 7 (Figure 5b). In this study, the proteins secreted in the medium were examined but the proteins inside the cells or attached to cells were not examined. Thus, there are still two possibilities. (a) The 34 kDa extracellular protein is specifically expressed at pH 6. (b) The protein is expressed at all pHs but secreted in the medium specifically only at pH 6.

The gene encoding the 34 kDa extracellular protein was successfully cloned according to the methods described in Methods section and its result is shown in Figure 6. This gene encodes a polypeptide of 219 amino acid residues with a total molecular mass of 24,217 Da. Its N-terminal sequence was blocked by a possible modification, which was confirmed as pyroglutamylation (pyroglutamate formation at the N-terminal located glutamine) using LC-MS/MS analysis (Figure 7a). The putative N-terminal signal peptide (1–25 corresponding to the underlined amino acids in Figure 6) was predicted using SOSUI-signal (Gomi, Akazawa, & Mitaku, 2000; Gomi, Sonoyama, & Mitaku, 2004). These first 25 residues are considered to serve as a transit peptide for transferring the protein across the cell membrane. After removal of the signal peptide by cleavage at the Ala25–Gln26 position, the mature protein comprised 194 amino acid residues with a total molecular mass of 21,488 Da. Its sequence highly corresponded with that of LOXL protein from *C. merolae* CMN144C (Figure 8). Thus, the 34 kDa extracellular protein secreted

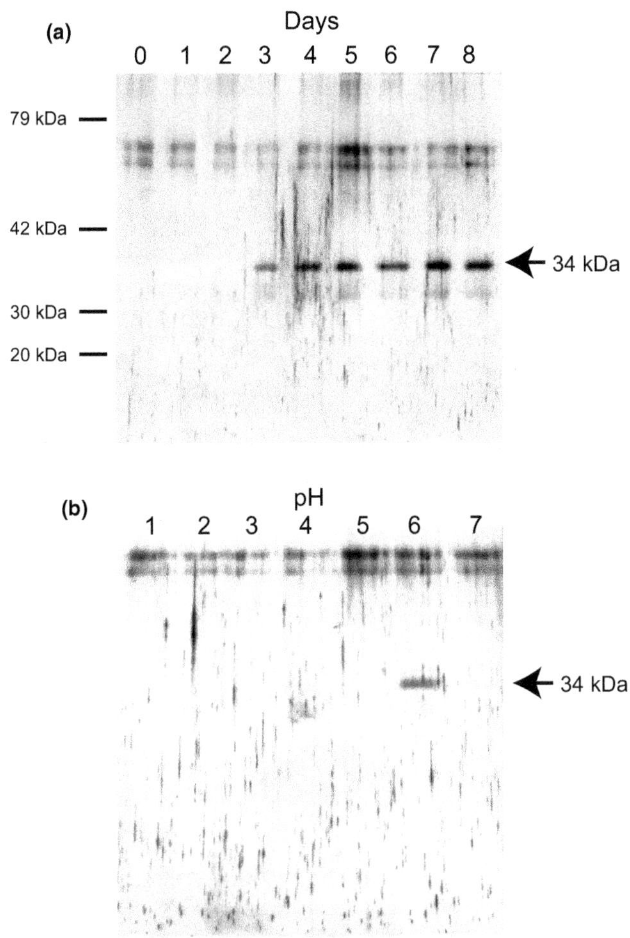

FIGURE 5 SDS-PAGE profiles of extracellular protein secreted from *Cyanidium caldarium.* (a) Extracellular protein secreted from *C. caldarium* cultured at pH 6. The algal cells were cultured at pH 6 for 0–8 days. (b) Extracellular protein secreted from *C. caldarium* cultured at various pHs. The algal cells were cultured at various pHs for 7 days. Each sample was centrifuged at 5,000 × *g* for 5 min, and then 20 μl of the supernatants was applied to SDS-PAGE

```
  1 TTCACATTCGTTGGTGAGAGCAGTCTGCATACACCACTGACTCTT  45
 46 AGAACAGTTTTCCTCATGTACAAACGTATTGCACTCGTGTTTCTG  90
  1                     M  Y  K  R  I  A  L  V  F  L   10
 91 TTTTGTTTGGTTTGCTTGGCGACCGCCGCATCTTCGGTGCGAGCA 135
 11 F  C  L  V  C  L  A  T  A  A  S  S  V  R  A    25
136 CAGGCCAAGTTTGATTTGGAGCCATCGCTGGAAATCCTCAAGGGC 180
 26 Q  A  K  F  D  L  E  P  S  L  E  I  L  K  G    40
181 AGTCTTCGGATCAAGACGGAGTCGTTCACGCCTCGCGATTGCGCG 225
 41 S  L  R  I  K  T  E  S  F  T  P  R  D  C  A    55
226 TTTGAGATGGGATGTGTGCTCGGAACTGGTCCGAGGAAGCTTTTG 270
 56 F  E  M  G  C  V  L  G  T  G  P  R  K  L  L    70
271 CAGCTCGCTGGCGCTATTGGAAACAACGGTCCCGATGACTTTGTT 315
 71 Q  L  A  G  A  I  G  N  N  G  P  D  D  F  V    85
316 GTTGGGAACCCGGCTGATCACCCCGATATCTACACCTGGGATACC 360
 86 V  G  N  P  A  D  H  P  D  I  Y  T  W  D  T   100
361 TGCACTGGGCAGTTCATCACGCGCAAGCCTATGTTGAACTACTCG 405
 91 C  T  G  Q  F  I  T  R  K  P  M  L  N  Y  S   115
406 ATCTGGGACTCTGCAGGCAACCTGGTGCGCAGTGGTAAGAAGGAG 450
106 I  W  D  S  A  G  N  L  V  R  S  G  K  K  E   130
451 AGTTTCTGCATGGAGGATTTGGCTCGTTTCCCTGAGTATACTGGG 495
121 S  F  C  M  E  D  L  A  R  F  P  E  Y  T  G   145
496 AACCCGGCGCTGGTGCCGGCTGAGCCTGTCCACACGTGCACGTTC 540
136 N  P  A  L  V  P  A  E  P  V  H  T  C  T  F   160
541 CAGGGTATCAGTGTCGGTTGGTGTGACATTTACGACGACTCTCTG 585
151 Q  G  I  S  V  G  W  C  D  I  Y  D  D  S  L   175
586 CCGTGCCAGTGGATCGATATCACCGGTGTCCCGTCAGGCCATTAC 630
166 P  C  Q  W  I  D  I  T  G  V  P  S  G  H  Y   190
631 ACGTTGCGTCTCGAGTGCAATGCGTATCATATCTTCAAGGAGACG 675
181 T  L  R  L  E  C  N  A  Y  H  I  F  K  E  T   205
676 ACGTATGCGAATAACGTTGTTGAGACGCCAGTGACGATCACCTAA 720
206 T  Y  A  N  N  V  V  E  T  P  V  T  I  T  *   219
721 GCGTGGACGCTGCGGCGCGTTGCGATGCACAGAATCATCTGTTTG 765
766 GAAAGTTTCGTGTGGTTTGGCGAGGGAAAAGTAGTCGTTACGCAG 810
815 CATTGATTGCGTAGAGCAGTTTAAAGTACGTCAACTATCCATTG  854
```

FIGURE 6 Nucleotide sequence of the gene encoding the 34 kDa extracellular protein secreted from *Cyanidium caldarium* cultured at pH 6. The deduced amino acid sequence is shown below the nucleotide sequence in the single-letter code. The putative N-terminal signal peptide (1–25; the underlined amino acid) was predicted using SOSUIsignal (Gomi et al., 2000, 2004)

from *Cyanidium* cells cultured at pH 6 was identified to be a lysyl oxidase-like protein (LOXL).

MS/MS analysis of *Achromobacter* protease I digests of the *Cyanidium* LOXL showed that high-mannose oligosaccharides were linked with the amide nitrogen of 113Asn in the peptide fragment of PMLNYSIWDSAGNLVRSGK in which Asn-X-Ser consensus sequence for N-linked glycosylation was present (Figure 7b). Discrepancy between the apparent molecular mass by SDS-PAGE (34 kDa) and the molecular mass calculated based on its amino acid sequence (21,488 Da) may, in part, be due to the N-linked high-mannose chains at 113Asn and may also be due to the presence of the cross-linked cofactor (lysyl tyrosyl quinine [LTQ]: described after) which would prevent the complete unfolding of the mature protein under the conditions of SDS-PAGE.

Lysyl oxidases encoding gene and LOX homologous genes were found to be widely distributed in the animal kingdom containing mammalian, bird, amphibian, fish, protochordate, echinoderm, arthropod, and hydra, as observed by basic local alignment search tool (BLAST) analysis using the DDBJ/EMBL/GenBank database (Table 1). In the plant kingdom, however, homologous genes were identified only in two acidophilic primitive red algae, *Cyanidium* and *Cyanidioschyzon*, but not in cyanobacteria, red algae, euglena, green algae, brown algae, diatoms, and all of higher plants. In bacteria, LOX homologous genes were found in myxobacteria (delta-proteobacteria) and beta-proteobacteria (Table 1) but not in alpha-proteobacteria which is considered to be the origin of mitochondria (Esser et al., 2004; Gray, Lang, & Burger, 2004). Furthermore, in archaea, homologous genes were found only in two halophilic archaeon (*Haloterrigena limicola* and *Haloterrigena turknenica*) but not the other archaea (Table 1). The genes were also found in *Acanthamoeba castellanii*. As

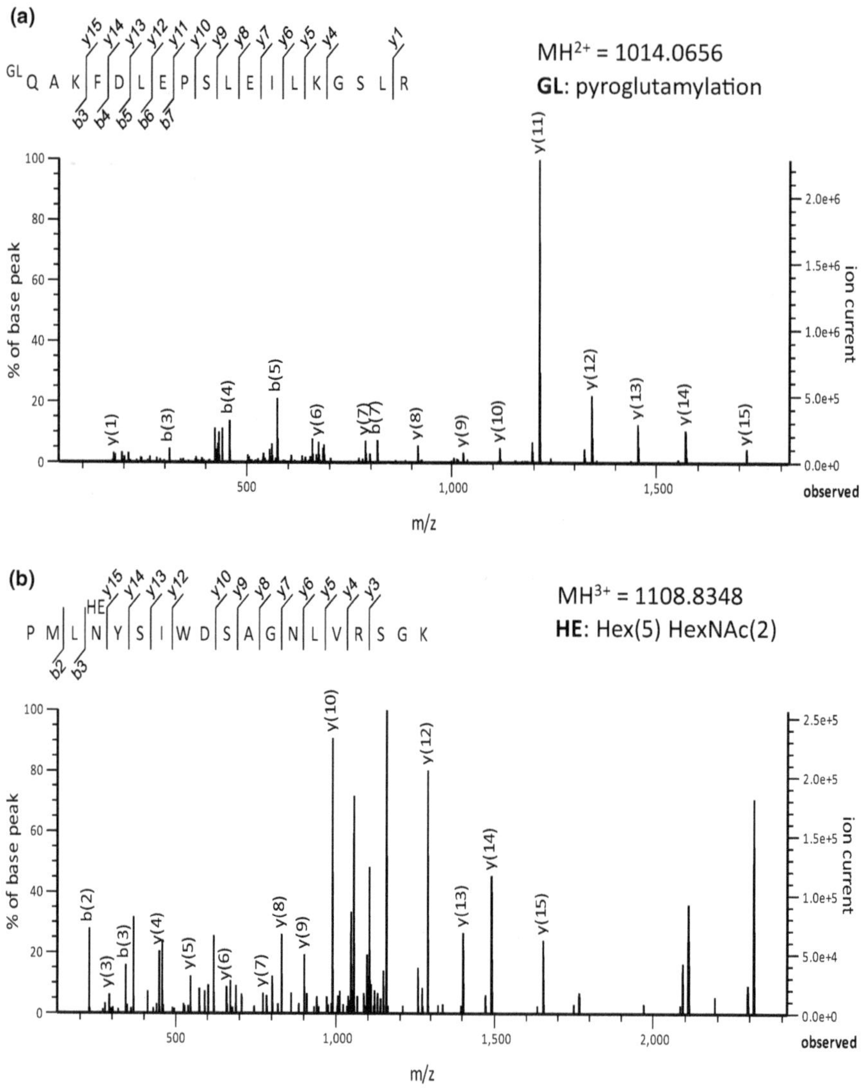

FIGURE 7 The MS/MS spectra corresponding to N-terminal peptide (26–39) with pyroglutamate (a) and the peptide (110–128) with N-linked high-mannose oligosaccharides (b). The identified peptides were satisfied the Expectation value of <0.05 in MASCOT Database search

shown in Table 1, sequence identities between the mature protein (C-domain) of *Cyanidium* LOXL and homologous proteins from the other organisms are relatively high: 97% with *Cyanidioschyzon*, 35%–38% with proteobacteria, and 29%–30% with mammalian LOX. These indicate that the C-terminal domain of *Cyanidium* LOXL has been conserved through the species, although the N-terminal regions involving signal sequences have been reported to be largely different among organisms (Csiszar, 2001; Kagan & Li, 2003; Lucero & Kagan, 2006).

Figure 8 compared the sequence of *Cyanidium* LOXL with that of other organism LOX(L). The presence of LTQ, a unique peptidyl quinine functioning as a carbonyl cofactor, has been reported in human and rat LOX (Csiszar, 2001; Kagan & Li, 2003; Lucero & Kagan, 2006). The two amino acids, "K" and "Y," involved in LTQ formation, have been completely conserved in acidophilic red algae (*C. caldarium* and *C. merolae*), myxobacteria (*Plesiocystis pacifica* and *Stigmatelia aurantiaca*), beta-proteobacteria (*Sphaerotilus natans*), halophilic archaea (*H. limicola*), and acanthamoeba (*A. castellanii*), indicating the presence of LTQ in LOXL of these organisms. Ten cysteines, which form five disulfide bonds and then contribute to the protein stability, have been also reported in animal LOX (Csiszar, 2001; Kagan & Li, 2003; Lucero & Kagan, 2006). Six in these 10 cysteines were conserved in the acidophilic algae, myxobacteria, and acanthamoeba LOXL, while only three or two cysteines were conserved in beta-proteobacteria or archaea LOXL, respectively. Four histidine residues in amino acid sequences corresponding to the copper-binding site (WXWHXCHXHYH) in human and rat LOX (Csiszar, 2001; Kagan & Li, 2003; Lucero & Kagan, 2006) were conserved in myxobacteria, beta-proteobacteria, halophilic archaea, and acanthamoeba LOXL except for the first histidine, but were not conserved in acidophilic red algal LOXL, suggesting that acidophilic red algal LOXL may lack the copper-binding site.

3 | DISCUSSION

As sown in Figure 5a, *Cyanidium* LOXL secretion was detected after 3 days of the cultivation at pH 6. The large cells with four endospores also became to increase gradually after 3 days of the cultivation at pH 6 (Figure 3b). Thus, the large cell formation seems to be

```
                         10        20        30        40        50        60        70        80        90       100
                ....|....|....|....|....|....|....|....|....|....|....|....|....|....|....|....|....|....|....|....|
C. caldarium    MYKRIALVFLFCLVCLATAASSVRAQAKFDLEPS------------------------------LEILKGSLRIKTESFTPRDCAFEMGCVLG-------
C. merolae      MYKRISTLFALCLVCLAAVASSVRAQAKFDLEPS------------------------------LKILQGSLMIKTESFTPHDCAFEMGCVLG-------
P. pacifica     DPDNPAVACGEELLCCAEGPVFDGVYRCEHPVPAMGDTP-----PAEPGDPLTGPLGCDAPDLVVLDEDLIVYLQNLFVAPGSCTLYEGCVGG-------
S. aurantiaca   GCGGQCGFCPNDELCVPATGTCETFAACDHLRPSCTPSCGTSEFCGTDCACHPVSKQLPDLIVDEARLRDEILFDTIFVTENSCARVEECVEG-------
S. natans       RAGDYFLMLHGYRAYSGVTLKATQSGTASAGRPD----------------------------ITISMAAANPRITTETFAANACEIEEGTITA-------
H. limicola     EENAATPPPTTVESDPKPSDDHVEDRTGVNFVPG----------------------------------VENFNVSTEVFDESSPDVEDGFVTP-------
A. castellanii  GPSRLPLISRACVTGCSNESRCVPAQGSINKRQSS---------------------ALPDITVDANRAQSTMYIEQRSFLSTDCAVREGCAA--------
H. sapiens      ----MSKLFQEVLAAPTPQS----SRCLPDLVPD------------------------------PNYVQASTYVQRAHLYSLRCAAEEKCLASTAYAPEA
                                                                         .    .

                        110       120       130       140       150       160       170       180       190       200
                ....|....|....|....|....|....|....|....|....|....|....|....|....|....|....|....|....|....|....|
C. caldarium    --TGPRKLLQLAGAIGNNGPDDFVVGNPADHPDIYTWDTCTGQFITRKPMLNYSIWDSAG-NLVRSGKKESFCMEDLARFPEYTGNPALVPAEPVHTCTF
C. merolae      --TGPRKLLKLAGAIGNNGPDDFVVGNPADHPNIYTWDTCTGQFITRKPMLNYSIWDSAG-NLVRSGKKESFCMEDLARFPEYTGNPALVPAEPVHTCTF
P. pacifica     --TGWRRVLRFDTKTANLGARDLLLGVPANHPDLYSYSPCHNHHHFDN-YASYVLLDPADDSVVAEGHKQAFCLLDFESWAWEGMSRQEDG---TYTCIN
S. aurantiaca   --TGERRVLRFSVEAVNQGSATLTVPAPAERPDLFTFSPCHGHYHFSG-FATYALVDADG-RTVLAGRKQAYCMEDTQRVATGPDVP----CAKKFTCDD
S. natans       ---GTHRLLRFNTQTRNIGSADLVLGNPASN-SSFEWGGCHGHYHFRS-FAQYRLLDASG-AVVRTGKKVGFCLMDITRIDSGANPS------ARYTCSN
H. limicola     ---GEHRLLRFDMIIYNMGDADAELGRPENRPDLFEYSESHGHAHLKG-FNNYILLDESG-ERTGAVRKQTFCLRDLYQTRSTASSS------PQFDCEY
A. castellanii  --AGTRVLLRFDTATPNIGRADLTLGNPAAG-GGFVYDACHRHYHFSG-YANYRLYSYDR-TLTRVGRKQAFCLEDFEPISGWTGSRSLLP---KFDCNN
H. sapiens      TDYDVRVLLRFPQRVKNQGTADFLPNRPRHT---WEWHSCHQHYHSMDEFSHYDLLDAATGKKVAEGHKASFCLEDSTCDFGNLKR-------YACTSHT
                   . : :*::     * *        *        : :  .  :          * : .      .    :*  :*: *                 .

                        210       220       230       240       250       260       270       280
                ....|....|....|....|....|....|....|....|....|....|....|....|....|....|....|....|....
C. caldarium    QGISVGWCDIYDDSLPCQWIDITGVPSGHYTLRLECN------AYHIFKETTYANNVVETPVTIT-------------------
C. merolae      QGISVGWCDIYDDSLPCQWIDITGVPSGHYTLRLECN------AYHIFKETTYANNVVETPVTIT-------------------
P. pacifica     QGLSLGWSDTYAASVDCQWIDVTGVEPGEYTLRMEINLPPEGKAHSTLVELRYDNNVLEIPVEVN-------------------
S. aurantiaca   QGIQRGWSDLYGNTLDCQWLDITDVPPGDYRLQVTLN------PSRAFQETTLDNNTSSVPVTLPAR-----------------
S. natans       QGIQAGWADVYSSNLSGQWVDITGVPAGNYVLEIVMD------PMNLIDELDESNNTGRLNVTIP-------------------
H. limicola     QGISAGWADEYDASLPGQYIVIDDLPDGEYTLQATTN------AAGTINETCDGDNTVRVDLSINNDTVTVHTPQSHYVRPSAC
A. castellanii  QGISVGWQDVYVANLDCQWIDITGITLGSYWLYVEINPVV--NGARAFTESDYTNNVFWIPFRIVNP-----------------
H. sapiens      QGLSPGCYDTYNADIDCQWIDITDVQPGNYILKVHVN------PKYIVLESDFTNNVVRCNIHYTGRYVSATNCKIVQS-----
                **:. *  * *   :  *:: : .:  * * *    :           . *    :*.    .
```

FIGURE 8 Alignment of the sequence of lysyl oxidase-like protein from *Cyanidium caldarium*, with that of homologous proteins from other acidophilic alga (*Cyanidioschyzon merola*), myxobacteria (*Plesiocystis pacifica* and *Stigmatelia aurantiaca*), beta-proteobacteria (*Sphaerotilus natans*), halophilic archaea (*Haloterrigena limicola*), acanthamoeba (*Acanthamoeba castellanii*), and mammalian (*Homo sapiens*). The conserved "K" and "Y," and "C" were highlighted in orange and red, respectively. The amino acid sequence corresponding to the copper-binding site was marked by underline

relevant to LOXL secretion. LOX in human and rat has been reported to function as the extracellular catalysis of lysine-derived cross-links in fibrillar collagens and elastin, playing a role in the formation and stabilization of extracellular matrix (Csiszar, 2001; Kagan & Li, 2003; Lucero & Kagan, 2006). *Cyanidium* cells have been also reported to have a thick cell wall, which is largely (50%–55%) proteinaceous (Bailey & Staehelin, 1968). Thus, it is likely that some proteins in the cell wall are cross-linked with LOXL which results in stabilization of the cell wall, thereby preventing the endospores from splitting.

Another acidophilic primitive red alga, *C. merolae*, has also a similar LOXL gene. However, *C. merolae* are considered not to have the survival mechanisms similar to *Cyanidium* cells, because *C. merolae* lacks a rigid proteinaceous cell wall and reproduces by binary fission without formation of endospore (Matsuzaki et al., 2004; Miyagishima et al., 2001).

In this study, to the best of our knowledge, *Cyanidium* LOXL protein was identified for the first time in the plant kingdom. It should be noted that the *Cyanidium* LOXL protein consists of a simple amino acid sequence just of a highly conserved C-terminal domain with only 25 residues of signal peptide, whereas mammal LOX and LOXL contain long N-terminal sequences (about 200–500 residues) in addition to a highly conserved C-terminal sequence (Csiszar, 2001; Kagan & Li, 2003; Lucero & Kagan, 2006). The diversifications in various mammal LOX and LOXL are considered to have been constructed during evolution of LOX by addition of various amino acids, such as signal peptides, proline-rich region, and scavenger receptor cysteine-rich domain to N-terminal sequence regions (Csiszar, 2001; Kagan & Li, 2003; Lucero & Kagan, 2006). Thus, it is likely that the *Cyanidium* LOXL may be one of the origins of mammalian LOX.

In the plant kingdom, the LOX homologous genes are present only in two acido- and thermo-philic primitive red algae, *Cyanidium* and *Cyanidioschyzon*. The genes had been lost in the third acido- and thermo-philic red alga, *G. sulphuraria* which shows advance features and is closer to the lower *Rhodophyta* (Seckbach, 1987), and then completely disappeared in the plant kingdom. These facts suggest that *Cyanidioschyzon* and *Cyanidium* may be the first eukaryotic algae.

4 | METHODS

4.1 | Cultures

Cyanidium caldarium strain RK-1 was cultured at 40°C under continuous illumination at 30 μmol/m^2/s in an inorganic medium (2.64 g $(NH_4)_2SO_4$, 0.544 g KH_2PO_4, 0.49 g $MgSO_4 \cdot 7\ H_2O$, 0.0148 g $CaCl_2 \cdot 2H_2O$/L, and traces of microelements involving Fe, Mn, Co,

TABLE 1 Comparison of the sequence identities and e-values of lysyl oxidase and its homologous proteins between *Cyanidium caldarium* and various organisms

	Species	Identity (%)	e-Value
Proteobacteria	*Sorangium cellulosum* So157-2	38	3e-32
	Rhodoferax sp. DCY110	35	8e-29
	Stigmatelia aurantiaca DVV4/3-1	38	7e-28
	Plesiocystis pacifica	35	8e-26
Archaea	*Haloterrigena limicola*	32	1e-20
Acanthamoeba	*Acanthamoeba castellanii*	37	3e-29
Red algae	*Cyanidioschyzon merolae*	97	9e-142
Hydra	*Hydra vulgaris (magnipapillata)*	29	7e-16
Arthropod	*Culex quinquefasciatus*	28	8e-18
	Acyrthosiphon pisum	29	3e-15
Echinoderm	*Strongylocentrotus purpuratus*	34	1e-21
Protochordata	*Branchiostoma floridae*	32	8e-22
	Ciona intestinalis	31	1e-16
Fish	*Tetraodon nigroviridis*	31	7e-16
	Salmo salar	29	2e-17
	Oncorhynchus mykiss	31	5e-17
Amphibian	*Xenopus laevis*	30	2e-14
	Xenopus tropicalis	30	5e-15
Bird	*Gallus gallus*	28	2e-15
	Taeniopygia guttata	29	2e-14
Mammalian	*Ornithorhynchus anatinus*	30	2e-16
	Mus musculus	29	5e-18
	Bos taurus	29	1e-19
	Homo sapien	29	5e-17

Note. The sequence identities and e-values were obtained based on ClustalW (Larkin et al., 2007) and BLAST. The sequence identities (%) are expressed the amino acid sequence of mature part of *Cyanidium* lysyl oxidase-like protein as 100%.

Mo, Cu, EDTA). The pH was adjusted to each objective pH with the addition of 0.1 M H_2SO_4 or 1 M NaOH. During growth studies, the pH was checked at frequent intervals, and where necessary, 0.1 M H_2SO_4 or 1 M NaOH was added to bring the pH back to the initial values, and pH 1–3 were maintained within ±0.05 unit, and pH 4–7 within ±0.2 unit. *Cyanidium* cells cultured at pH 2.5 were used for subcultures in all experiments.

4.2 | Measurements of chlorophyll fluorescence

The Chl fluorescence yield using single turn over flash analysis was measured by a double modulation fluorometer (Photon Systems Instruments). Cells were concentrated by low-speed centrifugation and resuspended in fresh medium at a Chl *a* concentration of 5 μg/ml. The cell suspension was incubated in the dark for 10 min prior to the measurements.

4.3 | Measurements of cell number and ratio of the cells with four endospores

Cyanidium cells cultured at pH 2.5 and pH 6 were observed under a differential interference contrast microscope, and number of cells and ratio of the cells with four endospores were counted using hemocytometer. Because cells cultured at pH 6 were significantly aggregated, the cells were shaken for 30 min in culture medium containing 0.01% Triton X-100 to separate the aggregated cells before the microscope observation.

4.4 | SDS-PAGE

Samples were solubilized with 2.5% SDS and 3% 2-merkaptoethanol for 10 min at room temperature. Solubilized samples were applied to a gel containing 12% acrylamide and 6 M urea. After electrophoresis at a constant current of 5 mA for 17 hr, gels were stained with a silver reagent and photographed.

4.5 | Cloning and sequencing of the gene encoding the 34 kDa extracellular protein secreted from *Cyanidium caldarium*

Total RNA of the alga was isolated using Isogen reagent (Nippon Gene, Tokyo, Japan) by treatments with glass beads. Subsequently, full-length cDNA appended with 5′ and 3′ adapter sequences was synthesized by RNA ligase-mediated rapid amplification of cDNA ends (GeneRacer kit; Invitrogen, Carlsbad, CA). For the first PCR amplifications, primers were designed based on the sequences that are conserved among the LC-MS/MS results and the sequence of LOXL protein from *C. merolae* CMN144C, which corresponded to partial sequences of the LC-MS/MS results of the 34 kDa extracellular protein from *C. caldarium*. In the second PCR, sets of primer pairs were designed on the basis of sequence information obtained from the first PCR product and adapter sequences. Resultant sequences were combined to obtain the full-length sequence of the cDNA. To confirm the gene sequence corresponding to the 34 kDa extracellular protein, the gene was amplified using specific primers from the 5′- and 3′-untranslated regions. The PCR products obtained were inserted into the plasmid pCR2.1-TOPO (Invitrogen, Carlsbad, CA), and the DNA sequences were determined using an ABI 3130 genetic analyzer (Applied Biosystems, Foster City, CA). At least three different clones were analyzed for each PCR product. Resulting sequences have been deposited at DDBJ/EMBL/GenBank under accession number LC314427.

4.6 | Mass spectrometry

The silver-stained protein band was cut out and de-stained. The protein band was reduced with DTT (Wako Pure Chemical Industries), followed by carboxymethylation with iodoacetate (Sigma-Aldrich). The band was digested with TPCK-treated trypsin (Worthington Biochemical Corporation) or Achromobacter protease I

(API; a gift from Takeharu Masaki, University of Ibaraki, Ibaraki, Japan) in 10 mM Tris (pH 8.0) at 37°C for 12 hr. The resulting peptide solution was separated on a nanoflow LC (Easy nLC 1000; Thermo Fisher Scientific) with a spray column (NTCC analytical column, C18, φ75 μm × 100 mm, 3 μm; Nikkyo Technos, Tokyo, Japan) with a linear gradient of 0%–35% buffer B (100% acetonitrile and 0.1% formic acid) at a flow rate of 300 nl/min over 10 min. The eluent was analyzed by nano LC-MS/MS using a Q Exactive™ Hybrid Quadrupole-Orbitrap Mass Spectrometer (Thermo Fisher Scientific) in a data-dependent TOP 10 MS/MS method. The acquired data were searched against the in-house database using an in-house MASCOT server (version: 2.5; Matrix Science).

ACKNOWLEDGMENTS

The present work was financially supported in part by Grant-in-Aids for Scientific Research (No. 26220801, 17K07453, 18H05177) and a Grant from JST PRESTO (T.T.).

AUTHOR CONTRIBUTIONS

T.T. and I.E. designed the research; A.O., T.S., O.M., R.K., N.I., and R.N. performed research; T.T., A.O., T.S., R.N., M.I., N.D., and I.E. analyzed data. T.T. and I.E. wrote the manuscript.

REFERENCES

Allen, M. B. (1959). Studies with *Cyanidium caldarium*, an anomalously pigmented chlorophyte. *Archives of Microbiology*, *32*, 270–277. https://doi.org/10.1007/BF00409348

Bailey, R., & Staehelin, L. (1968). The chemical composition of isolated cell walls of *Cyanidium caldarium*. *Microbiology*, *54*, 269–276.

Clarke, A. K., Hurry, V. M., Gustafsson, P., & Oquist, G. (1993). Two functionally distinct forms of the photosystem II reaction-center protein D1 in the cyanobacterium *Synechococcus* sp. PCC 7942. *Proceedings of the National Academy of Sciences of the United States of America*, *90*, 11985–11989. https://doi.org/10.1073/pnas.90.24.11985

Csiszar, K. (2001). Lysyl oxidases: A novel multifunctional amine oxidase family. *Progress in Nucleic Acid Research and Molecular Biology*, *70*, 1–32.

Doemel, W. N., & Brock, T. (1971). The physiological ecology of *Cyanidium caldarium*. *Microbiology*, *67*, 17–32.

Enami, I., Akutsu, H., & Kyogoku, Y. (1986). Intracellular pH regulation in an acidophilic unicellular alga, *Cyanidium caldarium*: ^{31}P-NMR determination of intracellular pH. *Plant and Cell Physiology*, *27*, 1351–1359. https://doi.org/10.1093/oxfordjournals.pcp.a077233

Enami, I., & Kura-Hotta, M. (1984). Effect of intracellular ATP levels on the light-induced H^+ efflux from intact cells of *Cyanidium caldarium* [Rhodophyceae]. *Plant and Cell Physiology*, *25*, 1107–1113. https://doi.org/10.1093/oxfordjournals.pcp.a076817

Enami, I., Nagashima, H., & Fukuda, I. (1975). Mechanisms of the acido-and thermo-phily of *Cyanidium caldarium* Geitler II. Physiological role of the cell wall. *Plant and Cell Physiology*, *16*, 221–231.

Esser, C., Ahmadinejad, N., Wiegand, C., Rotte, C., Sebastiani, F., Gelius-Dietrich, G., … Leister, D. (2004). A genome phylogeny for mitochondria among α-proteobacteria and a predominantly eubacterial ancestry of yeast nuclear genes. *Molecular Biology and Evolution*, *21*, 1643–1660. https://doi.org/10.1093/molbev/msh160

Ford, T. (1984). A comparative ultrastructural study of *Cyanidium caldarium* and the unicellular red alga *Rhodosorus marinus*. *Annals of Botany*, *53*, 285–294. https://doi.org/10.1093/oxfordjournals.aob.a086690

Gomi, M., Akazawa, F., & Mitaku, S. (2000). SOSUIsignal: Software system for prediction of signal peptide and membrane protein. *Genome Informatics*, *11*, 414–415.

Gomi, M., Sonoyama, M., & Mitaku, S. (2004). High performance system for signal peptide prediction: SOSUIsignal. *Chem-Bio Informatics Journal*, *4*, 142–147.

Grau-Bové, X., Ruiz-Trillo, I., & Rodriguez-Pascual, F. (2015). Origin and evolution of lysyl oxidases. *Scientific Reports*, *5*, 10568. https://doi.org/10.1038/srep10568

Gray, M. W., Lang, B. F., & Burger, G. (2004). Mitochondria of protists. *Annual Review of Genetics*, *38*, 477–524. https://doi.org/10.1146/annurev.genet.37.110801.142526

Gross, W. (2000). Ecophysiology of algae living in highly acidic environments. *Hydrobiologia*, *433*, 31–37. https://doi.org/10.1023/A:1004054317446

Herbert, S. K., Samson, G., Fork, D. C., & Laudenbach, D. E. (1992). Characterization of damage to photosystems I and II in a cyanobacterium lacking detectable iron superoxide dismutase activity. *Proceedings of the National Academy of Sciences of the United States of America*, *89*, 8716–8720. https://doi.org/10.1073/pnas.89.18.8716

Kagan, H. M., & Li, W. (2003). Lysyl oxidase: Properties, specificity, and biological roles inside and outside of the cell. *Journal of Cellular Biochemistry*, *88*, 660–672. https://doi.org/10.1002/(ISSN)1097-4644

Kura-Hotta, M., & Enami, I. (1981). Light-induced H^+ efflux from intact cells of *Cyanidium caldarium*. *Plant and Cell Physiology*, *22*, 1175–1183.

Kura-Hotta, M., & Enami, I. (1984). Respiration-dependent H^+ efflux from intact cells of *Cyanidium caldarium*. *Plant and Cell Physiology*, *25*, 1115–1122.

Kvíderová, J. (2012). Photochemical performance of the acidophilic red alga *Cyanidium* sp. in a pH gradient. *Origins of Life and Evolution of the Biosphere*, *42*, 223–234. https://doi.org/10.1007/s11084-012-9284-3

Larkin, M. A., Blackshields, G., Brown, N., Chenna, R., McGettigan, P. A., McWilliam, H., … Lopez, R. (2007). Clustal W and Clustal X version 2.0. *Bioinformatics*, *23*, 2947–2948. https://doi.org/10.1093/bioinformatics/btm404

Lucero, H., & Kagan, H. (2006). Lysyl oxidase: An oxidative enzyme and effector of cell function. *Cellular and Molecular Life Sciences*, *63*, 2304–2316. https://doi.org/10.1007/s00018-006-6149-9

Matsuzaki, M., Misumi, O., Shin-i, T., Maruyama, S., Takahara, M., Miyagishima, S., … Kuroiwa, T. (2004). Genome sequence of the ultrasmall unicellular red alga *Cyanidioschyzon merolae* 10D. *Nature*, *428*, 653–657. https://doi.org/10.1038/nature02398

Miyagishima, S.-Y., Takahara, M., Mori, T., Kuroiwa, H., Higashiyama, T., & Kuroiwa, T. (2001). Plastid division is driven by a complex mechanism that involves differential transition of the bacterial and eukaryotic division rings. *Plant Cell*, *13*, 2257–2268. https://doi.org/10.1105/tpc.13.10.2257

Ohta, H., Shirakawa, H., Uchida, K., Yoshida, M., Matuo, Y., & Enami, I. (1997). Cloning and sequencing of the gene encoding the plasma membrane H^+-ATPase from an acidophilic red alga, *Cyanidium caldarium*. *Biochimica et Biophysica Acta*, *1319*, 9–13. https://doi.org/10.1016/S0005-2728(97)00007-8

Reeb, V., & Bhattacharya, D. (2010). The thermo-acidophilic cyanidiophyceae (Cyanidiales). In *Red algae in the genomic age* (pp. 409–426). Berlin, Germany: Springer.

13

FarmCPUpp: Efficient large-scale genomewide association studies

Aaron Kusmec[1] | **Patrick S. Schnable**[1,2]

[1]Department of Agronomy, Iowa State University, Ames, IA, USA

[2]Plant Sciences Institute, Iowa State University, Ames, IA, USA

Correspondence
Patrick S. Schnable, Department of Agronomy, Iowa State University, Ames, IA, USA.
Email: schnable@iastate.edu

Funding information
National Science Foundation, Grant/Award Number: 1027527

Abstract

Genomewide association studies (GWAS) are computationally demanding analyses that use large sample sizes and dense marker sets to discover associations between quantitative trait variation and genetic variants. FarmCPU is a powerful new method for performing GWAS. However, its performance is hampered by details of its implementation and its reliance on the R programming language. In this paper, we present an efficient implementation of FarmCPU, called FarmCPUpp, that retains the R user interface but improves memory management and speed through the use of C++ code and parallel computing.

KEYWORDS
bioinformatics, genomewide association study, quantitative trait, software

1 | INTRODUCTION

Genomewide association studies (GWAS) use statistical tests to identify associations between genetic markers and phenotypes. Next-generation sequencing and improved phenotyping methods have enabled the use of increasingly large datasets in GWAS, posing challenges for the statistical models used and their computational efficiencies.

Two challenges arise from the biology underlying quantitative trait variation. One challenge is false negative marker-trait associations, which are often attributable to small sample sizes and the small effect sizes expected from most quantitative trait nucleotides (QTN). This challenge is met primarily by increasing sample sizes and through experimental designs that increase the precision of phenotypes. The other challenge is false positive marker-trait associations that can arise due to linkage between markers and QTN induced by population and family structure. This challenge has been addressed through more sophisticated single-marker models that incorporate subpopulation assignments (STRUCTURE) (Pritchard, Stephens, & Donnelly, 2000), principal components of the marker matrix (Price et al., 2006), and kinship matrices (Yu et al., 2006). Through this work, the general linear model (GLM) and mixed linear model (MLM) (Yu et al., 2006) became the dominant statistical paradigms in GWAS.

Significant new ideas were introduced by the multilocus mixed model (MLMM) (Segura et al., 2012) and SUPER (Wang, Tian, Pan, Buckler, & Zhang, 2014) methods. The MLMM addressed the issue posed by linkage disequilibrium between nearby markers by using model selection and multiple regression in conjunction with the MLM. SUPER introduced a new formulation of the kinship matrix based on pseudo-QTNs that are selected from a preliminary GWAS step. However, more complex statistical methods also impacted the computational efficiency of these methods.

The sizes of GWAS datasets also pose a challenge for the computational efficiency of the statistical methods. Various statistical and computational techniques, such as EMMAX (Kang et al., 2010), P3D (Zhang et al., 2010), and Fast-LMM (Lippert et al., 2011), have steadily improved the computational efficiency—but not the statistical power—of newer software relative to the original implementation of the MLM. While efficient, these methods provided approximate parameter estimates. GEMMA improved the statistical power of MLM-based methods by providing exact parameter estimates with an efficiency matching that of EMMAX/P3D (Zhou & Stephens, 2012).

FarmCPU (Liu, Huang, Fan, Buckler, & Zhang, 2016) is an innovative method that combines the model selection strategy of MLMM with the restricted kinship matrix of SUPER. Briefly, FarmCPU implements the GLM with optional population structure covariates to scan single markers. The results of this scan are used to select multiple sets of pseudo-QTNs similar to SUPER. The best set is selected based on its log-likelihood estimated from a random effects model. Markers are scanned again using the GLM plus pseudo-QTNs as covariates, and the process is repeated until the set of pseudo-QTNs does not change.

FarmCPU has been reported to be not only a statistically powerful but also a computationally efficient method for GWAS. However, the performance tests reported in the main text of the paper tested at most 60,000 markers with a sample size of 11,000 (660 million data points) (figure 6 in Liu et al., 2016). Tests on larger datasets were reported in the supplementary materials. Of particular, interest to us is the test on 10,000 individuals and 1,000,000 markers (10 billion data points) reported in Figure S26. This figure reports that FarmCPU completed the analysis in ~4 hr on a laptop with 4 GB of RAM. We initially tested FarmCPU with a dataset of 4,890 individuals and 2,452,207 markers (~12 billion data points) and found that this required 9 hr and 265 GB of RAM. This disparity is not due to the use of inferior hardware.

We had a large number of GWAS to perform such that existing software (e.g., MLMM, SUPER, FarmCPU) would have taken prohibitively long. Therefore, we investigated the source code of FarmCPU to identify memory- and time-related bottlenecks that could potentially be improved.

2 | MATERIALS AND METHODS

2.1 | Computational considerations for GWAS

The sizes of GWAS datasets pose memory management and computational speed challenges for software development. Univariate GWAS requires at minimum a $n \times 1$ vector of phenotypes and $n \times m$ marker matrix, where n is the number of samples and m is the number of markers. A $n \times q$ matrix of covariates, where q is the number of covariates, containing potential confounding variables, such as population structure, may also be included. The values of both n and m have increased rapidly as larger samples are collected to increase statistical power and whole-genome resequencing identifies more variants. Thus, how the marker matrix is represented and accessed in memory is important for all GWAS methods.

While FarmCPU shares the challenge of memory management with other GWAS models, its efficiency concerns differ. Popular early methods, such as the general linear model (GLM) and mixed linear model (MLM), tested each marker once in an analysis. Researchers devised multiple algorithms to improve the efficiency of the MLM, although none of these improved its statistical power. While FarmCPU uses the efficient GLM, it also uses an iterative, model selection approach to increase statistical power and decrease false positives. This requires testing each marker multiple times, adding to the computational burden of the analysis.

Subsequent sections discuss three ways in which FarmCPUpp improves on the original implementation of FarmCPU through the use of better memory management techniques, compiled code, and parallel computing.

2.2 | Memory management

R's design makes working with even modest numbers of samples and markers relatively inefficient. To address this issue, the original implementation of FarmCPU suggested, but did not require, the use of the `big.matrix` data structure from the `bigmemory` package (Kane, Emerson, & Weston, 2013) as an alternative to the base R `data.frame` for the storage and manipulation of the marker matrix.

`big.matrix` is a C++ data structure that returns a pointer to the R process. The structure and its pointer can also be stored in binary files and shared across processes to facilitate reuse and parallel computing. While `big.matrix` allows the user to exceed R's internal limit on object size, *sufficient RAM to store the dataset must still be available* (Kane et al., 2013). Thus, `bigmemory` is a useful tool for enabling the analysis of large datasets but is still subject to the physical limitations of the user's hardware.

The original implementation of FarmCPU has two shortcomings with respect to its handling of a `big.matrix` due to the details of the C++ implementation of the `big.matrix`.

First, a `big.matrix` can store only one of four (`char` [1 byte], `short` [2 bytes], `int` [4 bytes], or `double` [8 bytes]) data types at once. Marker scores for GWAS are typically represented as 0, 1, and 2, counting the number of minor alleles that an individual possesses at a single locus. Missing marker scores are often imputed as the mean nonmissing marker score (or twice the minor allele frequency). Because minor allele frequency is in the interval (0, 0.5], imputed marker scores will be in the interval (0, 1]. Therefore, marker scores are a mix of integers and real numbers. Base R seamlessly converts integers into real numbers and, in a `data.frame`, can store both types. However, real numbers improperly stored in a `char`, `short`, or `int` type `big.matrix` will be truncated, causing an inflated number of homozygous major individuals. The manual for FarmCPU recommends the use of a `char` type `big.matrix`; this is only appropriate in the absence of missing data—a condition that is not met in real data without imputation. Because storage of a `double` requires 8× the space of a `char`, in the presence of missing marker data, FarmCPU requires 8× more memory than reported.

Second, FarmCPU takes subsets of the `big.matrix` to handle samples with missing phenotypic data. There are two methods for subsetting a `big.matrix` that also return a `big.matrix`, but they have very different consequences for memory usage. The first option uses the `sub.big.matrix()` function to extract a continuous subset (or block) of the `big.matrix`. This can be handled efficiently using pointers, requiring no copying of the original matrix. The second option uses the `deepcopy()` function to extract a discontinuous subset, creating a new, potentially smaller `big.matrix`.

In the worst case that the entire `big.matrix` is copied, `deepcopy()` doubles the memory used to store the data if the original `big.matrix` is not discarded. FarmCPU uses `deepcopy()` to remove marker scores for samples with missing phenotypes without releasing the memory used by the original marker matrix.

FarmCPUpp exclusively uses the `big.matrix` as the data structure to represent the marker matrix. This allows the marker matrix to exceed R's internal limits on object size, rapidly reload a `big.matrix` into RAM, and easily share the `big.matrix` between different processes. Due to the implementation of single marker regression, FarmCPUpp only accepts a `double`-type `big.matrix` and will throw an error if given any other type. This ensures accurate representation of the marker scores. FarmCPUpp also declines to make any copies of the marker matrix using `sub.big.matrix()` or `deepcopy()`. Instead, missing data are handled by removing the appropriate marker scores using compiled code when that marker is tested. While this strategy would be inefficient in R, it is efficient in compiled code and eliminates the need to make expensive copies of the marker matrix either in RAM or on disk.

2.3 | Compiled single-marker regression

As mentioned above, FarmCPU uses the computationally efficient GLM, which has an asymptotic running time of $O(nm)$ (Liu et al., 2016). However, because FarmCPU optimizes the set of pseudo-QTNs used as covariates and iterates until this set converges, the actual runtime of FarmCPU will be proportional to the number of iterations which is a function of the dataset. FarmCPU splits the model matrix into two parts: a constant part including covariates and pseudo-QTNs and a variable part containing the marker scores for the current marker. FarmCPU implements this in R, and it is sufficient to perform GWAS on reasonably large datasets in less than one day. However, this does not scale well with sample size or number of markers. FarmCPUpp uses the same strategy but implements it in C++ using the `Rcpp` (Eddelbuettel & Francois, 2011) and `RcppEigen` (Bates & Eddelbuettel, 2013) packages.

2.4 | Parallelization

Single-marker regression assumes that tests for each marker are mutually independent. Therefore, another complementary strategy for increasing the speed of single-marker regression is parallel computing. The version of FarmCPU (v1.02) used for benchmarks has code for parallelized single-marker regression and an option to choose the number of cores used for analysis, but use of this code will throw an error when the first single-marker regression loop is reached. FarmCPUpp uses the `RcppParallel` package (Allaire et al., 2016) and the `ncores.glm` parameter to provide parallelized single-marker regression.

FarmCPU's model selection chooses the optimal number of pseudo-QTNs to add as covariates in each iteration. This optimization is optional but recommended by the authors. This step is not as large of a bottleneck as single-marker regression, but it can take a significant amount of time when the number of models to test, sample size, and/or number of pseudo-QTNs is large. FarmCPUpp uses the `foreach` package (Revolution Analytics & Weston 2015) and the `ncores.reml` parameter to parallelize this step.

2.5 | Feature comparison

FarmCPUpp implements only the core GWAS-related functionality of FarmCPU. Some functionality has been simplified, such as the choice of functions to use for single-marker regression because FarmCPUpp has only one option. Other functions, such as power and FDR tests, are not implemented by FarmCPUpp. Full information on each parameter can be found in the documentation.

FarmCPUpp also implements functions to generate manhattan and QQ plots. Unlike FarmCPU, these are separate from the GWAS function. This allows the user to control base R plotting variables. Sensible default values have been selected for each function.

2.6 | Simulations

Markers were simulated in PLINK v1.9 (Chang et al., 2015; Purcell & Chang, 2017) using the `-simulate-qt` flag. All marker sets were simulated with reference allele frequencies in [0.05, 0.5], $D' = 0.75$, and dominance deviations of 0. The per-QTL additive variance was 1/ m, where m is the number of markers, according to PLINK's internal upper bound. Two series of datasets were simulated. The first series had 60,000 markers and sample sizes of 500, 1,000, 2,500, 5,000, 7,500, or 10,000. The second series had 10,000, 50,000, 100,000, 500,000, 1,000,000, or 5,000,000 markers and a sample size of 1,000.

Following (Bernardo & Yu, 2007), phenotypes with a heritability of 0.3 were simulated for each dataset using 3,000 QTNs with effect sizes of 0.9^k, where k is the kth QTN. Phenotypes were constructed by summing up the QTN effects for each individual and adding i.i.d. normal deviates with mean 0 and variance $(1-h^2)\sigma_a^2/h^2$, where h^2 is the heritability, and σ_a^2 is the variance of the genetic effects.

2.7 | Benchmarks

All benchmarks were run on a Ubuntu 14.04 server with 44 Intel Xeon E5-2697 v3 @ 2.60 GHz processors, 377 GB of RAM, and a 4.6 TB HDD. All timing runs were performed with the optimum option for pseudo-QTN selection, the default options for number of pseudo-QTNs, a maximum of 20 iterations, and one core unless otherwise stated. The size of bins varied with the number of markers according to Table 1.

3 | RESULTS

3.1 | Correctness

The final results from all 12 simulated datasets were compared using the `all.equal()` function in R. FarmCPU reports p-values and effect estimates for each SNP, so we compared only these two measures even though FarmCPUpp also reports the standard errors and

TABLE 1 Bin sizes used for model selection at different marker densities. Sizes are given in base-pairs

Simulation	Bin sizes
10,000 markers	500; 1,000; 1,500; 2,000
50,000 markers	2,500; 5,000; 7,500; 10,000
60,000 markers	2,500; 5,000; 7,500; 10,000
100,000 markers	5,000; 10,000; 15,000; 20,000
500,000 markers	25,000; 50,000; 75,000; 100,000
1,000,000 markers	50,000; 100,000; 150,000; 200,000
5,000,000 markers	250,000; 500,000; 750,000; 1,000,000

t-values for each SNP. FarmCPU and FarmCPUpp produced identical results for effect estimates and *p*-values.

3.2 | Benchmarks

The performance of FarmCPU and FarmCPUpp was tested first on simulated datasets of 60,000 markers with 3,000 markers chosen as QTNs and a heritability of 0.3. Sample size varied from 500 to 10,000. Both packages scale well with increasing sample size (Figure 1a); at a sample size of 10,000, FarmCPUpp is faster by 159 s. In each iteration, the FarmCPU algorithm adds pseudo-QTNs to the model as covariates for single-marker regression. Increased model size has a stronger effect on FarmCPU than on FarmCPUpp (Figure 1b). At a sample size of 10,000, the longest single-marker regression time (most complex model) for FarmCPUpp is as fast as the shortest single-marker regression time (least complex model) for FarmCPU.

By contrast, the number of markers has a much stronger effect on the runtimes of both packages. These datasets were simulated with a sample size of 1,000 and numbers of markers that varied from 10,000 to 5,000,000. Phenotypes were simulated with 3,000 QTN and heritabilities of 0.3. With small samples sizes, the performance of both packages is similar at low numbers of markers (Figure 2a). However, FarmCPU scales much more poorly as the number of markers increases. With 5,000,000 markers, FarmCPU takes almost 2 hr to complete, while FarmCPUpp takes only 30 min. This reduction in runtime is largely attributable to reductions in the time spent on single-marker regression (Figure 2b). With 5,000,000 markers, FarmCPU performs single-marker regression in 517–832 s, while FarmCPUpp performs the same analysis in 137–185 s, a 74%–78% reduction in runtime.

Finally, we tested the performance of both packages on a real dataset described in (Kusmec, Srinivasan, Nettleton, & Schnable, 2017) that consisted of measurements for days to silk on 4,890 recombinant inbred lines (RILs) in the maize nested association mapping (NAM) population (Yu, Holland, McMullen, & Buckler, 2008) (Figure 3). The first three principal components of the marker matrix were used to control for population structure, and 2,452,207 markers were tested. We also used different numbers of cores in FarmCPUpp. As already shown in Figure 1, FarmCPUpp scales slightly better with increasing sample sizes, and using the full dataset FarmCPUpp runs ~3.5× faster. This difference represents the advantage of using C++ rather than R for single-marker regression. The use of multiple cores (Figure 3) further decreases the runtime of FarmCPUpp to a minimum observed runtime of ~51 min with 16 cores.

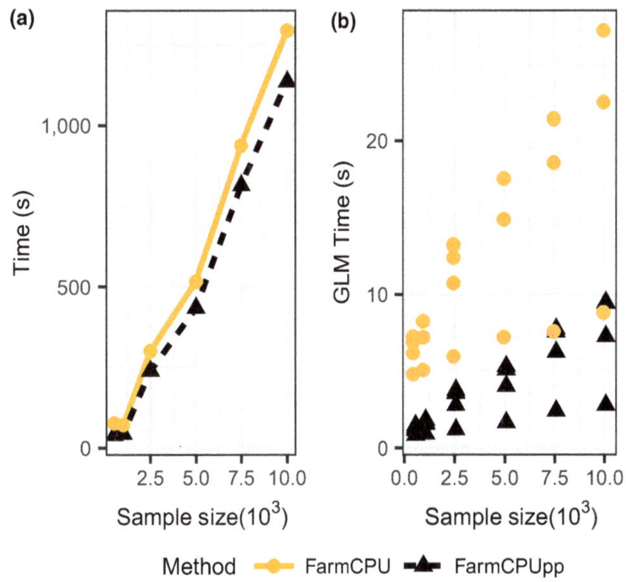

FIGURE 1 Effect of sample size on the runtime of FarmCPU and FarmCPUpp. (a) Total runtime in seconds. (b) Time for single-marker regression in seconds. Dots represent the time spent in each iteration at the indicated sample size

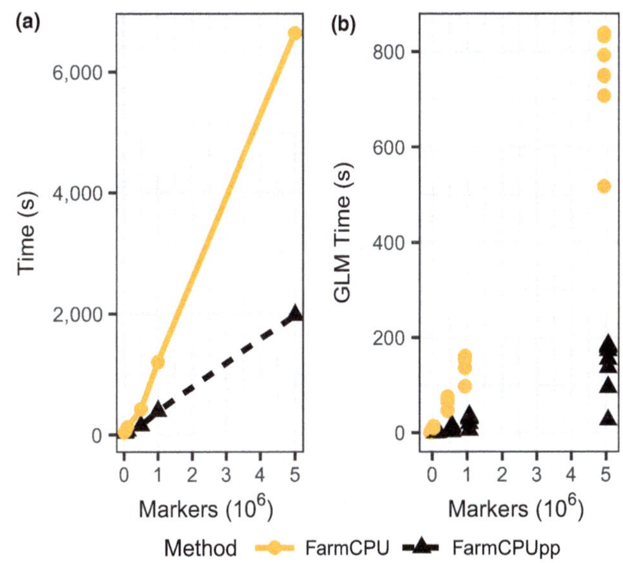

FIGURE 2 Effect of number of markers on the runtime of FarmCPU and FarmCPUpp. (a) Total runtime in seconds. (b) Time for single-marker regression in seconds. Dots represent the time spent in each iteration at the indicated number of markers

4 | DISCUSSION

GWAS datasets are rapidly growing due to improved phenotyping and genotyping methods, and it is now common to perform GWAS on datasets consisting of thousands of samples and millions of markers.

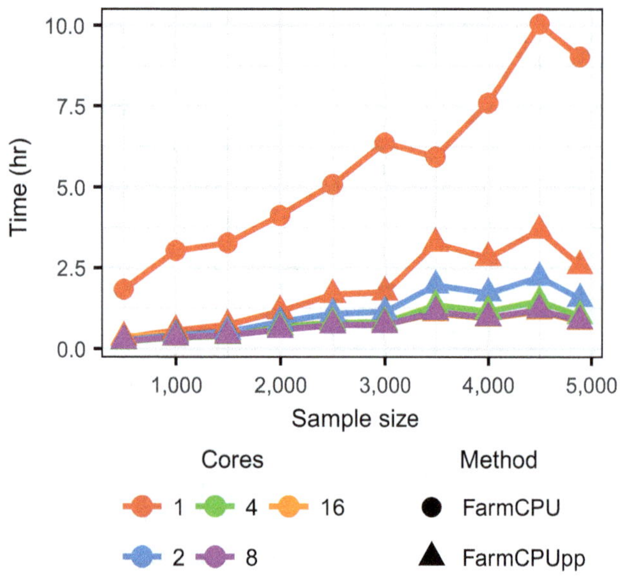

FIGURE 3 Effects of sample size and number of CPU cores on the runtime of FarmCPU and FarmCPUpp in hours. The full dataset has 4,890 samples, 2,452,207 markers, and three population structure covariates

Even though the GLM used in FarmCPU is linear with respect to number of samples and markers, runtimes rapidly increase when both dimensions of the data are increasing. This greater runtime may be acceptable when performing a single GWAS, but GWAS on multiple phenotypes are also common and long runtimes may be prohibitive in these situations. This growth poses a computational challenge to existing methods and software. We have presented FarmCPUpp, an improved version of the FarmCPU method, that uses compiled single-marker regression code and parallel computing to reduce the memory requirements and runtime, especially for large datasets.

The performance of FarmCPU and FarmCPUpp is comparable for datasets up to 10,000 samples by 60,000 markers or 1,000 samples by 1,000,000 markers. Runtimes may still be quite different below these thresholds depending on the dimensions of the data. FarmCPUpp achieves this speedup without parallel processing, although further speedups can be achieved if multiple cores are available. Because the interface of FarmCPUpp is nearly identical to FarmCPU's, it is not difficult to integrate FarmCPUpp into an existing analysis.

In summary, FarmCPUpp makes improvements on the original FarmCPU R code through the use of C++ and parallel computing. This decreases the memory requirements and runtime while maintaining the R interface of the original software. FarmCPUpp is available as an R package on GitHub at https://github.com/amkusmec/FarmCPUpp.

ACKNOWLEDGMENTS

This material is based upon work supported by the National Science Foundation (grant number 1027527) to P.S.S.

AUTHOR CONTRIBUTIONS

A.K. designed and wrote the modified software. A.K. and P.S.S. wrote the manuscript.

REFERENCES

Allaire, J., Francois, R., Ushey, K., Vandenbrouck, G., Geelnard, M., & Intel. (2016). *RcppParallel: Parallel programming tools for "Rcpp."* v4.3.20. Retrieved from https://CRAN.R-project.org/package=RcppParallel

Bates, D., & Eddelbuettel, D. (2013). Fast and elegant numerical linear algebra using the RcppEigen package. *Journal of Statistical Software*, *52*(5), 1–24.

Bernardo, R., & Yu, J. (2007). Prospects for genomewide selection for quantitative traits in maize. *Crop Science*, *47*(3), 1082–1090. https://doi.org/10.2135/cropsci2006.11.0690

Chang, C. C., Chow, C. C., Tellier, L. C., Vattikuti, S., Purcell, S. M., & Lee, J. J. (2015). Second-generation PLINK : Rising to the challenge of larger and richer datasets. *GigaScience*, *4*(7), 1–16.

Eddelbuettel, D., & Francois, R. (2011). Rcpp: Seamless R and C ++ integration. *Journal of Statistical Software*, *40*(8), 1–18.

Kane, M. J., Emerson, J. W., & Weston, S. (2013). Scalable strategies for computing with massive data. *Journal of Statistical Software*, *55*(14), 1–19.

Kang, H. M., Sul, J. H., Zaitlen, N. A., Kong, S. Y., Freimer, N. B., Sabatti, C., & Eskin, E. (2010). Variance component model to account for sample structure in genome-wide association studies. *Nature Genetics*, *42*(4), 348–354. https://doi.org/10.1038/ng.548

Kusmec, A., Srinivasan, S., Nettleton, D., & Schnable, P. S. (2017). Distinct genetic architectures for phenotype means and plasticities in Zea mays. *Nature Plants*, *3*(9), 715–723. https://doi.org/10.1038/s41477-017-0007-7

Lippert, C., Listgarten, J., Liu, Y., Kadie, C. M., Davidson, R. I., & Heckerman, D. (2011). FaST linear mixed models for genome-wide association studies. *Nature Methods*, *8*(10), 833–835. https://doi.org/10.1038/nmeth.1681

Liu, X., Huang, M., Fan, B., Buckler, E. S., & Zhang, Z. (2016). Iterative usage of fixed and random effect models for powerful and efficient genome-wide association studies. *PLoS Genetics*, *12*(2), e1005767. https://doi.org/10.1371/journal.pgen.1005767

Price, A. L., Patterson, N. J., Plenge, R. M., Weinblatt, M. E., Shadick, N. A., & Reich, D. (2006). Principal components analysis corrects for stratification in genome-wide association studies. *Nature Genetics*, *38*(8), 904–909. https://doi.org/10.1038/ng1847

Pritchard, J. K., Stephens, M., & Donnelly, P. (2000). Inference of population structure using multilocus genotype data. *Genetics*, *155*(2), 945–959.

Purcell, S. M., & Chang, C. (2017). PLINK v1.90b3.42. Retrieved from https://www.cog-genomics.org/plink/1.9

Revolution Analytics & Weston, S. (2015). foreach: Provides foreach looping construct for R. v1.4.3. Retrieved from https://CRAN.R-project.org/package=foreach

Segura, V., Vilhjálmsson, B. J., Platt, A., Korte, A., Seren, Ü., Long, Q., & Nordborg, M. (2012). An efficient multi-locus mixed-model approach for genome-wide association studies in structured populations. *Nature Genetics*, *44*(7), 825–830. https://doi.org/10.1038/ng.2314

Wang, Q., Tian, F., Pan, Y., Buckler, E. S., & Zhang, Z. (2014). A SUPER powerful method for genome wide association study. *PLoS ONE*, *9*(9), e107684. https://doi.org/10.1371/journal.pone.0107684

Yu, J., Holland, J. B., McMullen, M. D., & Buckler, E. S. (2008). Genetic design and statistical power of nested association mapping in maize. *Genetics*, *178*(1), 539–551. https://doi.org/10.1534/genetics.107.074245

Yu, J., Pressoir, G., Briggs, W. H., Bi, I. V., Yamasaki, M., Doebley, J. F., ... Kresovich, S. (2006). A unified mixed-model method for association mapping that accounts for multiple levels of relatedness. *Nature Genetics*, *38*(2), 203–208. https://doi.org/10.1038/ng1702

Zhang, Z., Ersoz, E., Lai, C. Q., Todhunter, R. J., Tiwari, H. K., Gore, M. A., ... Buckler, E. S. (2010). Mixed linear model approach adapted for genome-wide association studies. *Nature Genetics*, *42*(4), 355–360. https://doi.org/10.1038/ng.546

Zhou, X., & Stephens, M. (2012). Genome-wide efficient mixed-model analysis for association studies. *Nature Genetics*, *44*(7), 821–824. https://doi.org/10.1038/ng.2310

14

Noncanonical interactions between serpin and β-amylase in barley grain improve β-amylase activity *in vitro*

Maja Cohen | Robert Fluhr

Department of Plant Sciences, Weizmann Institute of Science, Rehovot, Israel

Correspondence
Robert Fluhr, Department of Plant Sciences, Weizmann Institute of Science, Rehovot, Israel.
Email: Robert.Fluhr@weizmann.ac.il

Funding information
US-Israel Binational Agricultural Research and Development Fund for Research IS-4915-16.

Abstract

Serpin protease inhibitors and β-amylase starch hydrolases are very abundant seed proteins in the endosperm of grasses. β-amylase is a crucial enzyme in the beer industry providing maltose for fermenting yeast. In animals and plants, inhibitory serpins form covalent linkages that inactivate their cognate proteases. Additionally, in animals, noninhibitory functions for serpins are observed such as metabolite carriers and chaperones. The function of serpins in seeds has yet to be unveiled. In developing endosperm, serpin Z4 and β-amylase showed similar *in vivo* spatio-temporal accumulation properties and colocalize in the cytosol of transformed tobacco leaves. A molecular interaction between recombinant proteins of serpin Z4 and β-amylase was revealed by surface plasmon resonance and microscale thermophoresis yielding a dissociation constant of 10^{-7} M. Importantly, the addition of serpin Z4 significantly changes β-amylase enzymatic properties by increasing its maximal catalytic velocity. The presence of serpin Z4 stabilizes β-amylase activity during heat treatment without affecting its critical denaturing temperature. Oxidative stress, simulated by the addition of $CuCl_2$, leads to the formation of high molecular weight polymers of β-amylase similar to those detected *in vivo*. The polymers were cross-linked through disulfide bonds, the formation of which was repressed when serpin Z4 was present. The results suggest an unprecedented function for a plant seed serpin as a β-amylase-specific chaperone-like partner that could optimize β-amylase activity upon germination. This report is the first to describe a noninhibitory function for a serpin in plants.

KEYWORDS
amylase, barley, enzyme stability, *Hordeum vulgare*, oxidative stress, protein interaction, serpin

1 | INTRODUCTION

Serpin proteins are part of a well-conserved superfamily of proteins in plants and animals. All serpin types share a similar secondary structure, although their specific amino acids sequence can differ. Serpins are metastable proteins and undergo conformational changes, from "stressed" to "relaxed" transition to complete their various functions (Huntington, Read, & Carrell, 2000). They are mostly known as serine protease inhibitors in animal cells (Gettins, 2002) and as cysteine protease inhibitors controlling cell death in plant cells (Fluhr, Lampl, & Roberts, 2012; Lampl, Alkan, Davydov, & Fluhr, 2013; Lampl et al., 2010; Lema Asqui et al., 2017; Vercammen et al., 2006). The secondary structure of a serpin includes a reactive center loop (RCL) domain that acts as bait for its cognate protease by mimicking the preferred P1-P1′ digestion site (Khan et al., 2011). Other functions for serpins were discovered in animal systems. The noninhibitory serpins do not exhibit protease inhibitory properties and instead have a panel of activities including: hormone transport by the serpin thyroxine-binding globulin (Pemberton, Stein, Pepys, Potter, & Carrell, 1988); chaperone-like functions by SERPINH1 binding to collagen (Sauk, Nikitakis, & Siavash, 2005) and as storage proteins for ovalbumin, in egg white (Stein et al., 1990). In contrast, the myeloid and erythroid nuclear termination (MENT) serpin retains protease inhibition but also functions in chromatin remodeling (Irving et al., 2002). In plants, noninhibitory functions for serpins have yet to be characterized.

Serpins are present in seeds of cereals such as barley, wheat, oat, and rye (Hejgaard, 2001; Hejgaard & Hauge, 2002; Østergaard, Finnie, Laugesen, Roepstorff, & Svennson, 2004; Rasmussen, Dahl, Nørgård, & Hejgaard, 1996; Rosenkrands, Hejgaard, Rasmussen, & Bjørn, 1994). In barley grains, the two major types, serpin Z4 and serpin Z7, represent up to 5% of the total seed proteome (Evans & Hejgaard, 1999). The amounts of the two proteins vary among the various cultivars (Evans & Hejgaard, 1999; Finnie et al., 2004). In wheat and oat, the RCL sequence of the grain serpins contains glutamines, a very abundant amino acid in storage proteins (Ostergaard, Rasmussen, Roberts, & Hejgaard, 2000). Therefore, it was suggested that serpins could protect grain storage proteins by inhibiting predator's proteases adapted to the proteolysis of glutamine-rich substrates. However in barley, the RCL sequence of serpin Z4 and Z7 is not similar to any of the main storage protein sequences, and, although both serpins are protease inhibitors of cathepsin G and trypsin, their inhibitory activity appears to be less potent than other plant serpins (Dahl, Rasmussen, & Hejgaard, 1996). Alternatively, considering their abundance in the endosperm, serpins were suggested to have a role as storage proteins. However, serpins in barley grains are relatively resistant to degradation by proteases during germination. Indeed, the level of the soluble serpin continues to increase 2 days after seed imbibition (Bønsager, Finnie, Roepstorff, & Svensson, 2007). Hence, the function of barley grains serpins remains an open question.

In seeds, serpins exist as two distinct forms, the free serpins, extractable with water-based buffer and the bound fraction of serpins, extractable only with thiol enriched buffers (Evans & Hejgaard, 1999). β-amylase, another very abundant endosperm protein, also exists in these two forms (Evans, Wallace, Lance, & MacLeod, 1997; Evans, MacLeod, et al., 1997). The ratio between bound and free β-amylase can be modulated by the presence of a cysteine at position 115 (Li et al., 2002). This position is also critical for conferring distinct IEF band patterns (Ma, Langridge, Logue, & Evans, 2002) and protein thermostability (Ma, Evans, Logue, & Langridge, 2001; Ma, Stewart, et al., 2000). β-amylase accumulates as a water-insoluble protein during grain ripening but during germination or malting becomes readily extractable (Finnie, Melchior, Roepstorff, & Svensson, 2002; Finnie et al., 2004). It is thought that cysteines play a key role in these changes. For example during germination, the C-terminal region of β-amylase that contains a cysteine is removed by limited proteolysis (Guerin, Lance, & Wallace, 1992; Lundgard & Svensson, 1986). In addition, two cysteines are targeted and reduced by thioredoxins (Hagglund et al., 2010). The combined action of proteases and reducing agent released by the aleurone layer convert the bound form of β-amylase in the endosperm to its free form. Serpin and β-amylase are both found in the endosperm tissue (Finnie & Svensson, 2003; Roberts, Marttila, Rasmussen, & Hejgaard, 2003) adhering to the border of starch granules (Borén, Larsson, Falk, & Jansson, 2004; Hara-Nishimura, Nishimura, & Daussant, 1986; Wang et al., 2011). Upon gel filtration of salt extract from barley dry grain, serpin and β-amylase eluted in high molecular weight fractions prompting speculation of their tight association (Hejgaard & Carlsen, 1977).

Although the exact function of β-amylase in natural seed physiology is not known, it is a critical enzyme in the beer industry; it produces highly fermentable maltose disaccharides through the hydrolysis of the α-1,4-glycosidic bond of polysaccharide chains (MacGregor, Bazin, Macri, & Babb, 1999). Moreover, as one of the most abundant barley protein and as one of the only residual beer proteins, serpin has been investigated for its impact on beer quality (Iimure, Nankaku, Kihara, Yamada, & Sato, 2012; Iimure & Sato, 2013; Specker, Niessen, & Vogel, 2014). In the results presented here, we show that serpin and β-amylase interact in a quantifiable manner. The interaction enhances β-amylase enzymatic properties and increases β-amylase stability to heat and oxidative stress. This research suggests an unprecedented noncanonical function for serpin in plants as a β-amylase-specific molecular chaperone.

2 | MATERIALS AND METHODS

2.1 | Plant material

Barley cv. Harrington was grown in the greenhouse under 24/20°C (day/night) at natural day length conditions. For sampling of developing grain, flowering kernels were marked and sampled at the indicated times. For germination experiments, barley grains were imbibed in water for 24 h and placed on wet filter paper at 22°C under natural day length conditions. The grains were sampled at the indicated time without the coleoptile and the radicle.

2.2 | Gel fractionation and immunoblots

Grains at different growth stages were ground in liquid nitrogen. Tissue (100 mg) was dissolved in 1 ml of Dulbecco's phosphate-buffered saline (DPBS without calcium and magnesium) and plant protease inhibitor cocktail (Sigma P9599) with or without 20 mM DTT. The extracted tissue was left shaking for 15 min at room temperature and centrifuged for 15 min at 17,000 ***g*** at 4°C. Gel fractionation on denaturing or nondenaturing gels was as described (Lampl et al., 2010). The membrane was developed with β-amylase antibodies (1:1000) and secondary anti-rabbit horseradish peroxidase (1:3000) or with serpin Z4 antibodies (1:1000), and secondary anti-rat horseradish peroxidase (1:3000). For purified proteins, the proteins were first dialyzed in PBS and then mixed with reducing or oxidizing buffers, left at room temperature for 10 min, and then fractionated onto 10% SDS-PAGE. Purified serpin Z4 was used to generate antibodies from rat serum. Antibodies against β-amylase were purchased from antibody-online (cat # ABIN285322).

2.3 | Cloning and transient expression in *Nicotiana benthaniama*

Serpin Z4 and β-amylase were cloned from cDNA of barley cv. Harrington. The serpin and the amylase were fused to an RFP on the N terminus and a GFP on the C-terminus, respectively, under a 35S promoter in vector pART27 and introduced into the plant transformation vector pML-BART (Pogorelko, Fursova, Ogarkova, & Tarasov, 2007). The primers used for the cloning are shown in Table S1. The plasmid was expressed in Agrobacterium GV301 and infiltrated into 1-month-old leaves of *N. benthamiana*. All images were taken with a model no. A1 Confocal Microscope (Nikon, Melville, NY) GFP emission/excitation 525/488 and RFP emission/excitation 595/561.

2.4 | Protein expression and purification

A construct encoding β-amylase (WT and mutants) or serpin Z4 isolated from barley cv. Harrington was cloned into the pET28-(a) vector (Novagen) with a C-terminal hexahistidine tag. The primers used for the cloning are shown in Table S1. Expression was induced in *Escherichia coli* SHuffle cells (Lobstein et al., 2012) grown to an OD_{600} of 0.6–0.7 at 28°C then supplied with 1 mM isopropyl-β-d-thiogalactoside and cultured for 16 h at 16°C. Cells were resuspended in lysis buffer (0.2 M NaH_2PO_4, 500 mM NaCl, 10 mM imidazole, 1 mM DTT pH 7.4) and lysed by sonication. Cleared lysate was purified by Ni-NTA affinity chromatography. The eluate was concentrated and purified further by gel filtration (Superdex 200 Increase 10/300GL; GE Healthcare) in PBS buffer supplied with 2 mM DTT. The purified proteins were aliquoted and stored at −80 in 5% glycerol. Wild-type β-amylase was analyzed offline on a Q-Trap 4000 (Sciex) operated in the positive ion mode. All samples were diluted into 50% methanol/5% formic acid and loaded into Nano spray borosilicate emitters. Typical instrument settings for Q1 MS were applied.

2.5 | Chymotrypsin inhibition assay

Peptidase activity of 70 ng of chymotrypsin (Sigma C4129) was tested on casein substrate with serpin BsZx or Z4 at a molar ratio of 2:1, serpin: protease. Proteolytic activity was measured using Pierce™ Fluorescent Protease Assay Kit (cat# 23266).

2.6 | Biacore and Microscale thermophoresis assays

Purified proteins (serpin Z4 and β-amylase) were first immobilized via amine coupling to the free carboxyl groups on the CM5 chip surface following standard BIAcore protocols. Next, the reciprocal protein (either serpin Z4 or β-amylase) was injected at various concentrations to generate kinetics sensorgrams. The experiments were carried out in PBS buffer on a Biacore T200 instrument (GE Healthcare Life Sciences).

Microscale thermophoresis analysis was performed on a NanoTemper Monolith NT.115 instrument (NanoTemper Technologies). β-amylase and serpin Z4 were fluorescently labeled with a NHS-Ester Fluorescent Dye. Labeled β-amylase (70 nM) was mixed with varying concentration of unlabeled serpin Z4 (from 40 μM to 1 nM) and 70 nM of labeled serpin Z4 with 16 μM to 0.5 nM of unlabeled β-amylase. Each solution was transferred into a standard glass capillary soon after mixing and immediately analyzed using settings of 20% power and 50% excitation power. For each set of binding experiments, three independent MST measurements were carried out. Data analysis was performed by the Nanotemper analysis software.

2.7 | β-amylase activity assays

Purified β-amylase was first dialyzed in PBS to remove DTT using centrifugal filter (Amicon ultra, millipore). β-amylase (2 or 2.5 pmol) in 10 μl PBS was mixed with 20 μl of 2% starch in 25 mM NaAc pH 4.8 and incubated for 10 min at 25°C. The reaction was terminated by the addition of 30 μl of 3,5-dinitrosalicylic acid (DNS) solution (Miller, 1959) then boiled for 5 min. After cooling, 250 μl of distilled water was added, and the absorbance was measured at 540 nm (microplate reader TECAN Infinite 200) with maltose as the standard. Amylase activity is expressed in nmol maltose per minutes. For additives, 2 or 2.5 pmol of β-amylase in 5 μl was mixed with 5 μl of 2× solution containing either serpin Z4, DTT, or $CuCl_2$ and left at room temperature for 10 min. Results were analyzed by linear regression performed by GraphPad Prism software. The specific measurement of β-amylase activity in grains was measured using a megazyme® β-amylase assay kit (Betamyl-3; Megazyme, Australia) following the manufacturer's instructions.

2.8 | Thermal shift assay

β-amylase (5–0.5 μM) in 2.5 μl was mixed with 2.5 μl of a 1:5 molar ratio of serpin Z4. PBS buffer (10 μl) and 5 μl of 4× sypro-orange solution (S5692; Sigma) were added to the protein and inserted into a real-time polymerase chain reaction device (StepOne™ Real-time PCR System) using 96-well plate in four replicates. The settings were as follow: Starting temperature was 25°C, and the fluorescence data were collected every 2°C until 99°C. The results were analyzed using protein thermal shift software provided by Thermo Fisher Scientific.

3 | RESULTS

3.1 | Serpin and β-amylase coexpress during development and colocalize to the cytoplasm

Specific antibodies were used to concurrently analyze by immunoblot the accumulation of serpin and β-amylase during grain development and germination in barley cv. Harrington. In developing grain, under fractionation in reducing conditions (1,4-dithiothreitol, DTT), polypeptides of approximately 40 and 60 kDa for serpin Z4 and β-amylase (Figure 1a and b lower blots) were detected, respectively. A gradual increase in both proteins was observed during the first 30 days followed by a plateau phase lasting until the grain was dry. In nonreducing conditions, in which pre-existing disulfide bonds are maintained, the monomeric forms of both serpin Z4 and β-amylase (Figure 1a and b upper blots) accumulated from 8 DAP to 28 DAP. They decreased from 28 DAP to the stage of "dry grains" while higher molecular weight forms of these polypeptides accumulated (100–200 kDa). Enzymatic activity of β-amylase was determined in the free (water soluble) and total (thiol extractable) fraction of the developing grains and is represented by the ratio of free β-amylase activity over total β-amylase activity. A decrease in the free/total β-amylase activity was recorded during grain development that parallels the accumulation of bound β-amylase and the appearance of higher molecular weight complexes.

In a similar manner, the content of serpin and β-amylase was investigated during germination. Under reducing conditions (Figure 1d and e; lower blots), the full-length protein and a processed form are observed. The processed size of serpin Z4 corresponds to the size of a spent serpin that has interacted with an unknown protease at the RCL (Gettins, 2002). The size of the processed β-amylase (55 kDa) coincides with the size of β-amylase after partial proteolysis of the C-terminal portion. Indeed, proteases released from the aleurone layer are known to target the β-amylase C-terminus (Guerin et al., 1992; Lundgard & Svensson, 1986). Fractionation of proteins without reducing agent show that the 200 kDa size migrating polypeptides disappeared between 5 and 24 HAI (Figure 1d and e; upper blots). The disappearance of high molecular weight proteins is consistent with reports of increasing accumulation of thioredoxin reductive activity (Hagglund et al., 2010; Lundgard & Svensson, 1986). In a reciprocal manner to what was observed in developing seeds, the ratio of free over total enzymatic activity of β-amylase increased in parallel with the disappearance of higher molecular weight forms (Figure 1f).

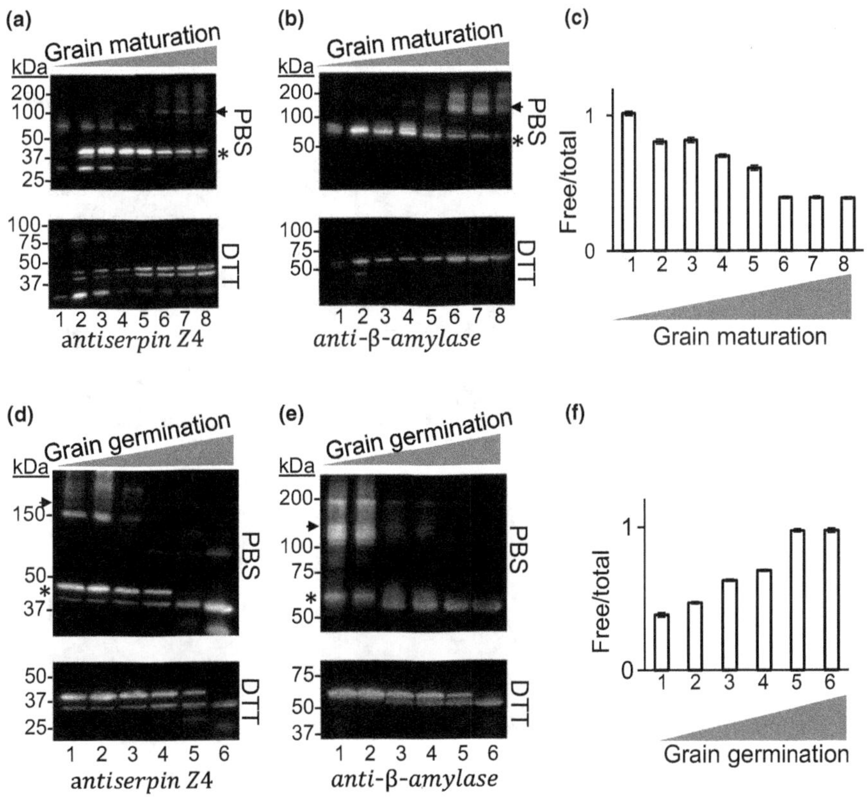

FIGURE 1 Accumulation of β-amylase and serpin Z4 during grain maturation and germination. Each Western blot was replicated at least three times using independent biological replicates. (a–c) Proteins were extracted from developing grains of barley cv. Harington with either PBS (upper blots) or PBS with 20 mM DTT (lower blots). In a and b, the proteins were fractionated on a nonreducing SDS-PAGE gel transferred to a PVDF membrane and blotted against serpin Z4 in a or against β-amylase in b. Lanes are days after pollination (DAP) Lanes 1–8; 8, 13, 18, 22, 28, 32, 39, dry grains. (c) β-amylase activity was assayed on the same samples using the Betamyl-3 method, shown is the ratio of free β-amylase activity over total β-amylase activity. (d–f) Proteins were extracted from 100 mg of germinating grains of barley cv. Harington with 1 ml of either PBS (upper blots) or PBS with 20 mM DTT (lower blots). Lanes are hour after imbibition (HAI) from left to right 1–6; 0, 5, 24, 30, 48, and 120 HAI. (d and e) Proteins were loaded on a nonreducing SDS-PAGE gel transferred to a PVDF membrane and blotted against serpin Z4 in d and against β-amylase in e. f, β-amylase activity was assayed on the same grain samples as carried out in (c). The sizes of the proteins are indicated in kDaltons (kDa). The asterisk indicates the monomer size of serpin and β-amylase, and the arrow indicates high molecular weight aggregates of the proteins

The results underscore correlation between the accumulation patterns of serpin Z4 and β-amylase during seed development and germination. Therefore, it was of interest to examine their cellular localization as well. Plant serpins are known to reside either in the cytoplasm or are secreted to extracellular spaces (Fluhr et al., 2012), while β-amylase can localize to the chloroplast (amyloplast) or cytoplasm in Arabidopsis leaves (Mita, Suzuki-Fujii, & Nakamura, 1995; Sparla, Costa, Schiavo, Pupillo, & Trost, 2006). Agrobacterium harboring constructs with serpin Z4 fused to RFP and β-amylase fused to GFP were coinfiltrated into *N. benthamiana* plants. To verify the reliability of this assay, immunoblot analysis showed that the expressed protein sizes were commiserated with their expected fused size (Fig. S1). As shown in Figure 2, the confocal images of both serpin Z4 and β-amylase colocalized to the cytoplasm surrounding the chloroplasts of spongy mesophyll (Figure 2a). They are also expressed in the cytoplasmic peripheral regions of leaf epidermal cells (Figure 2b).

3.2 | Serpin Z4 and ß-amylase exhibit quantifiable protein–protein interactions

To examine for possible interaction between serpin Z4 and β-amylase, their recombinant proteins were first purified by affinity chromatography and size exclusion chromatography. The gel image of the proteins is shown in Fig. S2. The proteins were further tested for conventional activity. The activity of serpin as protease inhibitor was established by calculating its inhibition factor, X_{inh}. This factor measures the fraction of complex forming serpin in equilibrium with the protease as determined by the residual peptidase activity. As shown in Fig. S3a, recombinant serpin Z4 inhibits chymotrypsin with a X_{inh} of 0.35 while a different barley recombinant serpin, BsZx inhibited the protease with a ratio of 0.88. The results are consistent with that measured previously for isolated barley serpins (Dahl et al., 1996). Recombinant amylase activity was measured using two independent assays; the DNSA reducing sugar method (Miller, 1959) and the Betamyl-3 method as described in the Material and Methods. Figure S3c and d shows that recombinant β-amylase activity in both methods was linear from 1 to 8 pmol of enzyme, similar to what was reported previously (Lundgard & Svensson, 1987).

Serpin Z4 and β-amylase activities were reported to comigrate during gel filtration of barley seed extracts (Hejgaard & Carlsen, 1977). This could be a result of their interaction or stem from similar molecular weights due to self-oligomerization. To examine this directly, we employed microscale thermophoresis as shown in Figure 3. Microscale thermophoresis measures the motion of molecules along a microscopic temperature gradient (Jerabek-Willemsen,

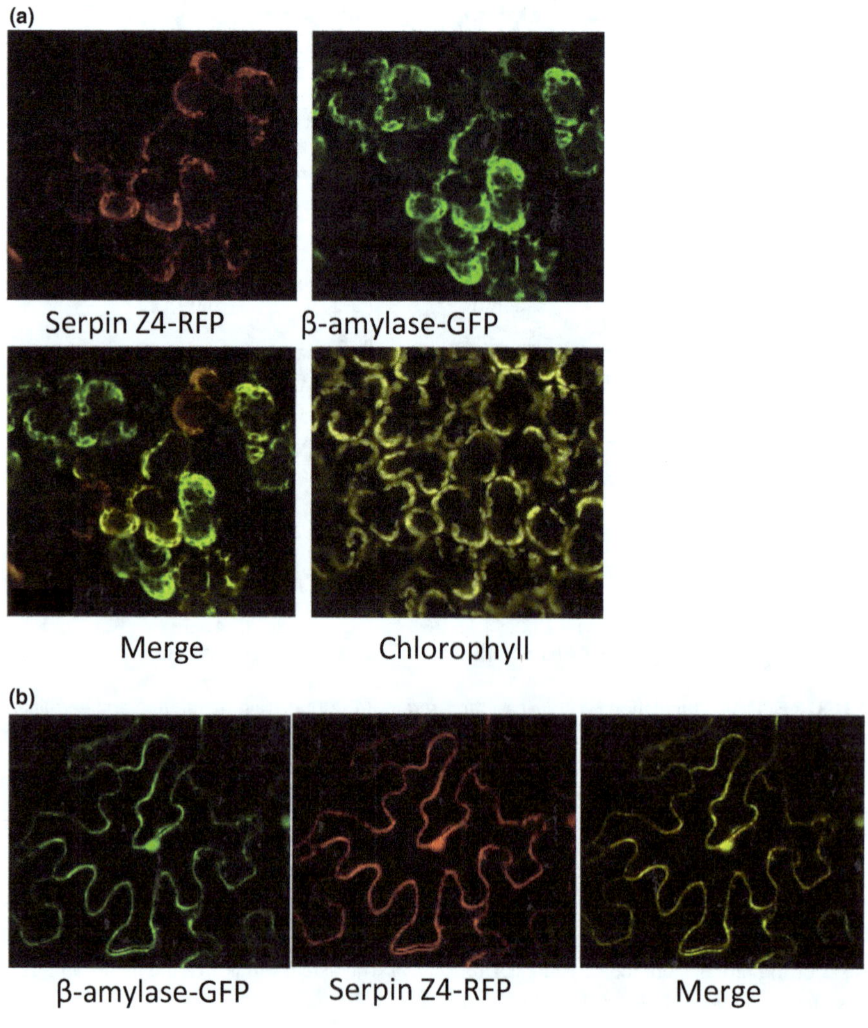

FIGURE 2 Transient expression of serpin Z4-RFP and β-amylase-GFP in leaves of *Nicotiana benthaniama*. Confocal microscope observation of a, spongy mesophyll cells and b, epidermal cells

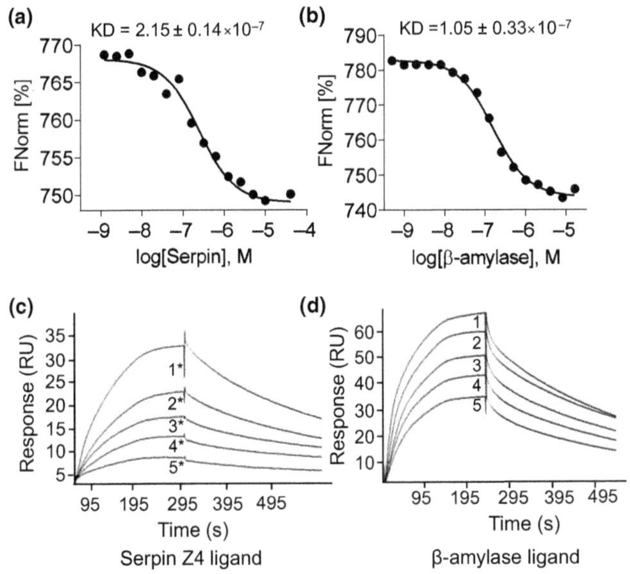

(e)

Ligand	KD	Ka (1/Ms)	Kdiss (1/s)
Z4	3.16×10^{-7}	0.77×10^{4}	2.43×10^{-3}
β-amylase	1.42×10^{-7}	1.86×10^{4}	2.63×10^{-3}

FIGURE 3 Interaction between serpin Z4 and β-amylase established by microscale thermophoresis and by surface plasmon resonance. (a) β-amylase was bound to a fluorescent probe, and increasing amounts of serpin Z4 were added. (b) Fluorescent serpin Z4 was probed with increasing β-amylase as in a. The assays were conducted in PBS buffer with 0.1% Tween-20. For estimation of KD, the results of three replicate experiments were combined. (c) Serpin Z4 was attached to a Biacore compatible sensor chip, and β-amylase was injected at the following concentrations: 2 μM, 0.9 μM, 0.4 μM, 0.2 μM, and 0. 1 μM (1* to 5*, respectively). (d) β-amylase was attached to the sensor chip and serpin Z4 injected at the following concentrations: 0.6 μM, 0.5 μM, 0.4 μM, 0.3 μM, and 0.2 μM (1–5, respectively). (e) Equilibrium constants of the interaction measured by surface plasmon resonance

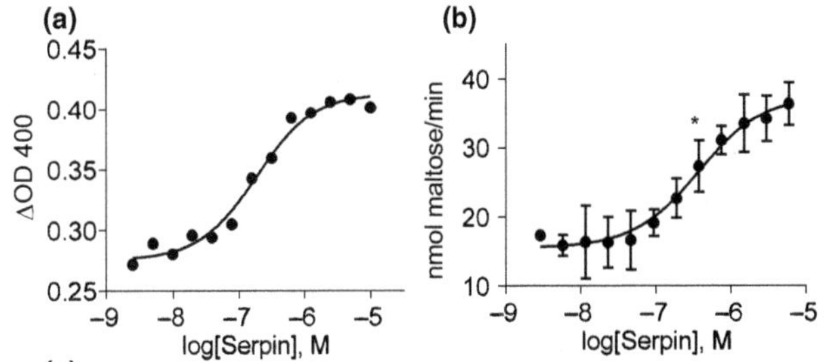

(c)

	V_{max} nmol maltose/min	K_m (mg starch/ml)
β-amylase	12.8 ± 0.98	0.008
β-amylase+ Z4*	28.3 ± 3.32	0.012
β-amylase + DTT 5 mM	16.6 ± 1.4	0.004
β-amylase+ Z4 +DTT 5 mM**	33.1 ± 2.8	0.009

FIGURE 4 Activity of β-amylase in the presence of serpin Z4. (a) The activity of 2 pmol of β-amylase was assayed using the Betamyl substrate from Megazyme in the presence of different concentrations of serpin. The graphs are representative of three independent experiments. Graph (a) is presented as change in O.D. (b) As in (a) using the reducing sugar DNSA method. The enzymatic activity of β-amylase in the presence of serpin beginning at the asterisk is significantly different from the no serpin control with a p-value < .03. (c) Kinetic properties of β-amylase with serpin Z4 and DTT. The maximal velocity (V_{max}) and the substrate affinity (Km) are derived from the Lineweaver–Burk model. V_{max} is expressed in nmol maltose per minute, and Km is in mg/ml of starch. * The V_{max} of β-amylase + serpin Z4 is significantly different from the Vmax of the β-amylase control with a p-value = .001. ** The V_{max} of β-amylase+ Z4 + DTT 5 mM is significantly different from the Vmax of the β-amylase + DTT 5 mM control with a p-value = .001. The kinetic properties of β-amylase were assayed in three independent experiments each one comprising three technical replicates

Wienken, Braun, Baaske, & Duhr, 2011). A fluorescent label was attached covalently, to one of the proteins to be tested and added to a serial dilution of the unlabeled cognate protein (X axis in log [M]; see Materials and Methods; Figure 3). The change in motion is presented as normalized fluorescence intensity (Y axis). Typical MST traces and capillary shapes are shown in Fig. S4. The KD of this interaction was found to be $2 \cdot 10^{-7}$ M or $1 \cdot 10^{-7}$ M depending on the direction of the reciprocal labeling (Figure 3a and b). The results quantify a physical interaction between serpin and β-amylase in native nonreducing conditions with an affinity of approximately 10^{-7} M.

The interaction of β-amylase and serpin Z4 was further measured by surface plasmon resonance using a BIAcore sensor chip (see Materials and Methods; Figure 3c and d). When serpin was immobilized on the surface and β-amylase was the analyte the recorded sensorgram yielded a binding affinity constant (KD) of $3.16 \cdot 10^{-7}$ M. Likewise, when the proteins were interchanged and β-amylase was immobilized the affinity measured was $1.42 \cdot 10^{-7}$ M. As shown in Figure 3e, the KD is the result of the ratio of the two rate constants, where Ka and Kdiss were ~10^4 and 10^{-3}, respectively. The relatively low Ka indicates a slow association rate and the relatively high Kdiss, a fast dissociation process. Thus, two diverse biophysical methods of measurement show that β-amylase and serpin Z4 display a significant measurable interaction with a binding affinity constant greater than 10^{-7} M. Although the motivation for this experiment relied on the observation that serpin Z4 and β-amylase comigrate during gel filtration, the consensus KD of 10^{-7} M measured here would be too low to form stable complexes during gel filtration. Hence, it is possible that previous observation of comigration emanated from the formation of homomultimers yielding similar filtration qualities (see below).

3.3 | Serpin interaction with β-amylase impacts on its enzymatic properties

To investigate possible ramifications of the interaction between β-amylase and serpin Z4, a dose–response activity curve for β-amylase in the presence of increased serpin concentration was performed. Figure 4a shows the effect of serpin on β-amylase enzymatic activity assayed with the Betamyl-3 method. As the amount of serpin increases (abscissa axis), β-amylase activity in the presence of excess substrate is significantly enhanced. The Betamyl-3 method uses short-chain substrate (p-nitrophenyl-β-D-maltotrioside) that requires auxiliary β-glucosidase activity. Hence, we also examined the effect of serpin on β-amylase using high MW starch substrate coupled to a direct assay of released reducing sugars. Figure 4b confirms the results using the DNSA technique. The concentration of serpin that was required for half of the maximal effect (EC50) is similar in both assays and was $1.8 \cdot 10^{-7}$ M and $3 \cdot 10^{-7}$ M for the Betamyl-3 and DNSA methods, respectively. The EC50 value was obtained at a molar ratio of 1 β-amylase to 1 serpin. The maximal increase in activity was obtained at a molar ratio of 1 β-amylase to 10 serpins (Figure 4). Similar experiments carried out with bovine serum albumin (BSA) did not yield any change in β-amylase activity indicating specificity (Fig. S5).

Notably, inspection of the dose–response of β-amylase activation upon addition of serpin yields a curve that is consistent with the KD values obtained for the direct protein–protein interactions (Figure 3). To further dissect the effect of serpin on the kinetic characteristics of β-amylase, Vmax and Km were estimated. As shown in Figure 4c, the addition of serpin at a 1:5 molar ratio led to a twofold increase in the Vmax independent of the presence of the reducing agent, DTT. Serpin Z4 had a slight negative effect on the affinity of β-amylase to its substrate as indicated by the increase in Km.

Temperature is a critical environmental parameter (e.g., during beer mashing) that affects the enzymatic activity of β-amylase (Evans, van Wegen, Ma, & Eglinton, 2003). β-amylase activity was assessed over a temperature gradient ranging from 37°C to 62°C, supplemented or not, with serpin Z4 (ratio of 1:5, β-amylase: Z4; Figure 5a). Over a critical range of 37–48°C, the presence of serpin significantly improved β-amylase activity. At 50°C, the activity of β-amylase alone dropped by 40%, but with the addition of serpin Z4, the enzyme activity was enhanced by 60% compared with β-amylase alone at optimal temperatures. Above 53°C, β-amylase lost more than 80% of its activity with or without serpins. Thus, the presence of serpin elevates β-amylase activity over a critical range of temperatures. However, the denaturing temperature for protein stability as determined by the "Thermal Shift" assay in Figure 5b showed that the intrinsic melting temperature of β-amylase does not change in the presence of serpin Z4 and remains 55°C.

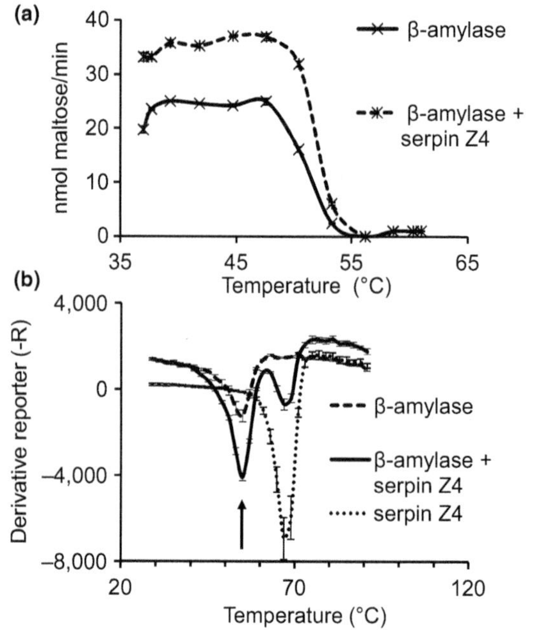

FIGURE 5 The effect of temperature on β-amylase activity. (a) β-amylase (2.5 pmol) was assayed by the DNSA method at different temperatures with or without serpin Z4 at a molar ratio of 1:5. The activity is presented in nmol of maltose released per minute of reaction. From 37°C to 53.3°C, the enzymatic activity of β-amylase in the presence of serpin is significantly different from the control with a p-value $\leq .03$. The experiment was repeated three times with three technical replicates. (b) The thermal stability of β-amylase with and without serpin Z4 was measured by thermal shift assay using sypro-orange dye. The arrow indicates the melting temperature of β-amylase. The graph is representative of three independent replicates

3.4 | Impact of serpin on the redox-related activity and polymerization state of β-amylase

Oxidative conditions were shown to play a role in β-amylase activity (Seung et al., 2013; Sparla et al., 2006). It was therefore of interest to examine whether the presence of serpin impacts on the redox reactivity of β-amylase. Oxidative conditions were simulated by the use of $CuCl_2$ that oxidizes reactive cysteines inducing the formation of cysteine–cysteine bonds (Cavallini, De Marco, Duprè, & Rotilio, 1969) (Figure 6a). Without serpins, β-amylase activity starts decreasing from 10 μM, until 50 μM at which point, less than 20% of the initial activity remains. The presence of serpins improves β-amylase stability to oxidation, as after treatment with 50 μM $CuCl_2$, β-amylase was as active as the control β-amylase without the oxidative agent. However, a gradual decrease in activity, although less steep, was observed due to $CuCl_2$ even in the presence of serpin.

Grain development and germination seem to affect the polymerization state of β-amylase (Figure 1), and these changes have been ascribed to shifts in the cellular redox milieu (Niku-Paavola, Skakoun, Nummi, & Daussant, 1973). Therefore, it was of interest to examine the impact of serpins on the polymerization status of β-amylase upon oxidation. To visualize this, recombinant proteins were combined at molar ratio of 1:0, 1:2, and 1:5 (β-amylase:serpin Z4). The mixtures were then subjected to treatment with $CuCl_2$. The proteins were fractionated on a SDS-PAGE with or without reducing agent and developed by immunoblot with anti-β-amylase or antiserpin Z4 antibody (Figure 6b–e). Oxidative treatment caused β-amylase and serpin Z4 to migrate in multiple higher molecular weight forms (Figure 6b and c). β-amylase migrated in apparent dimer and trimer size at approximately 120 and 180 kDa, respectively; however, higher molecular weight migration forms also appeared. In the case of serpin Z4, dimeric (80 kDa) and higher order forms were visible, although to a lesser extent. Interestingly, the addition of serpins increased the amount of β-amylase monomers, inhibiting the formation of high molecular weight β-amylase. Notably, all polypeptides

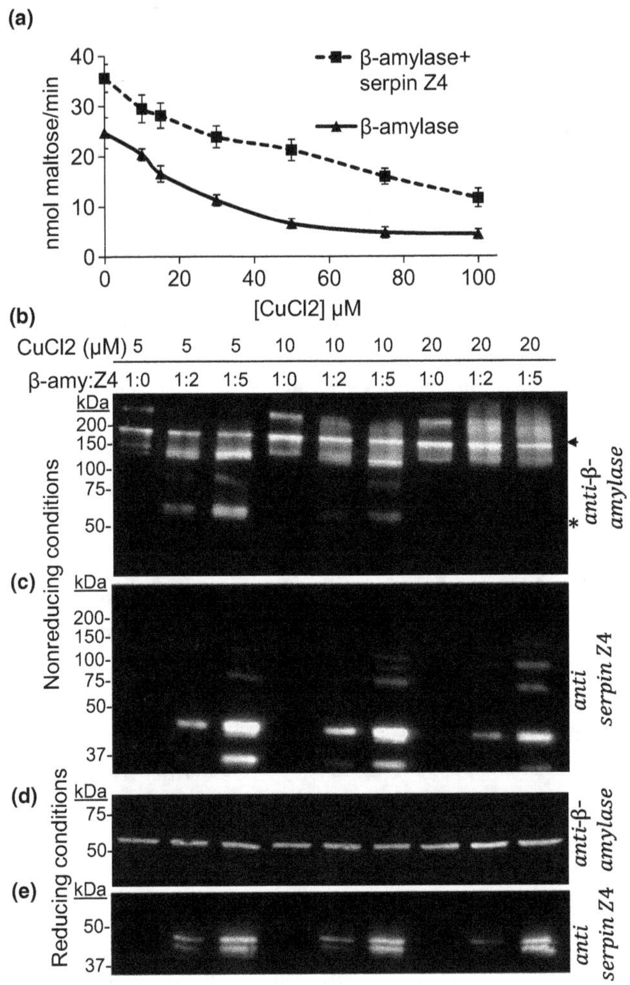

FIGURE 6 Enzymatic activity and immuno-detection of recombinant β-amylase and serpin under oxidizing conditions. (a) The activity of 2.5 pmol β-amylase was assayed using the DNSA method with or without the addition of serpin Z4 at a 1:5 ratio upon increasing $CuCl_2$ concentrations. (b–d, e) β-amylase and serpin Z4 were mixed at the indicated ratios and treated with 5, 10, and 20 μM $CuCl_2$ for 10 min. The proteins were fractionated in 10% SDS-PAGE gel with (b, c) or without the addition of reducing agent, β-mercaptoethanol (d, e) and blotted with antibody against β-amylase (b, d) or against serpin Z4 (c, e). The asterisk indicates β-amylase monomer size and the arrow polymeric forms. Western blots and enzymatic assays were replicated at least three times

migrated at their monomeric size after the addition of the reducing agent, β-mercaptoethanol (Figure 6d and e). Hence, β-amylase and serpin Z4 polymerization are due to the formation of cysteine bonds in the oxidative environment. The results show that the multiple polypeptide states of β-amylase are dictated by the oxidation level and that increasing levels of serpin Z4 reduce the formation of higher molecular weight β-amylase.

3.5 | β-amylase mutants exhibit a differential response to oxidation

To investigate the participation of specific cysteines residues in β-amylase contributing to its polymerization, two modified β-amylases were generated. The β-amylase WT used in this study originates from cv. Harrington and contains a cysteine at position 115 (Chiapparino, Donini, Reeves, Tuberosa, & O'Sullivan, 2006). It was modified by a C115R replacement. The second modification was the removal of the C-terminus of β-amylase from the glycine at position 489 (Trunc β-amylase). This removes a cysteine at position C503. These cysteines were chosen as they are known to impact on the thermostability and enzymatic properties of β-amylase (Ma, Eglinton, Evans, Logue, & Langridge, 2000; Ma et al., 2001). Moreover, C503 was found to be the target of thioredoxins released in the endosperm upon germination (Hagglund et al., 2010).

The effect of serpin Z4 was evaluated on β-amylase activity of WT, C115R, and Trunc β-amylase (Figure 7a). As shown in Figure 7a, WT and all the variants displayed similar activity enhancement with no significant statistical difference as result of the presence of serpin Z4. For example, at a 5:1 ratio (arrow; Figure 7a), the addition of serpin Z4 improves all β-amylase to the same degree. Hence, the action of serpin on β-amylase activity was unaffected by modifications of these cysteines.

To investigate the ramifications of the mutations on the polymerization patterns, 5–20 μM of $CuCl_2$ was applied to the β-amylase variants. Without $CuCl_2$, each polypeptide migrated at the expected molecular weight size of its monomeric form (Fig. S6). However, as observed in Figure 7b, while β-amylase WT rapidly polymerized due to increasing $CuCl_2$, the C115R variant remains mainly in its monomeric form even at the highest concentration of $CuCl_2$ (20 μM). Although WT β-amylase appears smaller than expected, LC/MS/MS analysis was conducted on the protein and showed the expected size value (Fig. S7). The Trunc β-amylase variant also showed polymerization patterns that were distinct from the WT. Beside the monomeric form, one additional band is spotted on the Western blot as a result of oxidation. However, its polymeric migration size as determined by nondenaturing gel fractionation is 140 kDa, a value that is larger than the predicted dimer size (117 kDa) but smaller than the expected trimer size (176 kDa). In addition, the Trunc β-amylase variant appeared to be less susceptible to oxidation than the WT as it remained as a monomer at 5 μM $CuCl_2$. When the percent of residual amylase activity in the presence of $CuCl_2$ was measured, both WT and Trunc β-amylase showed a decrease in activity but not the C115R variant (Figure 7c). Thus, the C115R β-amylase, that is less inclined to form polymers, shows less sensitivity to oxidation indicating that the decrease in enzymatic activity might be a consequence of the polymerization process. Although the three forms of β-amylase exhibited distinct and differential responses to $CuCl_2$, they all responded to serpin as the WT, with enhanced enzymatic activity as measured in Figure 7a. Hence, serpin can improve β-amylase enzymatic activity independent of its ability to modify its oligomerization state.

4 | DISCUSSION

Serpin Z4 and β-amylase are highly abundant seed proteins in many barley cultivars (Finnie et al., 2004). Similar to storage proteins, their

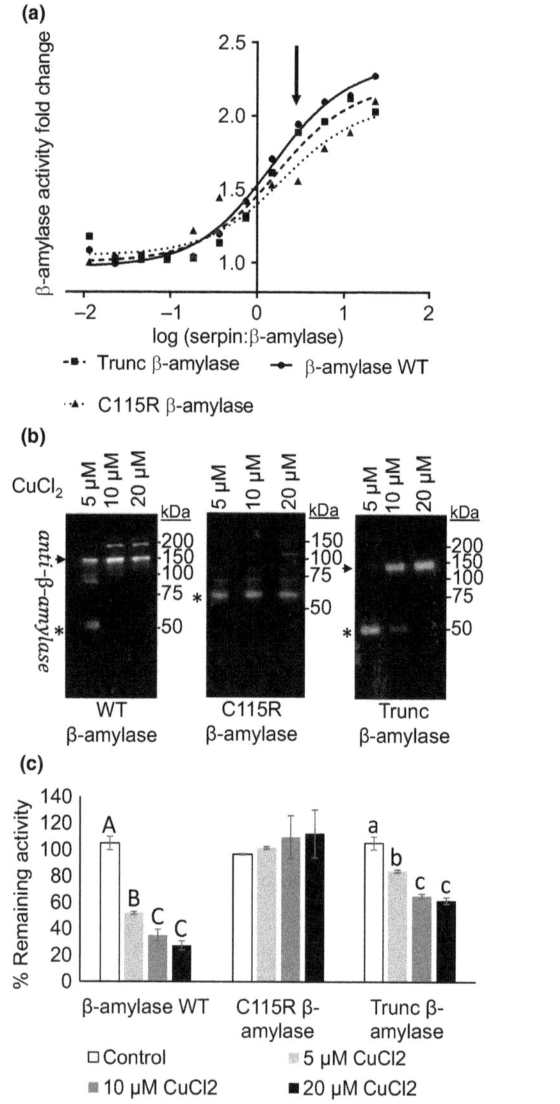

FIGURE 7 Enzymatic activity and immuno-detection of wild type and mutant forms of β-amylase in the presence of serpin Z4. (a) Activity of 2 pmol of β-amylase with increasing concentrations of serpin Z4 assayed by the DNSA method. Assays were on WT, C-terminus truncated β-amylase (Trunc β-amylase) and the C115R amino acid replacement. The X axis represents the ratio of serpin Z4 to the β-amylases. The Y axis expressed the fold change in the β-amylase activity with serpin over the activity of the β-amylase without serpin. Serpin Z4: β-amylase ratios range between 20:1 and 0.04:1. The arrow indicates a ratio of 1:5. (b) Immunoblot analysis of recombinant β-amylase wild type and mutants as in a. Proteins (0.6 μg) were treated with 5, 10, and 20 μM of $CuCl_2$, for 10 min and fractionated on a 10% SDS nonreducing PAGE gel and immunobloted with β-amylase antibody. (c) The residual activity of β-amylase compared to no treatment after addition of $CuCl_2$ as in b assessed by the DNSA method. Capital letters indicate significant differences in β-amylase WT-treated enzyme, lower cases letters, significant differences in Trunc β-amylase-treated enzyme, no statistical differences were found in the C115R β-amylase-treated enzyme. The results were analyzed by performing Tukey-Kramer HSD tests. The asterisk indicates the monomeric forms of the three β-amylases and the arrow, the polymers of β-amylase WT and Trunc β-amylase. Western blots and enzymatic assays were replicated at least three times

content increases in the endosperm of developing grains (Sreenivasulu et al., 2006) and accumulate as matrix-bound proteins. Here, we show that both proteins display comparable accumulation profiles during seed maturation and germination and colocalize to the cytosol in a heterologous expression system (Figures 1 and 2). As the grain ripens and cell death proceeds in the endosperm (Young & Gallie, 2000), cytosolic proteins agglutinate to coat the amyloplast-localized starch granules (Borén et al., 2004; Wang et al., 2011). Consistent with our results, this coating would include serpin Z4 and β-amylase. During the later stages of grain filling and the early stages of germination, both β-amylase and serpin Z4 are detected mostly as a bound nonfree fraction or as high molecular weight protein in the free fraction (Figure 1). As they accumulate, the polymerization profiles of these two barley proteins are consistent with the creation of disulfides bridges to form homopolymers as has been reported for proteins in ripening wheat grains (De Gara, de Pinto, Moliterni, & D'Egidio, 2003). Significantly, serpin Z4 and β-amylase were detected among the multiple putative targets of thioredoxins, that is, reducing agents released from the aleurone layer during germination (Hagglund et al., 2010). Thus, serpin Z4 and β-amylase share common spatio-temporal localization characteristics that would promote interaction during grain development and germination.

Using different techniques, serpin Z4 was shown here to interact with β-amylase with a binding constant (KD) of approximately 10^{-7} M (Figure 3). This interaction is in the range measured for a variety of chaperones (Schweiger, Soll, Jung, Heermann, & Schwenkert, 2013). Importantly, in the presence of serpin Z4, the enzymatic activity of β-amylase is improved significantly. The improvement was specific as the addition of BSA had no effect (Fig. S3). Thus, β-amylase activity exhibits a sigmoidal curve of activity upon the addition of serpin (Figure 4); a curve that typifies agonistic interactions between proteins (DeLean, Munson, & Rodbard, 1978). A ratio of 1 β-amylase to 5 serpins increased the maximal velocity by twofold and protected β-amylase activity from heat or oxidation stress (Figures 5 and 6). It was previously reported that a serpin Z4-deficient barley line showed more diastatic power, that is, was more efficient in converting starch (Iimure, Kimura, et al., 2012). However, as diastatic power is a gross measurement taking into consideration multiple enzymatic activities including β-amylase, α-amylases and limit dextrinase activity, it may be that, in that case, other qualities compensated for the reduced level of serpin.

Under oxidizing conditions, β-amylase WT rapidly polymerized and lost significant activity (Figure 7). The kinetics of polymerization shows that while the C115R mutation stabilized β-amylase in its monomeric form, the Trunc β-amylase mutant still polymerized but in a manner distinct from the WT. A number of β-amylase variants exist in barley. The SD1 and SD2 β-amylase isotypes are commonly found in malting-type barley (Allison & Swanston, 1974; Evans, Wallace, et al., 1997). The substitution of cysteine at position 115 by an arginine results in the conversion of SD1 to the SD2 type (Ma et al., 2002). SD1 is more thermostable than SD2 but has lower affinity to its substrate (Ma et al., 2001). Moreover, a SD1 type β-amylase from cv. Pirkka was shown to polymerize spontaneously (Niku-Paavola et al., 1973) in a manner similar to the WT described in this work (also a SD1 type). The C115R mutation transforms the SD1 β-amylase into a SD2 type and as shown here prevents the formation of high molecular weight moieties (Figure 7) identifying a cysteine critical for higher order polymerization. In this scenario, C115 could serve as a hub for cysteine bonds formation. Engineered removal of the C-terminus eliminates C503 yielding a more stable β-amylase (Figure 7). This is consistent with the finding that reduction of the C503 by thioredoxins (Hagglund et al., 2010) or its removal by proteases (Guerin et al., 1992; Lundgard & Svensson, 1987) efficiently solubilizes the bound β-amylase. The resulting processed β-amylase was found to be more thermostable but otherwise retained the same kinetic parameters (Ma, Eglinton, et al., 2000). The decrease in β-amylase activity due to oxidation reported here differs from that reported for the redox-sensitive Arabidopsis, trBAMY. The Arabidopsis trBAMY, a thioredoxin regulated β-amylase, does not form high molecular weight complexes, suggesting that in that case intramolecular disulfide bonds account for the control of enzymatic activity (Sparla et al., 2006).

Under oxidizing conditions, serpin Z4 interaction impacts on the enzymatic activity of WT β-amylase in two ways. Firstly, the interaction between β-amylase and serpin Z4 represses the formation of multimeric β-amylase (Figure 6b and c). Presumably such polymers are less active (Figure 7c). Secondly, its intrinsic ability to activate β-amylase is evidenced in the mutant C115 that lacks the ability to polymerize but is nevertheless enhanced by serpin. How serpin and β-amylase interactions interfere with polymerization is not known. Based on β-amylase structure (PDB ID: 2XFR), C115 is located on the outer surface far from the active site. One can speculate that in the topography of its interaction, serpin Z4 specifically interferes with the ability of C115 to form cysteine bonds perhaps by physical contact in that area.

The function for serpin Z4 in grains of barley as a chaperone-like stabilizer of β-amylase that enhances enzyme activity and favors monomeric forms differs from classical chaperones that operate in protein folding pathways. Yet, in mammals, the serpin family was found to participate in a wide range of activity beside protease inhibition (Gettins, 2002). For example, Hsp47 is a serpin whose function is to prevent the aggregation of protocollagen, a collagen precursor particularly abundant in the ER (Ito & Nagata, 2017; Natsume, Koide, S-i, Hirayoshi, & Nagata, 1994). Hsp47 stabilizes their cognate protein partners by binding to the triple helix of type I collagen to prevent further aggregation (Koide et al., 2006; Widmer et al., 2012).

Despite the importance of starch breakdown for providing a source of energy, β-amylase and serpin Z4 are not absolutely required for germination of barley seeds, and barley cultivars with variable amounts of both proteins can be found in nature (Iimure, Kimura, et al., 2012; Kihara, Kaneko, Ito, Aida, & Takeda, 1999). However, barley strains that are dedicated to the production of beer require functional α-amylase and β-amylase to be present in the mashing step for the release of maltose units favored by fermenting yeast. Serpin Z4 was also found to contribute to beer quality by improving foam stability along with other beer proteins such as LTPs (Evans & Hejgaard, 1999). Hence, selection pressure by thousands of years of human use has likely contributed to shaping the attributes of today's barley varieties.

This work shows that serpin Z4 interacts with β-amylase and optimizes its activity. In the presence of serpin Z4, the maximal catalytic velocity of β-amylase is doubled, and β-amylase was found to stabilized by serpin Z4 during heat and oxidative treatments. Thus, serpin Z4 shows novel noninhibitory attributes by displaying chaperone-like activity toward β-amylase.

ACKNOWLEDGMENTS

The authors thank Dr Irina Shin for her help in the microscale thermophoresis experiments, Dr. Aharon Rabinkov for his expertise in the Surface Plasmon Resonance technology, and Dr Olga Davidov for her support in general laboratory work. R.F. is grateful to the US-Israel Binational Agricultural Research and Development Fund for Research Project IS-4915-16.

AUTHOR CONTRIBUTION

M.C. and R.F. conceived the experiments. M.C. carried them out. M.C. and R.F. wrote the manuscript.

REFERENCES

Allison, M. J., & Swanston, J. S. (1974). Relationships between β-amylase polymorphisms in developing, mature and germinating grains of barley. *Journal of the Institute of Brewing*, *80*, 285–291. https://doi.org/10.1002/j.2050-0416.1974.tb03618.x

Bønsager, B. C., Finnie, C., Roepstorff, P., & Svensson, B. (2007). Spatio-temporal changes in germination and radical elongation of barley seeds tracked by proteome analysis of dissected embryo, aleurone layer, and endosperm tissues. *Proteomics*, *7*, 4528–4540. https://doi.org/10.1002/(ISSN)1615-9861

Borén, M., Larsson, H., Falk, A., & Jansson, C. (2004). The barley starch granule proteome internalized granule polypeptides of the mature endosperm. *Plant Science*, *166*, 617–626. https://doi.org/10.1016/j.plantsci.2003.10.028

Cavallini, D., De Marco, C., Duprè, S., & Rotilio, G. (1969). The copper catalyzed oxidation of cysteine to cystine. *Arch Biochem Biophys.*, *130*, 354–361. https://doi.org/10.1016/0003-9861(69)90044-7

Chiapparino, E., Donini, P., Reeves, J., Tuberosa, R., & O'Sullivan, D. M. (2006). Distribution of β-amylase I haplotypes among European cultivated barleys. *Molecular Breeding*, *18*, 341–354. https://doi.org/10.1007/s11032-006-9035-0

Dahl, S. W., Rasmussen, S. K., & Hejgaard, J. (1996). Heterologous expression of three plant serpins with distinct inhibitory specificities. *Journal of Biological Chemistry*, *271*, 25083–25088. https://doi.org/10.1074/jbc.271.41.25083

De Gara, L., de Pinto, M. C., Moliterni, V. M. C., & D'Egidio, M. G. (2003). Redox regulation and storage processes during maturation in kernels of *Triticum durum*. *Journal of Experimental Botany*, *54*, 249–258. https://doi.org/10.1093/jxb/erg021

DeLean, A., Munson, P. J., & Rodbard, D. (1978). Simultaneous analysis of families of sigmoidal curves: Application to bioassay, radioligand assay, and physiological dose-response curves. *American Journal of Physiology*, *235*, 97–102.

Evans, D. E., & Hejgaard, J. (1999). The impact of malt derived proteins on beer foam quality. Part I. The Effect of germination and kilning on the level of protein Z4, Protein Z7 and LTP1. *Journal of the Institute of Brewing*, *105*, 159–170. https://doi.org/10.1002/(ISSN)2050-0416

Evans, D. E., MacLeod, L. C., Eglinton, J. K., Gibson, C. E., Zhang, X., Wallace, W., ... Lance, R. C. M. (1997). Measurement of beta-amylase in malting Barley (*Hordeum vulgare* L.). I. Development of a quantitative ELISA for beta-amylase. *Journal of Cereal Science*, *26*, 229–239. https://doi.org/10.1006/jcrs.1997.0119

Evans, E., van Wegen, B., Ma, Y., & Eglinton, J. (2003). The impact of the thermostability of alpha-amylase, beta-amylase, and limit dextrinase on potential wort fermentability. *Journal-American Society of Brewing Chemists*, *61*, 210–218.

Evans, D. E., Wallace, W., Lance, R. C. M., & MacLeod, L. C. (1997). Measurement of beta-amylase in malting barley (*Hordeum vulgare* L.). II. The effect of germination and kilning. *Journal of Cereal Science*, *26*, 241–250. https://doi.org/10.1006/jcrs.1997.0120

Finnie, C., Maeda, K., Østergaard, O., Bak-Jensen, K. S., Larsen, J., & Svensson, B. (2004). Aspects of the barley seed proteome during development and germination. *Biochemical Society Transactions*, *32*, 517–519. https://doi.org/10.1042/bst0320517

Finnie, C., Melchior, S., Roepstorff, P., & Svensson, B. (2002). Proteome analysis of grain filling and seed maturation in barley. *Plant Physiology*, *129*, 1308–1319. https://doi.org/10.1104/pp.003681

Finnie, C., & Svensson, B. (2003). Feasibility study of a tissue-specific approach to barley proteome analysis: Aleurone layer, endosperm, embryo and single seeds. *Journal of Cereal Science*, *38*, 217–227. https://doi.org/10.1016/S0733-5210(03)00033-X

Fluhr, R., Lampl, N., & Roberts, T. H. (2012). Serpin protease inhibitors in plant biology. *Physiologia Plantarum*, *145*, 95–102. https://doi.org/10.1111/j.1399-3054.2011.01540.x

Gettins, P. G. W. (2002). Serpin structure, mechanism, and function. *Chemical Reviews*, *102*, 4751–4804. https://doi.org/10.1021/cr010170+

Guerin, J. R., Lance, R. C. M., & Wallace, W. (1992). Release and activation of barley beta-amylase by malt endopeptidases. *Journal of Cereal Science*, *15*, 5–14. https://doi.org/10.1016/S0733-5210(09)80052-0

Hagglund, P., Bunkenborg, J., Yang, F., Harder, L. M., Finnie, C., & Svensson, B. (2010). Identification of thioredoxin target disulfides in proteins released from barley aleurone layers. *Journal of Proteomics*, *73*, 1133–1136. https://doi.org/10.1016/j.jprot.2010.01.007

Hara-Nishimura, I., Nishimura, M., & Daussant, J. (1986). Conversion of free beta amylase to bound beta amylase on starch granules in the barley endosperm during desiccation phase of seed development. *Protoplasma*, *134*, 149–153. https://doi.org/10.1007/BF01275713

Hejgaard, J. (2001). Inhibitory serpins from rye grain with glutamine as P1 and P2 residues in the reactive center. *FEBS Letters*, *488*, 149–153. https://doi.org/10.1016/S0014-5793(00)02425-X

Hejgaard, J., & Carlsen, S. (1977). Immunoelectrophoretic identification of a heterodimer β-amylase in extracts of barley grain. *Journal of the Science of Food and Agriculture*, *28*, 900–904. https://doi.org/10.1002/(ISSN)1097-0010

Hejgaard, J., & Hauge, S. (2002). Serpins of oat (*Avena sativa*) grain with distinct reactive centres and inhibitory specificity. *Physiologia Plantarum*, *116*, 155–163. https://doi.org/10.1034/j.1399-3054.2002.1160204.x

Huntington, J. A., Read, R. J., & Carrell, R. W. (2000). Structure of a serpin-protease complex shows inhibition by deformation. *Nature*, *407*, 923–926. https://doi.org/10.1038/35038119

Iimure, T., Kimura, T., Araki, S., Kihara, M., Sato, M., Yamada, S., ... Sato, K. (2012). Mutation analysis of barley malt protein Z4 and protein Z7 on beer foam stability. *Journal of Agriculture and Food Chemistry*, *60*, 1548–1554. https://doi.org/10.1021/jf2044718

Iimure, T., Nankaku, N., Kihara, M., Yamada, S., & Sato, K. (2012). Proteome analysis of the wort boiling process. *Food Research International*, *45*, 262–271. https://doi.org/10.1016/j.foodres.2011.10.033

Iimure, T., & Sato, K. (2013). Beer proteomics analysis for beer quality control and malting barley breeding. *Food Research International*, *54*, 1013–1020. https://doi.org/10.1016/j.foodres.2012.11.028

Irving, J. A., Shushanov, S. S., Pike, R. N., Popova, E. Y., Brömme, D., Coetzer, T. H. T., ... Whisstock, J. C. (2002). Inhibitory activity of a heterochromatin-associated serpin (MENT) against papain-like cysteine proteinases affects chromatin structure and blocks cell proliferation. *Journal of Biological Chemistry*, *277*, 13192–13201. https://doi.org/10.1074/jbc.M108460200

Ito, S., & Nagata, K. (2017). Biology of Hsp47 (Serpin H1), a collagen-specific molecular chaperone. *Seminars in Cell & Developmental Biology*, *62*, 142–151. https://doi.org/10.1016/j.semcdb.2016.11.005

Jerabek-Willemsen, M., Wienken, C. J., Braun, D., Baaske, P., & Duhr, S. (2011). Molecular interaction studies using microscale thermophoresis. *Assay and Drug Development Technologies*, *9*, 342–353. https://doi.org/10.1089/adt.2011.0380

Khan, M. S., Singh, P., Azhar, A., Naseem, A., Rashid, Q., Kabir, M. A., & Jairajpuri, M. A. (2011). Serpin inhibition mechanism: A delicate balance between native metastable state and polymerization. *Journal of Amino Acids*, *2011*, 606797.

Kihara, M., Kaneko, T., Ito, K., Aida, Y., & Takeda, K. (1999). Geographical variation of β-amylase thermostability among varieties of barley (*Hordeum vulgare*) and β-amylase deficiency. *Plant Breed*, *118*, 453–455. https://doi.org/10.1046/j.1439-0523.1999.00397.x

Koide, T., Nishikawa, Y., Asada, S., Yamazaki, C. M., Takahara, Y., Homma, D. L., ... Kitagawa, K. (2006). Specific recognition of the collagen triple helix by chaperone HSP47: II. The HSP47-binding structural motif in collagens and related proteins. *Journal of Biological Chemistry*, *281*, 11177–11185. https://doi.org/10.1074/jbc.M601369200

Lampl, N., Alkan, N., Davydov, O., & Fluhr, R. (2013). Set-point control of RD21 protease activity by AtSerpin1 controls cell death in Arabidopsis. *Plant Journal*, *74*, 498–510. https://doi.org/10.1111/tpj.12141

Lampl, N., Budai-Hadrian, O., Davydov, O., Joss, T. V., Harrop, S. J., Curmi, P. M., ... Fluhr, R. (2010). *Arabidopsis* AtSerpin1, crystal structure and *in vivo* interaction with its target protease Responsive to Desication-21 (RD21). *Journal of Biological Chemistry*, *285*, 13550–13560. https://doi.org/10.1074/jbc.M109.095075

Lema Asqui, S., Vercammen, D., Serrano, I., Valls, M., Rivas, S., Van Breusegem, F., ... Coll, N. S. (2017). AtSerpin1 is an inhibitor of the metacaspase AtMC1-mediated cell death and autocatalytic processing *in planta*. *New Phytologist*. https://doi.org/10.1111/nph.14446

Li, C. D., Langridge, P., Zhang, X. Q., Eckstein, P. E., Rossnagel, B. G., Lance, R. C. M., ... Scoles, G. J. (2002). Mapping of barley (*Hordeum vulgare L.*) beta-amylase alleles in which an amino acid substitution determines beta-amylase isoenzyme type and the level of free beta-amylase. *Journal of Cereal Science*, *35*, 39–50. https://doi.org/10.1006/jcrs.2001.0398

Lobstein, J., Emrich, C. A., Jeans, C., Faulkner, M., Riggs, P., & Berkmen, M. (2012). SHuffle, a novel *Escherichia coli* protein expression strain capable of correctly folding disulfide bonded proteins in its cytoplasm. *Microbial Cell Factories*, *11*, 753. https://doi.org/10.1186/1475-2859-11-56

Lundgard, R., & Svensson, B. (1986). Limited proteolysis in the carboxy-terminal region of barley β-amylase. *Carlsberg Research Communications*, *51*, 487–491. https://doi.org/10.1007/BF02906890

Lundgard, R., & Svensson, B. (1987). The four major forms of barley β-amylase. Purification, characterization and structural relationship. *Carlsberg Research Communications*, *52*, 313–326. https://doi.org/10.1007/BF02907173

Ma, Y. F., Eglinton, J. K., Evans, D. E., Logue, S. J., & Langridge, P. (2000). Removal of the four C-terminal glycine-rich repeats enhances the thermostability and substrate binding affinity of barley β-amylase. *Biochemistry*, *39*, 13350–13355. https://doi.org/10.1021/bi000688s

Ma, Y., Evans, D., Logue, S., & Langridge, P. (2001). Mutations of barley β-amylase that improve substrate-binding affinity and thermostability. *Molecular Genetics and Genomics*, *266*, 345–352.

Ma, Y., Langridge, P., Logue, S. J., & Evans, D. E. (2002). A Single amino acid substitution that determines IEF band pattern of barley beta-amylase. *Journal of Cereal Science*, *35*, 79–84. https://doi.org/10.1006/jcrs.2001.0421

Ma, Y., Stewart, D. C., Eglinton, J. K., Logue, S. J., Langridge, P., & Evans, D. E. (2000). Comparative enzyme kinetics of two allelic forms of barley (*Hordeum vulgare* L.) beta-amylase. *Journal of Cereal Science*, *31*, 335–344. https://doi.org/10.1006/jcrs.2000.0303

MacGregor, A., Bazin, S., Macri, L., & Babb, J. (1999). Modelling the contribution of alpha-amylase, beta-amylase and limit dextrinase to starch degradation during mashing. *Journal of Cereal Science*, *29*, 161–169. https://doi.org/10.1006/jcrs.1998.0233

Miller, G. L. (1959). Use of dinitrosalicylic acid reagent for determination of reducing sugar. *Analytical Chemistry*, *31*, 426–428. https://doi.org/10.1021/ac60147a030

Mita, S., Suzuki-Fujii, K., & Nakamura, K. (1995). Sugar-inducible expression of a gene for beta-amylase in *Arabidopsis thaliana*. *Plant Physiology*, *107*, 895–904. https://doi.org/10.1104/pp.107.3.895

Natsume, T., Koide, T., S-i, Yokota, Hirayoshi, K., & Nagata, K. (1994). Interactions between collagen-binding stress protein HSP47 and collagen. Analysis of kinetic parameters by surface plasmon resonance biosensor. *Journal of Biological Chemistry*, *269*, 31224–31228.

Niku-Paavola, M. L., Skakoun, A., Nummi, M., & Daussant, J. (1973). The polymorphism of barley β-amylase. *BBA*, *322*, 181–184.

Østergaard, O., Finnie, C., Laugesen, S., Roepstorff, P., & Svennson, B. (2004). Proteome analysis of barley seeds: Identification of major proteins from two-dimensional gels (pI 4–7). *Proteomics*, *4*, 2437–2447. https://doi.org/10.1002/(ISSN)1615-9861

Ostergaard, H., Rasmussen, S. K., Roberts, T. H., & Hejgaard, J. (2000). Inhibitory serpins from wheat grain with reactive centers resembling glutamine-rich repeats of prolamin storage proteins. Cloning and characterization of five major molecular forms. *Journal of Biological Chemistry*, *275*, 33272–33279. https://doi.org/10.1074/jbc.M004633200

Pemberton, P. A., Stein, P. E., Pepys, M. B., Potter, J. M., & Carrell, R. W. (1988). Hormone binding globulins undergo serpin conformational change in inflammation. *Nature*, *336*, 257–258. https://doi.org/10.1038/336257a0

Pogorelko, G. V., Fursova, O. V., Ogarkova, O. A., & Tarasov, V. A. (2007). New vector system for induction of gene expression in dicotyledonous plants. *Russian Journal of Genetics (Translation of Genetika (Moscow, Russian Federation))*, *43*, 136–142.

Rasmussen, S. K., Dahl, S. W., Nørgård, A., & Hejgaard, J. (1996). A recombinant wheat serpin with inhibitory activity. *Plant Molecular Biology*, *30*, 673–677. https://doi.org/10.1007/BF00049343

Roberts, T. H., Marttila, S., Rasmussen, S. K., & Hejgaard, J. (2003). Differential gene expression for suicide-substrate serine proteinase inhibitors (serpins) in vegetative and grain tissues of barley. *Journal of Experimental Botany*, *54*, 2251–2263. https://doi.org/10.1093/jxb/erg248

Rosenkrands, I., Hejgaard, J., Rasmussen, S. K., & Bjørn, S. E. (1994). Serpins from wheat grain. *FEBS Letters*, *343*, 75–80. https://doi.org/10.1016/0014-5793(94)80610-1

Sauk, J. J., Nikitakis, N., & Siavash, H. (2005). Hsp47 a novel collagen binding serpin chaperone, autoantigen and therapeutic target. *Frontiers in Bioscience*, *10*, 107–118. https://doi.org/10.2741/1513

Schweiger, R., Soll, J., Jung, K., Heermann, R., & Schwenkert, S. (2013). Quantification of interaction strengths between chaperones and tetratricopeptide repeat domain-containing membrane proteins. *Journal of Biological Chemistry*, *288*, 30614–30625. https://doi.org/10.1074/jbc.M113.493015

Seung, D., Thalmann, M., Sparla, F., Hachem, M. A., Lee, S. K., Issakidis-Bourguet, E., ... Santelia, D. (2013). *Arabidopsis thaliana* AMY3 is a unique redox-regulated chloroplastic α-amylase. *Journal of Biological Chemistry*, *288*, 33620–33633. https://doi.org/10.1074/jbc.M113.514794

Sparla, F., Costa, A., Schiavo, F. L., Pupillo, P., & Trost, P. (2006). Redox regulation of a novel plastid-targeted β-amylase of Arabidopsis. *Plant Physiology*, *141*, 840–850. https://doi.org/10.1104/pp.106.079186

Specker, C., Niessen, L., & Vogel, R. F. (2014). *In vitro* studies on the main beer protein Z4 of *Hordeum vulgare* concerning heat stability, protease inhibition and gushing. *Journal of the Institute of Brewing*, *120*, 85–92. https://doi.org/10.1002/jib.118

Sreenivasulu, N., Radchuk, V., Strickert, M., Miersch, O., Weschke, W., & Wobus, U. (2006). Gene expression patterns reveal tissue-specific signaling networks controlling programmed cell death and ABA-regulated maturation in developing barley seeds. *Plant Journal*, *47*, 310–327. https://doi.org/10.1111/j.1365-313X.2006.02789.x

Stein, P. E., Leslie, A. G. W., Finch, J. T., Turnell, W. G., McLaughlin, P. J., & Carrell, R. W. (1990). Crystal structure of ovalbumin as a model for the reactive centre of serpins. *Nature*, *347*, 99–102. https://doi.org/10.1038/347099a0

Vercammen, D., Belenghi, B., van de Cotte, B., Beunens, T., Gavigan, J.-A., De Rycke, R., ... Van Breusegem, F. (2006). Serpin1 of *Arabidopsis thaliana* is a suicide inhibitor for Metacaspase 9. *Journal of Molecular Biology*, *364*, 625–636. https://doi.org/10.1016/j.jmb.2006.09.010

Wang, C. P., Pan, Z. F., Nima, Z. X., Tang, Y. W., Cai, P., Liang, J. J., ... Yu, M. Q. (2011). Starch granule-associated proteins of hull-less barley (*Hordeum vulgare* L.) from the Qinghai-Tibet Plateau in China. *Journal of the Science of Food and Agriculture*, *91*, 616–624. https://doi.org/10.1002/jsfa.4223

Widmer, C., Gebauer, J. M., Brunstein, E., Rosenbaum, S., Zaucke, F., Drögemüller, C., ... Baumann, U. (2012). Molecular basis for the action of the collagen-specific chaperone Hsp47/SerpinH1 and its structure-specific client recognition. *Proceedings of the National Academy of Sciences*, *109*, 13243–13247. https://doi.org/10.1073/pnas.1208072109

Young, T. E., & Gallie, D. R. (2000). Programmed cell death during endosperm development. In E. Lam, H. Fukuda, & J. Greenberg (Eds.), *Programmed cell death in higher plants* (pp. 39–57). Netherlands, Dordrecht: Springer. https://doi.org/10.1007/978-94-010-0934-8

15

Developmental dynamics of stem starch accumulation in *Sorghum bicolor*

Brian A. McKinley[1] | Anna L. Casto[1] | William L. Rooney[2] | John E. Mullet[1]

[1]Department of Biochemistry and Biophysics, Texas A&M University, College Station, Texas

[2]Department of Soil and Crop Sciences, Texas A&M University, College Station, Texas

Correspondence
John E. Mullet, Department of Biochemistry and Biophysics, Texas A&M University, College Station, TX 77845.
Email: jmullet@tamu.edu

Funding information
This research was supported by the Perry Adkisson Chair.

Abstract

Sweet sorghums were identified that accumulate up to ~9% of their total stem dry weight as starch. Starch accumulated preferentially in stem pith parenchyma in close proximity to vascular bundles. Stem starch accumulated slowly between floral initiation and anthesis and more rapidly between anthesis and 43 days post-anthesis before declining in parallel with tiller outgrowth. Genes involved in stem starch metabolism were identified through phylogenetic approaches and RNA-seq analysis of Della stem gene expression during the starch accumulation phase of development. Genes differentially expressed in stems were identified that are involved in starch biosynthesis (i.e., AGPase SS/LS, starch synthases, starch-branching enzymes), degradation (i.e., glucan-water dikinase, β-amylase, disproportionating enzyme, alpha-glucan phosphorylase) and amyloplast sugar transport (glucose-6-P translocator). Transcripts encoding AGPase SS and LS subunits with plastid localization were differentially induced during stem starch accumulation indicating that ADP-glucose for starch biosynthesis is primarily generated in stem plastids. Cytosolic heteroglucan metabolism may play a role in stem sucrose/starch accumulation because genes encoding cytosolic forms of the disproportionating enzyme and alpha-glucan phosphorylase were induced in parallel with stem sucrose/starch accumulation. Information on the stem starch pathway obtained in this study will be useful for engineering sorghum stems with elevated starch thereby improving forage quality and the efficiency of biomass conversion to biofuels and bio-products.

KEYWORDS
bioenergy, pith parenchyma, RNA-seq, transcriptome analysis

1 | INTRODUCTION

Plant stems transport water and nutrients to and from leaves and roots and contribute to diverse canopy architectures that influence light interception and leaf photosynthesis. Stems also serve as temporary depots for nonstructural carbohydrates such as sucrose, monosaccharides, starch, fructans, and isoprenoids (Jensen & Wilkerson, 2017; Slewinski, 2012). The C4 grasses sugarcane and sorghum are capable of accumulating large amounts and high concentrations of sucrose (Lingle, 1987; Mcbee & Miller, 1982; McKinley, Rooney, Wilkerson, & Mullet, 2016b; Murray et al., 2008). For example, the sweet sorghum Della accumulated ~31% of its stem dry weight as sucrose between floral initiation and grain maturity as well as lower levels of glucose, fructose, and starch (McKinley et al., 2016b). The large capacity for stem sucrose accumulation provides these plants with a sink for storing photosynthetically derived sucrose when requirements for metabolism, growth, and/or seed filling are satisfied. This allows continued photosynthesis with minimal feedback inhibition reducing the cost/damage associated with interception of light by leaves in the absence of a sink for energy utilization. The non-structural carbohydrates that accumulate in stems provide a reservoir of sugars that enhance plant resilience and enable grain filling under adverse environmental conditions that cause inhibition of photosynthesis (Czedik-Eysenberg et al., 2016). In sorghum, stem sucrose and starch levels decline in parallel with onset of tillering post-grain maturity indicating that stem reserves also contribute to traits such as ratooning that are often associated with perennial sorghum (McKinley, Rooney, Wilkerson, & Mullet, 2016a).

Sugarcane breeders have focused on increasing the purity, concentration, and yield of sucrose in sugarcane stems (Waclawovsky, Sato, Lembke, Moore, & Souza, 2010). As a consequence of strong selection, the sucrose concentration in stems of elite sugarcane has reached an apparent plateau (~0.7 M) and recent gains in yield are correlated with increases in plant size and biomass (Waclawovsky et al., 2010; Wu & Birch, 2007). The stem sucrose concentration plateau has been attributed to constraints caused by elevated turgor associated with high sucrose concentrations in pith parenchyma cells, sucrose-mediated feedback inhibition of photosynthesis, and carbohydrate sensing and signaling mechanisms (Wu & Birch, 2007). Sucrose concentrations in sweet sorghum are similar to sugarcane; however, as sorghum is not a large commercial source of sucrose, selection for sucrose purity has been less intense which may explain why significant levels of stem starch are observed in sweet sorghum genotypes (McKinley et al., 2016b).

Our group has been developing and designing high-biomass sorghum hybrids that can be used for forage and production of biofuels and bio-products (Mullet et al., 2014; Rooney, Blumenthal, Bean, & Mullet, 2007). Stems account for approximately 80% of this crop's harvested biomass; therefore, stem composition is an important attribute of biomass value and utilization. High stem sucrose yield is a useful trait for biofuels production because conversion of sucrose/sugars to biofuels and bio-products is low cost and highly efficient (Gnansounou & Dauriat, 2010; Wu et al., 2010). However, stem sugars are unstable following plant harvest due to metabolic and microbial degradation, therefore, commercial production of sucrose from sugarcane stems requires efficient harvesting, transport, and processing (Wu et al., 2010). The instability of stem sugars prevents long-term storage of harvested biomass restricting mill operation to times of the year when these crops can be harvested. These logistical obstacles could be reduced if sugars that are transported to stems were used to synthesize more stable polysaccharides such as starch or mixed linkage glucans. Higher stem starch could also be useful in forage crops to increase biomass digestibility. Therefore, one motivation for this study was to characterize the dynamics of stem starch accumulation during sorghum development and to increase knowledge of this pathway to enable selection and engineering.

During periods of carbohydrate surplus, leaves shunt hexoses into the synthesis of starch and starch is often the primary storage carbohydrate in tubers and grain. Starch metabolism has been extensively studied in leaves, potato tubers, rice and maize endosperms, as well as other systems (Sonnewald & Kossmann, 2013; Stitt & Zeeman, 2012). In many dicot species, starch synthesis in heterotrophic tissues involves the import G-6-P into plastids via the glucose-6-phosphate/phosphate translocator (GPT) followed by isomerization to glucose-1-phosphate by phosphoglucomutase (PGM). The heterotetrameric enzyme ADP-glucose pyrophosphorylase (AGPase) catalyzes the conversion of glucose-1-phosphate to ADP-glucose, the committed step in starch biosynthesis (Hädrich et al., 2011). In leaves, this reaction is rendered irreversible by the action of plastidial alkaline pyrophosphatase which hydrolyzes inorganic pyrophosphate into orthophosphate (Stitt & Zeeman, 2012). In grasses, a cytosolic AGPase has been identified that is responsible for the synthesis of 95% of the ADP-glucose in the grain of maize (Denyer, Dunlap, Thorbjornsen, Keeling, & Smith, 1996). The cytosolic ADP-glucose is then imported into the plastid by the action of Brittle-1, an adenylate translocator (Shannon, Pien, Cao, & Liu, 1998). Plastid localized ADP-glucose is then utilized by starch synthases (SS) to generate starch polymers with α-1,4-glycosidic linkages (Stitt & Zeeman, 2012). Following linear polysaccharide synthesis, starch-branching enzymes (BEs) hydrolyze an oligosaccharide from the terminal end of the starch polymer just below the granule surface and attaches it deeper into the growing granule, catalyzing the formation of an α-1,6-linked branch point (Tetlow & Emes, 2014). This action leads to the formation of semi-crystalline amylopectin and increases granule density, crystallinity, and stability, and the number of nonreducing ends available for chain elongation by starch synthase (Tetlow & Emes, 2014). The action of starch synthase in combination with glucan trimming by debranching enzymes (DBE)/isoamylases ISA1 and ISA2 fine tunes the structure and crystallinity of the granule and its physical properties which are compatible with species specific enzymes involved in starch degradation (Geigenberger & Fernie, 2014; Lorberth, Ritte, Willmitzer, & Kossmann, 1998; Tetlow & Emes, 2014).

The synthesis of starch creates a molecule that is highly stable (Hejazi et al., 2008). Because of this stability, industrial processing and degradation of starch require high heat to gelatinize starch. The addition of heat adds thermal energy to the polysaccharide chains of the molecule increasing the physical space between them allowing for solvation followed by hydrolysis by starch hydrolytic enzymes (Biliaderis, Maurice, & Vose, 1980). As plants cannot utilize heat to disrupt granule crystallinity, destabilization in vivo is thought to be achieved by increasing the negative charge density of the granule surface through phosphorylation thereby disrupting crystallinity and separating the polymer chains which increases enzyme accessibility (Hejazi et al., 2008). Phosphorylation of the granule surface occurs initially by glucan-water dikinase (GWD) which phosphorylates the C-6 position of terminal glucose monomers, followed by phosphorylation by phosphoglucan-water dikinase (PWD) which phosphorylates the C-3 positions of different glucosyl residues (Ritte et al., 2006). Phosphorylation by PWD is strictly dependent on prior phosphorylation by GWD (Ritte et al., 2006). After the localized swelling introduced by GWD and PWD, a phosphoglucan phosphatase encoded by *SEX4* subsequently removes the phosphate groups as these groups impede the accessibility of starch degradation enzymes to the granule surface (Oa et al., 2009). Hydrolysis begins once the reducing ends of the amylopectin chains are accessible to β-amylases (exo-amylases that liberate maltose). One member of the debranching enzyme family, ISA3, seems to be involved in hydrolysis of branch points facilitating degradation (Delatte et al., 2006). Additional degradation of oligosaccharides liberated from the granule is performed by the α-amylases (endo-amylases) and disproportionating enzymes (DPE1 and DPE2) (Critchley, Zeeman, Takaha, Smith, & Smith, 2001). Transport of glucose and maltose released by starch degradation from the amyloplast can occur through glucose transporters (pGlcT) and maltose exporters (MEX1) (Stitt & Zeeman, 2012).

The results of this study showed that sweet sorghum genotypes can accumulate up to 9% of their stem biomass as starch and that starch accumulates primarily in stem pith parenchyma. Stem starch accumulated slowly between floral initiation and anthesis then more rapidly until peak levels were reached 43 days post-anthesis. RNA-seq analysis of stem gene expression during this developmental time course in conjunction with homology and phylogenetic analysis of gene families involved in starch metabolism enabled the identification of genes that are likely involved in stem starch accumulation and turnover.

2 | MATERIALS AND METHODS

2.1 | Harvest of diverse sorghums post-grain maturity for starch quantification

The survey of starch in sorghum [*Sorghum bicolor* (L.) Moench] stems after grain maturity was conducted from plants grown in the experimental field plots in College Station, Texas in 2014. Before planting, ammonium nitrate (32-0-0) was applied at a rate of 140 kg N/ha. Samples were harvested when all plants had passed grain maturity. The genotypes in this study were at variable stages of post-grain maturity development. Three plants per plot were harvested, avoiding the end of the plots to avoid edge effects, and these three plants were used in subsequent analyses. After harvesting, leaves and leaf sheaths were removed from the stems. A stem section spanning three internodes of which the middle internode was located at 50% of the total stem length was removed by excision at the nodes. These internode sections were cut into smaller sections and dried in a forced air oven at 60°C for 3 days; the tissue was coarsely ground in a Wiley Mill (Thomas Scientific, Inc.) and subsequently finely ground in a Cyclone Sample Mill (Udy Corporation, Fort Collins, Colorado, USA) until the tissue exhibited the consistency of a powder. Then, this powered tissue was used for starch quantification.

2.2 | Harvest of the sweet sorghum Della during development for NSC quantification

Della (Reg. no. CV-130, PI1566819), a sweet sorghum developed from a cross of Dale and ATx622 (Harrison & Miller, 1993), was planted at the Texas A&M field station near College Station, Texas in 2012 and 2013. Before planting, ammonium nitrate (32-0-0) was applied at a rate of 140 kg N/ha. In both years, the experimental design consisted of two ranges of 10 plots each of Della planted in adjacent regions of the field. Additionally, two boarder rows were planted on either side of the experiment to prevent edge effects. The spacing between rows was 76 cm, and plants were thinned to 10-cm spacing at the five-leaf stage. Basal tillers were removed until anthesis when basal tillering ceased. Tiller removal was performed to maintain a planting density of one main culms per 10 cm. During the experiment, plants were irrigated as needed. Five plants were harvested from each experimental replication for a total of 10 plants harvested per time point for a total of 10 plants per sample. Plants were harvested a minimum of two feet from the end of the plot to avoid edge effects. After harvesting, leaves and leaf sheaths were removed from the stems. A three internode section, of which the middle internode was located at 50% of the total stem length, was removed by excision at the node. These internode sections were cut into smaller sections and dried in a forced air oven at 60°C for 3 days. Next, the internode tissue was coarsely ground in a Wiley Mill (Thomas Scientific, Inc.) and subsequently finely ground in a Cyclone Sample Mill (Udy Corporation, Fort Collins, Colorado, USA) until the tissue exhibited the consistency of a powder. Then, this powered tissue was used for NSC quantification.

2.3 | Della stem tissue harvest for RNA-seq

To obtain tissue for RNA-seq, Della was grown in a greenhouse in College Station during the fall of 2011. The day length was maintained at 14 hr, and the mid-day PAR of the green house was ~1,100–1,200 μE/m^2/s. Plants were germinated in five gallon pots containing Metromix potting soil. Osmocote was added to the soil at the beginning of the experiment, and additional fertilization was supplemented with Peter's nutrient solution every 30 days. The plants were watered with the use of an automated watering system to eliminate the possibility of drought stress. Leaves were numbered as they appeared insuring that internode 10 was harvested at each of the eight sampling dates. Three biological replicates were harvested in the morning on each harvest date. To harvest internode samples, leaves and leaf sheaths were removed from the stem and Internode 10 was excised at the nodes to insure that a complete internode was harvested for analysis. The internode was quickly sectioned into smaller pieces with a razor for storage and ease of processing after freezing. The internode sections were placed into a 50-ml conical tube, frozen in liquid nitrogen, and stored at −80°C until processing. To prepare tissue for RNA extraction, the frozen tissue of each internode was ground to a fine powder using a mortar and pestle and stored in 50-ml conical tube until extraction.

2.4 | Soluble carbohydrate quantitation

To quantify the abundance of stem nonstructural carbohydrates, previously, oven dried samples from field experiments were redried overnight using a forced air oven at 60°C to insure accuracy of dry weight measurements. Next, 200 mg of finely ground biomass was weighed (±0.5 mg) using an analytical balance and transferred to a 15-ml conical tube. Alternatively, to quantify nonstructural carbohydrates in tissue that gave rise to the RNA-seq data, approximately 0.5 g of frozen, pulverized tissue was placed into a previously weighed 15-ml conical tube and lyophilized until desiccated. The tube and tissue were reweighed, and tube weight subtracted to obtain the tissue weight. Both tissue types were subjected to the following extraction protocol. Water-soluble NSCs were extracted in 10 ml of water/Na azide (200 mg/L) solution at 50°C for 48 hr with agitation. This length of incubation was experimentally determined to be optimal for this experimental set-up. This extraction time allowed the extraction solvent to fully penetrate all of the biomass particles and extract the soluble carbohydrates. Aliquots of 20 μl were diluted 50 × into 980 μl of deionized water. All HPLC samples were filtered using 0.45-μm cellulose acetate sterile filters. Sucrose, glucose, and fructose concentrations were quantified using high performance anion exchange-pulsed amperometric detection (HPAE-PAD) with a Carbopac PA1 analytical column (Dionex, Sunnyvale, CA) as well as a Borate trap (Dionex) and an Aminopac column (Dionex) to reduce borate and amino acid interference (Murray et al., 2008). A solution of 75 mM NaOH was made from 50% NaOH solution (Sigma-Aldrich, St. Louise, MO, USA) instead of pellets to eliminate carbonate contamination. This running buffer was vacuum degassed overnight and stored under a helium bed for the duration of the chromatographic run. The standard curve was validated by the incorporation of curve validation samples of known concentration throughout the experiment in accordance with the NREL LAP (Sluiter, Ruiz, Scarlata, Sluiter, & Templeton, 2008).

2.5 | Starch quantification

The starch in stem biomass was quantified using the NREL Laboratory Analytical Procedure for extracting starch from solid biomass (Sluiter & Sluiter, 2005). After the soluble nonstructural carbohydrate extraction, samples were washed four times in methanol-chloroform-water extraction solvent at room temperature for 15 minutes then centrifuged at 1,640 *g* for 15 min to sediment structural biomass and starch granules. After the MCW washes, the samples were oven dried at 65°C for 48 hr. Next, DMSO was added to the dry samples and incubated overnight. Overnight imbibition insures that the DMSO, the gelatinization solvent, has fully penetrated the biomass and is capable of fully gelatinizing the starch which is essential to fully digest all the starch. After imbibition, the samples were incubated at 100°C for 10 min with agitation every two minutes to gelatinize the starch. Then, MOPS buffer and thermostable α-amylase were added to digest the starch to maltose at 95°C for 1 hr with agitation to insure full digestion. Subsequently, sodium acetate buffer and amyloglucosidase were added to hydrolyze the maltose to glucose at 55°C for 24 hr. To insure that starch was fully digested, digestion completion samples were included in the assay and sampled after digestion with amylase and amyloglucosidase and stained with Lugol's iodine solution. Glucose was quantified with high performance anion exchange-pulsed amperometric detection (HPAE-PAD) chromatography with the use of a Carbopac PA1 analytical column (Dionex, Sunnyvale, CA). Curve validation samples were randomly interspersed among the analytical samples during chromatography to validate the glucose, fructose, and sucrose standard curves. The percentage of starch in the dry biomass was calculated from the glucose released during digestion.

2.6 | Stereomicroscopy of internode cross sections

To visualize the location of starch in the stem by stereomicroscopy, field grown plants were harvested at anthesis and post-grain maturity. A 1-mm slice of stem tissue was excised with a razor blade from the middle of internode 11. These 1-mm internode sections were infused with Lugol's solution in a vacuum. After infusion, the internode sections were stored in this solution overnight. Excess staining solution was removed by incubating the tissue slice in water for 24 hr, changing the solvent every 6 hr. The sections were imaged under white light using a Zeiss M^2Bio Fluorescence Combination Zoom Stereo/Compound Microscope coupled with a Zeiss AxioCam digital camera (Kramer Scientific). Photographs were captured using the Zeiss AxioVision 3.0.6 software.

2.7 | Template preparation and sequencing

RNA extraction was performed using the Trizol extraction method (https://www.mrcgene.com/rna-isolation/tri-reagent). The high salt precipitation step was included because of the potential high concentration of soluble carbohydrates in the sample. Total RNA was further purified using the RNeasy kit (Qiagen, Venlo, the Netherlands) which isolates RNAs >200 nucleotides in length. An on column DNase1 digestion was performed using the RNase-free DNase set (Qiagen, Venlo, Netherlands) to remove any contaminating DNA. The quality of the RNA libraries was assessed using an Agilent Bioanalyzer. The mean RNA integrity number (RIN) from this quality control step was 8.9, and the minimum RIN was 8.3 indicating that the quality of the RNA libraries was high. RNA libraries were prepped for sequencing using the Illumina TruSeq® V2 mRNA Sample Preparation Kit (Illumina, San Diego, California, USA), which selects for polyadenylated RNA and excludes rRNA and non-polyadenylated RNAs from the sample. The 24 libraries were sequenced on one lane of an Illumina HiSeq 2500 in single-end mode. The average sequence depth was 5.8 M reads.

2.8 | Transcriptome analysis

The 70 bp reads were aligned to the *Sorghum bicolor* V3.1 genome using the HISAT2 aligner (Daehwan, Langmead, & Salzberg, 2015; McCormick et al., 2018). Expression was quantified using the StringTie version 1.3 software (Pertea et al., 2015). Analysis of gene expression was performed on TPM normalized data. The prepDE.py script was used to convert nucleotide coverage data from StringTie into reads that could be used by differential expression statistical packages that use conventional raw reads. Differential expression and the FDR-adjusted *p*-values were calculated using the edgeR package. The fold-change displayed in the tables is the absolute value of the maximum fold-change value between any two time points. The FDR-adjusted *p*-value represents the *p*-value associated with the maximum fold-change values shown in the gene tables. In some cases, zero expression at one time point, when being compared to another time point, led to a very high differential expression value. To minimize this effect in instances with very high DE values, the lowest time point was considered to be a TPM = 1 and, therefore, the fold-change was recalculated as FC = maxTPM/1 (McKinley et al., 2016b). These instances are indicated with the ">"sign in front of the fold-change values in the gene tables as the actual fold-change is assumed to be some value greater than the value displayed. Functional analysis of gene families utilized gene functional annotations generated by DOE-JGI for the *Sorghum bicolor* V3.1 genome available through Phytozome (https://phytozome.jgi.doe.gov/pz/portal.html#!info?alias=Org_Sbicolor) (McCormick et al., 2018).

2.9 | Phylogenetic analysis of gene families to identify orthologs

Phylogenetic analysis was used increase the accuracy of gene family annotation. This approach was based on the identification of genes in other species with known previously validated function. The primary protein sequences were obtained Phytozome or NCBI. The protein sequences of these validated genes and the sorghum candidate homologs were then clustered using the CLUSTAL Omega algorithm using default settings.

2.10 | Identification of chloroplast targeting peptide sequences

Peptide sequences were obtained from Biomart, hosted by the Phytozome database. The primary transcript was used for the analysis. The ChloroP algorithm (http://www.cbs.dtu.dk/services/ChloroP/) was used to predict the presence of chloroplast targeting sequences in peptides sequences.

3 | RESULTS

3.1 | Variation in sorghum stem starch accumulation in diverse sorghum accessions

The level of starch present in stems of field grown grain and sweet sorghum genotypes was analyzed after grain maturity. A stem segment spanning three internodes located approximately mid-stem was assayed for levels of starch and other nonstructural carbohydrates. The sorghum genotypes analyzed accumulated 0.2% to ~9% of their internode biomass as starch (Figure 1). Stem starch levels were generally higher in sweet sorghum genotypes (Figure 1, green bars) compared to grain sorghum genotypes (Figure 1, orange bars).

3.2 | Developmental dynamics of starch accumulation in Della sorghum stems

Sorghum progresses through a juvenile and vegetative phase following germination (Figure 2a). The reproductive phase begins with floral initiation, followed by anthesis approximately 25–30 days later, grain maturity and a post-grain maturity phase where tillering often occurs (Figure 2a). The sweet sorghum Della grown in the field in College Station reached anthesis 69 days after emergence (DAE), and grain maturity was reached approximately one month after anthesis. The developmental time course of nonstructural carbohydrate accumulation in the stems of the sweet sorghum Della was characterized in field grown plants during the summers of 2012 and 2013 (Figure 2b). The analysis revealed that Della stems accumulated starch post-anthesis and that peak stem starch levels were reached after grain maturity. Stem starch accumulated at its highest rate from 26 days post-anthesis to ~53 days post-anthesis (0.26% of internode dry biomass/day). Starch accounted for ~9.5% of Della's stem dry weight 53 days after anthesis. In contrast, stem soluble sugar levels reached a maximum ~26 days post-anthesis and remained high until 53 days post-anthesis (Figure 2b). Between 53 and 80 days after anthesis, levels of stem sucrose and starch decreased in parallel with leaf senescence and apical tiller outgrowth. Similar results were obtained when Della was grown in the greenhouse (Figure 2c). When Della was grown in the greenhouse to facilitate RNA-seq analysis, the developmental time course of soluble sugar and starch accumulation in stems was similar to plants grown in the field (Figure 2c). However, in greenhouse grown plants, soluble sugar and starch levels in Della stems were lower than in field grown plants likely due to lower light levels in the greenhouse.

3.3 | Distribution of starch in stems

The amount of starch and soluble carbohydrates in each of the elongated internodes of Della stems was assayed at grain maturity (Figure 3). Most of the stem internodes had similar nonstructural carbohydrate concentration profiles. However, sucrose concentrations were lower (and monosaccharides somewhat higher) in internodes located near the base of the plant (internode 8), internodes just below the peduncle (internodes 16, 17), and the peduncle. Excluding the peduncle, starch accounted for >5% of the internode dry biomass of every internode at grain maturity and internodes 9–11 has the highest starch levels (~12.5% of dry weight) (Figure 3).

The radial distribution of starch across stem internodes was assayed by infusing cross sections with Lugol's iodine solution which stains starch dark blue prior to visualization using a stereo-microscope (Figure 4). This analysis revealed that starch accumulated at relatively high levels in pith parenchyma cells in regions of high vascular bundle density (Figure 4a). Starch levels were also elevated in pith parenchyma cells that surround vascular bundles in the central

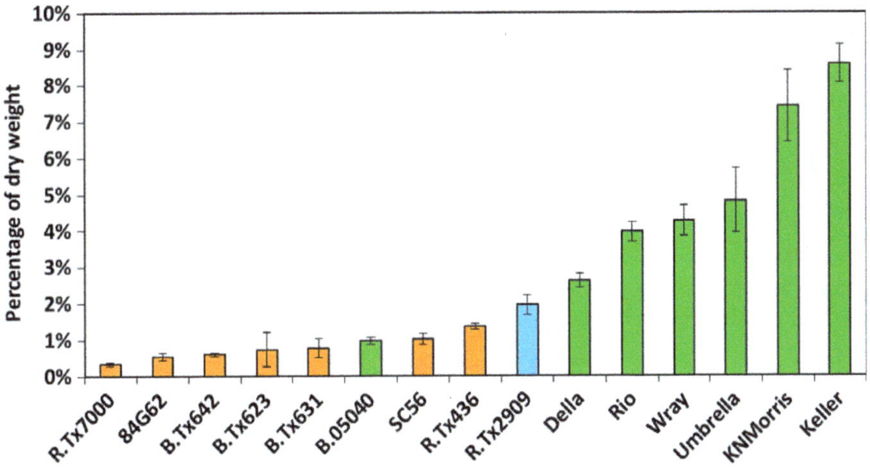

FIGURE 1 A survey of the percent composition of starch in the stem of diverse sorghum genotypes grown in the field in College Station, TX and assayed after grain maturity. Genotypes represented by green bars are sweet sorghums, orange bars represent grain sorghums, and the blue bar represents a forage sorghum. Error bars represent *SEM*

portion of the stem and decreased with increasing distance from the vascular bundles (Figure 4a and b). Starch granules were not observed within cells of the vascular bundles; however, this tissue was tinted blue after staining likely because of the scattering of blue light through these cell types from surrounding starch-containing cells (Figure 4c and d). High levels of starch were not observed in the epidermis of the stem, but starch accumulated in cells just beneath the epidermis (Figure 4a).

3.4 | Expression of genes involved in stem starch metabolism during plant development

One overall goal of this study was to identify and characterize the expression of genes involved in stem starch metabolism. The increase in starch accumulation in stems between floral initiation and 53 days post-anthesis provided an opportunity to identify genes in starch metabolism that show increased expression in parallel with stem starch accumulation. Data on stem gene expression were obtained by extracting RNA for RNA-seq analysis from stem internodes of Della plants grown in the greenhouse at eight times between floral initiation (~46 DAE) through 80 days post-anthesis (~175 DAE). Genes previously identified in other plants that are involved in starch synthesis, turnover, and plastid sugar transport were used to identify sorghum homologs based on sequence similarity and phylogenetic analysis of gene families (Supporting information Tables S1–S3). Sorghum encodes a family of genes for most steps in starch metabolism, and specific genes in each family were differentially expressed in stems in parallel with starch accumulation.

To help visualize patterns of gene expression in the context of the entire starch pathway, a working model of stem starch metabolism was developed based on prior knowledge of this pathway obtained in other plants/organs (Smith, Denyer, & Martin, 1997; Stitt & Zeeman, 2012). In the model, genes encoding enzymes that are likely involved in stem starch synthesis are highlighted in green, genes encoding proteins involved in starch turnover are highlighted in blue or red, and those encoding proteins involved in sugar transport through the outer plastid membrane are shown in colored boxes (Figure 5). Genes expressed in stems at >20 TPM and that were induced >fivefold in parallel with stem starch accumulation between floral initiation and grain maturity have information on fold-induction presented in yellow-highlighted boxes adjacent to each gene. This overview shows that genes encoding proteins involved in starch biosynthesis (i.e., plastid *AGPase, SS, BE, DBE*), turnover (i.e., *GWD, DBE, DPE, PHS*), and transport (i.e., *GPT, pGlcT*) were upregulated between floral initiation and 43 days post-anthesis in parallel with stem starch accumulation.

3.5 | Expression of genes encoding AGPases

Two sorghum genes encoding AGPase small subunits (SSs) and four genes encoding AGPase large subunits (LSs) were identified in the sorghum genome (Supporting information Figure S1; Supporting information Table S1). Several different transcripts were produced from these genes. One gene encoding AGPase SS (Sobic.002G160400, *APS1*) and one gene encoding AGPase LS (Sobic.009G245000; *APL2*) were expressed at lower levels in stems at floral initiation and at gradually increasing levels during development with peak expression between 25 and 43 days post-anthesis (Figure 6a; Supporting information Table S1). Sobic.009G245000 (*APL2*) was differentially expressed 47-fold higher in stems post-anthesis and Sobic.002G160400 (*APS1*) eightfold higher (Figure 6a; Supporting information Table S1). Both of these genes were expressed at ~five-tenfold lower levels in leaves compared to stems; however, expression in panicles was similar to stems (Supporting information Figure S1).

Many grasses synthesize some portion of the ADP-glucose used for starch synthesis in grain in the cytosol (Beckles, Smith, & Ap Rees, 2001). As the stem post-anthesis is a heterotrophic carbohydrate storage organ similar to grain (Emanuelsson, Nielsen, & von Heijne, 1999), an analysis was conducted to characterize the expression of genes/transcripts involved cytosolic ADP-glucose synthesis during development (Supporting information Figures S1 and S2). Alignment and phylogenetic analysis of sorghum and maize AGPase gene homologs that are localized in the cytoplasm or plastid (Huang, Hennen-Bierwagen, & Myers, 2014) and ChoroP prediction of plastid transit sequences (Xuan et al., 2013) were used to identify sorghum transcripts that encode AGPase subunits that would localize in the cytoplasm. This analysis

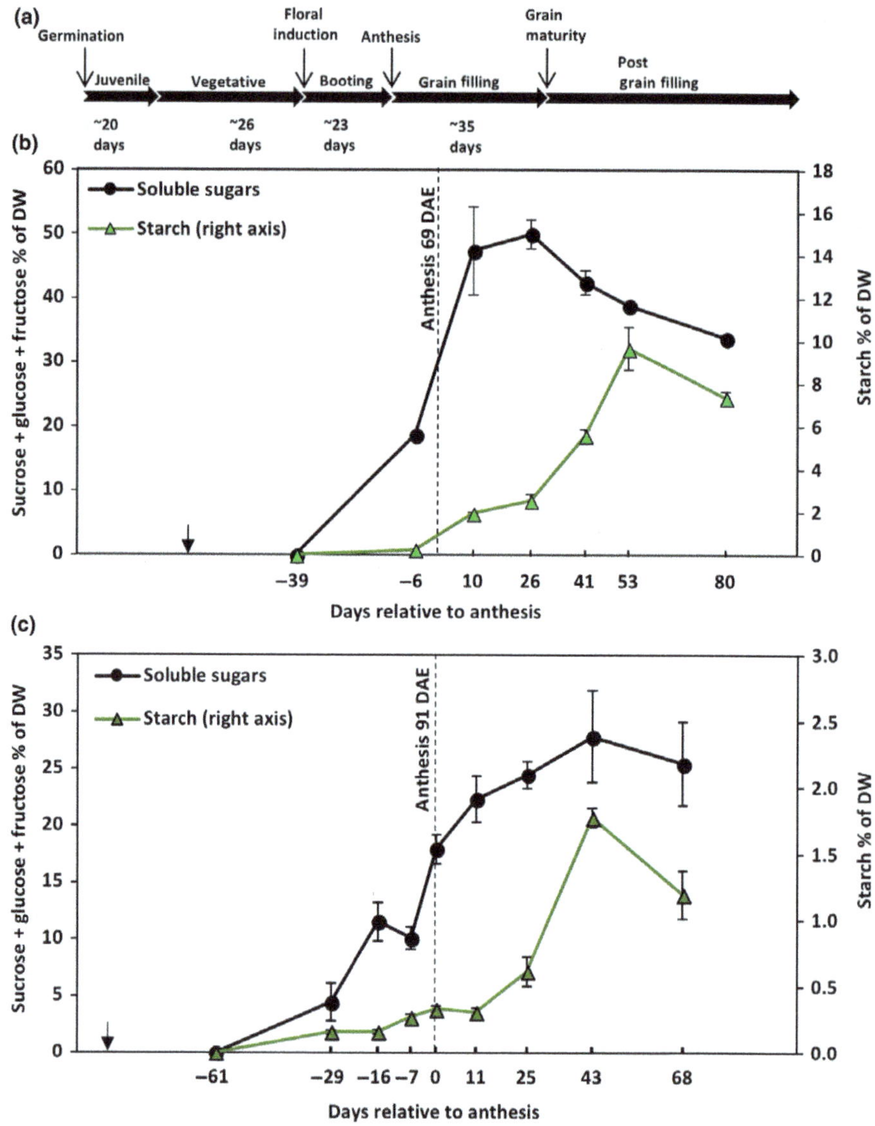

FIGURE 2 Temporal pattern of starch accumulation in the stem of the sweet sorghum Della in field and green house environments. (a) Timeline of Della development in the field in College Station, TX, in 2011. (b) The accumulation profile of total soluble sugars (sucrose + glucose + fructose) and starch in the Della stem in the field in College Station, TX, 2011. The dotted line represents anthesis. The *x*-axis represents days relative to anthesis. Negative numbers are days prior to anthesis, and positive numbers are days after anthesis. The black arrow represents seedling emergence, that is, the start of the experiment. (c) The accumulation profile of total soluble sugars (sucrose + glucose + fructose) and starch in the Della stem in the greenhouse in College Station, TX, Fall, 2011. Error bars represent*SEM*

identified two transcripts (Sobic.007G101500.2, Sobic.007G101500.3) that encode AGPase small subunits and three transcripts (Sobic.003G230500.1/.2/.3) that encode AGPase large subunits predicted to produce proteins that localize in the cytoplasm (Supporting information Figure S1). These transcripts are expressed at lower levels in stems compared to other genes/transcripts that encode AGPase subunits with predicted plastid localization. Taken together, this indicates that a plastid localized AGPase is the primary source of ADP-glucose for starch synthesis in stem amyloplasts.

3.6 | Genes encoding stem starch synthases and starch-branching/debranching enzymes

ADP-glucose is the substrate of starch synthases, the enzyme family responsible for adding glucose monomers to the reducing ends of growing amylose and amylopectin chains (Hirose & Terao, 2004). The starch synthase gene family consists of 10 genes in rice and maize (Supporting information Table S1). This family can be further divided into the soluble and granule bound starch synthases, with each of these groups containing unique subtypes (Supporting information Figure S3; Supporting information Table S1) (Hirose & Terao, 2004). Due to the functional complexity of the family, a phylogenetic analysis was conducted to align sorghum gene annotations with gene family members identified in rice and maize (Supporting information Figure S3) (Hirose & Terao, 2004). This analysis identified ten genes in the sorghum, the same number of genes in the rice starch synthase gene family (Supporting information Figure S3) (Hirose & Terao, 2004). Eight of the 10 genes in the family were expressed in sorghum stems. Most of these family members showed peak expression in stems post-anthesis (Supporting information Table S1). A gene encoding *GRANULE BOUND STARCH SYNTHASE* (*GBSS*), Sobic.002G116000.1, was the most highly expressed gene in the family (max TPM = 737), and the gene was differentially expressed in stems between floral initiation and 43 days post-anthesis (DE = 11-fold) (Supporting information Table S1). The second most abundantly expressed gene of this gene family was a *STARCH SYNTHASE I* (max TPM = 229, DE = 29-fold), and the third was a *STARCH SYNTHASE II* (max TPM = 162, DE = 15-fold) (Figure 6a).

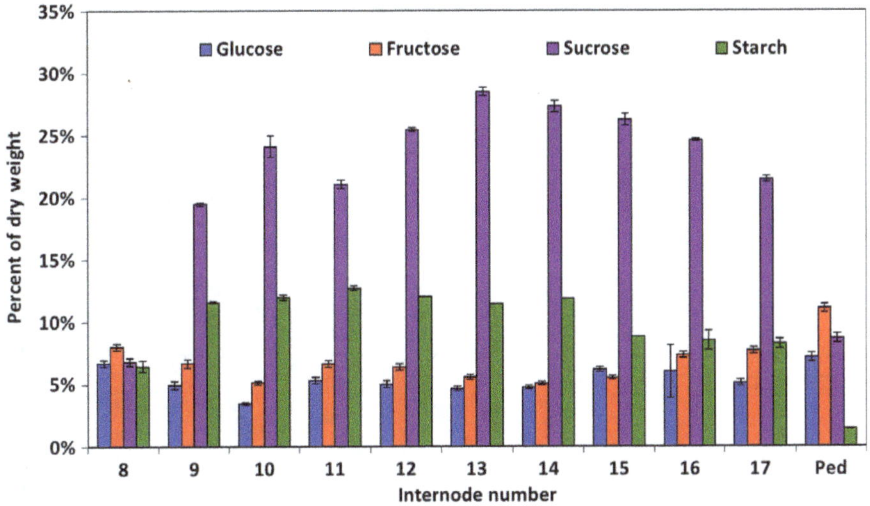

FIGURE 3 The distribution of nonstructural carbohydrates in the sweet sorghum stem as a percentage of the total dry biomass of each internode. Internode number increases proceeding up the stem. Ped, peduncle. Error bars represent*SEM*

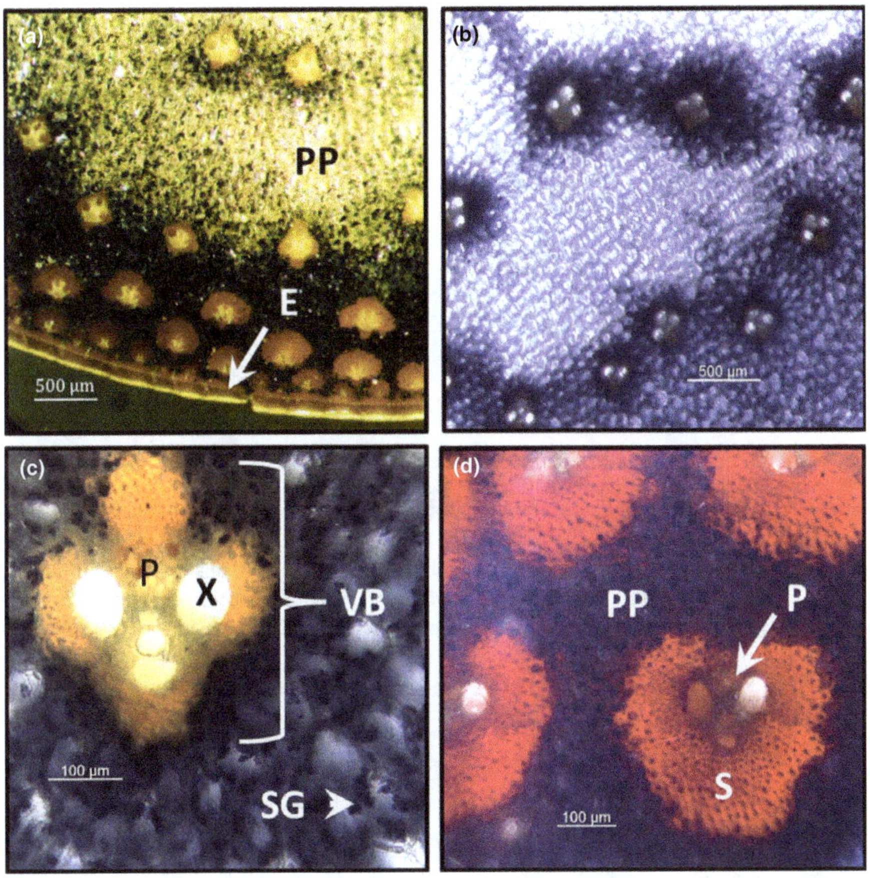

FIGURE 4 Localization of starch within sweet sorghum Della internodes visualized using stereomicroscopy. Starch was stained with Lugol's solution through vacuum infiltration. (a) Low resolution image of an internode cross section that shows the distribution of starch in the internode. (b) Image of vascular bundles of the central region of the Della internode. (c) High-resolution image of a vascular bundle located near the central region of the internode. (d) High-resolution image of a vascular bundle located in the rind region at the periphery of the internode near the epidermis. PP, pith parenchyma; SG, starch granule; E, epidermis; S, sclerenchyma sheath; P, phloem; X, xylem

The starch-branching enzymes and debranching enzymes tune starch granule structure to achieve specific physicochemical properties of the granule (Smith et al., 1997). Four genes encoding starch-branching enzymes were expressed in the sorghum stem with peak expression after anthesis (Supporting information Table S1). The most highly expressed gene/transcript was Sobic.006G066800.2 (max TPM = 865, 43 days after anthesis) a gene that was also differentially expressed during plant development (DE = 25-fold) (Figure 6a). Isoamylase 1-3 are starch-debranching enzymes. Isoamylases 1 and 2 are involved in starch biosynthesis (Supporting information Table S1), tuning the structure of amylopectin to facilitate crystallization (Hussain et al., 2003). Evidence suggests that isoamylase 3 is involved in starch degradation (Supporting information Table S2) (Delatte et al., 2006). Three transcripts of *ISA1* and *ISA2* were expressed in stems with peak expression occurring 25 days after anthesis (Figure 6a; Supporting information Table S1). The gene encoding isoamylase 3 (Sobic.002G233600) produced three transcripts that varied in abundance in stems with peak expression occurring post-anthesis (Supporting information Table S2).

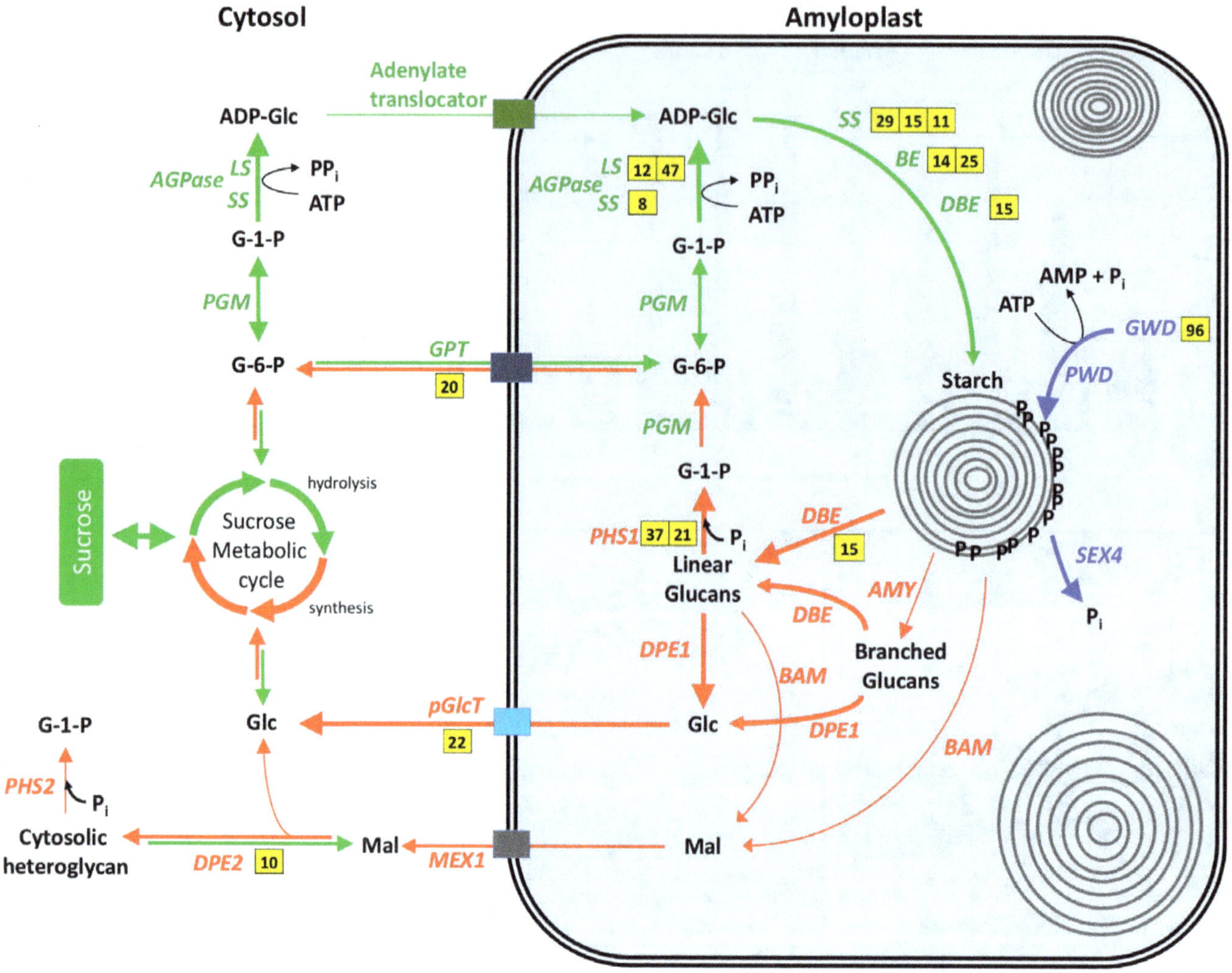

FIGURE 5 Differentially expressed genes of starch metabolism in the context of their pathways. Yellow boxes represent the fold-change of differentially expressed genes of the pathway. Selection criteria: 10-fold induction after anthesis with maximum expression >20 TPM. Thick lines represent hypothetical pathway flux. AGPase; ADP-glucose pyrophosphorylase, LS; large subunit (of AGPase), SS; small subunit (of AGPase), PGM; phosphoglucomutase, GPT; glucose-6-phosphate/phosphate translocator, SS; starch synthase, BE; branching enzyme, DBE; debranching enzyme, GWD; glucan-water dikinase, PWD; phosphoglucan-water dikinase, SEX4; phosphoglucan phosphatases, AMY; α-amylase, BAM; β-amylase, DPE; disproportionating enzyme, PHS; α-glucan phosphorylase, MEX1; maltose transporter, pGlcT; glucose transporter

3.7 | Genes involved in stem starch degradation

The expression of several genes that encode enzymes involved in starch turnover increased between floral initiation and grain maturity (Supporting information Table S2). To enhance starch degradation, the crystallinity of the starch granule is reduced through the action of GWD (Figure 5). The expression of *GWD* (Sobic.010G143500.1) was induced ~96-fold between floral initiation and grain maturity in parallel with the accumulation of starch in the sorghum stem (Figure 6b; Supporting information Table S2). There were three splice variants derived from a gene expressed in the stem that encodes PWD (Supporting information Table S2). All the *PWD* splice variants were expressed at ~10-fold lower levels compared to *GWD*, and only one transcript was differentially expressed (~fivefold) with peak expression post-anthesis (Table S2). A gene encoding phosphoglucan phosphorylase (*SEX4*, Sobic.001G53880) showed twofold higher expression in stems post-anthesis (Supporting information Table S2).

Amylases and isoamylase help degrade starch grains to maltose and branched and linear glucans (Figure 5). Three genes encoding α-amylase and ten genes encoding β-amylase were expressed in sorghum stems. The α-amylase gene with the highest expression was expressed at similar levels from floral initiation through grain maturity. Two other α-amylase genes were expressed at low levels until 68 days post-anthesis when starch levels were decreasing in stems (Supporting information Table S2). Two of the genes encoding β-amylase were expressed at high levels in stems at floral initiation. These genes may be associated with chloroplasts located in the outer cell layers of the stem. Four members of the gene family were expressed at approximately the same level in stems during the developmental stages analyzed. In addition, four other members of the β-*AMYLASE* gene family were expressed at relatively low levels prior

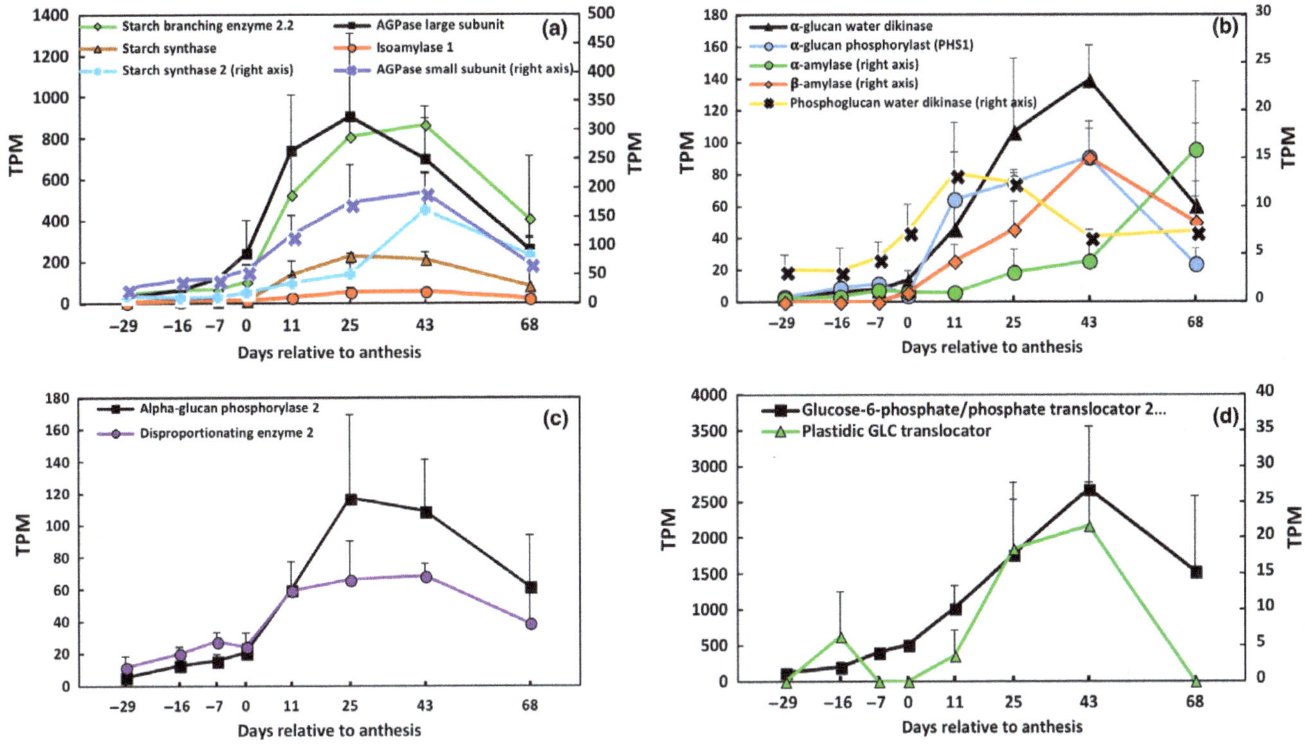

FIGURE 6 Expression profiles of starch metabolism genes in the stem that are induced after anthesis. (a) Starch biosynthesis genes. (b) Starch degradation genes. (c) Expression of starch metabolism genes localized outside of the amyloplast. (d) Transporters potentially involved in import and export of sugar into the amyloplast. Error bars represent *SEM*

to anthesis with their maximum expression peaking post-anthesis (Figure 6b; Supporting information Table S2).

Two genes encoding DPE1 and DPE2 were significantly upregulated in stems between floral initiation and grain maturity (Figure 6c; Supporting information Table S2). *DPE2*, which encodes an enzyme with predicted cytosolic localization, was more highly expressed and upregulated ~10-fold between floral initiation and 43 days post-anthesis. *DPE1*, which encodes an enzyme with predicted plastid localization, was expressed at lower levels, but also induced during this phase of plant development. Two genes encoding heteroglucan phosphorylase (*PHS1*, *PHS2*) were expressed at similar levels, and gene expression was induced ~21-fold and ~37-fold, respectively, with peak expression 43 days post-anthesis (Figure 6b and c; Supporting information Table S2). *SbPHS1* has a predicted localization in plastids whereas *SbPHS2* has a predicted localization in the cytosol (Figure 6b and c; Supporting information Table S2) consistent with analysis of these proteins in other plants (Beck & Ziegler, 1989; Fettke, Malinova, Eckermann, & Steup, 2009).

3.8 | Differential expression of starch pathway genes in stems and leaves

The expression of genes involved in starch metabolism in leaves and stems was compared to see if members of gene families were regulated in an organ-specific manner. The expression of pairs of genes from the same gene family was compared using RNA isolated from leaves and stems at the same time in the morning during four stages of plant development (Figure 7 and Supporting information Figure S4). The analysis showed that expression of Sobic.009G245000.2 (*AGL2*) was higher in stems vs. leaves during the phase of development analyzed whereas the related family member Sobic.001G100000.1 (*AGL2*) was expressed at higher levels in leaves compared to stems (Figure 7a and b; Supporting information Figure S4). Similarly, Sobic.002G160400.1 (*AGS1*) was expressed at higher levels in stems vs. leaves whereas the family member Sobic.007G101500.1 (*AGS1*) was expressed at higher levels in leaves compared to stems during these same developmental stages (Figure 7c and d; Supporting information Figure S4). A pair of genes encoding starch synthases (Sobic.010G093400.1, Sobic.010G047700.1) showed similar divergent patterns of differential expression in leaves and stems (Figure 7e and f; Supporting information Figure S4).

3.9 | Genes involved in stem amyloplast sugar transport

The increase in stem starch accumulation between floral initiation and grain maturity and remobilization post-grain maturity could require higher expression of genes encoding the plastid localized glucose-6-phosphate/phosphate translocator (*GPT*), the maltose transporter (*MEX1*), the plastidial glucose transporter (*pGlcT*), and/or the adenylate translocator. RNA-seq analysis showed that a gene encoding a member of the *GLUCOSE-6-PHOSPHATE/PHOSPHATE TRANSLOCATOR 2 (GPT2)* family was induced ~20-fold in parallel with increased stem starch accumulation during development (Figure 6d; Supporting information Table S3). This gene was highly expressed (max TPM = 2,689) and showed strong differential expression during development (FDR-adjusted *p*-value = 0.00021) (Figure 6d; Supporting information Table S3). Induction of this highly expressed gene suggests that glucose-6-phosphate is a predominant metabolite transported to and from stem amyloplasts. A gene *PLASTIDIAL GLUCOSE TRANSPORTER* (*pGlcT*) encodes a protein that transports glucose released from starch degradation out of the plastid. This gene showed increased expression in stems post-anthesis; however, the relative expression of *pGlcT* (max TPM = 150) was much lower than *GPT2* (TPM = 2,689) (Figure 6c, Supporting information Table S3). The plastid ATP/ADP-transporter translocates adenylates across plastid membranes thus maintaining plastidial ATP supply (Figure 6c; Supporting information Table S3) (Tjaden, Mohlmann, Kampfenkel, Henrichs, & Neuhaus, 1998). One member of this gene family was expressed in stems (max TPM = 120) although expression did not increase in parallel with starch accumulation during development (Supporting information Table S3). The gene *MALTOSE EXCESS 1* (*MEX1*) encodes a protein that transports maltose out of plastids. This gene was expressed at low levels in stems (max TPM = 8) and was not differentially expressed between floral initiation and 43 days post-anthesis (Supporting information Table S3).

4 | DISCUSSION

The overall goal of this study was to learn more about the regulation of starch accumulation in sorghum stems because increasing stem starch levels could improve biomass yield and the composition of high-biomass sorghum that is used for forage and production of biofuels and bio-products. While prior studies have noted the presence of starch in sorghum stems (Gutjahr et al., 2013; McBee, Waskom, & Creelman, 1983; McKinley et al., 2016b), the current study showed that a number of sweet sorghum genotypes can accumulate up to ~9% of their stem dry weight as starch. The capacity for significant stem starch accumulation in sweet sorghum genotypes that also accumulate high levels of stem sucrose indicates that even higher levels starch could be generated by altering the partitioning between sucrose and starch. Moreover, nearly all of the internodes that comprise Della stems accumulated high levels of sucrose and ~10%-12% starch. Therefore, high-biomass sweet sorghums that accumulate 80% of their biomass in stems that grow 4–5 m in length and 20–40-mm diameter have a large potential capacity for storing stem starch. The stems of the sweet sorghum Della have a nonstructural carbohydrate to lignocellulose ratio of ~2.8 at harvest (McKinley et al., 2016b). Increasing stem starch further could potentially improve yield by reducing feedback inhibition of photosynthesis that occurs when stems reach maximum sucrose holding capacity (Gutjahr et al., 2013; Paul & Foyer, 2001) and improve the efficiency and lower the cost of biomass conversion to biofuels and bio-products.

Starch accumulation was highest in stem pith parenchyma cells adjacent to vascular bundles in the subepidermal region of the stem where vascular bundle density was highest. The observation that starch accumulates in this region of the stem suggests that sucrose/sugar availability is high near vascular bundles. Starch also accumulated in pith parenchyma cells immediately adjacent to vascular bundles located in the central region of the stem. Interestingly, in the center of the stem where vascular bundle density is lower, pith parenchyma cells located furthest from vascular bundles accumulated lower levels of starch suggesting that sucrose/sugars derived from the phloem are used preferentially by pith parenchyma cells closest to the phloem and that starch accumulation in stem pith cells is responsive to sucrose availability.

Leaves are the primary source of sucrose that is used for synthesis of starch in stem pith parenchyma cells. Prior studies have examined the transport of sucrose from sorghum leaf mesophyll cells to stems (Sowder, Tarpley, Vietor, & Miller, 1997) and identified transporters that are involved in this pathway (Bihmidine, Baker, Hoffner, & Braun, 2015; Bihmidine, Julius, Dweikat, & Braun, 2016; Milne et al., 2017; Mizuno, Kasuga, & Kawahigashi, 2016). In fully elongated internodes that are accumulating sucrose post-floral initiation, sucrose can move from the phloem to stem storage parenchyma cells via an apoplastic and/or symplastic pathway with subsequent sequestration in pith parenchyma cell vacuoles (Bihmidine et al., 2015; Milne, Offler, Patrick, & Grof, 2015; Sowder et al., 1997; Tarpley & Vietor, 2007). Sucrose taken up by storage pith parenchyma post-anthesis is then sequestered in vacuoles or hydrolyzed by cytoplasmic invertases or sucrose synthases generating hexoses that can be used for starch biosynthesis. The ~20-fold induction of *GPT2* expression in stems in parallel with increased starch accumulation during development indicates that transport of G-6-P to and from the plastid plays an important role in stem starch metabolism.

4.1 | Genes involved in sorghum stem starch metabolism

Sorghum genes potentially involved in stem starch metabolism were identified based on their sequence similarity to genes previously identified in other species, phylogenetic analysis of gene families, and increased expression during starch accumulation in sorghum stems between floral initiation and 43 days post-anthesis. This approach identified sorghum genes predicted to encode enzymes involved in starch metabolism that were significantly upregulated in parallel with the accumulation of stem starch during development (Figure 5). Several genes encoding steps in starch biosynthesis were differentially expressed relative to other gene family members in stems vs. leaves during monocot development. This suggests that stem starch accumulation is regulated in part through differential expression of specific gene family members involved in starch metabolism. The basis of differential regulation and confirmation that these genes are directly involved in stem starch metabolism will require additional studies involving targeted gene modification.

In monocot grain endosperm tissue, most of the ADP-glucose used for starch biosynthesis is synthesized in the cytoplasm and imported into amyloplasts (Beckles et al., 2001; Johnson et al.,

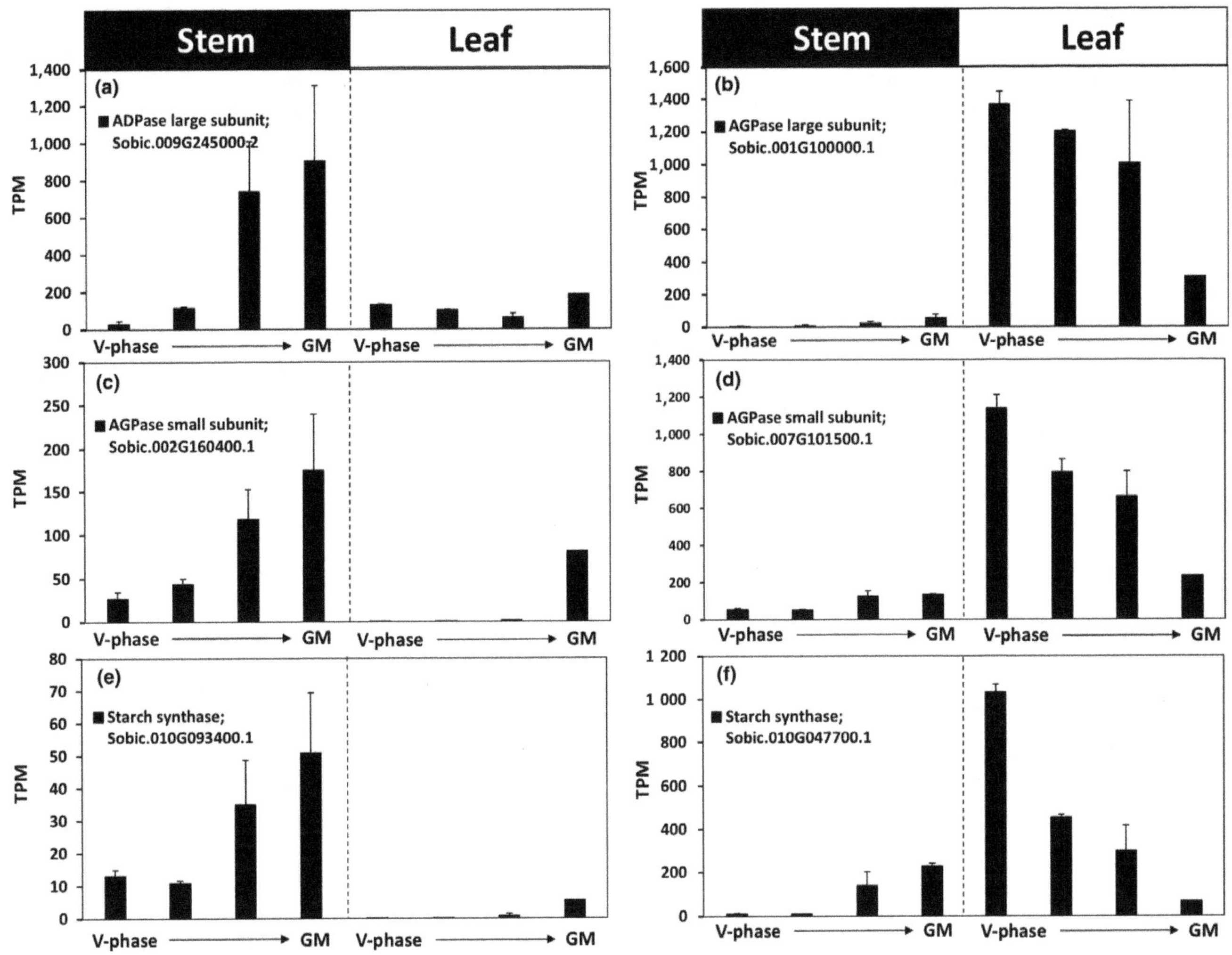

FIGURE 7 Tissue-specific gene expression of starch biosynthetic genes. (a and b) AGPase large subunit expressed in stems and leaves, respectively. (c and d) AGPase small subunit in stems and leaves, respectively. (e and f) Tissue-specific starch synthases expressed in stems and leaves, respectively. Error bars represent *SEM*

2003). In sorghum stems that were accumulating starch, the level of transcripts encoding cytosolic versions of AGPase subunits was low and abundance did not increase in parallel with stem starch accumulation. In contrast, transcripts encoding AGPase subunits with predicted plastid localization were present at higher levels and RNA abundance increased ~eightfold (Sobic.002G1600400.1; *APS1*) and ~30-fold (Sobic.009G245000.2; *APL2*) from floral initiation to 43 days post-anthesis. This indicates that generation of ADP-glucose for starch biosynthesis occurs in plastids similar to other monocot vegetative tissues. This conclusion is consistent with elevated expression of *GPT2* that could transport G-6-P into plastids thereby providing a source of G-1-P for ADP-glucose biosynthesis. Eight genes encoding starch synthases were expressed in stems, and three of these genes showed increased expression (11–29-fold) in parallel with stem starch accumulation between floral initiation and 43 days post-anthesis. Two of the three genes encoding starch-branching enzymes and a gene encoding isoamylase 1 showed similar dynamics of gene expression. Taken together, this study identified genes in gene families that encode enzymes for known steps in starch biosynthesis that are differentially expressed in parallel with stem starch accumulation.

As stem starch is a carbohydrate reserve, it was not surprising that genes encoding enzymes involved in starch turnover were induced in parallel with stem starch accumulation. A gene encoding alpha-glucan-water dikinase that mediates an early step in starch turnover showed the highest differential expression during the developmental time course (~96-fold). Expression of starch phosphorylase (*PHS1*), which generates G-1-P from linear malto-oligosaccharides released from the starch granule, was induced 37-fold. Subsequent conversion of G-1-P to G-6-P would facilitate export of hexoses from the plastid via GPT2. Genes encoding other enzymes involved in starch turnover were induced to lower extents. The expression patterns of the ten gene family members encoding β-amylase that were expressed in stems was especially complex, with some family members differentially expressed in stems at floral initiation followed by decreased expression, and others showing differential expression post-anthesis. Further studies on the biological function of this complexity are warranted.

Genes encoding the disproportionating enzyme (DPE) and glucan phosphorylase (PHS) were expressed in stems. These genes encoded proteins with predicted plastid localization (DPE1, PHS1) as well as proteins with predicted cytosolic localization (DPE2, PHS2). Prior studies have shown that the cytosolic forms of these enzymes are involved in the synthesis/turnover of cytosolic heteroglucans (Fettke et al., 2005, 2009). The function of the cytosolic heteroglucans has been investigated using knock-outs/RNAi constructs targeting *PHS2* and *DPE2* (Chia et al., 2004; Duwenig, Steup, Willmitzer, & Kossmann, 1997; Schopper et al., 2015). Mutants that lack DPE2 are compromised in starch turnover in darkness and show a starch excess phenotype (Chia et al., 2004). In potato, reduction of *PHS2* expression results in early flowering and increased sprouting, possibly indicating modification of sugar signaling (Schopper et al., 2015). In sorghum stems, genes encoding DPE2 and PHS2 show 37-fold and 10-fold increases in expression in parallel with stem sucrose and starch accumulation, DPE and PHS may be involved in synthesis and remobilization of sucrose from the stem, where they catalyze intermediate steps between maltose metabolism and sucrose synthesis (Lu & Sharkey, 2004).

4.2 | Regulation of stem starch gene expression and accumulation

The sucrose transported to stems is utilized for many purposes that depend on growth and development. During rapid stem elongation, sucrose is used principally for cell wall biosynthesis and only low levels of sucrose and starch accumulate in elongating internodes under good growing conditions. Floral initiation reprograms the shoot apical meristem from production of nascent internodes to production of the peduncle and panicle. Nascent internodes produced prior to floral transition grow out between floral transition and anthesis. Stem growth and secondary cell wall formation continues until ~7 days before anthesis maintaining a sink for sucrose transported to stems. However, during the transition from rapid stem growth/secondary cell wall formation and rapid grain filling, there is a reduction in these sink activities that is correlated with the accumulation of stem sucrose and starch. An early step in the transition of stems to becoming heterotrophic storage depots involves the downregulation of vacuolar invertase gene expression starting at floral initiation so that sucrose can accumulate in vacuoles (McKinley et al., 2016a). Stem sucrose sometimes begins to accumulate prior to anthesis but usually reaches maximal levels between anthesis and grain maturity. Some internodes accumulate as a percentage of their biomass, ~10%–12% starch, and ~30% sucrose. Preferential accumulation of sucrose is logical because sequestration and remobilization of sucrose are energetically less costly compared to starch. However, as starch is sequestered as a large polymer with low osmotic activity, accumulation of starch can occur in pith cells that already have high levels of vacuolar sucrose providing an additional, longer term store of sugar for grain filling or tillering (ratooning) post-anthesis.

As the induction of the starch biosynthesis pathway in stems is correlated with the accumulation of sucrose, it is likely that the expression of genes involved in stem starch metabolism is regulated by stem sucrose levels directly or indirectly. Follow on studies will be needed to determine whether sucrose/glucose-sensing pathways (i.e., T6P, SnRK1, TOR, hexokinase) regulate the expression of genes in the stem starch pathway. Additional levels of regulation that occur in chloroplasts and amyloplasts (i.e., metabolic feedback, redox) (Cho et al., 1999; Mikkelsen et al., 2005; Valerio et al., 2011) may also be important modulators of starch accumulation and turnover in stems.

4.3 | Summary

In this study the temporal dynamics and spatial distribution of starch accumulation in stems of the sweet sorghum Della was characterized. Transcriptome and phylogenetic analyses helped identify differentially expressed genes involved in stem starch metabolism and sugar transport through the outer membrane of amyloplasts. Specific gene family members involved in stem starch metabolism were differentially expressed in stems and leaves.

ACKNOWLEDGMENTS

We would like to thank Vlad Panin for use of his HPLC for carbohydrate quantification. We would also like to thank Keerti Rathore for the use his microscope. This research was supported by the Perry Adkisson Chair in Agricultural Biology and the Great Lakes Bioenergy Research Center (DE-FC02-07ER64494 and DE-SC0018409).

AUTHOR CONTRIBUTIONS

B.A.M. and J.E.M. designed the research; B.A.M. and W.L.R. cultivated plant material; B.A.M. and A.L.C. performed experiments; B.A.M. and J.E.M. performed the analysis of data; B.A.M. and J.E.M. wrote the manuscript.

REFERENCES

Beck, E., & Ziegler, P. (1989). Biosynthesis and degredation of starch in higher plants. *Annual Review of Plant Physiology and Plant Molecular Biology*, *40*, 95–117. https://doi.org/10.1146/annurev.pp.40.060189.000523

Beckles, D. M., Smith, A. M., & Ap Rees, T. (2001). A cytosolic ADP-glucose pyrophosphorylase is a feature of graminaceous endosperms, but not of other starch-storing organs. *Plant Physiology*, *125*, 818–827. https://doi.org/10.1104/pp.125.2.818

Bihmidine, S., Baker, R. F., Hoffner, C., & Braun, D. M. (2015). Sucrose accumulation in sweet sorghum stems occurs by apoplasmic phloem unloading and does not involve differential Sucrose transporter expression. *BMC Plant Biology*, *15*, 186. https://doi.org/10.1186/s12870-015-0572-8

Bihmidine, S., Julius, B. T., Dweikat, I., & Braun, D. M. (2016). *Tonoplast Sugar Transporters (SbTSTs)* putatively control sucrose accumulation in sweet sorghum stems. *Plant Signaling & Behavior*, *11*, e1117721. https://doi.org/10.1080/15592324.2015.1117721

Biliaderis, C. G., Maurice, T. J., & Vose, J. R. (1980). Starch gelatinization phenomena studied by differential scanning calorimetry. *Journal of Food Science*, *45*, 1669–1674. https://doi.org/10.1111/j.1365-2621.1980.tb07586.x

Chia, T., Thorneycroft, D., Chapple, A., Messerli, G., Chen, J., Zeeman, S. C., … Smith, A. M. (2004). A cytosolic glucosyltransferase is required for conversion of starch to sucrose in *Arabidopsis* leaves at night. *Plant Journal*, *37*, 853–863. https://doi.org/10.1111/j.1365-313X.2003.02012.x

Cho, M. J., Wong, J. H., Marx, C., Jiang, W., Lemaux, P. G., & Buchanan, B. B. (1999). Overexpression of thioredoxin h leads to enhanced activity of starch debranching enzyme (pullulanase) in barley grain. *Proceedings of the National Academy of Sciences of the United States of America*, *96*, 14641–14646. https://doi.org/10.1073/pnas.96.25.14641

Critchley, J. H., Zeeman, S. C., Takaha, T., Smith, A. M., & Smith, S. M. (2001). A critical role for disproportionating enzyme in starch breakdown is revealed by a knock-out mutation in Arabidopsis. *Plant Journal*, *26*, 89–100. https://doi.org/10.1046/j.1365-313x.2001.01012.x

Czedik-Eysenberg, A., Arrivault, S., Lohse, M. A., Feil, R., Krohn, N., Encke, B., … Stitt, M. (2016). The Interplay between carbon availability and growth in different zones of the growing maize leaf. *Plant Physiology*, *172*, 943–967.

Daehwan, K., Langmead, B., & Salzberg, S. L. (2015). HISAT: A fast spliced aligner with low memory requirements. *Nature Methods*, *12*, 357–360.

Delatte, T., Umhang, M., Trevisan, M., Eicke, S., Thorneycroft, D., Smith, S. M., & Zeeman, S. C. (2006). Evidence for distinct mechanisms of starch granule breakdown in plants. *Journal of Biological Chemistry*, *281*, 12050–12059. https://doi.org/10.1074/jbc.M513661200

Denyer, K., Dunlap, F., Thorbjornsen, T., Keeling, P., & Smith, A. M. (1996). The major form of ADP-glucose pyrophosphorylase in maize endosperm is extra-plastidial. *Plant Physiology*, *112*, 779–785. https://doi.org/10.1104/pp.112.2.779

Duwenig, E., Steup, M., Willmitzer, L., & Kossmann, J. (1997). Antisense inhibition of cytosolic phosphorylase in potato plants (*Solanum tuberosum* L.) affects tuber sprouting and flower formation with only little impact on carbohydrate metabolism. *Plant Journal*, *12*, 323–333. https://doi.org/10.1046/j.1365-313X.1997.12020323.x

Emanuelsson, O., Nielsen, H., & von Heijne, G. (1999). ChloroP, a neural network-based method for predicting chloroplast transit peptides and their cleavage sites. *Protein Science*, *8*, 978–984. https://doi.org/10.1110/(ISSN)1469-896X

Fettke, J., Malinova, I., Eckermann, N., & Steup, M. (2009). Cytosolic heteroglycans in photoautotrophic and in heterotrophic plant cells. *Phytochemistry*, *70*, 696–702. https://doi.org/10.1016/j.phytochem.2009.03.016

Fettke, J., Poeste, S., Eckermann, N., Tiessen, A., Pauly, M., Geigenberger, P., & Steup, M. (2005). Analysis of cytosolic heteroglycans from leaves of transgenic potato (*Solanum tuberosum* L.) plants that under- or overexpress the Pho 2 phosphorylase isozyme. *Plant and Cell Physiology*, *46*, 1987–2004. https://doi.org/10.1093/pcp/pci214

Geigenberger, P., & Fernie, A. R. (2014). Metabolic control of redox and redox control of metabolism in plants. *Antioxidants & Redox Signaling*, *21*, 1389–1421. https://doi.org/10.1089/ars.2014.6018

Gnansounou, E., & Dauriat, A. (2010). Techno-economic analysis of lignocellulosic ethanol: A review. *Bioresource Technology*, *101*, 4980–4991. https://doi.org/10.1016/j.biortech.2010.02.009

Gutjahr, S., Clement-Vidal, A., Soutiras, A., Sonderegger, N., Braconnier, S., Dingkuhn, M., & Luquet, D. (2013). Grain, sugar and biomass accumulation in photoperiod-sensitive sorghums. II. Biochemical processes at internode level and interaction with phenology. *Functional Plant Biology*, *40*, 355–368. https://doi.org/10.1071/FP12177

Hädrich, N., Gibon, Y., Schudoma, C., Altmann, T., Lunn, J. E., & Stitt, M. (2011). Use of TILLING and robotised enzyme assays to generate an allelic series of *Arabidopsis thaliana* mutants with altered ADP-glucose pyrophosphorylase activity. *Journal of Plant Physiology*, *168*, 1395–1405. https://doi.org/10.1016/j.jplph.2011.01.013

Harrison, R. L., & Miller, F. R. (1993). Registration of Della sweet sorghum. *Crop Science*, *33*, 1416. https://doi.org/10.2135/cropsci1993.0011183X003300060076x

Hejazi, M., Fettke, J., Haebel, S., Edner, C., Paris, O., Frohberg, C., … Ritte, G. (2008). Glucan, water dikinase phosphorylates crystalline maltodextrins and thereby initiates solubilization. *Plant Journal*, *55*, 323–334. https://doi.org/10.1111/j.1365-313X.2008.03513.x

Hirose, T., & Terao, T. (2004). A comprehensive expression analysis of the starch synthase gene family in rice (*Oryza sativa* L.). *Planta*, *220*, 9–16. https://doi.org/10.1007/s00425-004-1314-6

Huang, B., Hennen-Bierwagen, T. A., & Myers, A. M. (2014). Functions of multiple genes encoding ADP-glucose pyrophosphorylase subunits in maize endosperm, embryo, and leaf. *Plant Physiology*, *164*, 596–611.

Hussain, H., Mant, A., Seale, R., Zeeman, S., Hinchliffe, E., Edwards, A., … Bustos, R. (2003). Three isoforms of isoamylase contribute different catalytic properties for the debranching of potato glucans. *Plant Cell*, *15*, 133–149. https://doi.org/10.1105/tpc.006635

Jensen, J. K., & Wilkerson, C. G. (2017). Brachypodium as an experimental system for the study of stem parenchyma biology in grasses. *PLoS One*, *12*, e0173095. https://doi.org/10.1371/journal.pone.0173095

Johnson, P. E., Patron, N. J., Bottrill, A. R., Dinges, J. R., Fahy, B. F., Parker, M. L., … Denyer, K. (2003). A low-starch barley mutant, Risø 16, lacking the cytosolic small subunit of ADP-glucose pyrophosphorylase, reveals the importance of the cytosolic isoform and the identity of the plastidial small subunit. *Plant Physiology*, *131*, 684–696.

Lingle, S. E. (1987). Sucrose metabolism in the primary culm of sweet sorghum during development. *Crop Science*, *27*, 1214–1219. https://doi.org/10.2135/cropsci1987.0011183X002700060025x

Lorberth, R., Ritte, G., Willmitzer, L., & Kossmann, J. (1998). Inhibition of a starch-granule-bound protein leads to modified starch and repression of cold sweetening. *Nature Biotechnology*, *16*, 473–477. https://doi.org/10.1038/nbt0598-473

Lu, Y., & Sharkey, T. D. (2004). The role of amylomaltase in maltose metabolism in the cytosol of photosynthetic cells. *Planta*, *218*, 466–473. https://doi.org/10.1007/s00425-003-1127-z

Mcbee, G. G., & Miller, F. R. (1982). Carbohydrates in sorghum culms as influenced by cultivars, spacing, and maturity over a diurnal period. *Crop Science*, *22*, 381–385. https://doi.org/10.2135/cropsci1982.0011183X002200020042x

McBee, G. G., Waskom, R. M., & Creelman, R. A. (1983). Effect of senescene on carbohydrates in sorghum during late Kernel Maturity states. *Crop Science*, *23*, 372–376. https://doi.org/10.2135/cropsci1983.0011183X002300020044x

McCormick, R. F., Truong, S. K., Sreedasyam, A., Jenkins, J., Shu, S., Sims, D., … Mullet, J. (2018). The Sorghum bicolor reference genome: Improved assembly, gene annotations, a transcriptome atlas, and signatures of genome organization. *Plant Journal*, *93*, 338–354. https://doi.org/10.1111/tpj.13781

McKinley, B., Rooney, W., Wilkerson, C., & Mullet, J. (2016a). Dynamics of biomass partitioning, stem gene expression, cell wall biosynthesis, and sucrose accumulation during development of Sorghum bicolor. *Plant Journal*, *88*, 662–680. https://doi.org/10.1111/tpj.13269

McKinley, B., Rooney, W., Wilkerson, C., & Mullet, J. (2016b). Dynamics of biomass partitioning, stem gene expression, cell wall biosynthesis, and sucrose accumulation during development of *Sorghum bicolor*. *Plant Journal*, *88*, 662–680. https://doi.org/10.1111/tpj.13269

Mikkelsen, R., Mutenda, K. E., Mant, A., Schü, P., Blennow, A., & Buchanan, B. B. (2005). α-Glucan, water dikinase (GWD): A plastidic enzyme with redox-regulated and coordinated catalytic activity and binding affinity. *Proceedings of the National Academy of Sciences of the United States of America*, *102*, 1785–1790. https://doi.org/10.1073/pnas.0406674102

Milne, R. J., Offler, C. E., Patrick, J. W., & Grof, C. P. L. (2015). Cellular pathways of source leaf phloem loading and phloem unloading in developing stems of Sorghum bicolor in relation to stem sucrose storage. *Functional Plant Biology*, *42*, 957. https://doi.org/10.1071/FP15133

Milne, R. J., Perroux, J. M., Rae, A. L., Reinders, A., Ward, J. M., Offler, C. E., … Grof, C. P. L. (2017). Sucrose transporter localization and function in phloem unloading in developing stems. *Plant Physiology, 173*, 1330–1341. https://doi.org/10.1104/pp.16.01594

Mizuno, H., Kasuga, S., & Kawahigashi, H. (2016). The sorghum SWEET gene family: Stem sucrose accumulation as revealed through transcriptome profiling. *Biotechnology for Biofuels, 9*, 127. https://doi.org/10.1186/s13068-016-0546-6

Mullet, J., Morishige, D., McCormick, R., Truong, S., Hilley, J., McKinley, B., … Rooney, W. (2014). Energy sorghum–a genetic model for the design of C4 grass bioenergy crops. *Journal of Experimental Botany, 65*, 3479–3489. https://doi.org/10.1093/jxb/eru229

Murray, S. C., Sharma, A., Rooney, W. L., Klein, P. E., Mullet, J. E., Mitchell, S. E., & Kresovich, S. (2008). Genetic improvement of sorghum as a biofuel feedstock: I. QTL for stem sugar and grain nonstructural carbohydrates. *Crop Science, 48*, 2165. https://doi.org/10.2135/cropsci2008.01.0016

Oa, W., Kö, O., Santelia, D., Edner, C., Eicke, S., Marthaler, T., Gentry, M. S.,… Zeeman, S. C. (2009). STARCH-EXCESS4 is a Laforin-like phosphoglucan phosphatase required for starch degradation in *Arabidopsis thaliana*. *Plant Cell, 21*, 334–346.

Paul, M. J., & Foyer, C. H. (2001). Sink regulation of photosynthesis. *Journal of Experimental Botany, 52*, 1383–1400. https://doi.org/10.1093/jexbot/52.360.1383

Pertea, M., Pertea, G. M., Antonescu, C. M., Chang, T.-C., Mendell, J. T., & Salzberg, S. L. (2015). StringTie enables improved reconstruction of a transcriptome from RNA-seq reads. *Nature Biotechnology, 33*, 290–295. https://doi.org/10.1038/nbt.3122

Ritte, G., Heydenreich, M., Mahlow, S., Haebel, S., Kötting, O., & Steup, M. (2006). Phosphorylation of C6-and C3-positions of glucosyl residues in starch is catalysed by distinct dikinases. *FEBS Letters, 580*, 4872–4876. https://doi.org/10.1016/j.febslet.2006.07.085

Rooney, W. L., Blumenthal, J., Bean, B., & Mullet, J. E. (2007). Designing sorghum as a dedicated bioenergy feedstock. *Biofuels, Bioprod Biorefining, 1*, 147–157. https://doi.org/10.1002/bbb.15

Schopper, S., Mühlenbock, P., Sörensson, C., Hellborg, L., Lenman, M., Widell, S., … Andreasson, E. (2015). *Arabidopsis* cytosolic alpha-glycan phosphorylase, PHS2, is important during carbohydrate imbalanced conditions. *Plant Biology, 17*, 74–80. https://doi.org/10.1111/plb.12190

Shannon, J. C., Pien, F.-M., Cao, H., & Liu, K.-C. (1998). Brittle-1, an adenylate translocator, facilitates transfer of extraplastidial synthesized ADP-glucose into amyloplasts of maize endosperms. *Plant Physiology, 117*, 1235–1252. https://doi.org/10.1104/pp.117.4.1235

Slewinski, T. L. (2012). Non-structural carbohydrate partitioning in grass stems: A target to increase yield stability, stress tolerance, and biofuel production. *Journal of Experimental Botany, 63*, 4647–4670. https://doi.org/10.1093/jxb/ers124

Sluiter, A., Ruiz, R., Scarlata, C., Sluiter, J., & Templeton, D. (2008). Determination of extractives in biomass: Laboratory analytical procedure. *NREL Laboratory analytical procedure*, Retrieved from https://www.nrel.gov/docs/gen/fy08/42619.pdf

Sluiter, A., & Sluiter, J. (2005). Determination of starch in solid biomass samples by HPLC: Laboratory Analytical Procedure. *NREL Laboratory Analytical Procedure*, Retrieved from http://devafdc.nrel.gov/pdfs/9360.pdf

Smith, A. M., Denyer, K., & Martin, C. (1997). The synthesis of the starch granule. *Annual Review of Plant Physiology and Plant Molecular Biology, 48*, 67–87. https://doi.org/10.1146/annurev.arplant.48.1.67

Sonnewald, U., & Kossmann, J. (2013). Starches—from current models to genetic engineering. *Plant Biotechnology Journal, 11*, 223–232. https://doi.org/10.1111/pbi.12029

Sowder, C. M., Tarpley, L., Vietor, D. M., & Miller, F. R. (1997). Leaf photoassimilation and partitioning in stress-tolerant sorghum. *Crop Science, 37*, 833. https://doi.org/10.2135/cropsci1997.0011183X003700030023x

Stitt, M., & Zeeman, S. C. (2012). Starch turnover: Pathways, regulation and role in growth. *Current Opinion in Plant Biology, 15*, 282–292. https://doi.org/10.1016/j.pbi.2012.03.016

Tarpley, L., & Vietor, D. M. (2007). Compartmentation of sucrose during radial transfer in mature sorghum culm. *BMC Plant Biology, 7*, 33. https://doi.org/10.1186/1471-2229-7-33

Tetlow, I. J., & Emes, M. J. (2014). A review of starch-branching enzymes and their role in amylopectin biosynthesis. *IUBMB Life, 66*, 546–558. https://doi.org/10.1002/iub.1297

Tjaden, J., Mohlmann, T., Kampfenkel, K., Henrichs, G., & Neuhaus, H. E. (1998). Altered plastidic ATP/ADP-transporter activity influences potato (*Solanum tuberosum* L.) tuber morphology, yield and composition of tuber starch. *Plant Journal, 16*, 531–540. https://doi.org/10.1046/j.1365-313x.1998.00317.x

Valerio, C., Costa, A., Marri, L., Issakidis-Bourguet, E., Pupillo, P., Trost, P., & Sparla, F. (2011). Thioredoxin-regulated β-amylase (BAM1) triggers diurnal starch degradation in guard cells, and in mesophyll cells under osmotic stress. *Journal of Experimental Botany, 62*, 545–555. https://doi.org/10.1093/jxb/erq288

Waclawovsky, A. J., Sato, P. M., Lembke, C. G., Moore, P. H., & Souza, G. M. (2010). Sugarcane for bioenergy production: An assessment of yield and regulation of sucrose content. *Plant Biotechnology Journal, 8*, 263–276. https://doi.org/10.1111/j.1467-7652.2009.00491.x

Wu, L., & Birch, R. G. (2007). Doubled sugar content in sugarcane plants modified to produce a sucrose isomer. *Plant Biotechnology Journal, 5*, 109–117. https://doi.org/10.1111/j.1467-7652.2006.00224.x

Wu, X. R., Staggenborg, S., Propheter, J. L., Rooney, W. L., Yu, J. M., & Wang, D. H. (2010). Features of sweet sorghum juice and their performance in ethanol fermentation. *Industrial Crops and Products, 31*, 164–170. https://doi.org/10.1016/j.indcrop.2009.10.006

Xuan, Y. H., Hu, Y. B., Chen, L.-Q., Sosso, D., Ducat, D. C., Hou, B.-H., & Frommer, W. B. (2013). Functional role of oligomerization for bacterial and plant SWEET sugar transporter family. *Proceedings of the National Academy of Sciences of the United States of America, 110*, E3685–E3694. https://doi.org/10.1073/pnas.1311244110

16

Domain swaps of Arabidopsis secondary wall cellulose synthases to elucidate their class specificity

Joseph Lee Hill Jr[1] | Ashley Nicole Hill[1] | Alison W. Roberts[2] | Candace H. Haigler[3] | Ming Tien[1]

[1]Department of Biochemistry and Molecular Biology, The Center for Lignocellulose Structure and Formation, Pennsylvania State University, University Park, Pennsylvania

[2]Department of Biological Sciences, University of Rhode Island, Kingston, Rhode Island

[3]Department of Crop and Soil Sciences and Department of Plant and Microbial Biology, North Carolina State University, Raleigh, North Carolina

Correspondence
Ming Tien, Department of Biochemistry and Molecular Biology, The Center for Lignocellulose Structure and Formation, Pennsylvania State University, 305 S. Frear Laboratory, University Park, PA 16802.
Email: mxt3@psu.edu

Present address
Joseph Lee Hill Jr, Department of Horticulture, Michigan State University, East Lansing, Michigan 48824.

Funding information
U.S. Department of Energy, Grant/Award Number: DE-SC0001090

Abstract

Cellulose microfibrils are synthesized by membrane-embedded cellulose synthesis complexes (CSCs), currently modeled as hexamers of cellulose synthase (CESA) trimers. The three paralogous CESAs involved in secondary cell wall (SCW) cellulose biosynthesis in Arabidopsis (CESA4, CESA7, CESA8) are similar, but nonredundant, with all three isoforms required for assembly and function of the CSC. The molecular basis of protein–protein recognition among the isoforms is not well understood. To investigate the locations of the interfaces that are responsible for isoform recognition, we swapped three domains between the Arabidopsis CESAs required for SCW synthesis (CESA4, CESA7, and CESA8): N-terminus, central domain containing the catalytic core, and C-terminus. Chimeric genes with all pairwise permutations of the domains were tested for in vivo functionality within knockout mutant backgrounds of *cesa4*, *cesa7*, and *cesa8*. Immunoblotting with isoform-specific antibodies confirmed the anticipated protein expression in transgenic plants. The percent recovery of stem height and crystalline cellulose content was assayed, as compared to wild type, the mutant background lines, and other controls. Retention of the native central domain was sufficient for CESA8 chimeras to function, with neither its N-terminal nor C-terminal domains required. The C-terminal domain is required for class-specific function of CESA4 and CESA7, and CESA7 also requires its own N-terminus. Across all isoforms, the results indicate that the central domain, as well as the N- and C-terminal regions, contributes to class-specific function variously in Arabidopsis CESA4, CESA7, and CESA8.

KEYWORDS
Arabidopsis thaliana, cellulose synthase, chimera, class specificity, domain swap, protein interaction, secondary cell wall

1 | INTRODUCTION

Cellulose is the most abundant biopolymer on earth. It plays a critical role in plant cell growth and morphogenesis, acting as one of the load-bearing components of the cell wall and helping to regulate anisotropic growth (Cosgrove, 2014). Much remains unknown about plant cellulose, including the details of its para-crystalline structure and biosynthesis. In contrast to the recently crystallized bacterial cellulose synthase (Morgan, Strumillo, & Zimmer, 2013), cellulose synthases from plants and some other organisms form cellulose synthesis complexes (CSCs) where multiple β-1,4-glucan chains are produced in close proximity as a prelude to microfibril formation (Giddings, Brower, & Staehelin, 1980; Kimura et al., 1999; Mueller & Brown, 1980). CSCs exist in a variety of structural configurations, which are thought to control cellulose microfibril structure (Itoh, Kimura, & Brown, 2007; Okuda, 2007; Tsekos, 1999). In higher plants, CSCs form hexameric, or "rosette," structures (Mueller & Brown, 1980). Rosettes were recently proposed to comprise 18 cellulose synthase (CESA) proteins arranged in a "hexamer of trimers" configuration (Hill, Hammudi, & Tien, 2014; Newman, Hill, & Harris, 2013; Nixon et al., 2016).

In *Arabidopsis thaliana*, the CSCs that synthesize secondary cell wall (SCW) cellulose are composed of CESA4, CESA7, and CESA8 (Taylor, Howells, Huttly, Vickers, & Turner, 2003). CESAs that synthesize primary cell wall (PCW) cellulose are composed of CESA1, CESA3, and CESA6 (or 6-like CESAs) (Desprez et al., 2007; Persson et al., 2007; Somerville, 2006). Recent studies show that in both cases, the three CESA isoforms are present in equimolar stoichiometries (Gonneau, Desprez, Guillot, Vernhettes, & Hofte, 2014; Hill et al., 2014). Characterization of cellulose-deficient phenotypes in numerous genotypes with mutations in only one CESA shows a stringent requirement for three distinct CESAs during PCW and SCW synthesis. For example, the loss of just one SCW CESA in Arabidopsis causes complete loss of detectable SCW cellulose, with no further effect in double or triple SCW *atcesako* lines (Kumar & Turner, 2015). In addition, when one SCW *AtCESA* is knocked out, protein levels of the remaining two interacting AtCESAs are lost or severely depleted (Hill et al., 2014). The pattern of two CSCs, each with three CESAs is broadly conserved in seed plants, for which characterized genomes contain members of six phylogenetic clades that each encompass one of the required Arabidopsis CESAs (Carroll & Specht, 2011; Kumar et al., 2009).

In an effort to identify regions of the CESA proteins that might be involved in CESA–CESA interaction within the CSC, Carroll and Specht analyzed 82 CESAs from 11 plant species to identify "class-specific regions," that is portions of the sequence alignment that have higher similarity within versus between CESA classes (Carroll & Specht, 2011). They concluded that regions of high sequence class specificity are similar among the clades with the exception of the far N-terminus, which is missing in the CESA3, CESA4, and CESA8 classes, and the far C-terminus, where the CESA1 and CESA8 classes are highly divergent. To empirically test the functional significance of this sequence class specificity, Kumar and coworkers performed a number of reciprocal domain swaps with AtCESA4, AtCESA7, and AtCESA8 (Kumar, Atanassov, & Turner, 2017). Their results suggest that no one individual region is responsible for functional class specificity and that features distributed throughout CESA proteins contribute to class-specific function.

Several regions within CESA sequences are absent from bacterial cellulose synthase and thus might contribute to the unique assembly and class specificity observed in plant CESAs. Zn-binding RING domains are implicated in protein–protein recognition and binding (Leon & Roth, 2000). When expressed heterologously, cotton CESA Zn-binding domains dimerize in a redox-dependent manner (Jacob-Wilk, Kurek, Hogan, & Delmer, 2006), suggesting one possible step of CSC assembly (Carpita, 2011). In the central domain, the Plant-Conserved Region (P-CR) (Pear, Kawagoe, Schreckengost, Delmer, & Stalker, 1996) is highly conserved in sequence and structure among plant CESAs (Carroll & Specht, 2011; Rushton et al., 2017; Sethaphong et al., 2016), but its functional role is not yet proven (Rushton et al., 2017; Sethaphong et al., 2016). In contrast, the Class-Specific-Region (CSR), as its name implies, is more similar within versus between orthologous CESA groups (Carroll & Specht, 2011; Vergara & Carpita, 2001). Similar to the N-terminal domain, heterologously expressed truncated CESA central domains can form multimers: rice CESA8 central domains dimerized (Olek et al., 2014), whereas AtCESA1 central domains trimerized (Vandavasi et al., 2016). In addition, the crystal structure of heterologously expressed rice CESA8 P-CR includes a 3-fold contact. However, a trimer modeled using this contact is thought to be impossible when the membrane domain is included (Rushton et al., 2017). To date, a heterologously expressed truncated CESA C-terminal domain has not been studied, but the chitin synthase SPSA of *Bacillus subtilis*, a homolog of CESA, requires a similar C-terminal tail for dimerization (Charnock, Henrissat, & Davies, 2001).

A recent study tested chimeric CESAs produced by swapping nine relatively short CESA regions (Kumar et al., 2017). In contrast, we tested fewer and generally larger protein regions (compare Figure 1a,b) for their ability to function within paralogous SCW CESAs. We reasoned that swapping larger domains could preserve functional regions within the tertiary structure of chimeric CESAs, including those that may cross the boundaries between smaller regions. As in the previous study (Kumar et al., 2017), we tested the ability of chimeric genes to rescue crystalline cellulose deficiency and short stems in the relevant knockout mutant background lines of *AtCESA4*, *AtCESA7*, and *AtCESA8* (the *cesa4ko*, *cesa7ko*, and *cesa8ko* lines, respectively). In our experiments, we swapped three domains (Figure 1): (a) a N-terminal region inclusive of a Zn-binding RING motif (Zn), a variable sequence region (VR1), and transmembrane helices (TMH1,2); (b) a large cytosolic/catalytic central domain, which includes the P-CR (within CR1) and CSR (within VR2); and (c) a relatively short C-terminal region composed mostly of TMHs (Pear et al., 1996; Saxena & Brown, 1997). Our results indicate that the central domain and C-terminus confer class specificity in AtCESAs involved in SCW cellulose synthesis. Although the N-terminal domain appears not to be important in class-specific interactions, this does not rule out the possibility that it participates in CESA–CESA interactions that are non-class-specific.

2 | MATERIALS AND METHODS

All chemicals were obtained from Sigma-Aldrich, St. Louis, MO unless otherwise specified.

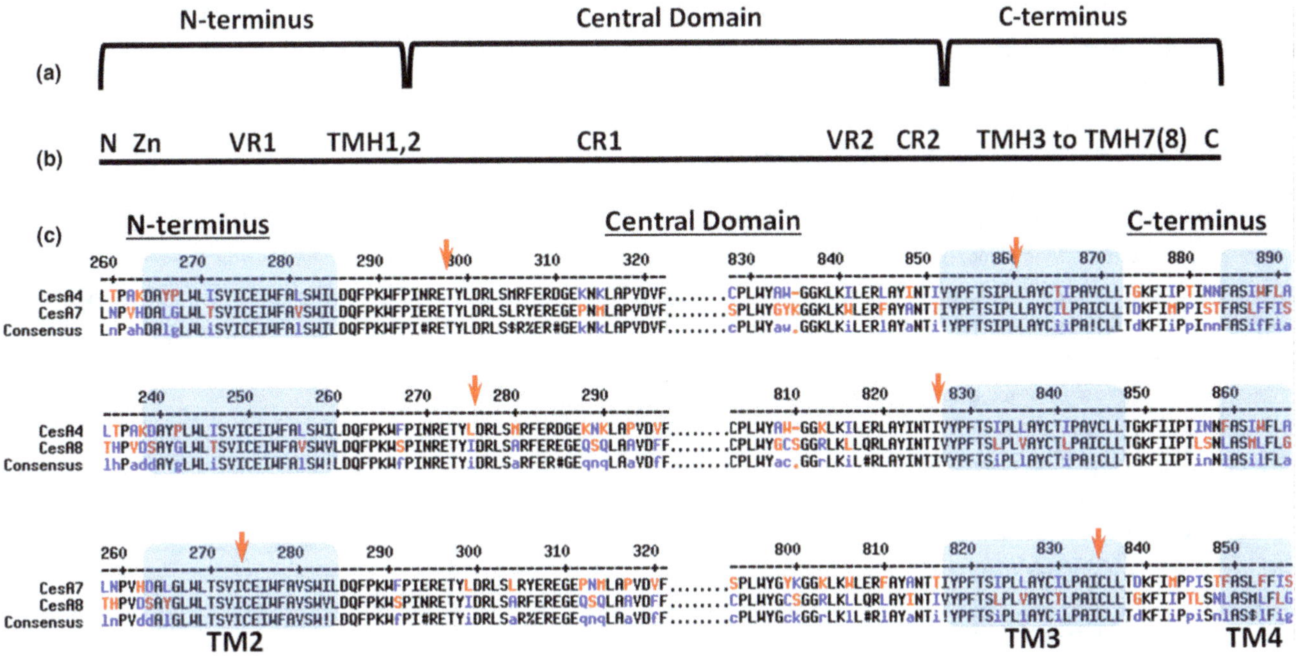

FIGURE 1 Chimeric CESA construction. (a) Three CESA segments assembled to make chimeric constructs: N-terminus; central domain containing the catalytic core and the peripheral P-CR and CSR domains; and the C-terminus. (b) Nine CESA segments swapped by Kumar and coworkers (Kumar et al., 2017) to make chimeric constructs: N = short N-terminus prior to the Zn-binding domain; Zn = Zinc-binding domain; VR1 = variable region 1; TMH1,2; CR1 = conserved region 1 with the P-CR in the middle; VR2 = variable region 2 composed mostly of the CSR; CR2 = conserved region 2; TMH3-7(8), and C = remaining protein after the last TM helix. (c) Trimmed MULTALIN alignments showing the junction (arrow) between segments assembled in chimeric constructs. These junctions were all in the vicinity of TM2 or TM3 (blue boxes), but were selected for each domain swap pair to maximize amino acid and nucleotide sequence identity adjacent to the junction

2.1 | Seed and DNA stocks

Seeds were obtained from the Arabidopsis Biological Research Center (ABRC, Ohio State) for wild-type Arabidopsis of the Columbia ecotype (CS70000), *cesa4ko* (*irx5-4*, SALK_084627), *cesa7ko* (*irx3-4*, SALK_029940C), *cesa8ko* (*irx1-5*, SALK_026812C) (Alonso et al., 2003). Only one of these, *cesa4ko* (*irx5-4*) with a T-DNA insertion in the third exon close to the N-terminus, was used as a background mutant line by Kumar and coworkers (Kumar et al., 2017). However, all three mutant lines were described previously (Brown, Zeef, Ellis, Goodacre, & Turner, 2005) and used by Carroll and coworkers (Carroll et al., 2012) for promoter-swap experiments and by Hill and coworkers (Hill et al., 2014) to demonstrate that they are complete null alleles. Corresponding to that, each one shows the well-known irregular xylem (*irx*) phenotype (Turner & Somerville, 1997). CESA4 and CESA7 cDNA clones were obtained from the ABRC (stock #U50150 and #U22199, respectively). The *CESA8* cDNA clone was a gift from Ying Gu (The Pennsylvania State University). The pORE-O3 plant transformation vector was obtained from the ABRC (stock #CD3-922).

2.2 | Generation of transgenic *Arabidopsis* plant lines

Promoter fragments comprising approximately 2.5 kb of sequence upstream of the CESA4 or CESA7 start sites were amplified by polymerase chain reaction with the primers listed in Supporting Information Table S1, digested with *SacII* and *NotI*, and then ligated into *SacII/NotI* digested pORE-O3 (Coutu et al., 2007) to generate pORE-O3[Pro4] and pORE-O3[Pro7].

CESA topology was predicted with TOPCONs web server (Tsirigos, Peters, Shu, Kall, & Elofsson, 2015) to define the transmembrane (TM) regions, TM2 and TM3. Pairwise amino acid (Corpet, 1988) and cDNA (Kumar, Tamura, & Nei, 1994) alignments were made between all combinations of AtCESA4, AtCESA7 and AtCESA8 with default parameters (BLOSUM-62 for amino acid, ClustalW with IUB matrix for cDNA). From these alignments, primers were designed to assemble chimeric CESA genes with regions of high amino acid and cDNA identity selected as the junctions between domains. CESA4 (AT5G44030; NM_123770), CESA7 (AT5G17420; NM_121748), and CESA8 (AT4G18780; NM_117994) cDNA fragments corresponding to the N-terminus, central domain, or C-terminus were amplified with the primers listed in Supporting Information Table S2. These purified insert pieces were then assembled via a SLiCE reaction into *NotI/PstI* digested pORE-O3[Pro4] (for CESA4/CESA8 domain swaps) or pORE-O3[Pro7] (for CESA4/CESA7 and CESA7/CESA8 domain swaps) (Zhang, Werling, & Edelmann, 2012).

For simplicity, the chimeric CESA genes are named according to the isomer origin of their constituent domains, for example CESA484 (or simply 484 in graphs) is composed of the N-terminus of CESA4, the central domain of CESA8, and the C-terminus of CESA4. The chimeric constructs were introduced by the floral dip method

(Clough & Bent, 1998) into two cognate knockout lines among three possibilities: *cesa4ko* (*irx5-4*), *cesa7ko* (*irx3-4*), or *cesa8ko* (*irx1-5*). For example, *CESA484* was transformed into both the *cesa4ko* and *cesa8ko* background lines to yield two novel genotypes designated as $cesa4ko^{CESA484}$ or $cesa8ko^{CESA484}$. Transgenic plants were selected by spraying soil-grown seedlings with 2 ml of 75 μg/ml Glufosinate-ammonium at 7, 10, and 13 days after planting. For biochemical analysis, plants at the T2 or homozygous T3 stage were typically used, and a consistent stem height phenotype was observed for at least three independent transformants of each genotype across multiple generations. The exceptions were $cesa8ko^{CESA484}$, where only a single transformant was recovered, and $cesa4ko^{CESA484}$, where a pool of two dozen T1 plants was analyzed.

To generate $CESA8^{\Delta NT}$, a fragment amplified from CESA8 cDNA with primers 5′ATCTCCGGCCGTCCCTGCGGCCGCCATGAGGAC AAAAATCACTTCATATAGG3′ and 5′TCACTAGTAAAAGGTACCGA GCTCCTTAGCAATCGATCAAAAGACAGTTC3′ was inserted in pORE-O3[Pro7] via a SLiCE reaction (Coutu et al., 2007; Zhang et al., 2012). This construct was transformed into *cesa8ko* (*irx1-5*), and two independent transgenic lines were selected as described above.

2.3 | Phenotype analysis

Plants were grown at 22–24°C with 18 hr days in 4″ square pots, 6 plants per pot on average, containing ProMix BX (Premier Tech Horticulture, Quakertown, PA), supplemented with Osmocote (14-14-14) slow release fertilizer at a rate of 3 g per liter of growing media (ScottsMiracle-Gro, Marysvile, OH). With the exception of $cesa8ko^{CESA484}$ and $cesa8ko^{CESA8\Delta NT}$, all plants were grown at the same time under identical conditions. The $cesa8ko^{CESA8\Delta NT}$ and $cesa8ko^{CESA484}$ lines were grown beside wild type and *irx1-5* for direct comparison. Normalization of cellulose content and stem height to the value of the wild type in the same experiment minimized the effects of confounding variables that could potentially have arisen from variances in different growing cycles. Stem height was measured with a ruler (minimum of 6 plants per line) as the full length of the primary inflorescence stem of 10-week-old plants.

Crystalline cellulose content was assessed in 10-week-old-stems after dissolving other components in strong acid (Updegraff, 1969). Primary inflorescence stems of 8–20 plants per line were stripped of branches, siliques, and leaves and cut into small pieces. Tissue was extracted with 70% ethanol then 100% acetone for at least 1 day each. After removal of acetone, the tissue was air-dried and ball milled to a fine powder at ambient temperature with a CryoMill (Retsch, Haan, Germany). Five 2–8 mg samples of each stem tissue pool were assayed independently, and standard deviations are reported for these technical replicates. Each sample was incubated in 1 ml of Updegraff reagent (8:2:1, acetic acid:H_2O:nitric acid) for 30 min in a boiling water bath. Cooled samples were pelleted by centrifugation, washed successively with 1 ml of H_2O and 1 ml of acetone, then air-dried. The pellets were resuspended and completely dissolved in 1 ml of 12 M H_2SO_4, requiring about 16 hr. A 20-50 μl aliquot of each sample was diluted with H_2O to a final volume of 350 μl prior to adding 650 μl of concentrated H_2SO_4 containing 0.2% anthrone. The samples were boiled (5 min) alongside a glucose standard curve, then cooled. The absorbance at 620 nm of 200 μl aliquot was determined in a microplate reader. The crystalline cellulose content was calculated from the linear standard curve of glucose and expressed as a percentage of the wild-type value. Genotypes that differed significantly in stem height and cellulose content from both the wild type and the background mutant (Mann–Whitney nonparametric test, $p < 0.01$) were interpreted as partial rescues. The Real Statistics Resource Pack (Release 5.1) for EXCEL was used for this analysis.

2.4 | Immunoblotting

For protein blotting, protein was extracted from 7-week-old primary inflorescence stems. Stems were ground in liquid nitrogen and acetone containing 10% trichloroacetic acid was added for protein extraction (Wang et al., 2006). The protein pellets were washed with acetone, dried, and then resuspended in phosphate-buffered saline containing 1% SDS. Protein content was determined as per (Peterson, 1977). Samples for immunoblotting were diluted to 2 mg/ml in SDS-PAGE loading buffers prior to immunoblotting as previously described (Hill et al., 2014). Antibodies were made by Covance (Denver, PA) using synthetic peptide antigens targeted to unique N-terminal or central domain regions for each CESA, and affinity purified in our lab as previously described (Hill et al., 2014). The specificity of the antibodies was tested by western blots against heterologously expressed CESAs and extracts of wild type and knockout lines of Columbia ecotype as previously described (Hill et al., 2014). Although the individual domains were not used for western blot analysis, the antibodies were synthesized to specific domains of each CESA.

Proteins were subjected to SDS-PAGE and then transferred to 0.1 μm pore nitrocellulose (Whatman, Piscataway, NJ). The membranes were then processed with primary antibody, and secondary antibody (Goat anti-rabbit horseradish peroxidase conjugate, KPL 95058-730) as described previously (Hill et al., 2014). Blots were visualized with SuperSignal West Pico Chemiluminescent Substrate and CL-Xposure Film (both from ThermoFisher Scientific, Waltham, MA). Cropped blot images are provided in the main text figures and full-length blots are shown in Supporting Information Figure S4.

3 | RESULTS

3.1 | Domain swap strategy and implementation

CESAs were divided into three large regions, the N-terminus, central domain, and C-terminus (Figure 1). To minimize possible complications arising from improper folding of chimeric constructs, domains were spliced at highly conserved locations in the TMH region. TMH2 was chosen as the transition between the N-terminus and central domain and TMH3 as the transition point between the central domain and C-terminus (Figure 1). Although it is uncertain whether particular predicted TMHs are authentic and consequently whether CESA has seven or eight TMHs (Slabaugh, Davis, Haigler, Yingling, & Zimmer, 2014), it is generally accepted that TMH2 and TMH3 form the boundaries of the large central/catalytic domain in the cytoplasm. The TMH2 transition point defines an N-terminal region of 218–295 amino acids, including the Zn-binding domain and a variable region. The central domain fragment corresponds to a 517–562 amino acid region containing the CESA catalytic core as well as the P-CR and CSR regions. At last, the C-terminal fragment consists of the remaining TMHs and the final 204–224 amino acids. Assembly of these domains in every pairwise combination between *CESA4*, *CESA7*, and *CESA8* produced 18 different constructs (Supporting Information Figure S2) named as described in the Material and Methods section. The native *CESA4* promoter was used to drive expression of the swaps between *CESA4* and *CESA8*, whereas the native *CESA7* promoter was used for the *CESA4/CESA7* and *CESA7/CESA8* swaps.

3.2 | Rescue of *cesa8ko*

Of the 18 chimeric constructs (Supporting Information Figure S2), four at least partially restored the wild-type phenotype in the *cesa8ko (irx1-5)* background: *CESA488*, *CESA788*, *CESA484*, and *CESA787* (Figure 2 and Supporting Information Figure S2). In addition, the positive control of *CESA8* driven by the *CESA4* promoter (*Pro*^{CESA4}*:CESA8*) validated the use of the *CESA4* promoter for driving *CESA8* expression. Rescue was revealed visually and quantitatively by the height of the plants (Figure 2b, Supporting Information Figure S3). The *cesa8ko*^{CESA484} plants were the same height as wild type, whereas *cesa8ko*^{CESA488}, *cesa8ko*^{CESA787}, and *cesa8ko*^{CESA788} plants had stem heights intermediate between wild type and the mutant background line. All of these chimeras retained the CESA8 central domain. The crystalline cellulose content in three of the transformants (*cesa8ko*^{CESA484}, *cesa8ko*^{CESA488}, *cesa8ko*^{CESA788}), was similar to wild type and the *cesa8ko*$^{ProCESA4:CESA8}$ control (Figure 2a). In contrast, the *cesa8ko*^{CESA787} chimera exhibited partial rescue (73% of the wild-type value, $p < 0.01$, Figure 2a).

To verify the genotypes of the rescued lines, we probed protein extracts (Figure 2c) with previously developed antibodies to the N-terminus of AtCESA4 (anti-CES4.1), AtCESA7 (anti-CESA7.1), and AtCESA8 (anti-CESA8.2) and central domain of AtCESA7 (anti-CESA7.3) (Hill et al., 2014). In the *cesa8ko*^{CESA488} and *cesa8ko*^{CESA484} lines, anti-CESA4.1 (N-terminus) identifies both endogenous CESA4 and a 4 kDa smaller band corresponding to the chimeric proteins (Figure 2c). This mass-shift is expected, as the CESA8 central domain

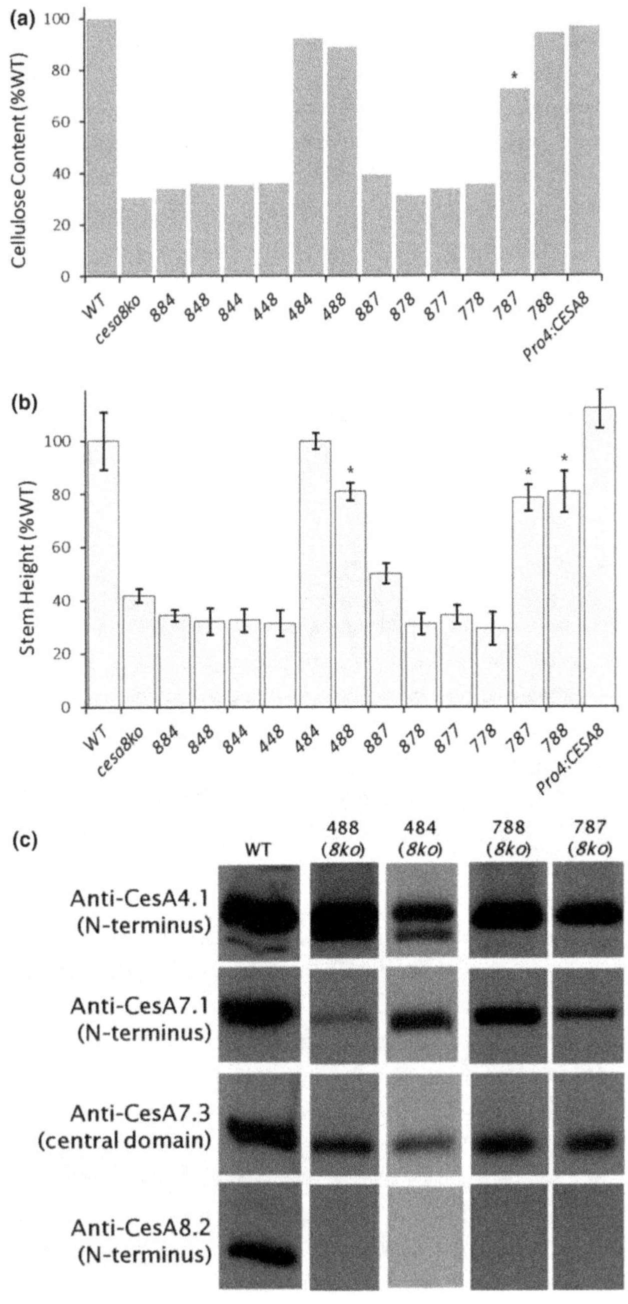

FIGURE 2 *cesa8ko* is rescued by four chimeric constructs containing the CESA8 central domain. Among 12 chimeric CESAs tested, CESA484, CESA488, CESA787, and CESA788 were able to rescue the *cesa8ko (irx1-5)* phenotype. Phenotypes were also rescued by the positive control (Pro4:CESA8, *CESA8* driven by a *CESA4* native promoter). (a) Crystalline cellulose content of each genotype as a percent of wild type (WT). Error bars are standard deviations (STD) from $n = 5$ technical replicates of a pooled sample composed of primary inflorescence stems harvested from 8–20 plants per line. (b) Stem height of each genotype. STD error bars derive from $n = 8$–25 stems from individuals from a single transgenic line. In (a) and (b), asterisks indicate partially rescued lines ($p < 0.01$ as determined by nonparametric Mann–Whitney test between values for the wild type and the transformed line). (c) Immunoblot analysis characterizing domain swap lines that rescued the mutant phenotype. Anti-CESA4.1 recognizes endogenous CESA4 in all lines and the 4 kDa smaller CESA488 and CESA484. Both anti-CESA7.1 and anti-CESA7.3 recognize endogenous CESA7, but CESA788 and CESA787 are too similar in molecular weight to CESA7 to be separated from the native protein. In all transgenic rescue lines, no signal from anti-CESA8.2 was observed, confirming the *cesa8ko* genetic background

of CESA488 and CESA484 lacks a CESA4-specific insertion within the CSR (Supporting Information Figure S1). Endogenous CESA7 protein is recognized by both anti-CESA7.1 and anti-CESA7.3 (Figure 2c). Although CESA788 and CESA787 are presumably recognized by anti-CESA7.1 (N-terminus), they cannot be distinguished from CESA7 based on molecular mass. However, in all four lines, the absence of immunoblot signal when probing with anti-CESA8.2 (N-terminus) confirms the lack of a CESA possessing the CESA8 N-terminal domain (Figure 2c).

3.3 | Rescue of *cesa4ko*

In the *cesa4ko* (*irx5-4*) background, only the *CESA744* chimera was able to partially rescue the mutant phenotypes, restoring 85% or 89% of the wild-type stem height or cellulose content, respectively. Both of these values are significantly different than wild type ($p < 0.01$) (Figure 3a,b). As controls, we also showed that *CESA4*, driven by *CESA4* or *CESA7* promoter fragments (Pro^{CESA4}:CESA4 and Pro^{CESA7}:CESA4, respectively), was able to restore both crystalline cellulose content and stem height to wild-type levels (Figure 3a,b). When the $cesa4ko^{CESA744}$ line was characterized by immunoblot analysis (Figure 3c), CESA8 was detected with anti-CESA8.2 and no signal arose from probing with anti-CESA4.1 (N-terminus), as expected in the *cesa4ko* genetic background. However, a doublet was observed with anti-CESA7.1 (N-terminus), where CESA744 has a 5 kDa higher mass due to the larger CESA4 CSR. In addition, the absence of a doublet when probing with anti-CESA7.3 (central domain) signifies the absence of the CESA7 central domain in the higher molecular weight species recognized by CESA7.1 (N-terminus), an immunoblot "fingerprint" identifying this line as $cesa4ko^{CESA744}$.

3.4 | Rescues of *cesa7ko*

In the *cesa7ko* (*irx3-4*) background, only the *CESA747* chimera was able to partially rescue the mutant phenotypes, restoring 79% or 67% of the wild-type stem height or cellulose content, respectively ($p < 0.01$) (Figure 4a,b). In the positive control of *CESA7* driven by a *CESA7* promoter (Pro^{CESA7}:CESA7), crystalline cellulose content was similar to wild type although stems were shorter ($p < 0.01$). As a negative control, we expressed *CESA4* driven by a *CESA7* promoter (Pro^{CESA7}:CESA4), which failed to rescue the mutant phenotypes (Figure 4a,b).

Again, immunoblotting confirmed the identity of the expressed CESA in the $cesa7ko^{CESA747}$ line (Figure 4c). Anti-CESA4.1 and anti-CESA8.2 confirmed the expected presence of the endogenous CESA4 and CESA8 proteins, respectively. When we probed for the CESA7.1 (N-terminus), we observed only a single band for $cesa7ko^{CESA747}$. This cross-reactive band is mass-shifted upward compared to native CESA7 protein, due to the presence of the CESA4 central domain adding 5 kDa. Furthermore, probing for the CESA7 central domain (anti-CESA7.3) gave the expected absence of signal, as CESA747 does not contain a CESA7 central domain and this is in a *cesa7ko* genetic background (Figure 2c).

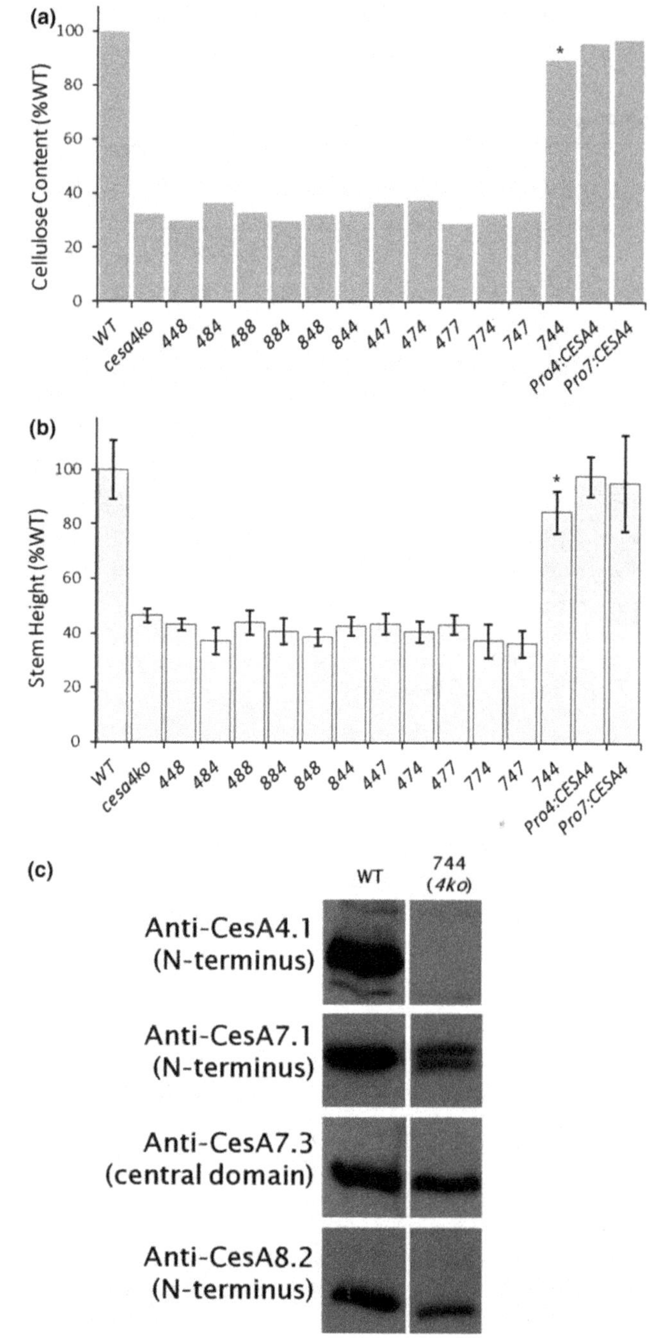

FIGURE 3 *cesa4ko* is rescued only by the CESA744 chimeric construct. Of 12 chimeric CESAs tested, only CESA744 provided phenotypic rescue of *cesa4ko* (*irx5-4*) along with the two positive controls of CESA4 driven by native promoter fragments of CESA4 or CESA7 (Pro4:CESA4 and Pro7:CESA4). (a) Cellulose content of lines (see Figure 2 for details). (b) Stem height measurements for each line (n = 7–26 stems with STD error bars). In (a) and (b), asterisks indicate partially rescued lines ($p < 0.01$ as determined by nonparametric Mann–Whitney test between values for the wild type and the transformed line). (c) Immunoblot analysis showed that, as expected, no signal was observed in the rescued CESA744 line with anti-CESA4.1, whereas a doublet was detected when probing with anti-CESA7.1. Anti-CESA7.3 did not recognize the 5 kDa larger CESA744, but endogenous CESA7 was recognized. Anti-CESA8.2 confirmed the presence of CESA8

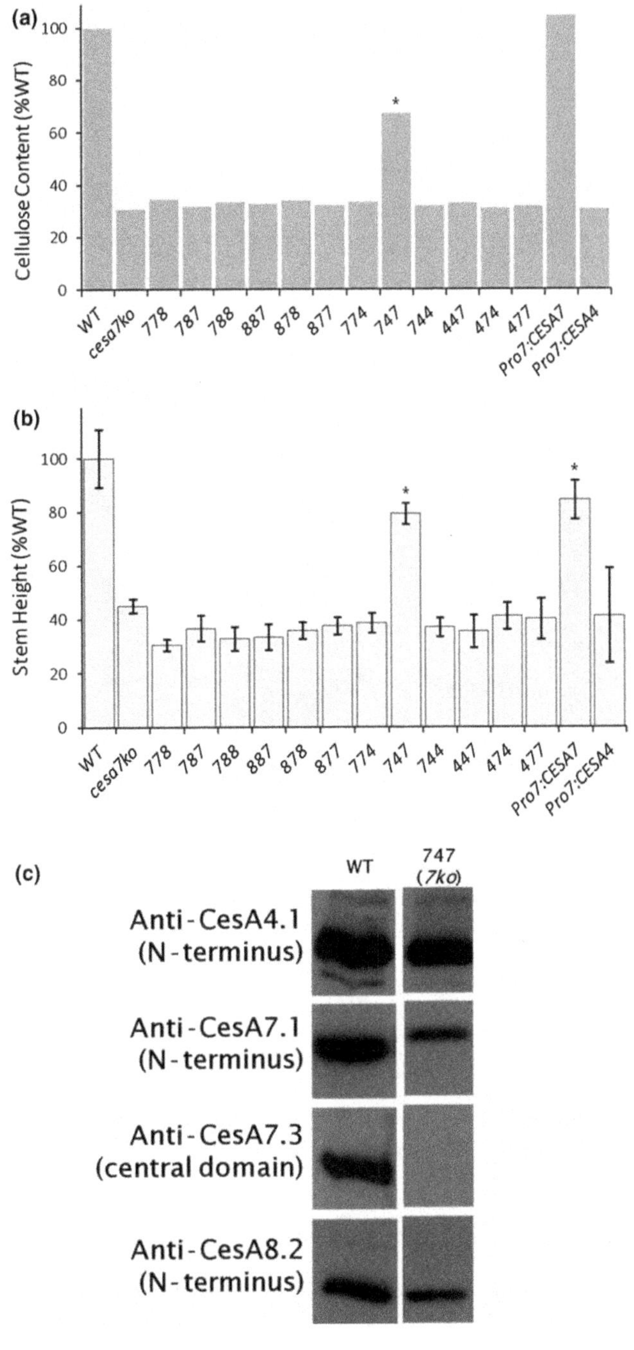

FIGURE 4 *cesa7ko* is rescued by the CESA747 chimeric construct. Of 12 chimeric CESAs, only CESA747 rescued the *cesa7ko* (*irx3-4*) mutant phenotype, along with the positive control of Pro7:CESA7. CESA4 driven by the native CESA7 promoter (Pro4: CESA4) did not rescue. (a) Cellulose content of lines (see Figure 2 for details). (b) Stem height measurements for each line (n = 6–26 stems with STD error bars). In (a) and (b), asterisks indicate partially rescued lines ($p < 0.01$ as determined by nonparametric Mann–Whitney test between values for the wild type and the transformed line). (c) Immunoblot analysis showed normal CESA4 and CESA8. Furthermore, as expected, no signal was observed in the rescued CESA747 line using anti-CESA7.3 that recognizes the central domain, while a 5 kDa larger protein is observed when probing with anti-CESA7.1

3.5 | An N-terminal deletion of CESA8 rescues the *cesa8ko*

Our domain swaps in the *cesa8ko* (*irx1-5*) background revealed some promiscuity in the CESA8 N-terminal domain. The N-terminal domain swaps using either CESA4 or CESA7 ($cesa8ko^{CESA488}$ and $cesa8ko^{CESA788}$) fully rescued crystalline cellulose content and partially rescued stem height in *cesa8ko* (Figure 2). To further determine the limits of this flexibility, we tested whether an N-terminally truncated version of CESA8 could rescue *cesa8ko*. A methionine residue 29 amino acids before the predicted start of TMH1 was used as the translational start site for $CESA8^{\Delta NT}$, which lacks its initial 153 amino acids.

$CESA8^{\Delta NT}$ provided substantial phenotypic rescue of the *cesa8ko*, as shown for two independent lines (Figure 5). Cellulose content and stem height were approximately 75% and 85% of wild-type values, respectively ($p < 0.01$). This level of recovery is similar to that of $cesa8ko^{CESA787}$ (Figure 2). The initial methionine of $CESA8^{\Delta NT}$ lies within the middle of the epitope used to generate anti-CESA8.2. In a fortunate way, we found that the remaining portion of the anti-CESA8.2 epitope was sufficient to visualize $CESA8^{\Delta NT}$ during immunoblotting. Figure 5d clearly shows the absence of full-length CESA8 and the presence of the substantially lower molecular weight $CESA8^{\Delta NT}$, as well as normal CESA4 and CESA7. Our results show that the N-terminus of CESA8 can be removed with only moderate to low impacts on CESA8 functionality.

4 | DISCUSSION

Using the metrics of stem length and crystalline cellulose content, our results demonstrate a limited ability of chimeric CESA genes to rescue mutant phenotypes in knockout lines of *AtCESA4*, *AtCESA7*, and *AtCESA8*. Of 18 chimeric genes we tested with swaps between isomers of the N-terminal, Central, and C-terminal domains (Figure 6a, Supporting Information Figure S2), only six were able to rescue the *cesa4ko* (Figure 6b), *cesa7ko* (Figure 6c) or *cesa8ko* (Figure 6d) mutant background lines (Supporting Information Figure S3). Linear regression demonstrated a strong positive correlation (R = 0.92, R^2 = 0.84) between stem length and cellulose content for mutant background and rescued lines (Supporting Information Figure S5). Among the rescued genotypes, 67% of wild-type crystalline cellulose was the lowest amount that facilitated partial rescue of stem height (79% of wild-type for CESA747 rescuing the *cesa7ko*).

The successful complementation experiments reported in our current study and the related study by Kumar and coworkers (Kumar et al., 2017) are summarized in Figure 6. The goal of the present study was to identify portions of the CESA proteins that are important in protein–protein interactions as revealed by successful complementation. Although failure of complementation is potentially informative, improper folding of a transgenic protein is difficult to distinguish from its inability to integrate into CSCs, because in both cases the CSCs will fail

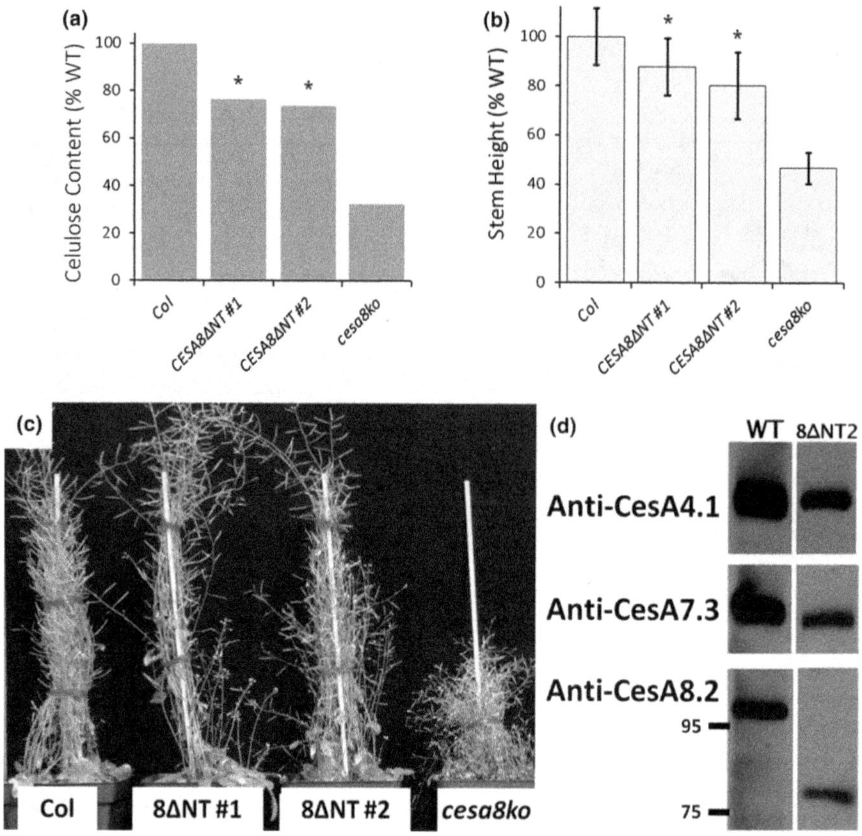

FIGURE 5 Truncation of the CESA8 N-terminus does not abolish function. A truncated version of CESA8 ($CESA8^{\Delta NT}$) lacking 153 amino acids, or 84% of the 182 amino acids in the N-terminus, rescued the phenotypes of *cesa8ko*. (a) Cellulose content of wild type, *cesa8ko*, and two independent $CESA8^{\Delta NT}$ transgenic lines (see Figure 2 for details). (b) Stem height measurements for each line (n = 24–39 stems with STD error bars). In (a) and (b), asterisks indicate partially rescued lines ($p < 0.01$ as determined by nonparametric Mann–Whitney test between values for the wild type and the transformed line). (c) Images of plants showing good phenotypic rescue with $CESA8^{\Delta NT}$ in comparison with the wild type and the cesa8ko. (d) Immunoblot analysis of $CESA8^{\Delta NT}$ showed normal CESA4 and CESA7 and that the lower MW truncated form of CESA8 can be visualized when probing with anti-CESA8.2

to arrive at the plasma membrane and all SCW CESAs will be rapidly degraded (Atanassov, Pittmann, & Turner, 2009; Hill et al., 2014).

The results of the current study and the one of Kumar and coworkers (Kumar et al., 2017) are both consistent and complementary. The *cesa4ko* line was nearly fully rescued in our study by the CESA744 chimera with the entire CESA7 N-terminus (Figure 3), whereas the prior results showed rescue of *cesa4ko* when four smaller N-terminal subdomains (N, ZN, VR1, or TM1) were swapped (Kumar et al., 2017) (Figure 6b). Similar to that, our results showed partial rescue of *cesa7ko* with CESA747 containing the entire central domain of CESA4 (Figure 4), which is consistent with results from testing swaps of the small VR2 and CR2 domains, as well as CR1 together with CR2 (Kumar et al., 2017) (Figure 6c). Rescue of *cesa8ko* by four chimeras (488, 484, 788, and 787) in our study (Figure 2) showed that CESA8 could function with the N-terminus from CESA4 or CESA7, and with either of these together with the C-terminus from the same isomer. Kumar and coworkers (Kumar et al., 2017) likewise showed that diverse chimeras containing parts of either CESA4 or CESA7 could replace CESA8 (Figure 6d). This is consistent with phylogenetic evidence that AtCESA8 is more specialized than AtCESA7 (Scavuzzo-Duggan et al., 2018). These results further suggest that AtCESA8 is more specialized than AtCESA4, as no chimeric constructs containing any portion of AtCESA8 were able to rescue *cesa4ko* or *cesa7ko*.

4.1 | Functionality of CESA N-terminal domains

Among the three CESA domains swapped in this study, the N-terminus is the least similar between isoforms in both sequence identity (Supporting Information Table S3) and size. CESA7 has a long N-terminus (Carroll & Specht, 2011), whereas the N-termini of CESA4 and CESA8 are shorter by 23 and 54 amino acids, respectively. The longer CESA7 N-terminus also shares the highest sequence similarity with the N-termini of CESAs from the moss *Physcomitrella patens* (Carroll & Specht, 2011), which represents a lineage that diverged from the other land plants prior to the diversification of the CESA family (Roberts & Bushoven, 2007). Within the lineage that includes ferns and seed plants, the CESA7 clade diverged first, followed by divergence of the CESA4 and CESA8 clades (Carroll & Specht, 2011; Kumar et al., 2016; Yin, Johns, Cao, & Rupani, 2014). This is consistent with shortening of the N-terminus within the lineage that includes CESA4 and CESA8. In an interesting manner, the short N-terminus of CESA8 could be replaced by an N-terminus from either

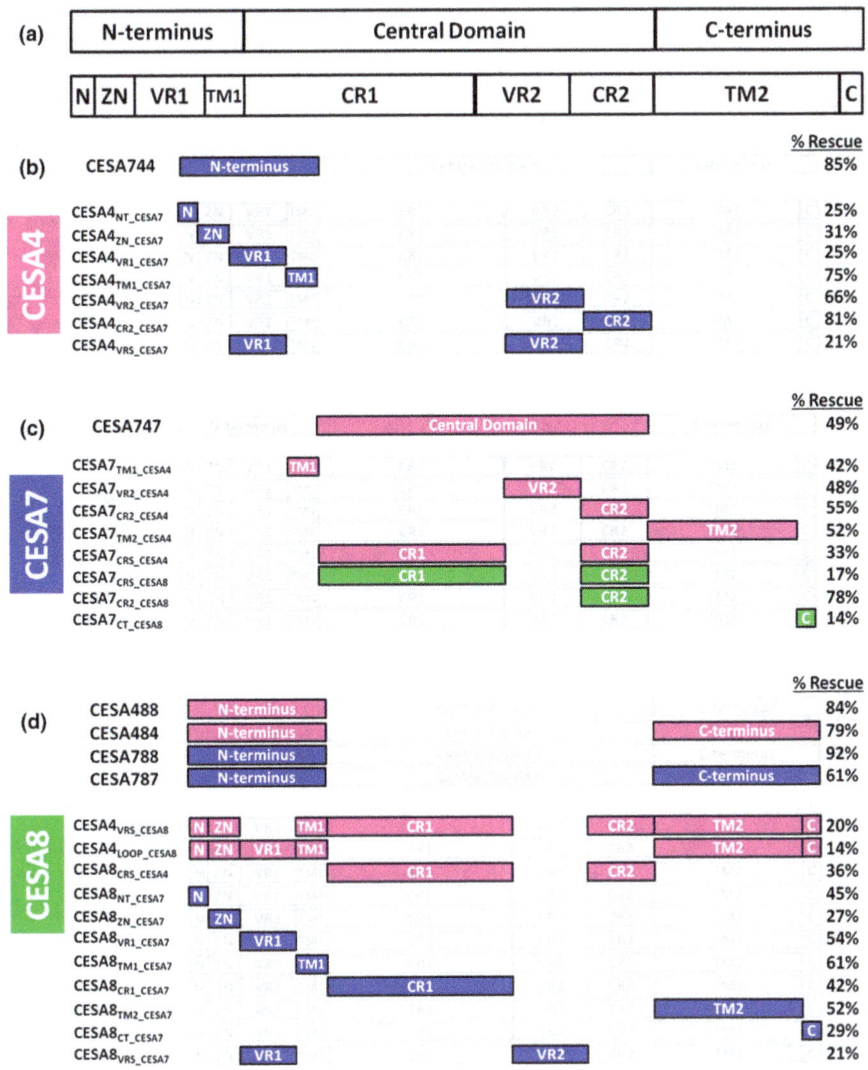

FIGURE 6 Diagram illustrating chimeric constructs complementing knockout mutants in two studies. The positive results of the current work are compared with those reported by Kumar and coworkers (Kumar et al., 2017). Panel A illustrates the three domains used in the current work and how they align with those swapped previously (Kumar et al., 2017). The % Rescue shown refers to crystalline cellulose content, using the previously published method of subtracting the residual amount in the knockout lines from the wild-type value before normalizing the values of other lines (Kumar et al., 2017). Swapped/donor CESAs are color coded: CESA4 (magenta), CESA7 (blue), or CESA8 (green), whereas unchanged regions are white. All constructs able to rescue each knockout line are shown in Panel b, for *cesa4ko*; Panel c for *cesa7ko*; and Panel d for *cesa8ko*. Note that this figure is conceptually useful, but the % rescue values are not precisely comparable because the transgenic lines studied by Kumar and coworkers (Kumar et al., 2017) were not homozygous

CESA4 or CESA7 (Figure 2), or even partially deleted (Figure 5), without abolishing the ability to rescue *cesa8ko*. This indicates that the CESA8 N-terminus is not critical for CSC function in vivo.

In an interesting manner, recent work on heterologously expressed poplar CESA8 demonstrated that the N-terminus could be deleted with little to no effect on catalytic activity (Purushotham et al., 2016). However, the in vitro-synthesized cellulose was more acid-labile (less crystalline) and globular particles (potential CESA complexes) that were seen in in vitro controls were no longer observed. This suggests that the N-terminus may indeed be involved in homomeric CESA8-CESA8 interactions in vitro. However, these in vitro results cannot be directly compared to *in planta* results because CESA4 and CESA7 were absent and rosette CSCs were not observed in vitro. In a potential way, the poplar CESA8 N-terminus facilitates homomeric CESA–CESA interactions in vitro that are not essential for the in vivo function of its Arabidopsis orthologue within a heteromeric CSC.

Reflecting differences between CESA isomers, bioinformatic analyses (Carroll & Specht, 2011) and other domain swapping experiments (Kumar et al., 2017) led to the conclusion that the N-terminus of CESA7 contributes substantially to class specificity. Our observation that the *cesa7ko* could not be rescued by any chimeric CESA lacking a CESA7 N-terminus (i.e., CESA477 or CESA877) supports the functional significance of this domain. However, the general ability of the N-terminus to act as a class-specific determinant is called into question by the results of *cesa4ko*^{CESA744}, where the CESA7 N-terminus does not prevent CESA744 from functioning as a CESA4 protein.

Although our results show that the CESA7 N-terminus is not itself sufficient to confer CESA7 class specificity, we hypothesize that it retains an ancestral domain or motif that is essential for function of the rosette CSC as a whole and that this domain/function has been lost, or partially lost, in CESA4 and CESA8. Phosphoproteomics has shown that CESA1 and CESA3 of the PCW CSC as well as CESA4 and CESA7 of the SCW CSC are phosphorylated in their N-terminal domains (Chen, Ehrhardt, & Somerville, 2010; Jones et al., 2016; Sanchez-Rodriguez et al., 2017; Taylor, 2007). The CESA5 N-terminus is also phosphorylated (Bischoff et al., 2011; Nuhse, Stensballe, Jensen, & Peck, 2004). Multiple experiments with PCW CESAs using site-directed mutagenesis to mimic the presence or absence of phosphorylation have shown that phosphorylation is able to regulate the activity of the CSC (Bischoff et al., 2011; Chen et al., 2010, 2016; Sanchez-Rodriguez et al., 2017). In addition, experiments involving phosphorylation of the CESA7 N-terminal domain indicate a possible role in protein stability (Taylor, 2007). Additional work will be required to precisely determine the function of the CESA N-terminal domain in different isomers.

4.2 | The central and C-terminal domains provide CESA class specificity

Of the six chimeric CESAs that rescued a mutant phenotype, one did so with a mismatched central domain: *cesa7ko*^{CESA747} was partially rescued (Figure 6c). Thirteen vectors failed to rescue when the central domain matched the CESA that was knocked out, including *cesa4ko*^{CESA747}. Although we cannot confidently interpret results for vectors that failed to rescue any mutant, results overall indicate that the central domain is not solely responsible for class-specific function of SCW CESAs. Indeed, partial rescue of the *cesa7ko* only by CESA747 supports the importance of the N- and C-terminal domains in the class specificity of AtCESA7 (Figure 4). Due to the nonrescue reported for a similar construct (CESA7$_{LOOP_CESA4}$) previously (Kumar et al., 2017), the results for three independent lines are shown in Supporting Information Figure S6. The differences between the two studies could be explained by factors such as details of chimeric gene splicing locations, as CESA747 contains additional CESA4 regions compared to CESA7$_{LOOP_CESA4}$. This finding was crucial for our conclusion that both the central and C-terminal domains confer class specificity between SCW AtCESAs.

CESA8 is clearly differentiated from CESA4 and CESA7 by its central domain, as both CESA484 and CESA787 rescue the *cesa8ko* (Figure 6d). But, the central domain cannot be responsible for determining class specificity between CESA4 and CESA7, as CESA747 rescued the *cesa7ko*. Whereas a combination of factors in the N- and C-termini could differentiate CESA4 and CESA7, we propose that the primary determinant(s) lies in the C-terminal domain because switching the C-terminal domain alters class specificity for chimeras of CESA4 and CESA7, with CESA747 able to replace only CESA7 (Figure 4) and CESA744 able to replace only CESA4 (Figure 3). Thus, our data support a model where the C-terminal domain is critical in differentiating between CESA4 and CESA7. However, this role of the C-terminus cannot be extended to CESA8, as both CESA484 and CESA787 function as CESA8 proteins.

Overall, the results indicate that the determinants of Arabidopsis CESA class specificity do not reside in a single region, consistent with conclusions of others (Carroll & Specht, 2011; Kumar et al., 2017) and illustrated for the two sets of domain swap experiments in Figure 6. Portions of CESA7 are able to replace the corresponding CESA4 region, with the exception of the CR1, TM2, and "C" domains as defined by Kumar and coworkers (Figure 6b) (Kumar et al., 2017). The central domain and TM regions of CESA7 can be replaced by those of CESA4, leaving only parts of the CESA7 N-terminus and "C" domain as reservoirs of class specificity. Select CR regions and the "C" domain of CESA8 are able to substitute within CESA7 (Figure 6c). Extensive flexibility is observed in the CESA8 N-terminal and C-terminal domains, with the analogous domains of either CESA4 or CESA7 able to substitute (Figure 6d). Within the central domain, the CR1 domain of CESA8 also lacks class specificity (Kumar et al., 2017). This leaves only VR2 and CR2, which harbor regions that differentiate CESA8 from CESA4 and CESA7, respectively.

Current hypotheses of protein complex evolution predict that CESA class specificity involves multiple interfaces that arose sequentially (Doolittle, 2012; Finnigan, Hanson-Smith, Stevens, & Thornton, 2012). According to that, the requirement for three CESA isomers to form typical PCW and SCW rosette CSCs is predicted to be the outcome of gene duplication followed by accumulation of neutral mutations that generated interfaces between paralogs and abolished interfaces between identical subunits (Doolittle, 2012; Finnigan et al., 2012). Therefore, single CESA isomers became unable to form homomeric CSCs, leading to the requirement of three nonredundant and class-specific CESA isomers, even while each of them could independently synthesize a ß-1,4-glucan chain. Our positive rescue results are consistent with a major role of the central domain in defining AtCESA8 class specificity, whereas the C-terminus differentiates AtCESA7 from AtCESA4. This variability between isomers in the regions (or residues) determining their class specificity is consistent with evolution of the hetero-oligomeric state through accumulation of neutral mutations that generate interfaces between distinct subunits. The results of Kumar and coworkers (Kumar et al., 2017) and those currently reported are complementary and provide a cumulative foundation for future work in mapping the CESA–CESA interaction interfaces in the Arabidopsis SCW CSC.

ACKNOWLEDGMENTS

We thank Dr. Ying Gu (Pennsylvania State University) for the gift of CESA8 cDNA. This research was supported as part of The Center for LignoCellulose Structure and Formation, an Energy Frontier Research Center funded by the U.S. Department of Energy, Office of Science, Basic Energy Sciences under Award # DE-SC0001090. The contribution of CHH was partially supported by the USDA National Institute of Food and Agriculture, Hatch project 1000932.

REFERENCES

Alonso, J. M., Stepanova, A. N., Leisse, T. J., Kim, C. J., Chen, H. M., Shinn, P., ... Ecker, J. R. (2003). Genome-wide insertional mutagenesis of *Arabidopsis thaliana*. *Science*, *301*, 653–657. https://doi.org/10.1126/science.1086391

Atanassov, I. I., Pittmann, J. K., & Turner, S. R. (2009). Elucidating the mechanisms of assembly and subunit interaction of the cellulose synthase complex of *Arabidopsis* secondary walls. *Journal of Biological Chemistry*, *284*, 3833–3841. https://doi.org/10.1074/jbc.M807456200

Bischoff, V., Desprez, T., Mouille, G., Vernhettes, S., Gonneau, M., & Hofte, H. (2011). Phytochrome regulation of cellulose synthesis in *Arabidopsis*. *Current Biology*, *21*, 1822–1827. https://doi.org/10.1016/j.cub.2011.09.026

Brown, D. M., Zeef, L. A. H., Ellis, J., Goodacre, R., & Turner, S. R. (2005). Identification of novel genes in *Arabidopsis* involved in secondary cell wall formation using expression profiling and reverse genetics. *Plant Cell*, *17*, 2281–2295. https://doi.org/10.1105/tpc.105.031542

Carpita, N. C. (2011). Update on mechanisms of plant cell wall biosynthesis: How plants make cellulose and other (1 -> 4)-beta-D-Glycans. *Plant Physiology*, *155*, 171–184. https://doi.org/10.1104/pp.110.163360

Carroll, A., Mansoori, N., Li, S. D., Lei, L., Vernhettes, S., Visser, R. G. F., ... Trindade, L. M. (2012). Complexes with mixed primary and secondary cellulose synthases are functional in *Arabidopsis* plants. *Plant Physiology*, *160*, 726–737. https://doi.org/10.1104/pp.112.199208

Carroll, A., & Specht, C. D. (2011). Understanding plant cellulose synthases through a comprehensive investigation of the cellulose synthase family sequences. *Frontiers in Plant Science*, *2*, 5.

Charnock, S. J., Henrissat, B., & Davies, G. J. (2001). Three-dimensional structures of UDP-sugar glycosyltransferases illuminate the biosynthesis of plant polysaccharides. *Plant Physiology*, *125*, 527–531. https://doi.org/10.1104/pp.125.2.527

Chen, S. L., Ehrhardt, D. W., & Somerville, C. R. (2010). Mutations of cellulose synthase (CESA1) phosphorylation sites modulate anisotropic cell expansion and bidirectional mobility of cellulose synthase. *Proceedings of the National Academy of Sciences of the United States of America*, *107*, 17188–17193. https://doi.org/10.1073/pnas.1012348107

Chen, S. L., Jia, H. L., Zhao, H. Y., Liu, D., Liu, Y. M., Liu, B. Y., ... Somerville, C. R. (2016). Anisotropic cell expansion is affected through the bidirectional mobility of cellulose synthase complexes and phosphorylation at two critical residues on CESA3. *Plant Physiology*, *171*, 242–250. https://doi.org/10.1104/pp.15.01874

Clough, S. J., & Bent, A. F. (1998). Floral dip: A simplified method for Agrobacterium-mediated transformation of *Arabidopsis thaliana*. *Plant Journal*, *16*, 735–743. https://doi.org/10.1046/j.1365-313x.1998.00343.x

Corpet, F. (1988). Multiple sequence alignment with hierarchical-clustering. *Nucleic Acids Research*, *16*, 10881–10890. https://doi.org/10.1093/nar/16.22.10881

Cosgrove, D. J. (2014). Re-constructing our models of cellulose and primary cell wall assembly. *Current Opinion in Plant Biology*, *22*, 122–131. https://doi.org/10.1016/j.pbi.2014.11.001

Coutu, C., Brandle, J., Brown, D., Brown, K., Miki, B., Simmonds, J., & Hegedus, D. D. (2007). pORE: A modular binary vector series suited for both monocot and dicot plant transformation. *Transgenic Research*, *16*, 771–781. https://doi.org/10.1007/s11248-007-9066-2

Desprez, T., Juraniec, M., Crowell, E. F., Jouy, H., Pochylova, Z., Parcy, F., ... Vernhettes, S. (2007). Organization of cellulose synthase complexes involved in primary cell wall synthesis in *Arabidopsis thaliana*. *Proceedings of the National Academy of Sciences of the United States of America*, *104*, 15572–15577. https://doi.org/10.1073/pnas.0706569104

Doolittle, W. F. (2012). Evolutionary biology A ratchet for protein complexity. *Nature*, *481*, 270–271. https://doi.org/10.1038/nature10816

Finnigan, G. C., Hanson-Smith, V., Stevens, T. H., & Thornton, J. W. (2012). Evolution of increased complexity in a molecular machine. *Nature*, *481*, 360–364. https://doi.org/10.1038/nature10724

Giddings, T. H., Brower, D. L., & Staehelin, L. A. (1980). Visualization of particle complexes in the plasma-membrane of micrasterias-denticulata associated with the formation of cellulose fibrils in primary and secondary cell-walls. *Journal of Cell Biology*, *84*, 327–339. https://doi.org/10.1083/jcb.84.2.327

Gonneau, M., Desprez, T., Guillot, A., Vernhettes, S., & Hofte, H. (2014). Catalytic subunit stoichiometry within the cellulose synthase complex. *Plant Physiology*, *166*, 1709–1712. https://doi.org/10.1104/pp.114.250159

Hill, J. L. Jr, Hammudi, M. B., & Tien, M. (2014). The Arabidopsis cellulose synthase complex: A proposed hexamer of CESA trimers in an equimolar stoichiometry. *Plant Cell*, *26*, 4834–4842. https://doi.org/10.1105/tpc.114.131193

Itoh, T., Kimura, S., & Brown, R. M. (2007). Immunogold labeling of cellulose-synthesizing terminal complexes. In R.M.B.A.I.M. Saxena (Ed.), *Cellulose: Molecular and structural biology: Selected articles on the synthesis, structure, and applications of cellulose* (pp. 237–255). Dordrecht, The Netherlands: Springer. https://doi.org/10.1007/978-1-4020-5380-1

Jacob-Wilk, D., Kurek, I., Hogan, P., & Delmer, D. P. (2006). The cotton fiber zinc-binding domain of cellulose synthase A1 from Gossypium hirsutum displays rapid turnover in vitro and in vivo. *Proceedings of the National Academy of Sciences of the United States of America*, *103*, 12191–12196. https://doi.org/10.1073/pnas.0605098103

Jones, D. M., Murray, C. M., Ketelaar, K. J., Thomas, J. J., Villalobos, J. A., & Wallace, I. S. (2016). The emerging role of protein phosphorylation as a critical regulatory mechanism controlling cellulose biosynthesis. *Frontiers in Plant Science*, *7*, 684.

Kimura, S., Laosinchai, W., Itoh, T., Cui, X. J., Linder, C. R., & Brown, R. M. (1999). Immunogold labeling of rosette terminal cellulose-synthesizing complexes in the vascular plant *Vigna angularis*. *Plant Cell*, *11*, 2075–2085. https://doi.org/10.1105/tpc.11.11.2075

Kumar, M., Atanassov, I., & Turner, S. (2017). Functional Analysis of Cellulose Synthase (CESA) protein class specificity. *Plant Physiology*, *173*, 970–983. https://doi.org/10.1104/pp.16.01642

Kumar, S., Tamura, K., & Nei, M. (1994). Mega – Molecular evolutionary genetics analysis software for microcomputers. *Computer Applications in the Biosciences*, *10*, 189–191.

Kumar, M., Thammannagowda, S., Bulone, V., Chiang, V., Han, K. H., Joshi, C. P., ... Ellis, B. E. (2009). An update on the nomenclature for the cellulose synthase genes in Populus. *Trends in Plant Science*, *14*, 248–254. https://doi.org/10.1016/j.tplants.2009.02.004

Kumar, M., & Turner, S. (2015). Protocol: A medium-throughput method for determination of cellulose content from single stem pieces of *Arabidopsis thaliana*. *Plant Methods*, *11*, 46. https://doi.org/10.1186/s13007-015-0090-6

Kumar, M., Wightman, R., Atanassov, I., Gupta, A., Hurst, C. H., Hemsley, P. A., & Turner, S. (2016). S-Acylation of the cellulose synthase complex is essential for its plasma membrane localization. *Science*, *353*, 166–169. https://doi.org/10.1126/science.aaf4009

Leon, O., & Roth, M. (2000). Zinc fingers: DNA binding and protein-protein interactions. *Biological Research*, *33*, 21–30.

Morgan, J. L. W., Strumillo, J., & Zimmer, J. (2013). Crystallographic snapshot of cellulose synthesis and membrane translocation. *Nature*, *493*, 181–192.

Mueller, S. C., & Brown, R. M. (1980). Evidence for an intramembrane component associated with a cellulose microfibril-synthesizing complex in higher-plants. *Journal of Cell Biology*, *84*, 315–326. https://doi.org/10.1083/jcb.84.2.315

Newman, R. H., Hill, S. J., & Harris, P. J. (2013). Wide-angle X-ray scattering and solid-state nuclear magnetic resonance data combined to test models for cellulose microfibrils in mung bean cell walls. *Plant Physiology, 163*, 1558–1567. https://doi.org/10.1104/pp.113.228262

Nixon, B. T., Mansouri, K., Singh, A., Du, J., Davis, J. K., Lee, J. G., . . . Haigler, C. H. (2016). Comparative structural and computational analysis supports eighteen cellulose synthases in the plant cellulose synthesis complex. *Scientific Reports, 6*, 28696. https://doi.org/10.1038/srep28696

Nuhse, T. S., Stensballe, A., Jensen, O. N., & Peck, S. C. (2004). Phosphoproteomics of the Arabidopsis plasma membrane and a new phosphorylation site database. *Plant Cell, 16*, 2394–2405. https://doi.org/10.1105/tpc.104.023150

Okuda, K. A. S. (2007). Cellulose-synthesizing complexes of a dinoflagellate and other unique algae. In R.M.B.A.I.M. Saxena (Ed.), *Cellulose: Molecular and structural biology: Selected articles on the synthesis, structure, and applications of cellulose* (pp. 199–215). Dordrecht, The Netherlands: Springer. https://doi.org/10.1007/978-1-4020-5380-1

Olek, A. T., Rayon, C., Makowski, L., Kim, H. R., Ciesielski, P., Badger, J., . . . Carpita, N. C. (2014). The structure of the catalytic domain of a plant cellulose synthase and its assembly into dimers. *Plant Cell, 26*, 2996–3009. https://doi.org/10.1105/tpc.114.126862

Pear, J. R., Kawagoe, Y., Schreckengost, W. E., Delmer, D. P., & Stalker, D. M. (1996). Higher plants contain homologs of the bacterial celA genes encoding the catalytic subunit of cellulose synthase. *Proceedings of the National Academy of Sciences of the United States of America, 93*, 12637–12642. https://doi.org/10.1073/pnas.93.22.12637

Persson, S., Paredez, A., Carroll, A., Palsdottir, H., Doblin, M., Poindexter, P., . . . Somerville, C. R. (2007). Genetic evidence for three unique components in primary cell-wall cellulose synthase complexes in Arabidopsis. *Proceedings of the National Academy of Sciences of the United States of America, 104*, 15566–15571. https://doi.org/10.1073/pnas.0706592104

Peterson, G. L. (1977). Simplification of protein assay method of Lowry Et Al - Which is more generally applicable. *Analytical Biochemistry, 83*, 346–356.

Purushotham, P., Cho, S. H., Diaz-Moreno, S. M., Kumar, M., Nixon, B. T., Bulone, V., & Zimmer, J. (2016). A single heterologously expressed plant cellulose synthase isoform is sufficient for cellulose microfibril formation in vitro. *Proceedings of the National Academy of Sciences of the United States of America, 113*, 11360–11365. https://doi.org/10.1073/pnas.1606210113

Roberts, A. W., & Bushoven, J. T. (2007). The cellulose synthase (CESA) gene superfamily of the moss *Physcomitrella patens*. *Plant Molecular Biology, 63*, 207–219.

Rushton, P. S., Olek, A. T., Makowski, L., Badger, J., Steussy, C. N., Carpita, N. C., & Stauffacher, C. V. (2017). Rice cellulose SynthaseA8 plant-conserved region is a coiled-coil at the catalytic core entrance. *Plant Physiology, 173*, 482–494. https://doi.org/10.1104/pp.16.00739

Sanchez-Rodriguez, C., Ketelaar, K., Schneider, R., Villalobos, J. A., Somerville, C. R., Persson, S., & Wallace, I. S. (2017). BRASSINOSTEROID INSENSITIVE2 negatively regulates cellulose synthesis in *Arabidopsis* by phosphorylating cellulose synthase 1. *Proceedings of the National Academy of Sciences of the United States of America, 114*, 3533–3538. https://doi.org/10.1073/pnas.1615005114

Saxena, I. M., & Brown, R. M. (1997). Identification of cellulose synthase (s) in higher plants: Sequence analysis of processive beta-glycosyltransferases with the common motif 'D, D, D35Q(R, Q)XRW'. *Cellulose, 4*, 33–49. https://doi.org/10.1023/A:1018411101036

Scavuzzo-Duggan, T. R., Chaves, A. M., Singh, A., Sethaphong, L., Slabaugh, E., Yingling, Y. G., . . . Roberts, A. W. (2018). Cellulose synthase 'class specific regions' are intrinsically disordered and functionally undifferentiated. *Journal of Integrative Plant Biology*. https://doi.org/10.1111.jipb.12637 [E-pub ahead of print].

Sethaphong, L., Davis, J. K., Slabaugh, E., Singh, A., Haigler, C. H., & Yingling, Y. G. (2016). Prediction of the structures of the plant-specific regions of vascular plant cellulose synthases and correlated functional analysis. *Cellulose, 23*, 145–161. https://doi.org/10.1007/s10570-015-0789-6

Slabaugh, E., Davis, J. K., Haigler, C. H., Yingling, Y. G., & Zimmer, J. (2014). Cellulose synthases: New insights from crystallography and modeling. *Trends in Plant Science, 19*, 99–106. https://doi.org/10.1016/j.tplants.2013.09.009

Somerville, C. (2006). Cellulose synthesis in higher plants. *Annual Review of Cell and Developmental Biology, 22*, 53–78. https://doi.org/10.1146/annurev.cellbio.22.022206.160206

Taylor, N. G. (2007). Identification of cellulose synthase AtCesA7 (IRX3) in vivo phosphorylation sites – A potential role in regulating protein degradation. *Plant Molecular Biology, 64*, 161–171. https://doi.org/10.1007/s11103-007-9142-2

Taylor, N. G., Howells, R. M., Huttly, A. K., Vickers, K., & Turner, S. R. (2003). Interactions among three distinct CesA proteins essential for cellulose synthesis. *Proceedings of the National Academy of Sciences of the United States of America, 100*, 1450–1455. https://doi.org/10.1073/pnas.0337628100

Tsekos, I. (1999). The sites of cellulose synthesis in algae: Diversity and evolution of cellulose-synthesizing enzyme complexes. *Journal of Phycology, 35*, 635–655. https://doi.org/10.1046/j.1529-8817.1999.3540635.x

Tsirigos, K. D., Peters, C., Shu, N., Kall, L., & Elofsson, A. (2015). The TOPCONS web server for consensus prediction of membrane protein topology and signal peptides. *Nucleic Acids Research, 43*, W401–W407. https://doi.org/10.1093/nar/gkv485

Turner, S. R., & Somerville, C. R. (1997). Collapsed xylem phenotype of *Arabidopsis* identifies mutants deficient in cellulose deposition in the secondary wall. *Plant Cell, 9*, 689–701. https://doi.org/10.1105/tpc.9.5.689

Updegraff, D. M. (1969). Semimicro determination of cellulose in biological materials. *Analytical Biochemistry, 32*, 420–424. https://doi.org/10.1016/S0003-2697(69)80009-6

Vandavasi, V. G., Putnam, D. K., Zhang, Q., Petridis, L., Heller, W. T., Nixon, B. T., . . . O'Neill, H. (2016). A structural study of CESA1 catalytic domain of Arabidopsis cellulose synthesis complex: Evidence for CESA trimers. *Plant Physiology, 170*, 123–135. https://doi.org/10.1104/pp.15.01356

Vergara, C. E., & Carpita, N. C. (2001). beta-D-Glycan synthases and the CesA gene family: Lessons to be learned from the mixed-linkage (1 -> 3), (1 -> 4)beta-D-glucan synthase. *Plant Molecular Biology, 47*, 145–160. https://doi.org/10.1023/A:1010631431620

Wang, W., Vignani, R., Scali, M., & Cresti, M. (2006). A universal and rapid protocol for protein extraction from recalcitrant plant tissues for proteomic analysis. *Electrophoresis, 27*, 2782–2786.

Yin, Y. B., Johns, M. A., Cao, H. S., & Rupani, M. (2014). A survey of plant and algal genomes and transcriptomes reveals new insights into the evolution and function of the cellulose synthase superfamily. *BMC Genomics, 15*, 260. https://doi.org/10.1186/1471-2164-15-260

Zhang, Y. W., Werling, U., & Edelmann, W. (2012). SLiCE: A novel bacterial cell extract-based DNA cloning method. *Nucleic Acids Research, 40*, e55. https://doi.org/10.1093/nar/gkr1288

17

Functional characterization of squalene synthase and squalene epoxidase in *Taraxacum koksaghyz*

Kristina Unland[1*] | Katharina M. Pütter[2*] | Kirsten Vorwerk[2] | Nicole van Deenen[2] | Richard M. Twyman[3] | Dirk Prüfer[1,2] | Christian Schulze Gronover[1]

[1]Fraunhofer Institute for Molecular Biology and Applied Ecology (IME), Muenster, Germany

[2]Institute of Plant Biology and Biotechnology, University of Muenster, Muenster, Germany

[3]TRM Ltd, Scarborough, UK

Correspondence
Christian Schulze Gronover, Fraunhofer Institute for Molecular Biology and Applied Ecology (IME), Schlossplatz 8, 48143 Muenster, Germany.
Email: christian.schulze.gronover@ime.fraunhofer.de

Abstract

The Russian dandelion *Taraxacum koksaghyz* produces high-value isoprenoids such as pentacyclic triterpenes and natural rubber in the latex of specialized cells known as laticifers. Squalene synthase (SQS) and squalene epoxidase (SQE) catalyze key steps in the biosynthesis of cyclic terpenoids, but neither enzyme has yet been characterized in *T. koksaghyz*. Genomic analysis revealed the presence of two genes (*TkSQS1* and *TkSQS2*) encoding isoforms of SQS, and four genes (*TkSQE1–4*) encoding isoforms of *SQE*. Spatial expression analysis in different *T. koksaghyz* tissues confirmed that TkSQS1 and TkSQE1 are the latex-predominant isoforms, with highly similar mRNA expression profiles. The TkSQS1 and TkSQE1 proteins colocalized in the endoplasmic reticulum membrane and their enzymatic functions were confirmed by in vitro activity assays and yeast complementation studies, respectively. The functions of TkSQS1 and TkSQE1 were further characterized in the latex of *T. koksaghyz* plants with depleted *TkSQS1* or *TkSQE1* mRNA levels, produced by RNA interference. Comprehensive expression analysis revealed the coregulation of *TkSQS1* and *TkSQE1*, along with a downstream gene in the triterpene biosynthesis pathway encoding the oxidosqualene cyclase TkOSC1. This indicates that the coregulation of *TkSQS1*, *TkSQE1*, and *TkOSC1* could be used to optimize the flux toward specific terpenoids during development.

KEYWORDS

latex, oxidosqualene cyclase, pentacyclic triterpene, RNA interference, squalene epoxidase, squalene synthase, *Taraxacum koksaghyz* (Russian dandelion), transcriptional regulation

1 | INTRODUCTION

The dandelion species *Taraxacum koksaghyz* (Tk) produces a plethora of secondary metabolites in specialized cells known as laticifers, including high-value isoprenoids such as the polyisoprene natural rubber, sterols, and pentacyclic triterpenes. The pentacyclic triterpenes in particular exhibit extraordinary biological activities against fungi and bacteria, making them ideal for agricultural and pharmaceutical applications (Ghosh, 2016).

The biosynthesis of pentacyclic triterpenes requires squalene synthase (SQS, EC 2.5.1.21) and squalene epoxidase (SQE, EC 1.14.14.7) as rate-limiting enzymes. SQS is a bifunctional membrane-bound enzyme that catalyzes the synthesis of squalene from two C_{15} allylic farnesyl diphosphate (FPP) molecules in two steps. First, presqualene diphosphate is produced via the head-to-head condensation of two FPP molecules, and this is subsequently reduced to squalene in an NADPH-dependent second step requiring divalent

*These authors contributed equally to the article

cations (Jarstfer, Zhang, & Poulter, 2002). Only a single *SQS* gene is found in yeast and humans (Jennings, Tsay, Fisch, & Robinson, 1991; Robinson, Tsay, Kienzle, Smith-Monroy, & Bishop, 1993), whereas there are 1–3 *SQS* genes in plants. Thus far, single *SQS* genes have been reported among others in *Oryza sativa* (Hata et al., 1997), *Lotus japonicus* (Akamine et al., 2003), *Taxus cuspidata* (Huang et al., 2007), and *Euphorbia tirucalli* (Uchida et al., 2009), whereas two paralogs are found in *Nicotiana tabacum* (Devarenne, Shin, Back, Yin, & Chappell, 1998), *Glycyrrhiza glabra* (Hayashi, Hirota, Hiraoka, & Ikeshiro, 1999), *Glycine max* (Nguyen et al., 2013), *Malus domestica* (Navarro Gallón et al., 2017), and *Arabidopsis thaliana*, although in the last case, one copy was shown to be a pseudogene (Busquets et al., 2008). *Panax ginseng* possesses three *SQS* paralogs (Kim, Han, Huh, & Choi, 2011). SQS enzymes contain a C-terminal hydrophobic transmembrane domain that anchors the enzyme into the endoplasmic reticulum (ER) membrane, whereas the large catalytic N-terminal domain is located in the cytosol (Stamellos et al., 1993). Because squalene is the first precursor of triterpenoids such as sterols, brassinosteroids, and pentacyclic triterpenes, SQS activity is an important switch and major branching point between triterpene and polyisoprene biosynthesis (Figure 1). As such, SQS represents a rate-limiting step in triterpene biosynthesis and is associated with the overall yield of these secondary metabolites. This has been demonstrated in several plant species: the overexpression of *SQS* in *P. ginseng* and *Eleutherococcus senticosus* increased the overall SQS activity and caused the accumulation of sterols and triterpenes (Lee et al., 2004; Seo et al., 2005), whereas the virus-induced gene silencing of *SQS* in *Withania somnifera* and *M. domestica* caused the sterol levels to decline (Navarro Gallón et al., 2017; Singh et al., 2015). Various RNA interference (RNAi) methods have also been used to silence the *SQS* gene and reduce sterol levels and this also caused the accumulation of artemisinin in *Artemisia annua* (Zhang et al., 2009). Moreover, tobacco cell suspension cultures treated with a fungal elicitor revealed that SQS is regulated at multiple transcriptional and post-translational levels (Devarenne, Ghosh, & Chappell, 2002; Devarenne et al., 1998).

Squalene epoxidase (SQE) catalyzes the epoxidation of squalene to 2,3-oxidosqualene, which is the first oxidation reaction in the triterpene biosynthesis pathway. Subsequently, 2,3-oxidosqualene is converted into various triterpene end-products by oxidosqualene cyclases (OSCs). The oxidation of squalene requires O_2 as well as NADPH and FAD cofactors (Abe & Prestwich, 1999; Nakamura & Sato, 1979; Ono, Ozasa, Hasegawa, & Imai, 1977). In *Saccharomyces cerevisiae*, SQE is located in the ER and is associated with lipid particles (Leber et al., 1998). In plants, SQE is anchored to the ER via transmembrane domains, although the number of predicted transmembrane domains can vary within a species (Han, In, Kwon, & Choi, 2010; He, Zhu, He, & Zhang, 2008; Laranjeira et al., 2015). Like *SQS*, only single *SQE* genes are present in yeast and humans, whereas some plants possess multiple paralogs, for example, *P. ginseng* and *A. thaliana* (Han et al., 2010; Rasbery et al., 2007). The different isoforms encoded by these genes may fulfill different functions; for example, only three of the six SQE isoforms in *A. thaliana* restore

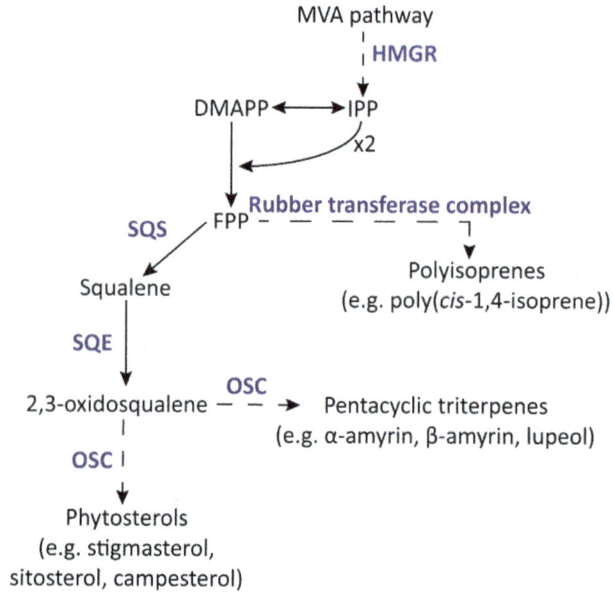

FIGURE 1 Proposed isoprenoid biosynthesis pathway in *Taraxacum koksaghyz* latex. Enzymes or enzyme complexes are shown in blue, and dashed arrows indicate multiple enzymatic steps. DMAPP, dimethylallyl diphosphate; FPP, farnesyl diphosphate; HMGR, 3-hydroxy-methyl-glutaryl-CoA reductase; IPP, isopentenyl diphosphate; MVA, mevalonate; OSC, oxidosqualene cyclase; SQE, squalene epoxidase; SQS, squalene synthase

wild-type activity to the yeast *erg1* mutant. And among them, only *AtSQE1* is essential for growth and development, given that *Atsqe1* mutant plants suffer from defects such as impaired stem elongation and infertility (Rasbery et al., 2007). In *P. ginseng*, two SQE isoforms with different regulatory roles were identified by treatment with methyl jasmonate, which suppresses *PgSQE2* expression but induces the accumulation of *PgSQE1* mRNA in roots. In addition, the knockdown of *PgSQE1* reduced the level of ginsenosides, whereas the overexpression of *PgSQE2* and *PNX* (encoding cycloartenol synthase) increased the accumulation of sterols (Han et al., 2010).

To gain deeper insight into the roles of SQS and SQE in the regulation of isoprenoid biosynthesis in *T. koksaghyz* latex, the isoforms of both enzymes were identified and functionally characterized *in planta*. The transcriptional regulation of the corresponding genes was analyzed at different developmental stages in wild-type dandelion plants and in *SQS* and *SQE* RNAi lines.

2 | MATERIALS AND METHODS

2.1 | Plant material and cultivation conditions

Taraxacum koksaghyz wild-type and transgenic plants were cultivated at 18°C and 20 klux (high pressure sodium lamp, HPS 600 W, Greenbud, enhanced yellow and red spectrum) with a 16-h photoperiod in controlled growth chambers or in the greenhouse. Plants were cultivated in a pre-fertilized 1:1 mixture of standard soil (ED73 Einheitserde, Fröndenberg, Germany) and garden mold (Botanical Garden Münster, Germany). They were fed every 4 weeks with a commercial fertilizer according to the manufacturer's recommendations (Hakaphos Plus, Compo GmbH, Münster, Germany). *Nicotiana benthamiana* seeds were obtained from the Sainsbury Laboratory (John Innes Centre, Norwich, UK) and cultivated as stated above.

2.2 | Total RNA extraction and cDNA synthesis

Total RNA was extracted from *T. koksaghyz* latex, root, leaf, peduncle and flower tissues using the innuPREP RNA Mini Kit (Analytik Jena, Jena, Germany) according to the manufacturer's instructions. Full-length cDNA was synthesized from 500 ng total RNA using PrimeScript RT Master Mix (TaKaRa, Clontech, Saint-Germain-en-Laye, France) according to the manufacturer's instructions.

2.3 | In silico analysis

TkSQS1 and *TkSQE1-2* full-length cDNA sequences were obtained from RNA-Seq data, whereas *TkSQS2* and *TkSQE3-4* full-length gDNA sequences were identified from draft genome data (Lin et al., 2017).

Full-length cDNA was translated using the ExPASy translate tool (http://web.expasy.org/translate/). The isoelectric point and molecular weight were predicted using the ProtParam tool (http://web.expasy.org/protparam/), and protein secondary structures were predicted using SOPMA (Geourjon & Deleage, 1995; https://npsa-prabi.ibcp.fr/cgi-bin/npsa_automat.pl?page=/NPSA/npsa_sopma.html). Conserved protein domains were identified using the NCBI conserved domain search (https://www.ncbi.nlm.nih.gov/Structure/cdd/wrpsb.cgi) and PROSITE (http://prosite.expasy.org/), whereas transmembrane domains were predicted using TMHMM software (http://www.cbs.dtu.dk/services/TMHMM-2.0/).

2.4 | Phylogenetic analysis

Amino acid sequences were obtained from GenBank (https://www.ncbi.nlm.nih.gov/genbank/) and are listed in Supporting Information Table S1. Multiple alignments of protein sequences were generated using the Clustal MUSCLE algorithm (Edgar, 2004; http://www.ebi.ac.uk/Tools/msa/muscle/). Phylogenetic trees were created using MEGA6 (Tamura, Stecher, Peterson, Filipski, & Kumar, 2013; http://www.megasoftware.net/).

2.5 | Quantitative RT-PCR

Quantitative RT-PCR analysis was carried out as previously described (Pütter, van Deenen, Unland, Prüfer, & Schulze Gronover, 2017). *Taraxacum koksaghyz* wild-type plants were grown for 12 weeks for spatial expression analysis and for 8–20 weeks for temporal expression analysis. RNA was extracted from nine individual plants, before pooling the cDNA from three plants. All oligonucleotide sequences for expression analysis are listed in Supporting Information Table S2. Primer efficiencies and amplification factors are summarized in Supporting Information Table S3.

2.6 | Isolation of genomic DNA

Genomic DNA was isolated using the NucleoSpin Plant II Kit (Macherey-Nagel, Düren, Germany) according to the manufacturer's instructions.

2.7 | Cloning and stable transformation procedures

For localization studies, the full-length *TkSQS1* cDNA was amplified using primers TkSQS1-fwd-blunt and TkSQS1-rev-XhoI, digested with XhoI and inserted into the XmnI/XhoI sites of the Gateway-compatible vector pENTR3C (Thermo Fisher Scientific Inc., Darmstadt, Germany), resulting in the final vector pENTR3C-TkSQS1. The full-length *TkSQE1* cDNA was amplified using primers TkSQE1-fwd-NcoI and TkSQE1-rev-NotI, and a similar fragment lacking the stop codon (*TkSQE1-wos*) was amplified using the primers TkSQE1-fwd-NcoI and TkSQE1-wos-rev-NotI. Both fragments were digested with NcoI and NotI and inserted into the NcoI/NotI sites of the Gateway-compatible vector pENTR4 (Thermo Fisher Scientific). For localization analysis, the full-length *TkSQS1* and *TkSQE1* sequences were introduced into the Gateway-compatible vector pBatTL-Cerulean-ccdB, whereas the *TkSQE1-wos* fragment was introduced into the vector pBatTL-ccdB-Cerulean (Epping et al., 2015) by LR recombination using LR Clonase (Thermo Fisher Scientific). The pBatTL-Cerulean-ccdB vector was cloned by amplifying the Cerulean fragment from a pUC18 Cerulean vector (GenBank: CAO79587) using the primer combination Cerulean-BglII-fwd and Cerulean-BglII-rev, digesting the plasmid with the corresponding restriction enzyme and inserting the product into the BglII site of pBaTL-ccdB (Jach, Pesch, Richter, Frings, & Uhrig, 2006).

For TkSQS1 activity assays, the *TkSQS1* full-length sequence was introduced into the Gateway-compatible vector pBatTL-ccdb to yield pBatTL-TkSQS1. The latter was subsequently introduced into *Agrobacterium tumefaciens* strain GV3103pMP90RK as previously described (Pütter et al., 2017).

For the TkSQE1 complementation assay, pENTR4-TkSQE1 was introduced into the Gateway-compatible destination vectors and pAG425GPD-ccdB-HA (Addgene, Cambridge, MA, USA). The expression vector pLab12.5-pREF containing the promoter of the rubber elongation factor (REF) was prepared as previously described (Epping et al., 2015). For the construction of pLab12.5-pREF-TkSQS1-RNAi and pLab12.5-pREF-TkSQE1-RNAi, a 178-bp TkSQS1-RNAi and a 285-bp TkSQE1-RNAi PCR fragment were amplified using the RNA dicer-optimized primers TkSQS1-RNAi-fwd-NcoI and TkSQS1-RNAi-rev-XhoI or TkSQE1-RNAi-fwd-NcoI and TkSQE1-RNAi-rev-XhoI, respectively. Both RNAi fragments were inserted into the NcoI and XhoI sites of the Gateway vector pBluescript II KS (+) (Addgene). Subsequently, the TkSQS1-RNAi and the TkSQE1-RNAi fragments were transferred to the pLab12.5-pREF vector by LR recombination to generate pLab12.5-pREF-TkSQS1-RNAi and pLab12.5-pREF-TkSQE1-RNAi. The transformation of *T. koksaghyz* by *A. tumefaciens* strain EHA105 was carried out as previously described (Stolze et al.,

2017). Oligonucleotide sequences are shown in Supporting Information Table S4.

2.8 | Methyl jasmonate assay

Taraxacum koksaghyz wild-type plants were cultivated under standard conditions for 6 weeks, subsequently the plants were repotted into cylindrical vessels (diameter 3.5 cm, height 5 cm) in vermiculite (Nestaan, Blandain, Belgium). Eight-week-old plants were watered with 25 ml of 0.8 mmol/L MeJA (Sigma-Aldrich, St. Louis, USA) or water as previously described (Cao et al., 2017). As MeJA (95%) was dissolved in absolute ethanol before dilution, the same amount of ethanol was added to the water used for the control plants. Six hours after the treatment latex was harvested from three plants each and used for mRNA expression analysis.

2.9 | Subcellular localization studies

Subcellular localization studies were carried out as previously described (Epping et al., 2015).

In silico analysis revealed the presence of a single TkSQS1 C-terminal transmembrane domain, so an N-terminal fusion with the blue fluorescent protein Cerulean (Cerulean-TkSQS1) was used for the localization studies. In contrast, TkSQE1 was shown to contain both N-terminal and C-terminal transmembrane domains, so we prepared and tested both N-terminal and C-terminal fusions (Cerulean-TkSQE1 and TkSQE1-Cerulean). All three fusion proteins were transiently expressed in *N. benthamiana* leaf cells, which were analyzed by confocal laser scanning microscopy (CLSM). We used the monomeric red fluorescent protein (mRFP) as a cytosolic marker and the N-terminal sequence of CYP51G1 fused to mRFP (NtermCYP51G1-mRFP) as an ER marker, to facilitate the localization of the Cerulean fusion proteins. NtermCYP51G1-mRFP presents mRFP on the cytosolic surface of the ER (Bassard, Mutterer, Duval, & Werck-Reichhart, 2012).

For TkSQS1 and TkSQE1 localization, the corresponding pBatTL constructs were infiltrated into *N. benthamiana* leaves in the following combinations: pBatTL-Cerulean-TkSQS1 + pBatTL-mRFP, pBatTL-Cerulean-TkSQS1 + pBatTL-NtermCYP51G1-mRFP, pBatTL-TkSQE1-Cerulean + pBatTL-mRFP, pBatTL-TkSQE1-Cerulean+ pBatTL-NtermCYP51G1-mRFP, pBatTL-Cerulean + pBatTL-NtermCYP51G1-mRFP.

2.10 | SQS activity assay

Nicotiana benthamiana leaves were infiltrated with the binary vector pBatTL-TkSQS1 in which *TkSQS1* was controlled by the constitutive Cauliflower mosaic virus 35S promoter. After incubation for 5 days, a crude protein extract from leaves was tested using an in vitro SQS assay with [1,2-^{14}C]-labeled FPP as the substrate. Protein extraction and quantification were carried out as previously described (Pütter et al., 2017). SQS activity was assessed in vitro followed by reversed-phase thin-layer chromatography analysis and quantification by scintillation counting as described by Busquets et al. (2008).

2.11 | Yeast complementation assay

For the yeast complementation assay of SQE activity, the *S. cerevisiae* ($\Delta erg1$) mutant stain KLN1 (*MAT (α), ERG1::URA3, leu2, ura3, trp1*) (Landl, Klönsch, & Turnowsky, 1996) and KLN1-pERG1 (a KLN1 derivative carrying a leucine-selectable expression plasmid for *ERG1*) were kindly provided by *F. Turnowsky* (University of Graz, Austria). KLN1 was transformed with the vector pAG425GPD-TkSQE1 or the empty vector by lithium acetate transformation as previously described (Hill, Donald & Griffiths, 1991). A drop test was performed on SD-Leu minimal medium under aerobic conditions and under anaerobic conditions with the supplement of 12 μg/μl ergosterol and 0.5% (v/v) Tween-80. The yeast were suspended in Tris-EDTA buffer, dropped out at $OD_{600\ nm}$ 1, $OD_{600\ nm}$ 0.1, $OD_{600\ nm}$ 0.01, and $OD_{600\ nm}$ 0.001 and incubated for 3 days under anaerobic conditions or 5 days under aerobic conditions.

2.12 | Chemical analysis

Whole *T. koksaghyz* roots were frozen in liquid nitrogen, freeze-dried, and crushed to a fine powder. For triterpene analysis, 100 mg of dry root material was used for saponification by adding 20 ml of methanol containing 6% potassium hydroxide and heating to 80°C for 2 hr. As an internal standard, we added 100 μl of betuline (2.5 mg/ml stock solution in acetone). Samples were extracted three times with 1 volume of hexane. The hexane phases were pooled and evaporated, and the samples were redissolved in 1 ml of acetone. Triterpene analysis was performed with a GC-MS-QP 210 Ultra system (Shimadzu, Duisburg, Germany) equipped with a Rxi®-5 ms column (Restek GmbH, Bad Homburg, Germany). We injected 0.5 μl of the extract using split modus (1:10) at an injector and interface temperature of 260°C. The GC temperature program was as follows: 120°C for 3 min, temperature gradient of 15°C per minute up to 330°C, 330°C for 10 min. Electron ionization (EI) in the MS was set to 70 eV. Peak integration and identification was performed with the LabSolution software (Shimadzu, Duisburg, Germany) using a NIST library (NIST = National Institute of Standards and Technology) or by analyzing the corresponding standards obtained from Extrasynthese (Genay, France). Quantification was performed in relation to the internal standard. Retention indices for the triterpene compounds were determined using the LabSolution software in relation to a C8–C40 alkane calibration standard (Sigma-Aldrich, Taufkirchen, Germany) using the same operating conditions.

3 | RESULTS

3.1 | Molecular cloning and sequence analysis of two *SQS* paralogs (*TkSQS1* and *TkSQS2*) and four *SQE* paralogs (*TkSQE1–TkSQE4*) in *T. koksaghyz*

RNA-Seq data obtained from latex, root, and leaf mRNA expression profiles and the recently published draft genome of *T. koksaghyz* (Lin et al., 2017) were screened for the presence of *SQS* and *SQE* cDNA sequences. One contig showed high similarity to known *SQS* sequences and was strongly expressed in *T. koksaghyz* latex. The primers TkSQS1-fwd and TkSQS1-rev were therefore designed to amplify the full-length cDNA sequence, subsequently designated *TkSQS1*. The resulting 1257-bp cDNA was predicted to encode a protein of 418 amino acids (GenBank: MG646369). A second SQS gene (*TkSQS2*) in Contig6767 was identified in the draft genome sequence, and the full-length cDNA was obtained from peduncle mRNA by PCR using primers TkSQS2-fwd and TkSQS2-rev. The 1254-bp cDNA was predicted to encode a protein of 417 amino acids (GenBank: MG646370). The TkSQS1 and TkSQS2 proteins had predicted isoelectric points of 8.18 and 8.57, respectively. The calculated molecular weights were 47.981 and 47.713 kDa, respectively. The secondary structure of TkSQS1 was predicted to consist mainly of α-helices (65%), followed by random coils (18%), extended strands (9%), and β-turns (8%), and was therefore quite similar to that of TkSQS2, with 60% α-helices, 22% random coils, 11% extended strands, and 7% β-turns. The TkSQS1 and TkSQS2 amino acid sequences were aligned in silico with known SQS proteins from other plant species (Supporting Information Figure S1a). TkSQS1 showed high identity to SQS from *Helianthus annuus* (91%) and *A. annua* (89%), with lower sequence identity to SQS from *Hevea brasiliensis* (81%), *N. tabacum* (80%), *A. thaliana* (78%), and *Zea mays* (70%). Interestingly, TkSQS1 showed only 72% sequence identity to TkSQS2, which in turn showed comparable identities to the SQS proteins from *H. annuus*, *H. brasiliensis* and *A. annua* (73%, 72% and 71%, respectively). TkSQS2 showed only 64% sequence identity to *Z. mays* SQS.

TkSQS1 and TkSQS2 belong to the isoprenoid biosynthesis enzymes class 1 superfamily (Conserved Protein Domain Family Accession cl00210) and possess three conserved SQS domains that are typical for mammalian SQS proteins and are known to be involved in catalysis (Supporting Information Figure S1a; Gu, Ishii, Spencer, & Shechter, 1998). Domain A, with its conserved tyrosine residue Y171, is presumably responsible for the first step of catalysis and is identical to the squalene and phytoene synthases signature 1 pattern (PROSITE Accession Number PS01044) at TkSQS1 residues 168–183 and TkSQS2 residues 169–184. For TkSQS1, domain B encompasses residues 199–226 and contains the essential conserved amino acids D213 and D217 involved in substrate binding via the magnesium salt bridge, whereas in TkSQS2, domain B encompasses residues 200–227 with D214 and D218 as the conserved amino acids (Gu et al., 1998). This domain also includes the characteristic signature 2 pattern of squalene and phytoene synthases (PROSITE Accession Number PS01045), which spans TkSQS1 amino acid residues 201–221 and TkSQS2 amino acid residues 202–222. Domain C is responsible for the second catalytic step (TkSQS1 residues 280–297 and TkSQS2 residues 281–298) with the conserved residues P280, F283, F285, and Q290 in TkSQS1, and P281, F284, F286, and Q291 in TkSQS2. All three essential catalytic domains, including the conserved residues responsible for enzyme activity, were present in TkSQS1 and TkSQS2, indicating that both *T. koksaghyz* SQS isoforms may be involved in squalene biosynthesis via the proposed mechanism. In addition, the kingdom-specific SQS hinge region was identified in all the plant sequences we analyzed (Supporting Information Figure S1a; Linscott, Niehaus, Zhuang, Bell, & Chappell, 2016) and a phylogenetic tree was constructed with *S. cerevisiae* SQS as the outgroup (Figure 2). Computer-based predictions of transmembrane domains revealed that TkSQS1 and TkSQS2 each possess one C-terminal transmembrane domain (encompassing amino acid residues 386–408 and 390–407, respectively), which are highly similar to those predicted for *H. annuus* and *A. annua* SQS (Supporting Information Figure S1a). Functionally, this transmembrane domain may anchor the enzyme into the ER membrane (Busquets et al., 2008; Linscott et al., 2016). Interestingly, a second transmembrane domain close to domain C has been predicted in *A. annua*, *M. truncatula*, and *H. brasiliensis* SQS, and identified in *P. ginseng* PgSS1, PgSS2, and PgSS3, as well as *G. max* SQS1 and SQS2, but this was not present in TkSQS1 or TkSQS2 (Kim et al., 2011; Nguyen et al., 2013).

The RNA-Seq data were also mined for *SQE* sequences, revealing two contigs with significant similarities to *SQE* genes. The specific primer pairs TkSQE1-fwd/TkSQE1-rev and TkSQE2-fwd/TkSQE2-rev were designed and used to amplify the corresponding cDNA sequences (*TkSQE1* and *TkSQE2*). The *TkSQE1* and *TkSQE2* cDNAs were 1596 and 1554 bp in length, respectively (GenBank: MG646371, MG646372), potentially encoding proteins of 513 (TkSQE1) and 517 (TkSQE2) amino acids. Two further *SQE* genes (*TkSQE3* and *TkSQE4*) were identified in the draft genome and were amplified using the primer combinations TKSQE3-fwd/TkSQE3-rev and TKSQE4-fwd/TkSQE4-rev. The full-length 1857-bp *TkSQE3* cDNA (GenBank: MG646373) potentially encodes a protein of 618 amino acids, whereas the full-length 1566-bp *TkSQE4* cDNA (GenBank: MG646374) potentially encodes a protein of 521 amino acids. The predicted isoelectric points of TkSQE1–4 were 8.81, 8.93, 9.39, and 8.88, respectively, and the calculated molecular weights were 58.080, 56.399, 68.107, and 56.550 kDa, respectively. The secondary structure of all four TkSQE proteins predominantly consisted of random coils (31%–37%) and α-helices (27%–32%) followed by extended strands (24%–27%) and β-turns (10%–12%). The in silico alignment of the four TkSQE polypeptide sequences revealed a maximum of 72% identity (TkSQE1 vs. TkSQE2) and a minimum of 62% (TkSQE3 vs. TkSQE4) (Supporting Information Figure S1b). TkSQE1, TkSQE3, and TkSQE4 showed the highest sequence identities to *H. annuus* SQE (89%, 69%, and 66%, respectively), whereas TkSQE2 showed greater identity to SQE1 and SQE2 from *P. ginseng* (77% in each case) compared to *H. annuus* SQE (75%). The phylogenic tree showed that TkSQE1–TkSQE4 cluster with diverse SQEs from other plants rather than together and that TkSQE3 shows the longest separation distance from other plant SQEs, indicating a putative functional difference (Figure 2). The *T. koksaghyz* SQEs also possess two conserved domains involved in enzyme activity (motif I) and substrate binding (motif II) that have already been identified in *S. cerevisiae* ERG1 (Uchida et al., 2015). Motif I, also termed the FAD I "fingerprint," consists of a Rossmann fold with alternating β-sheets and α-helices containing the sequence profile GXGXXG (TkSQE1–4 residues 73–78, 60–65, 150–155, and 64–69, respectively), which is

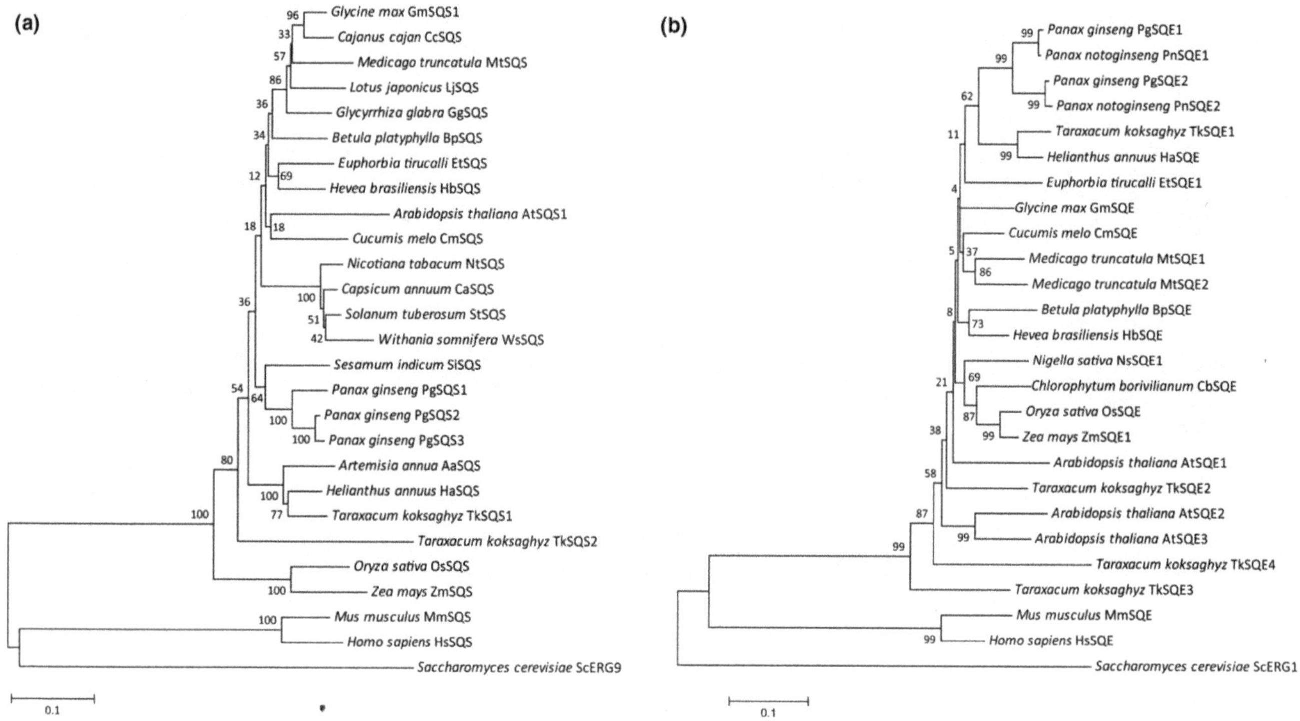

FIGURE 2 Phylogenetic tree of (a) SQS and (b) SQE amino acid sequences. The tree was constructed using the neighbor-joining method in MEGA6 software (Tamura et al., 2013; http://www.megasoftware.net/) with a bootstrap of 1000 replicates. Protein sequences were obtained from GenBank (https://www.ncbi.nlm.nih.gov/genbank/) and are provided in Table S1

needed to bind the ADP moiety of FAD (Ruckenstuhl, Eidenberger, Lang, & Turnowsky, 2005). In addition, the FAD II "fingerprint" contains the conserved GD motif (TkSQE1–4 residues 352–353, 339–340, 429–430, and 343–344, respectively), which binds NAD(P)H and the ribose moiety of FAD (Eppink, Schreuder, & van Berkel, 1997). Both domains are fully conserved in all *T. koksaghyz* SQE sequences (Supporting Information Figure S1b). Furthermore, the prediction of the transmembrane domains revealed that TkSQE1 contains four transmembrane domains, only one is present in TkSQE2, TkSQE3 contains six, and TkSQE4 contains two. The in silico data are summarized in Table S5. Finally, the *TkSQS1-2* and *TkSQE1-4* exon–intron structures were determined (Supporting Information Figure S2). The ubiquitous intron motif GT-AG was present in all sequences.

3.2 | Spatial and temporal expression profiles reveal the coregulation of *TkSQS1* and *TkSQE1* in dandelion latex

The spatial expression patterns of all *TkSQS* and *TkSQE* genes was determined by quantitative real time PCR (qRT-PCR) using total RNA from the latex, roots, leaves, peduncles and flowers of 12-week-old *T. koksaghyz* plants grown under greenhouse conditions. *TkSQS1* was preferentially expressed in the latex, with > threefold higher expression levels compared to leaves and 10-fold higher expression levels compared to peduncles. Only low levels of *TkSQS1* mRNA were detected in the roots and flowers (Figure 3a). In contrast, *TkSQS2* mRNA was not detected in the latex, roots or leaves, and only low levels were detected in the peduncles and flowers (Figure 3b). TkSQS1 therefore appears to be the key enzyme responsible for isoprenoid biosynthesis in the latex. *TkSQE1* was predominantly expressed in the latex, whereas only low or very low mRNA levels were detected in the leaves, roots, peduncles and flowers (Figure 3a). *TkSQE2*, *TkSQE3*, and *TkSQE4* were expressed at very low levels, particularly in the latex. *TkSQE2* expression was highest in peduncles, whereas *TkSQE3* was expressed most strongly in the roots and *TkSQE4* was expressed at similar levels in the latex, leaves, and peduncles (Figure 3b).

These results suggest that *TkSQS1* and *TkSQE1* are probably the only genes encoding key enzymes for the biosynthesis of cyclic terpenoids in latex, given the negligible expression of *TkSQS2* and *TkSQE2*–*TkSQE4* in this specialized cytoplasm. *TkSQS2* and *TkSQE2*–*TkSQE4* were therefore excluded from further studies. The potential coregulation of *TkSQS1* and *TkSQE1* expression in *T. koksaghyz* latex was investigated by growing wild-type plants in the greenhouse and isolating latex from 8- to 20-week-old plants. RNA was pooled from three individual plants and three pools were analyzed by qRT-PCR. *TkSQS1* and *TkSQE1* shared near-identical expression profiles with expression peaking in 12- and 16-week-old plants (Figure 3c). In addition to the accumulation of triterpenoids during this growth period (Figure 3d,f) we also analyzed the expression of the OSC gene *TkOSC1* (GenBank: MG646376), which was recently shown to be responsible for the synthesis of the major pentacyclic triterpenes in dandelion latex. Although the squalene content increased in 10- and

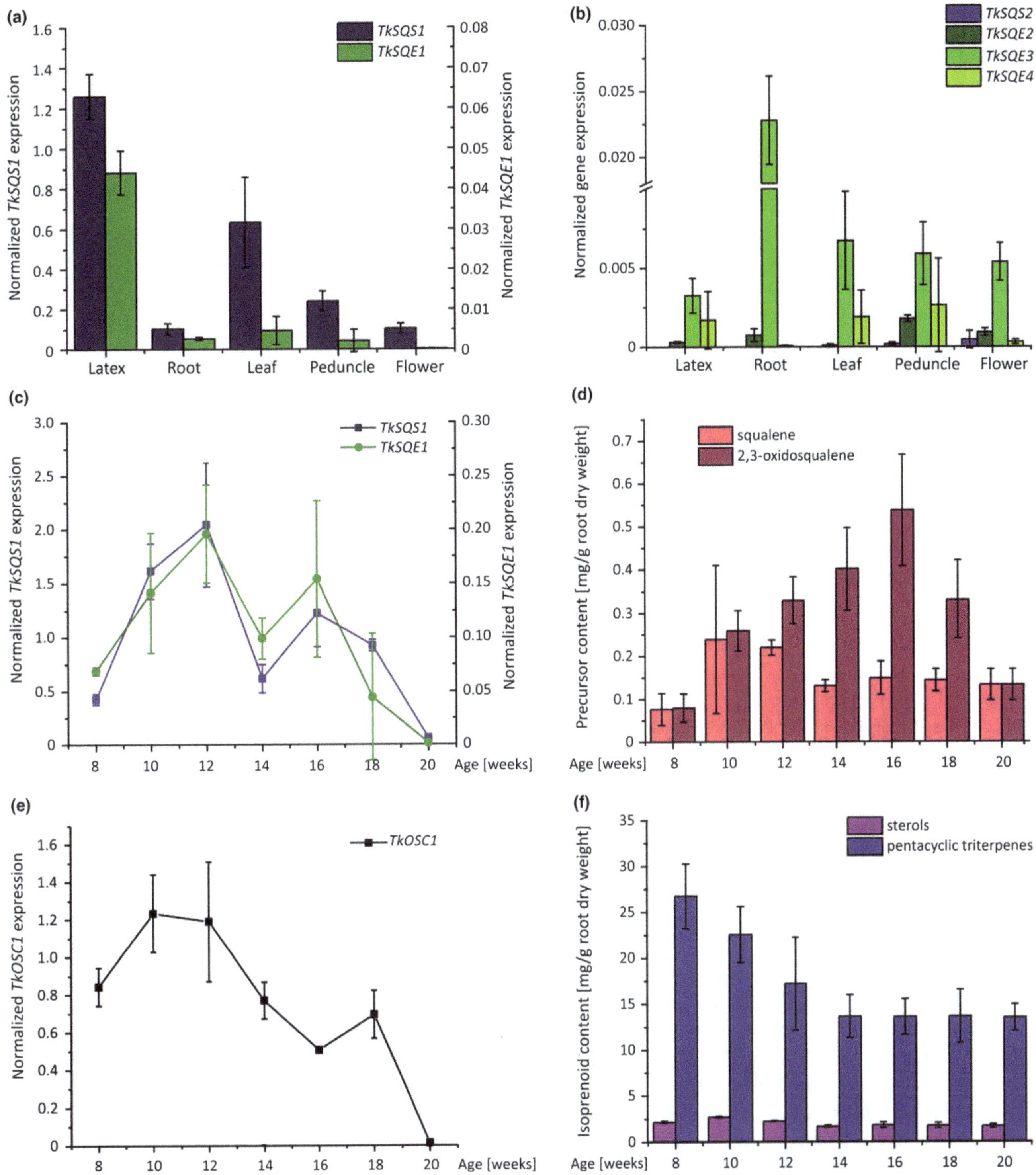

FIGURE 3 Spatial and temporal expression profiles of *TkSQS*, *TkSQE* and *TkOSC1* mRNA in wild-type *Taraxacum koksaghyz* plants determined by qRT-PCR and temporal isoprenoid levels in wild-type *T. koksaghyz* plants. The corresponding mRNA levels were normalized against the constitutive genes encoding elongation factor 1α (*TkEF1α*) and ribosomal protein L27 (*TkRP*) from *T. koksaghyz*. Data are means ± standard errors of three pools, each consisting of three independent wild-type plants. (a) *TkSQS1* and *TkSQE1* and (b) *TkSQS2* and *TkSQE2–4* mRNA levels in latex, roots, leaves, peduncles, and flowers of 12-week-old wild-type *T. koksaghyz* plants. (c) *TkSQS1* and *TkSQE1* mRNA levels in latex of 8- to 20-week-old wild-type *T. koksaghyz* plants. (d) Squalene and 2,3-oxidosqualene levels in roots of 8- to 20-week-old *T. koksaghyz* plants. (e) *TkOSC1* mRNA levels in the latex of 8- to 20-week-old wild-type *T. koksaghyz* plants. (f) Sterol and pentacyclic triterpene levels in roots of 8- to 20-week-old wild-type *T. koksaghyz* plants

12-week-old wild-type *T. koksaghyz* plants, 2,3-oxidosqualene continued to accumulate until the plants were 16 weeks old and then declined (Figure 3d). Moreover, although the sterol content remained nearly constant, the quantity of pentacyclic triterpenes continued to decline with plant age (Figure 4f). Interestingly, the pentacyclic triterpenes which decreased most were α-amyrin, β-amyrin and taraxasterol (Supporting Information Table S6). Accordingly, *TkOSC1* expression was highest in 10- and 12-week-old *T. koksaghyz* plants and declined thereafter (Figure 3e). In addition, we also investigated whether treatment with methyl jasmonate (MeJA) was able to induce *TkSQS1–TkSQE2* and *TkSQE1–TkSQE4* gene expression in 8-week-old *T. koksaghyz* wild-type plants (Supporting Information Table S4). Here, *TkSQS2* and *TkSQE2* showed no expression in *T. koksaghyz* latex, whereas *TkSQE3–TkSQE4* were expressed at low levels in MeJA-treated plants, but displayed no significant difference in gene expression compared to control plants. *TkSQS1* and *TkSQE1* gene expression was slightly elevated in MeJA-treated plants compared to the control plants, but again not significantly.

3.3 | Localization and enzymatic activity of TkSQS1 and TkSQE1

Next we investigated the subcellular localization of TkSQS1 and TkSQE1 by transient expression in *N. benthamiana* epidermal leaf cells followed by the functional analysis of enzyme activity using in vitro and yeast complementation assays for TkSQS1 and TkSQE1, respectively.

CLSM analysis revealed that the Cerulean-TkSQS1 fusion protein colocalized with NtermCYP51G1-mRFP, but not with the cytosolic mRFP marker (Figure 4a–b and Supporting Information Figure S3a–b). This is indicated by merged images and quantitative analysis of the fluorescent signals in regions of interest. As a negative control, cytosolic Cerulean was coinfiltrated with NtermCYP51G1-mRFP to ensure that the ER localization of TkSQS1 was not elicited by the fusion with Cerulean (Supporting Information Figure S3e). Likewise, the C-terminal TkSQE1 fusion protein (TkSQE1-Cerulean) colocalized with NtermCYP51G1-mRFP, but not with cytosolic mRFP (Figure 4c–d and Supporting Information Figure S3c–d).

The N-terminal TkSQE1 fusion protein (Cerulean-TkSQE1) aggregated in the tobacco cells, indicating that normal protein folding was disrupted by the fusion. Therefore, the potentially distinct role of the two TkSQE1 transmembrane domains in the localization of the enzyme could not be evaluated. However, our results clearly showed that both enzymes localize to the ER membrane in the heterologous *N. benthamiana* expression system.

To confirm the functionality of TkSQS1, an in vitro SQS assay was carried out with [1,2-^{14}C]-labeled FPP as the substrate in protein extracts from *N. benthamiana* leaves which was quantified by scintillation counting (Figure 5a). Proteins extracted from *TkSQS1*-infiltrated *N. benthamiana* leaves showed a sevenfold increase in squalene production compared to the infiltrated empty vector control, which confirmed the activity of TkSQS1 (Figure 5b).

The squalene epoxidase activity of TkSQE1 was confirmed in a yeast complementation assay. *S. cerevisiae* strain KLN1 (Δ*erg1*) (Lipinski et al., 2009) was transformed with the vector pAG425GPD-TkSQE1 or the empty vector pAG425GPD-ccdB as a negative control, and KLN1-pERG1 served as positive control. *ERG1* encodes the *S. cerevisiae* SQE, so that KLN1 (Δ*erg1*) is depleted for essential sterols and can only grow on ergosterol under anaerobic conditions. KLN1 complementation with TkSQE1 was feasible and resulted in growth under aerobic conditions, whereas the negative control was only able to grow under anaerobic conditions with ergosterol provided as a supplement (Figure 5c). These data confirmed the functionality of TkSQE1.

3.4 | *TkOSC1* expression is downregulated in TkSQS1-RNAi and TkSQE1-RNAi plants

Having identified the *TkSQS1* and *TkSQE1* genes and functionally characterized the corresponding enzymes, we investigated the effect of silencing the expression of either *TkSQS1* or *TkSQE1* in *T. koksaghyz* plants by RNAi.

We characterized 13 TkSQS1-RNAi and 4 TkSQE1-RNAi transgenic lines in terms of T-DNA insertion and gene expression. Transgenic T0 plants were pollinated using a wild-type plant for seed generation, and 10 plants from six different TkSQS1-RNAi lines with a strong silencing effect were selected and analyzed in the T1 generation. *TkSQS1* expression in the latex harvested from roots was measured by qRT-PCR in 12-week-old plants because this time point showed the highest expression level in wild-type plants (Figure 3c). Another seven TkSQS1-RNAi lines displaying no RNAi effect were used as negative controls. The 10 lines showing a strong RNAi effect and the seven controls were used for the further analysis of gene expression and metabolite levels. Putative regulatory effects at the transcriptional level were determined by analyzing *TkSQS2*, *TkSQE1*, and *TkOSC1* gene expression (Figure 6a). *TkSQS2* expression was not detected in latex samples from 12-week-old wild-type plants, indicating that *TkSQS2* does not compensate for the reduction of *TkSQS1* expression. Therefore, TkSQS1 and TkSQS2 do not appear to be functionally redundant.

TkSQE1, as the major *SQE* gene expressed in the latex, showed a slight but not statistically significant reduction in expression in the TkSQS1-RNAi lines compared to the control plants. Remarkably, we also observed a positive correlation between *TkOSC1* and *TkSQS1* expression in the TkSQS1-RNAi lines (Figure 6a). Moreover, gene expression in the root material of TkSQS1-RNAi lines was assessed for *TkSQS1-2*, *TkSQE1*, *TkSQE3*, and *TkOSC1* expression (Supporting Information Figure S4a). In all transgenic lines, the expression of *TkSQS1*, *TkSQE1*, *TkSQE3*, and *TkOSC1* was similar or lower compared to control plants, while *TkSQS2* showed no expression at all so that compensatory gene induction in root tissue can be excluded.

For the analysis of TkSQE1-RNAi lines, we selected a set of 11 T2 transgenic plants representing four independent transgenic lines. Because all the transgenic plants showed an RNAi effect, six non-transgenic plants from the same generation were used as controls.

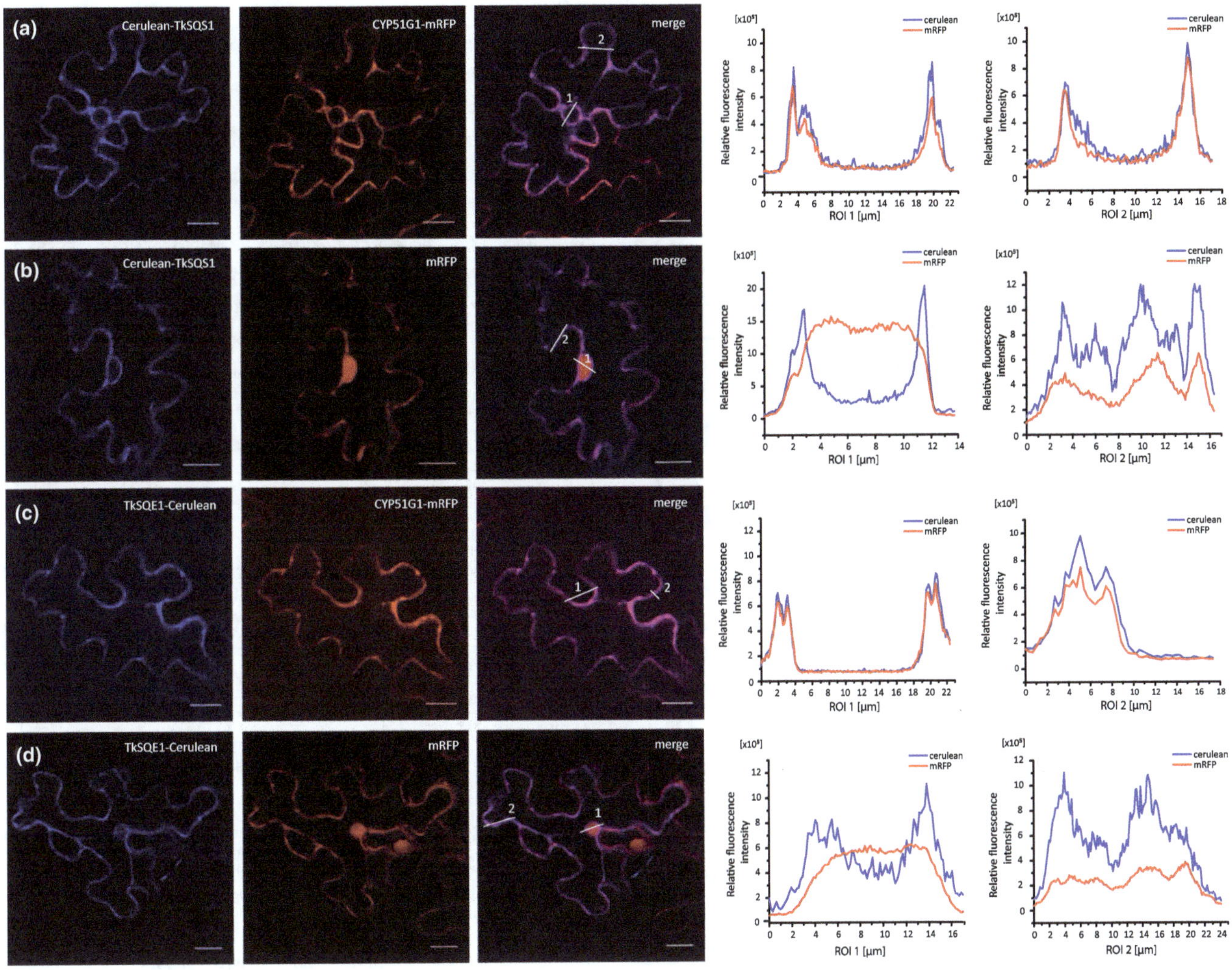

FIGURE 4 Localization of TkSQS1 and TkSQE1. (a–d) The CLSM images show red (mRFP, excitation 543 nm) and blue (Cerulean, excitation 458 nm) fluorescence of fusion proteins expressed in *Nicotiana benthamiana* epidermal cells. The relative fluorescence intensity in each region of interest (ROI) is marked in two regions. NtermCYP51G1-mRFP = ER-localized marker containing a 120-bp cDNA fragment representing the N-terminus of *Arabidopsis thaliana* CYP51G1 sterol 14-demethylase; mRFP = cytosolic marker; scale bar = 20 µm

Subsequent qPCR analysis revealed a significant reduction in *TkSQE1* expression to 20% of the level in wild-type plants (Figure 6b). At the same time, the expression levels of *TkSQS1* and *TkOSC1* were significantly reduced, which indicated the coregulation of *TkSQS1*, *TkSQE1*, and *TkOSC1* as suggested by the analysis of the TkSQS1-RNAi lines. The expression of *TkSQE2–TkSQE4* was also examined, and there were no differences from the control plants, indicating that *TkSQE2–TkSQE4* do not compensate for the reduction in *TkSQE1* expression. In addition, gene expression in root material of TkSQE1-RNAi lines was analyzed for *TkSQS1-2*, *TkSQE1*, *TkSQE3*, and *TkOSC1* expression (Supporting Information Figure S4b). In all transgenic lines, the expression of *TkSQS1*, *TkSQE3*, and *TkOSC1* was lower compared to control plants although not significantly, whereas *TkSQE1* displayed a significantly lower expression level in transgenic compared to wild-type plants. Moreover, *TkSQS2* could not be detected in either transgenic or in wild-type root material.

In addition, roots of 12-week-old TkSQS1-RNAi and TkSQE1-RNAi plants were cut directly below the rosette, freeze-dried, ground, and used for the extraction of metabolites for analysis by gas chromatography-mass spectrometry (GC-MS). Interestingly, the squalene and 2,3-oxidosqualene levels in 12-week-old TkSQS1-RNAi transgenic lines were not significantly different from the levels in the control lines (Figure 7A, Supporting Information Table S7). The same was true for the levels of pentacyclic triterpenes and sterols (Figure 7c, Supporting Information Table S7). However, GC-MS analysis of TkSQE1-RNAi roots revealed a 27-fold increase in squalene levels compared to control plants, but no significant differences in the levels of 2,3-oxidosqualene (Figure 7b, Supporting Information Table S8).

The *TkSQE1* expression profile and the accumulation of squalene confirmed the effectiveness of this RNAi approach. Surprisingly, neither the sterol content nor the amount of pentacyclic triterpenes was significantly reduced (Figure 7d, Supporting Information Table S8). Like the TkSQS1-RNAi lines, the TkSQE1-RNAi lines did not show significant differences in the accumulation of isoprenoid end-products compared to controls, although *TkSQE1* expression and the accumulation of squalene were knocked down successfully.

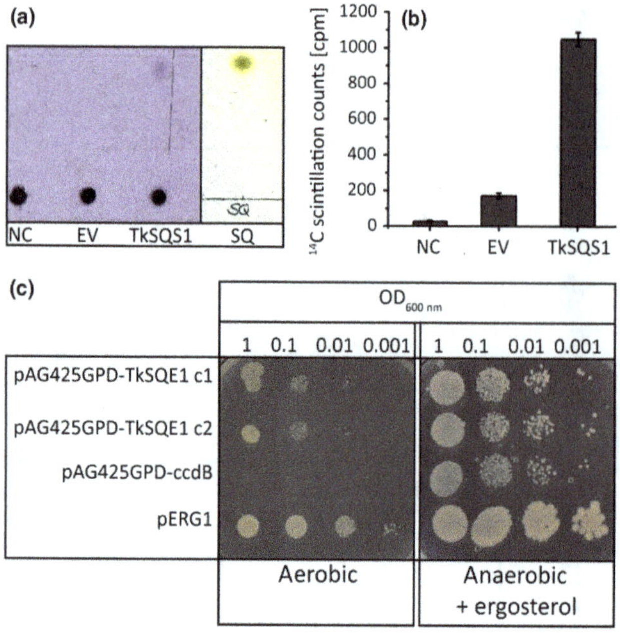

FIGURE 5 Functional analysis of TkSQS1 and TkSQE1. TkSQS1 (a–b) activity assay and TkSQE1 complementation (c). (a) Phosphor imaging with negative control (NC), empty vector control (EV) and TkSQS1 protein extracts and corresponding reversed-phase TLC plate showing squalene standard visualized with iodine vapor. (b) Silica was scraped from the TLC plate at the height of the squalene standard, and radioactivity was measured by scintillation counting. (c) Complementation of the yeast strain KLN1 (Δ*erg1*) to confirm the activity of TkSQE1. The yeast was transformed with the vector pAG425GPD-TkSQE1 or the negative control vector pAG425GPD-ccdB. KLN1-pERG1 complemented with the endogenous ERG1 served as a positive control. Two colonies of KLN1-TkSQE1 and one colony of KLN1-pERG1 and KLN1-ccdB were selected, dropped out, and incubated for 3 days under anaerobic conditions in the presence of ergosterol, or incubated for 5 days under aerobic conditions without supplements

4 | DISCUSSION

We have carried out a detailed analysis of the *SQS* and *SQE* genes in *T. koksaghyz* and a functional characterization of their products. In silico analysis of the *TkSQS1*, *TkSQS2*, and *TkSQE1–TkSQE4* cDNA sequences revealed high similarities to other known SQS and SQE sequences and the presence of known catalytically active domains indicating functional enzyme activity (Supporting Information Figure S1). Interestingly, TkSQS1 and TkSQS2 did not cluster together in the same branch of the SQS phylogenetic tree, suggesting that gene duplication may have predated speciation, hinting at a specialized tissue-specific role or sub-functionalization (Figure 2). Similar results have been obtained for SQS proteins from *G. glabra* in phylogenetic analyses (Hayashi et al., 1999; Navarro Gallón et al., 2017). However, this contrasts with the *M. domestica* MdSQS1 and MdSQS2 isoforms, which show higher intraspecific identities and may indicate more recent gene duplication events (Navarro Gallón et al., 2017).

Similarly, TkSQE3 and TkSQE4 did not cluster closely with other plant SQEs (Figure 2). This could indicate a specialized function which should be assessed in future studies, as already proposed for *A. thaliana SQE* genes (Laranjeira et al., 2015; Rasbery et al., 2007). Tissue-specific expression analysis revealed that *TkSQS1* and *TkSQE1* are the predominant latex isoforms (Figure 3a,b) and are therefore likely to play a key role in dandelion latex isoprenoid biosynthesis. The multitissue expression of *SQS* and *SQE* is common and has been described for *SQS* and *SQE* genes in *A. thaliana* (Busquets et al., 2008; Rasbery et al., 2007), *P. ginseng* (Han et al., 2010; Lee et al., 2004) and *W. somnifera* (Bhat et al., 2012; Razdan et al., 2013) among others, mirroring our results in *T. koksaghyz*.

Infiltration studies confirmed the localization of TkSQS1 and TkSQE1 on the ER membrane (Figure 4) as postulated for other SQS and SQE enzymes (Busquets et al., 2008; Kajikawa et al., 2015; Laranjeira et al., 2015; Leber et al., 1998). The in vitro TkSQS1 activity assay and the complementation of the impaired yeast strain KLN1 (Δ*erg1*) by TkSQE1 confirmed the activities of both *T. koksaghyz* enzymes (Figure 5).

Comparative expression analysis revealed that *TkSQS1* and *TkSQE1* share similar latex-preferred expression profiles during dandelion development (Figure 3c). Such coregulation may ensure that isoprenoid precursors are efficiently shuttled toward sterol or pentacyclic triterpene biosynthesis. In addition, minor effects of higher *TkSQS1* and *TkSQE1* gene expression in MeJA-treated plants compared to control plants could hint at an involvement in defense against herbivores or pathogens as has been reported for other *T. koksaghyz* genes involved in secondary metabolite biosynthesis (Cao et al., 2017). Moreover, expression peaks in 12- and 16-week-old plants indicated an increase in the demand for sterols and/or pentacyclic triterpenes, perhaps due to the rapid expansion of root volume (Schaller, 2003) or the rapid growth of laticiferous cells, in turn requiring the efficient shuttling of precursors via coordinated expression. Similar studies in *T. brevicorniculatum* revealed that genes in the mevalonate pathway also show similar expression patterns and expression peaks in 16-week-old plants (Pütter et al., 2017). Thus, membrane sterols seem to be required at this stage of development and are provided by the mevalonate pathway and subsequent SQS, SQE and OSC activity. Additional post-transcriptional mechanisms also seem to play a role in SQE activity. Although the squalene content during *T. koksaghyz* development increases in 10- to 12-week-old plants and then declines to a low level, 2,3-oxidosqualene levels increase until the plants are 16 weeks old before declining (Figure 3d). Therefore, although *TkSQS1* and *TkSQE1* expression levels remain similar at these times, further stabilization mechanisms or post-translational modifications may be required to increase the total 2,3-oxidosqualene content of the roots.

To gain insight into the role of TkSQS1 and TkSQE1 in dandelion isoprenoid biosynthesis, we generated RNAi lines in which these genes were silenced (Figures 6 and 7). TkSQS1-RNAi lines showed significantly lower *TkSQS1* expression, and surprisingly the expression levels of the genes *TkSQE1* and *TkOSC1* were simultaneously reduced, although the downregulation of *TkSQE1* was not statistically significant. A similar correlation was observed for the TkSQE1-

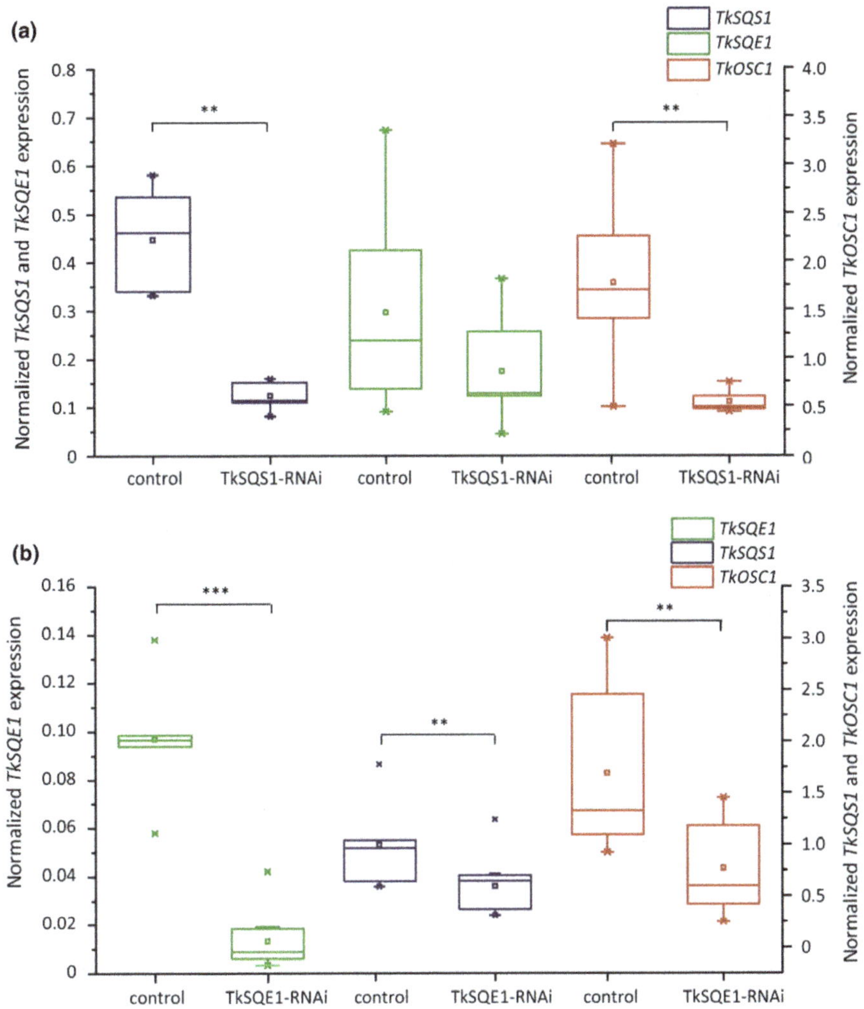

FIGURE 6 *TkSQS1*, *TkSQE1*, and *TkOSC1* mRNA expression analysis in latex from TkSQS1-RNAi and TkSQE1-RNAi *Taraxacum koksaghyz* plants. The corresponding mRNA levels were normalized against the constitutive genes encoding elongation factor 1α (*TkEF1α*) and ribosomal protein L27 (*TkRP*) from *T. koksaghyz*. The transgenic plants were analyzed 12 weeks after sowing. *TkSQS1*, *TkSQE1*, and *TkOSC1* mRNA expression analysis in (a) TkSQS1-RNAi plants ($n = 10$ from six independent lines) compared to control plants ($n = 7$) and (b) TkSQE1-RNAi plants ($n = 11$ from four independent lines) compared to control plants ($n = 6$). Asterisks denote statistical significance compared to control (two-tailed *t* test, ** = $p < 0.01$, *** = $p < 0.001$)

RNAi lines, in which the expression levels of *TkSQS1*, *TkSQE1*, and *TkOSC1* were significantly reduced compared to control lines. These findings indicate that *TkSQS1*, *TkSQE1*, and *TkOSC1* are coregulated in *T. koksaghyz* latex, as reflected by the similar developmental expression profiles of *TkSQS1* and *TkSQE1* (Figure 3c). Transcriptional coregulation may increase the flux toward isoprenoid biosynthesis and is one of several mechanisms that can be used to regulate the production of isoprenoids.

In *P. ginseng*, the overexpression of *SQS* resulted in higher expression levels of *SQE*, *βAS* (*OSC*) and cycloartenol synthase, which also support the positive coregulation of genes in the isoprenoid biosynthesis pathway (Lee et al., 2004). In addition, the virus-induced gene silencing of *W. somnifera SQS* reduced the expression of several downstream genes including *WsSQE*, whereas upstream genes were induced. Contrary to our results, the expression of *WsβAS1* was upregulated whereas cycloartenol synthase was downregulated (Singh et al., 2015). Therefore, although the regulation of OSCs may differ in a species-dependent manner, there appears to be a positive correlation between OSC and SQS/SQE expression in *T. koksaghyz* latex.

To investigate the regulation of isoprenoid biosynthesis in more detail, freeze-dried root material from the RNAi lines was evaluated to determine the content of triterpene precursors, sterols and pentacyclic triterpenes (Figure 7). The quantity of pentacyclic triterpenes was not affected by the silencing of *TkSQS1* or *TkSQE1*, perhaps reflecting the overall high-level accumulation of pentacyclic triterpenes in the early stages of dandelion development. The quantity of pentacyclic triterpenes peaked at 26.8 mg/g dry weight (dw) in 8-week-old *T. koksaghyz* wild-type plants before declining with age (Figure 3f), but high levels were still present after 12 weeks (17.2 mg/g dw). *TkSQS1* and *TkSQE1* silencing and the reduced expression of *TkOSC1* therefore do not have a profound effect on total pentacyclic triterpene levels in 12-week-old plants. In contrast, the remarkable impact of *TkSQE* silencing on squalene accumulation was clearly seen at this developmental stage, given that squalene is present at only low levels in 12-week-old wild-type plants (0.2 mg/g dw). This effect was previously reported in *A. thaliana Atsqe1* mutants (Rasbery et al., 2007) and demonstrates the reduced flux between squalene and 2,3-oxidosqualene elicited by TkSQE1. Squalene accumulation as a successful indication of SQE inhibition or knockdown has also been observed in other species: the treatment of tobacco BY-2 cells with terbinafine strongly increased squalene levels, as did artificial microRNA-mediated *SQE* silencing and terbinafine treatment in *Chlamydomonas reinhardtii* (Kajikawa et al., 2015;

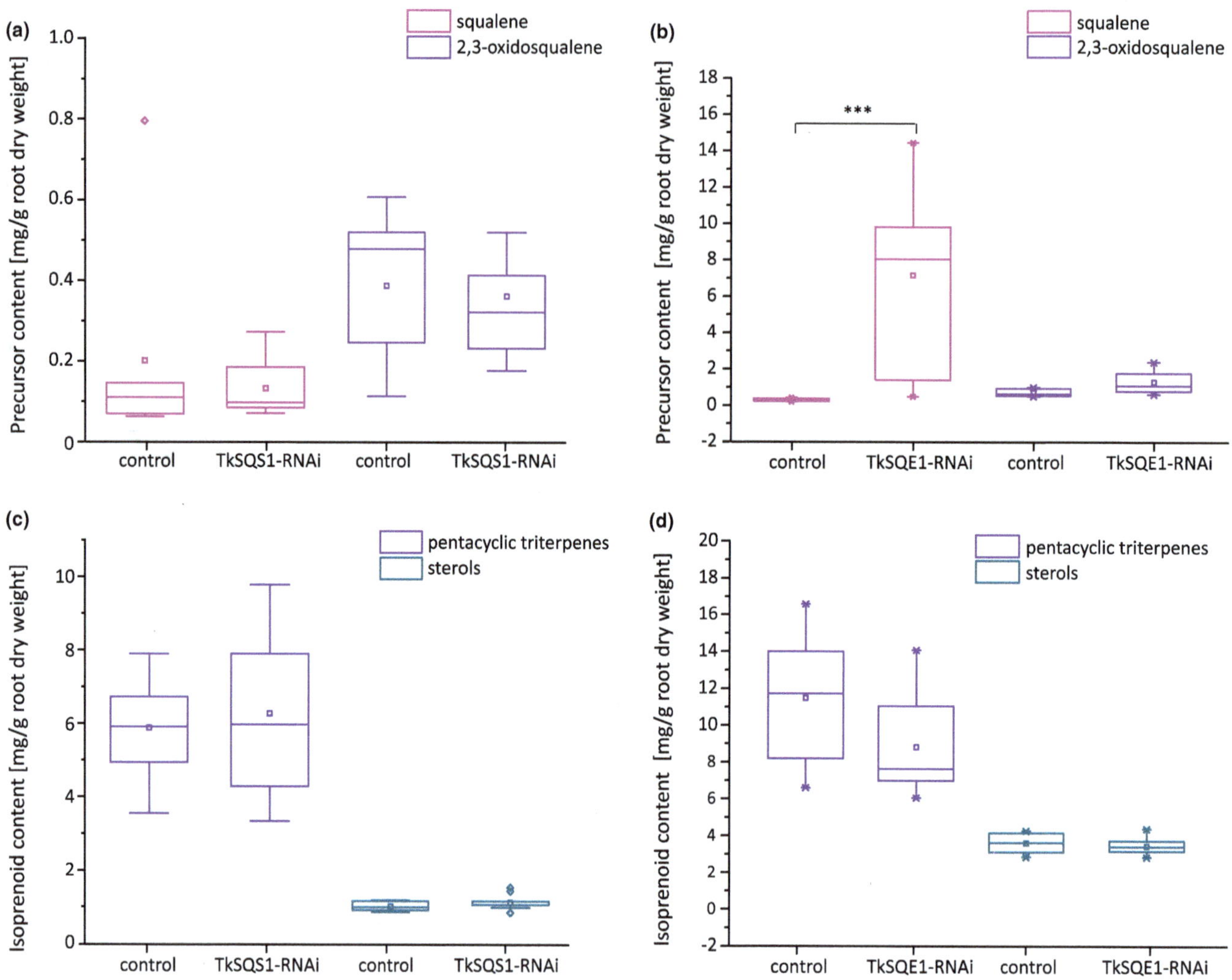

FIGURE 7 Metabolic analysis of TkSQS1-RNAi and TkSQE1-RNAi plants. Squalene and 2,3-oxidosqualene levels in freeze-dried root material were determined by GC-MS in (a) TkSQS1-RNAi (n = 10 from six independent lines) and corresponding control plants (n = 7) and in (b) TkSQE1-RNAi (n = 11 from four independent lines) and corresponding control plants (n = 6). Pentacyclic triterpene and sterol levels in freeze-dried root material determined by GC-MS in (c) TkSQS1-RNAi (n = 10 from six independent lines) and corresponding control plants (n = 7) and in (d) TkSQE1-RNAi (n = 11 from four independent lines) and corresponding control plants (n = 6). Pentacyclic triterpenes = taraxerol, α-amyrin, β-amyrin, lupeol and taraxasterol; sterols = campesterol, stigmasterol and γ-sitosterol. Asterisks denote statistical significance compared to control (two-tailed t test, ** = $p < 0.01$, *** = $p < 0.001$)

Wentzinger, Bach, & Hartmann, 2002). Interestingly, although the inhibition of SQE reduced sterol levels, the gene silencing approach failed to reduce sterol levels significantly, which was attributed to residual SQE activity in the *C. reinhardtii* knockdown lines (Kajikawa et al., 2015). Sufficient SQE activity in *T. koksaghyz* latex might also explain the unaltered isoprenoid end-product levels, possibly due to post-translational regulation. The expression of *TkSQS2* and *TkSQE2–4* did not change in the RNAi lines, indicating that other *SQS* and *SQE* genes are not upregulated in latex to compensate for the depletion of *TkSQS1* and *TkSQE1* (Han et al., 2010). Moreover, *SQS* and *SQE* gene expression in the root tissue of the RNAi lines also did not compensate for the corresponding reduced latex-predominant expression of *TkSQS1* and *TkSQE1*, so that compensatory expression in adjacent tissues can be excluded (Supporting Information Figure S4).

However, because the conversion of squalene to 2,3-oxidosqualene occurs at a constant rate, post-translational regulation of SQE might increase the overall stability of the enzyme to ensure that the levels of downstream isoprenoids remain constant.

Human SQE is regulated by cholesterol-dependent proteasomal N-terminal degradation on the ER membrane (Chua, Howe, Jatana, Thukral, & Brown, 2017; Gill, Stevenson, Kristiana, & Brown, 2011). In yeast, a similar feedback mechanism is known to be lanosterol-dependent (Foresti, Ruggiano, Hannibal-Bach, Ejsing, & Carvalho, 2013). A related mechanism may exist in plants to promote the degradation of SQEs when isoprenoid end-products such as sterols are present. When these end-products are depleted, the enzyme may be stabilized until a certain quantity of end-products accumulates again. Despite low *SQE* expression levels, this could stabilize the enzyme and would explain the lack of an end-product effect in our RNAi lines. The activation or stabilization of TkSQS1 and TkSQE1 through such a post-translational feedback mechanism at the ER membrane will be investigated in the future.

5 | CONCLUSIONS

We analyzed the *T. koksaghyz* enzymes SQS and SQE. The latex-preferred isoforms TkSQS1 and TkSQE1 were knocked down by RNAi but this was not sufficient to cause an effect on isoprenoid biosynthesis even though the *TkSQS1*, *TkSQE1* and *TkOSC1* genes were downregulated in both RNAi approaches. These findings indicate that pentacyclic triterpene biosynthesis is facilitated by efficient coregulation at the level of transcription.

ACKNOWLEDGMENTS

We kindly thank F. Turnowsky (University of Graz, Austria) for providing the yeast strains and we gratefully acknowledge the technical assistance of Daniela Ahlert (University of Muenster, Institute of Plant Biology and Biotechnology) and Gianina Ramos (Institute for Molecular Biology and Applied Ecology, Muenster).

CONFLICT OF INTEREST

The authors declare that they have no competing interests.

AUTHORS' CONTRIBUTIONS

N.v.D. and C.S.G. conceived and designed the experiments. K.U., K.M.P., and K.V. performed the experiments. C.S.G. and N.v.D. analyzed the data. K.U., K.M.P., N.v.D., and R.M.T. wrote the manuscript. D.P. and C.S.G supervised and complemented the writing.

ACCESSION NUMBERS

TkSQS1 (MG646369), TkSQS2 (MG646370), TkSQE1 (MG646371), TkSQE2 (MG646372), TkSQE3 (MG646373), TkSQE4 (MG646374), TkOSC1 (MG646376).

REFERENCES

Abe, I., & Prestwich, G. D. (1999). Squalene epoxidase and oxidosqualene : Lanosterol cyclase—key enzymes in cholesterol biosynthesis. *Comprehensive Natural Products Chemistry* Elsevier, *2*, 267–298.

Akamine, S., Nakamori, K., Chechetka, S. A., Banba, M., Umehara, Y., Kouchi, H., … Hata, S. (2003). cDNA cloning, mRNA expression, and mutational analysis of the squalene synthase gene of *Lotus japonicus*. *Biochimica et Biophysica Acta (BBA) - Gene Structure and Expression*, *1626*, 97–101. https://doi.org/10.1016/S0167-4781(03)00042-3

Bassard, J.-E., Mutterer, J., Duval, F., & Werck-Reichhart, D. (2012). A novel method for monitoring the localization of cytochromes P450 and other endoplasmic reticulum membrane associated proteins: A tool for investigating the formation of metabolons. *FEBS Journal*, *279*, 1576–1583. https://doi.org/10.1111/j.1742-4658.2011.08312.x

Bhat, W. W., Lattoo, S. K., Razdan, S., Dhar, N., Rana, S., Dhar, R. S., … Vishwakarma, R. A. (2012). Molecular cloning, bacterial expression and promoter analysis of squalene synthase from *Withania somnifera* (L.) Dunal. *Gene*, *499*, 25–36. https://doi.org/10.1016/j.gene.2012.03.004

Busquets, A., Keim, V., Closa, M., Del Arco, A., Boronat, A., Arró, M., & Ferrer, A. (2008). *Arabidopsis thaliana* contains a single gene encoding squalene synthase. *Plant Molecular Biology*, *67*, 25–36. https://doi.org/10.1007/s11103-008-9299-3

Cao, X., Yan, J., Lei, J., Li, J., Zhu, J., & Zhang, H. (2017). De novo Transcriptome Sequencing of MeJA-Induced *Taraxacum koksaghyz* Rodin to Identify Genes Related to Rubber Formation. *Scientific Reports*, *7*, 15697. https://doi.org/10.1038/s41598-017-14890-z

Chua, N. K., Howe, V., Jatana, N., Thukral, L., & Brown, A. J. (2017). A conserved degron containing an amphipathic helix regulates the cholesterol-mediated turnover of human squalene monooxygenase, a rate-limiting enzyme in cholesterol synthesis. *Journal of Biological Chemistry*, *292*(49), 19959–19973. jbc.M117.794230.

Devarenne, T. P., Ghosh, A., & Chappell, J. (2002). Regulation of squalene synthase, a key enzyme of sterol biosynthesis, in tobacco. *Plant Physiology*, *129*, 1095–1106. https://doi.org/10.1104/pp.001438

Devarenne, T. P., Shin, D. H., Back, K., Yin, S., & Chappell, J. (1998). Molecular characterization of tobacco squalene synthase and regulation in response to fungal elicitor. *Archives of Biochemistry and Biophysics*, *349*, 205–215. https://doi.org/10.1006/abbi.1997.0463

Edgar, R. C. (2004). MUSCLE: Multiple sequence alignment with high accuracy and high throughput. *Nucleic Acids Research*, *32*, 1792–1797. https://doi.org/10.1093/nar/gkh340

Epping, J., van Deenen, N., Niephaus, E., Stolze, A., Fricke, J., Huber, C., … Schulze Gronover, C. (2015). A rubber transferase activator is necessary for natural rubber biosynthesis in dandelion. *Nature Plants*, *1*, 15048. https://doi.org/10.1038/nplants.2015.48

Eppink, M. H., Schreuder, H. A., & van Berkel, W. J. (1997). Identification of a novel conserved sequence motif in flavoprotein hydroxylases with w putative dual function in FAD/NAD(P)H binding. *Protein Science*, *6*, 2454–2458.

Foresti, O., Ruggiano, A., Hannibal-Bach, H. K., Ejsing, C. S., & Carvalho, P. (2013). Sterol homeostasis requires regulated degradation of squalene monooxygenase by the ubiquitin ligase Doa10/Teb4. *Elife*, *2013*, 1–17.

Geourjon, C., & Deleage, G. (1995). Significant improvements in protein secondary structure prediction by prediction from multiple alignments. *Bioinformatics*, *11*, 681–684. https://doi.org/10.1093/bioinformatics/11.6.681

Ghosh, S. (2016). Biosynthesis of structurally diverse triterpenes in plants: The role of oxidosqualene cyclases. *Proceedings of the Indian National Science Academy*, *82*, 1189–1210.

Gill, S., Stevenson, J., Kristiana, I., & Brown, A. J. (2011). Cholesterol-dependent degradation of squalene monooxygenase, a control point in cholesterol synthesis beyond HMG-CoA reductase. *Cell Metabolism*, *13*, 260–273. https://doi.org/10.1016/j.cmet.2011.01.015

Gu, P., Ishii, Y., Spencer, T. A., & Shechter, I. (1998). Function-structure studies and identification of three enzyme domains involved in the catalytic activity in rat hepatic squalene synthase. *Journal of Biological Chemistry*, *273*, 12515–12525. https://doi.org/10.1074/jbc.273.20.12515

Han, J. Y., In, J. G., Kwon, Y. S., & Choi, Y. E. (2010). Regulation of ginsenoside and phytosterol biosynthesis by RNA interferences of squalene epoxidase gene in *Panax ginseng*. *Phytochemistry*, *71*, 36–46. https://doi.org/10.1016/j.phytochem.2009.09.031

Hata, S., Sanmiya, K., Kouchi, H., Matsuoka, M., Yamamoto, N., & Izui, K. (1997). cDNA cloning of squalene synthase genes from mono- and dicotyledonous plants, and expression of the gene in rice. *Plant and Cell Physiology*, *38*, 1409–1413. https://doi.org/10.1093/oxfordjournals.pcp.a029137

Hayashi, H., Hirota, A., Hiraoka, N., & Ikeshiro, Y. (1999). Molecular cloning and characterization of two cDNAs for *Glycyrrhiza glabra* squalene synthase. *Biological &/and Pharmaceutical Bulletin*, *22*, 947–950.

He, F., Zhu, Y., He, M., & Zhang, Y. (2008). Molecular cloning and characterization of the gene encoding squalene epoxidase in *Panax notoginseng*. *DNA Sequence*, *19*, 270–273. https://doi.org/10.1080/10425170701575026

Hill, J., Donald, K. A. I. G., & Griffiths, D. E. (1991). DMSO-enhanced whole cell yeast transformation. *Nucleic Acids Research*, *19*, 5791. https://doi.org/10.1093/nar/19.20.5791

Huang, Z., Jiang, K., Pi, Y., Hou, R., Liao, Z., Cao, Y., … Tang, K. (2007). Molecular cloning and characterization of the yew gene encoding squalene synthase from *Taxus cuspidata*. *Journal of Biochemistry and Molecular Biology*, *40*, 625–635.

Jach, G., Pesch, M., Richter, K., Frings, S., & Uhrig, J. F. (2006). An improved mRFP1 adds red to bimolecular fluorescence complementation. *Nature Methods*, *3*, 597–600. https://doi.org/10.1038/nmeth901

Jarstfer, M. B., Zhang, D. L., & Poulter, C. D. (2002). Recombinant squalene synthase. Synthesis of non-head-to-tail isoprenoids in the absence of NADPH. *Journal of the American Chemical Society*, *124*, 8834–8845. https://doi.org/10.1021/ja020410i

Jennings, S. M., Tsay, Y. H., Fisch, T. M., & Robinson, G. W. (1991). Molecular cloning and characterization of the yeast gene for squalene synthetase. *Proceedings of the National Academy of Sciences*, *88*, 6038–6042. https://doi.org/10.1073/pnas.88.14.6038

Kajikawa, M., Kinohira, S., Ando, A., Shimoyama, M., Kato, M., & Fukuzawa, H. (2015). Accumulation of squalene in a microalga *Chlamydomonas reinhardtii* by genetic modification of squalene synthase and squalene epoxidase genes. *PLoS ONE*, *10*, 1–21.

Kim, T.-D., Han, J.-Y., Huh, G. H., & Choi, Y.-E. (2011). Expression and functional characterization of three squalene synthase genes associated with saponin biosynthesis in *Panax ginseng*. *Plant and Cell Physiology*, *52*, 125–137. https://doi.org/10.1093/pcp/pcq179

Landl, K. M., Klönsch, B., & Turnowsky, F. (1996). ERG1, encoding squalene epoxidase, is located on the right arm of chromosome VII of *Saccharomyces cerevisiae*. *Yeast*, *12*, 609–613. https://doi.org/10.1002/(SICI)1097-0061(199605)12:6<609::AID-YEA949>3.0.CO;2-B

Laranjeira, S., Amorim-Silva, V., Esteban, A., Arró, M., Ferrer, A., Tavares, R. M., … Azevedo, H. (2015). Arabidopsis squalene epoxidase 3 (SQE3) complements SQE1 and is important for embryo development and bulk squalene epoxidase activity. *Molecular Plant*, *8*, 1090–1102. https://doi.org/10.1016/j.molp.2015.02.007

Leber, R., Landl, K., Zinser, E., Ahorn, H., Spök, A., Kohlwein, S. D., … Daum, G. (1998). Dual localization of squalene epoxidase, Erg1p, in yeast reflects a relationship between the endoplasmic reticulum and lipid particles. *Molecular Biology of the Cell*, *9*, 375–386. https://doi.org/10.1091/mbc.9.2.375

Lee, M.-H., Jeong, J.-H., Seo, J.-W., Shin, C.-G., Kim, Y.-S., In, J.-G., … Choi, Y.-E. (2004). Enhanced Triterpene and Phytosterol Biosynthesis in *Panax ginseng* Overexpressing Squalene Synthase Gene. *Plant and Cell Physiology*, *45*, 976–984. https://doi.org/10.1093/pcp/pch126

Lin, T., Xu, X., Ruan, J., Liu, S., Wu, S., Shao, X., … Li, J. (2017). Genome analysis of *Taraxacum kok-saghyz* Rodin provides new insights into rubber biosynthesis. *National Science Review*, *5*(1), 78–87. https://doi.org/10.1093/nsr/nwx101

Linscott, K. B., Niehaus, T. D., Zhuang, X., Bell, S. A., & Chappell, J. (2016). Mapping a kingdom-specific functional domain of squalene synthase. *Biochimica et Biophysica Acta (BBA) - Molecular and Cell Biology of Lipids*, *1861*, 1049–1057. https://doi.org/10.1016/j.bbalip.2016.06.008

Lipinski, M., Scholz, M., Pieper, K., Fischer, R., Prüfer, D., & Müller, K. (2009). A squalene epoxidase from *Nigella sativa* participates in saponin biosynthesis and mediates terbinafine resistance in yeast. *Central European Journal of Biology*, *4*, 163–169.

Nakamura, M., & Sato, R. (1979). The roles of soluble factors in squalene epoxidation. *Biochemical and Biophysical Research Communications*, *89*, 900–906. https://doi.org/10.1016/0006-291X(79)91863-1

Navarro Gallón, S. M., Elejalde-Palmett, C., Daudu, D., Liesecke, F., Jullien, F., Papon, N., … Besseau, S. (2017). Virus-induced gene silencing of the two squalene synthase isoforms of apple tree (*Malus* × *domestica* L.) negatively impacts phytosterol biosynthesis, plastid pigmentation and leaf growth. *Planta*, *246*, 45–60.

Nguyen, H. T. M., Neelakadan, A. K., Quach, T. N., Valliyodan, B., Kumar, R., Zhang, Z., & Nguyen, H. T. (2013). Molecular characterization of *Glycine max* squalene synthase genes in seed phytosterol biosynthesis. *Plant Physiology and Biochemistry*, *73*, 23–32. https://doi.org/10.1016/j.plaphy.2013.07.018

Ono, T., Ozasa, S., Hasegawa, F., & Imai, Y. (1977). Involvement of NADPH-cytochrome c reductase in the rat liver squalene epoxidase system. *Biochimica et Biophysica Acta (BBA) - Lipids and Lipid Metabolism*, *486*, 401–407. https://doi.org/10.1042/BA20080068

Pütter, K. M., van Deenen, N., Unland, K., Prüfer, D., & Schulze Gronover, C. (2017). Isoprenoid biosynthesis in dandelion latex is enhanced by the overexpression of three key enzymes involved in the mevalonate pathway. *BMC Plant Biology*, *17*, 88. https://doi.org/10.1186/s12870-017-1036-0

Rasbery, J. M., Shan, H., LeClair, R. J., Norman, M., Matsuda, S. P. T., & Bartel, B. (2007). *Arabidopsis thaliana* squalene epoxidase 1 is essential for root and seed development. *Journal of Biological Chemistry*, *282*, 17002–17013. https://doi.org/10.1074/jbc.M611831200

Razdan, S., Bhat, W. W., Rana, S., Dhar, N., Lattoo, S. K., Dhar, R. S., & Vishwakarma, R. A. (2013). Molecular characterization and promoter analysis of squalene epoxidase gene from *Withania somnifera* (L.) Dunal. *Molecular Biology Reports*, *40*, 905–916. https://doi.org/10.1007/s11033-012-2131-9

Robinson, G. W., Tsay, Y. H., Kienzle, B. K., Smith-Monroy, C. A., & Bishop, R. W. (1993). Conservation between human and fungal squalene synthetases: Similarities in structure, function, and regulation. *Molecular and Cellular Biology*, *13*, 2706–2717.

Ruckenstuhl, C., Eidenberger, A., Lang, S., & Turnowsky, F. (2005). Single amino acid exchanges in FAD-binding domains of squalene epoxidase of *Saccharomyces cerevisiae* lead to either loss of functionality or terbinafine sensitivity. *Biochemical Society Transactions*, *33*, 1197. https://doi.org/10.1042/BST0331197

Schaller, H. (2003). The role of sterols in plant growth and development. *Progress in Lipid Research*, *42*, 163–175. https://doi.org/10.1016/S0163-7827(02)00047-4

Seo, J. W., Jeong, J. H., Shin, C. G., Lo, S. C., Han, S. S., Yu, K. W., … Choi, Y. E. (2005). Overexpression of squalene synthase in *Eleutherococcus senticosus* increases phytosterol and triterpene accumulation. *Phytochemistry*, *66*, 869–877. https://doi.org/10.1016/j.phytochem.2005.02.016

Singh, A. K., Dwivedi, V., Rai, A., Pal, S., Reddy, S. G. E., Rao, D. K. V., … Nagegowda, D. A. (2015). Virus-induced gene silencing of *Withania somnifera* squalene synthase negatively regulates sterol and defence-related genes resulting in reduced withanolides and biotic stress tolerance. *Plant Biotechnology Journal*, *13*, 1287–1299. https://doi.org/10.1111/pbi.12347

Stamellos, K. D., Shackelford, J. E., Shechter, I., Jiang, G., Conrad, D., Keller, G. A., & Krisans, S. K. (1993). Subcellular localization of squalene synthase in rat hepatic cells. Biochemical and immunochemical evidence. *Journal of Biological Chemistry*, *268*, 12825–12836.

Stolze, A., Wanke, A., van Deenen, N., Geyer, R., Prüfer, D., & Schulze Gronover, C. (2017). Development of rubber-enriched dandelion varieties by metabolic engineering of the inulin pathway. *Plant Biotechnology Journal*, *15*, 740–753. https://doi.org/10.1111/pbi.12672

Tamura, K., Stecher, G., Peterson, D., Filipski, A., & Kumar, S. (2013), MEGA6: Molecular evolutionary genetics analysis version 6.0. *Molecular Biology and Evolution*, *30*, 2725–2729.

Uchida, H., Sumimoto, K., Ferriols, V. M. E., Imou, K., Saga, K., Furuhashi, K., ... Okada, S. (2015). Isolation and characterization of two squalene epoxidase genes from *Botryococcus braunii*, race B. *PLoS ONE*, *10*, 1–13.

Uchida, H., Yamashita, H., Kajikawa, M., Ohyama, K., Nakayachi, O., Uchida, H., Yamashita, H., Kajikawa, M., Ohyama, K., Nakayachi, O., Sugiyama, R., ... Ohyama, K. (2009). Cloning and characterization of a squalene synthase gene from a petroleum plant, *Euphorbia tirucalli* L. *Planta*, *229*, 1243–1252.

Wentzinger, L. F., Bach, T. J., & Hartmann, M.-A. (2002). Inhibition of squalene synthase and squalene epoxidase in tobacco cells triggers an up-regulation of 3-hydroxy-3-methylglutaryl coenzyme a reductase. *Plant Physiology*, *130*, 334–346. https://doi.org/10.1104/pp.004655

Zhang, L., Jing, F., Li, F., Li, M., Wang, Y., Wang, G., ... Tang, K. (2009). Development of transgenic *Artemisia annua* (Chinese wormwood) plants with an enhanced content of artemisinin, an effective antimalarial drug, by hairpin-RNA-mediated gene silencing. *Biotechnology and Applied Biochemistry*, *52*, 199–207.

18

MicroRNA *MIMIC* binding sites: Minor flanking nucleotide alterations can strongly impact *MIMIC* silencing efficacy in Arabidopsis

Gigi Wong | Maria Alonso-Peral | Bingjun Li | Junyan Li | Anthony A. Millar

Division of Plant Science, Research School of Biology, The Australian National University, Canberra, Australian Capital Territory, Australia

Correspondence
Anthony A Millar, Division of Plant Science, Research School of Biology, The Australian National University, Canberra, ACT, Australia.
Email: tony.millar@anu.edu.au

Funding information
G.W. was supported by an Australian Government Research Training Program RTP Scholarship. This study was funded by an Australian Research Council Discovery Grant (DP130103697) to A.A.M.

Abstract

In plants, microRNA (miRNA) target *MIMICs* (*MIMs*) have been widely used to inhibit miRNA function. They are based on the Arabidopsis *INSENSITIVE TO PHOSPATE STARVATION 1* (*IPS1*) gene that corresponds to a non-coding RNA containing a miR399 binding site that can be modified to sequester and inhibit any miRNA of interest. However, the efficacy of miRNA inhibition of these different *MIMs* can vary greatly. Using *MIMs* that have strong efficacy (*MIM159*) and poor efficacy (*MIM165*), we investigate the underlying cause of this variation. Firstly, sequence alignments of *IPS1* homologs from the *Brassicaceae* identified a highly conserved sequence immediately downstream of the miRNA binding site. Mutating this sequence in the context of the *MIM159* attenuates its strong efficacy. This conserved flanking region contains a predicted stem-loop structure that is also predicted to be present in most modified *MIMs* that appear to have a strong efficacy, but not in *MIM165* that has a poor efficacy. Restoring this predicted stem-loop in *MIM165* via mutation of only three or five nucleotides within the conserved flanking region resulted in *MIM165* variants that have very strong efficacies of miRNA inhibition. However, specifically mutating this predicted stem-loop in the *MIM159* context failed to significantly reduce efficacy, and additional mutations to restore this predicted stem-loop weakened efficacy further. Although this shows there is no simple correlation between this predicted stem-loop and efficacy, these results add to the growing evidence that the sequence context of miRNA binding sites is important, and that minor nucleotide substitutions to flanking sequences of miRNA binding sites can strongly enhance or attenuate the miRNA-target interaction.

KEYWORDS
Arabidopsis, *MIMICs*, miR159, miR165/166, miRNAs

1 | INTRODUCTION

Plant microRNAs (miRNAs) are a class of small RNAs (20–24 nts) that are key negative regulators of gene expression, controlling diverse traits such as development, response to environmental stimuli, and plant-microbe interactions (Li, Reichel, Li, & Millar, 2014). They operate by guiding the RNA induced silencing complex (RISC) to target mRNAs of high complementarity, where they repress expression through mRNA degradation (mRNA cleavage) or via a translational repression mechanism (Iwakawa & Tomari, 2015). Unlike animal systems, where miRNA-target pairs can tolerate many mis-matches, most canonical plant miRNA-target pairs only contain a small number of mis-matches, implying high complementarity is a strict requirement for repression in plants (Liu, Wang, & Axtell, 2014; Schwab et al., 2005). This constraint of high complementarity has remained a corner stone of plant miRNA biology, being the driving parameter of numerous bioinformatic approaches related to plant miRNAs (Dai, Zhuang, & Zhao, 2018).

In animal systems, it is clear that factors in addition to complementarity, such as RNA secondary structure or RNA-binding proteins, can strongly impact the silencing outcome of miRNA-target pair interactions (Kertesz, Iovino, Unnerstall, Gaul, & Segal, 2007; van Kouwenhove, Kedde, & Agami, 2011). By contrast, such evidence in plants is limited. However, we recently demonstrated that RNA secondary structure plays a role in controlling the regulation of the Arabidopsis *MYB33* gene, a conserved target of miR159. In this study, the flanking sequences of the miR159-binding site of *MYB33* were shown to form conserved RNA secondary structures that were required for strong silencing (Li, Reichel, & Millar, 2014; Zheng et al., 2017). This demonstrates that sequence context of a miRNA binding site can strongly impact the silencing outcome, possibly through favorable RNA secondary structure that promotes miRNA-target recognition.

We have previously argued that such factors beyond complementarity may underlie a much narrower functional specificity of plant miRNAs than that of bioinformatically predicted targets that are based primarily on complementarity (Li, Reichel, Li et al., 2014). Such a principle may also impact the efficacy of artificial miRNAs (amiRNA) or miRNA decoys, such as target *MIMICs* (*MIMs*), *SPONGEs* (*SPs*), and *Short Tandem Target MIMICs* (*STTMs*). For instance, different amiRNAs that have analogous complementarities to their target mRNAs work with considerable variability that cannot be explained by complementarity alone (Deveson, Li, & Millar, 2013; Li et al., 2013). Similarly, identical miRNA binding sites embedded within different decoy backbones, inhibit miRNA function with highly variable efficacies (Reichel, Li, Li, & Millar, 2015; Yan et al., 2012). In appraising the efficacies of *MIMs*, *SPs*, and *STTMs* to inhibit two different miRNA families, miR159 and miR165, the efficacy of the approaches was highly variable (Reichel et al., 2015). For example, *MIM159* could inhibit miR159 much more strongly than either a *STTM159* or *SP159* approach. Conversely, *SP165* and *STTM165* inhibited miR165 much more efficiently than *MIM165* (Reichel et al., 2015). Therefore, no one single decoy approach could guarantee strong miRNA inhibition. Not only does the same binding site within different backbones result in variable efficacies, but also different binding sites within the same backbone, such as the contrasting strong efficacy of *MIM159* to the weak efficacy of *MIM165* (Reichel et al., 2015). Given the current broad use of different miRNA decoy approaches to modulate and understand the miRNA function (Zhang et al., 2017; Zhao et al., 2016, 2017), understanding the underlying mechanism(s) responsible for decoy efficacy will be important for their optimal use. Here, we focus on the widely used *MIMs*, and find their efficacy of miRNA inhibition can be strongly impacted by subtle changes to flanking nucleotide composition.

2 | MATERIALS AND METHODS

2.1 | Plant material and growth conditions

All plant lines were in the *Arabidopsis thaliana* Columbia-0 (Col-0) background. Seeds were vapor sterilized using chlorine gas by mixing 100 ml of 100% sodium hypochlorite with 3 ml of 36% HCl. Seeds were sown on MS medium agar plates with hygromycin selection. At 7–10 days old, seedlings were transferred to soil (Debco plugger soil mix with 3.5 g/L Osmocote fertiliser). All seeds were stratified for 24 hr at 4°C in the dark prior to growth in "long day" conditions (16 hr light/8 hr dark cycle) under fluorescent illumination of 150–200 μmol/m^2/s at 22°C.

2.2 | Generation of *MIM* constructs and transformation into Arabidopsis

All *MIM159* and *MIM165* constructs were synthesized (Integrated DNA Technologies) with Gateway attB1 and attB2 sites (Invitrogen) harboring *MIM* binding sites as designed by Todesco, Rubio-Somoza, Paz-Ares, and Weigel (2010). Sequences were cloned into the Gateway entry vector, pDONR/Zeo (Invitrogen), using Gateway BP Clonase II (Invitrogen) and verified using restriction enzyme digestion analysis and sequencing. Resulting plasmids were then recombined into the Gateway compatible binary vector, pMDC32 (Curtis & Grossniklaus, 2003) using Gateway LR Clonase II (Invitrogen). Binary vectors were verified using restriction enzyme digestion analysis and were then transformed into *Agrobacteria tumefaciens* strain GV3101 (Hellens, Mullineaux, & Klee, 2000) by electroporation. Plasmids were extracted for verification using restriction enzyme digestion analysis. *MIM159* and *MIM165* constructs were transformed into Col-0 using the floral dipping method (Clough & Bent, 1998).

2.3 | qRT-PCR

Total RNA was extracted from 4-week-old rosettes using Trizol reagent (Invitrogen). For all *MIM159* constructs, 14 rosettes from independent lines were pooled together for RNA extraction to represent one biological replicate. Three biological replicates were used per construct. Twenty microgram of total RNA was RQ1 DNase treated (Promega) with the addition of 40 units/20 μl RNaseOUT Recombinant RNase inhibitor (Invitrogen). DNase treated RNA was purified using RNeasy Mini kit (QIAGEN) and RNA quality analyzed on 1% agarose gel. 500 ng–2 μg of treated RNA was used for reverse transcription with Superscript III Reverse Transcriptase (Invitrogen) using the oligo dT primer (Invitrogen). cDNA was diluted to 50-fold in nuclease-free water and used for qRT-PCR. *CYCLOPHILIN 5* (At2 g29960) was also measured to normalize the total amount of cDNA input allowing for comparative quantitation using the Rotor-

Gene Q software package. The following gene specific primers were used to amplify *MIM159* and *MIM165* constructs for qRT-PCR [5′-CATTATGTTTGGGTTGTACC][5′-GCACTGGTCTGACT ATTCTCC]. *OsMIM159* was amplified using the following primers [5′-TCTCAAAGAGGCACCAATAC][5′-ATAATGTGGAGTGTGCCCTG]. qRT-PCR was performed on a Rotor-Gene Q real-time PCR machine (QIAGEN) using three technical replicates. The following cycle conditions were used: 1 cycle of 95°C/5 min, 45 cycles of 95°C/15 s, 60°C/15 s, and 72°C/20 s of fluorescence was acquired at the 72°C step. A 55–99°C melting cycle was then performed.

2.4 | Bioinformatic alignments and prediction of RNA secondary structure

Gene sequences were obtained using the Basic Local Alignment Search Tool (BLAST), and aligned using the alignment software MAFFT (Katoh, Misawa, Kuma, & Miyata, 2002). Bioinformatic prediction of RNA secondary structure of *MIMs* was calculated using the Vienna RNAfold WebServer, (http://rna.tbi.univie.ac.at/cgi-bin/RNAWebSuite/RNAfold.cgi; Hofacker, 2003). All RNAfold analyses were performed using the Turner model, 2004 energy parameter (Mathews et al., 2004) at 22°C. Whole sequences were used for all RNA secondary structure prediction.

2.5 | Statistical analysis

Plant morphological phenotyping was analyzed using Pearson Chi-Squared tests. qRT-PCR data for *MIM159* were analyzed using analysis of variance (ANOVA) on log transformed relative expression followed by a post hoc multiple comparisons using Tukey's HSD. Data were log transformed to homogenize the standard deviations.

3 | RESULTS

3.1 | Nucleotides flanking the miR399 binding sites of *IPS1* homologs are highly conserved

Target *MIM*s are based on the non-coding RNA, *INSENSITIVE TO PHOSPHATE STARVATION 1* (*IPS1*) from Arabidopsis. It contains a miR399 binding site with a central 3 nt bulge to prevent slicing, where it is thought that miR399 irreversibly binds *IPS1* resulting in miR399 sequestration and functional inhibition (Franco-Zorrilla et al., 2007). Replacing the *IPS1* miR399 binding site with sequences that target other miRNAs is able to inhibit a wide range of different miRNAs, making this *MIM* approach a versatile functional tool (Franco-Zorrilla et al., 2007; Todesco et al., 2010).

IPS1 homologs are present in diverse monocot and dicot species of plants, and although *IPS1* homologs are non-coding RNAs, their length is consistently over 500 nt (Table S1), suggesting the maintenance of such a lengthy backbone signifies functional importance. However, other than the 23-nt miR399 binding site, very little sequence conservation is shared amongst *IPS1* homologs from these diverse species of plants (Franco-Zorrilla et al., 2007; Huang, Shirley, Genc, Shi, & Langridge, 2011; Supporting Information Figure S1). By contrast, when alignments were performed with *IPS1* homologs from just the *Brassicaceae* family, three strongly conserved regions were identified (Supporting Information Figure S2). Most notably was a fully conserved 32-nt region immediately downstream of the miR399 binding site in the *Brassicaceae IPS1* homologs (Figure 1a). Similarly, an alignment of monocot *IPS1* homologs identified a 12-nt region immediately upstream of the miR399 binding site that was strongly conserved (Figure 1b), when compared to the other regions of *IPS1* homologs from monocot species (Supporting Information Figure S3). The proximity of these conserved sequences to the miR399 binding site suggests they may be playing a role in the *IPS1*–miR399 interaction.

3.2 | The *IPS1* backbone sequence impacts the efficacy of miRNA inhibition

To test whether these three conserved regions impact the ability of Arabidopsis *IPS1* to inhibit miRNA function, we utilized *MIM159*, a modified *IPS1* in which the binding site has been changed to target miR159 (Todesco et al., 2010). Expression of *MIM159* in Arabidopsis results in characteristic phenotypic defects of the rosette whose severity can be readily assessed to gauge the extent of miR159 inhibition (Reichel et al., 2015). We mutated the nucleotide region immediately adjacent to the miRNA binding site of *MIM159* to a different sequence (but maintaining overall GC content) to result in the variant, *MIM159-1m* (Figure 2a). Additional mutations were then made to the other two conserved regions to generate the variant *MIM159-3m*, so that it contained three different mutated regions (Figure 2a). Finally, a *MIM159* binding site was incorporated into *IPS1* from *Oryza sativa*, designated as *OsMIM159*. *OsIPS1* has little sequence identity to *Arabidopsis IPS1*, where the miR399 binding site is the only strongly conserved region between these two *IPS1* homologs (Supporting Information Figure S4). The performance of these three *MIMs* were then compared against the Arabidopsis *MIM159* transgene.

All *MIMs* were cloned into the vector pMDC32, so to be under the transcriptional control of the 2×35S promoter (Curtis & Grossniklaus, 2003). After transformation into Arabidopsis, the rosette phenotype of multiple independent transformants was scored. Phenotypes were classified as either No Phenotypic (N; indistinguishable from wild-type), Mild (M; presence of leaf curvature), and Severe (S; the plant has strongly curled leaves with the abaxial side of two or more leaves being visible from an aerial view) (Figure 2b). The relative proportions of these phenotypic categories were used as a measure of silencing efficacy of the different *MIM* transgenes. A *35S:GUS* construct was used as a transgenic negative control.

As previously reported, *MIM159* displays a strong efficacy, as the majority of plants [102/114 (89%)] display "severe" phenotypic defects (Figure 2c). By contrast, *MIM159-1m* transformants display "severe" phenotypes at a significantly lower frequency [57/81 (70%), $p = 0.001144$]. Furthermore, *MIM159-3m* transformants displayed even fewer plants with "severe" phenotypes [34/92 (37%)], being significantly different from *MIM159-1m* ($p = 7.476e-06$). Consistent

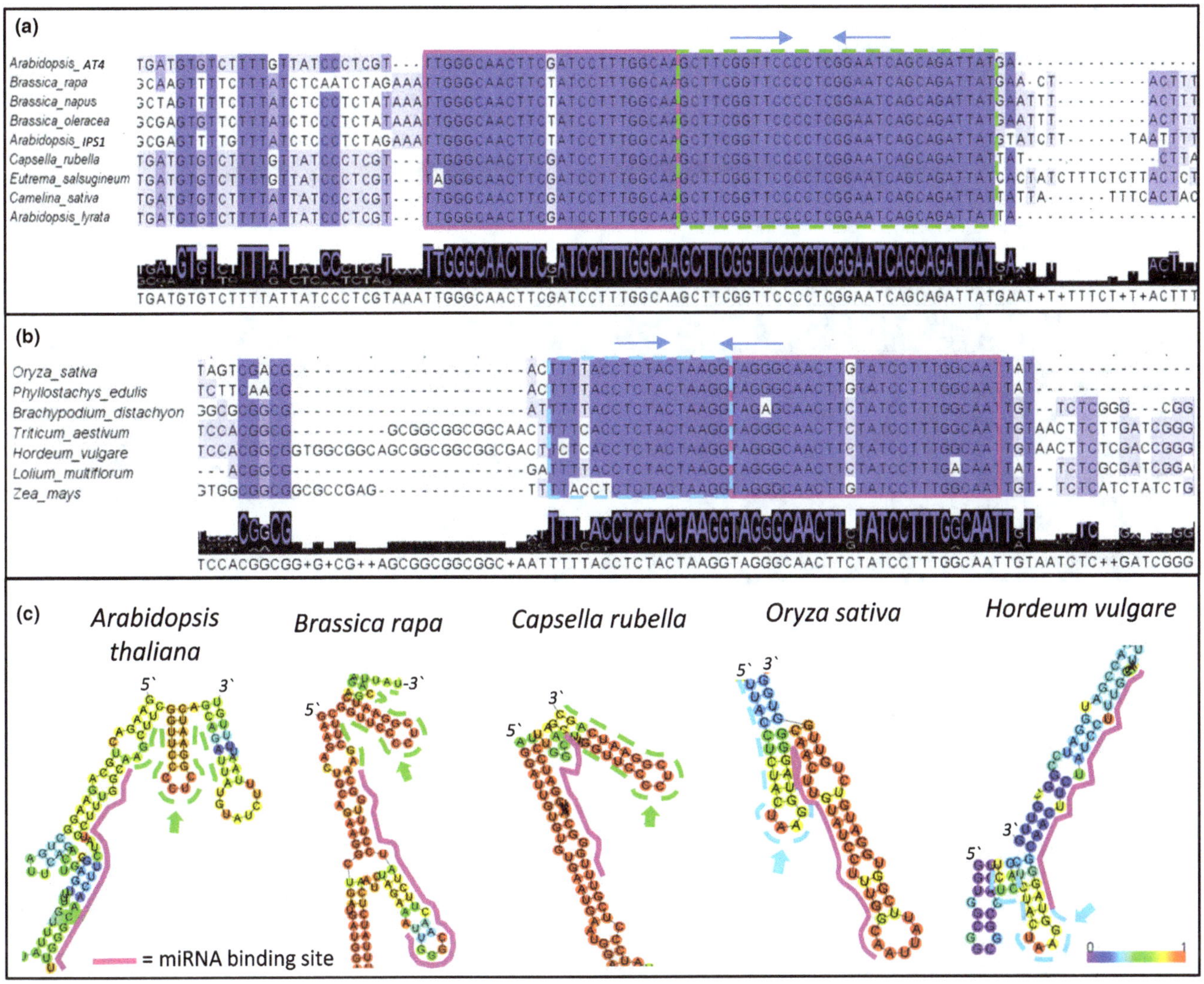

FIGURE 1 *IPS1* homologs have conserved nucleotides flanking the binding site that contain putative stem-loop regions. *IPS1* homologs were aligned using MAFFT. Conserved nucleotide sequences are observed in (a) *Brassicaceae* family members (green dashed box) and (b) various monocotyledonous plants (light blue dashed box). The binding site is indicated in the red box. Putative RNA stems in conserved sequences are indicated with arrows. (c) Predicted RNA secondary structures of *IPS1* homologs from the *Brassicaceae* family and monocotyledonous plants as determined using the RNAfold WebServer. The miR399 binding site is indicated by the pink line and the conserved sequence in green dashed lines for the *Brassicaceae IPS1* homologs and blue dashed lines for the monocotyledonous homologs. Arrows of the respective colors indicate the predicted stem-loop. Color of nucleotides represents the probability of structure formation

with our previous data (Reichel et al., 2015), this demonstrates that sequences outside of this miRNA binding site impacts the silencing efficacy of the *MIM159* construct. Furthermore, the frequency of *OsMIM159* transformants that displayed severe phenotypes was significantly lower than *MIM159* [42/79 (53%), p = 3.988e-08] (Figure 2c). Therefore, changing the *IPS1* backbone changes the silencing efficacy of the *MIM*.

To further analyze these plants, we used qRT-PCR to measure the transcript levels of a major target gene of miR159, *MYB33*, as well as a *MYB33* downstream gene, *CP1* (Alonso-Peral et al., 2010). RNA was prepared from multiple randomly selected transformants and qRT-PCR was performed. Generally, both the *MYB33* and *CP1* mRNA levels correlated with the strength of the phenotypic defects, with *MIM159* plants having the highest *MYB33/CP1* mRNA levels and *MIM159-3m* having the lowest (Figure 2e–f). *OsMIM159* plants that had intermediate efficacy compared to *MIM159* and *MIM159-3m*, contained intermediate levels of *MYB33* and *CP1* transcript levels (Figure 2e–f).

3.3 | The mRNA levels of the *MIMs* inversely correlate with silencing efficacy

Although all the *MIMs* in the above experiment are all under the control of an identical promoter (2x 35S), it is possible that the different *MIM* transcripts have different RNA stabilities (steady-state levels) which may explain their different efficacies. To investigate this, the RNA levels of the different *MIMs* were measured using qRT-PCR. The primers used were identical for *MIM159*, *MIM159-1m*,

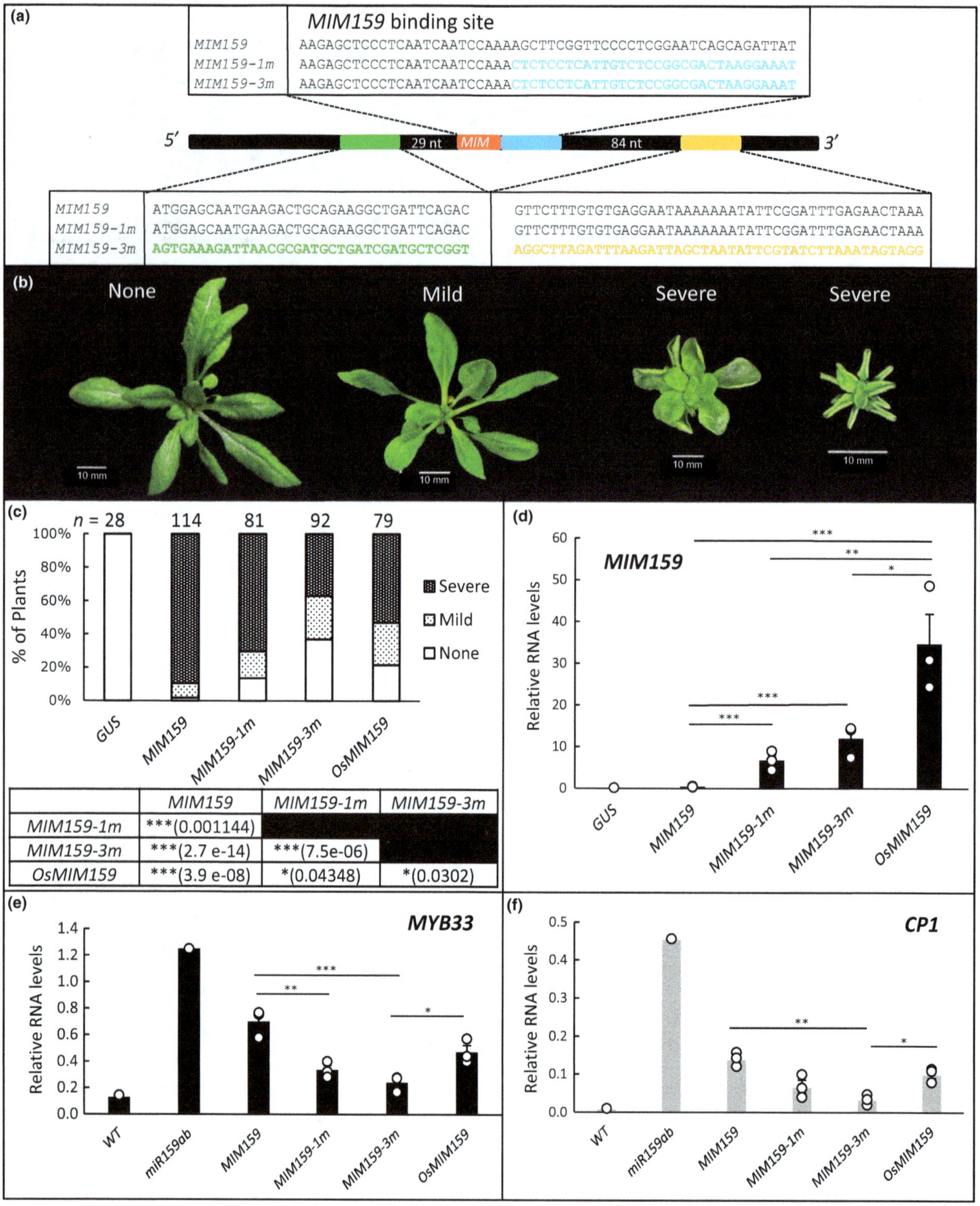

FIGURE 2 Generation and analysis of different *MIM159* transgenes. (a) *MIM159* was mutated in the downstream conserved flanking region of the miRNA binding site (in red) to create *MIM159-1m* (in blue). Two further mutations were made in the conserved regions further downstream and upstream of the miR159 binding site (in orange and green, respectively) to create *MIM159-3m*. (b) Rosette phenotypes were scored from 4-wk-old transformed plants. Phenotypic categories of "None" (phenotype indistinguishable from WT), "Mild" (presence of leaf curvature), and "Severe" (Leaves strongly curled so that the abaxial side is showing of two or more leaves) plant phenotypes. (c) Percentage of plants displaying each category of phenotypes as shown in B. Significance values are shown, calculated using Pearson Chi-Squared tests. (d-f) Transcript profiling of (d) the different *MIMs*, (E) *MYB33*, and (e) *CP1*. Measurements are composed of three biological replicates (shown as dots), with each replicate being composed of 14 randomly selected primary transformants. Statistical analysis was performed using analysis of variance where significant differences between means are indicated by '*' $p < 0.05$, '**' $p < 0.01$ and '***' $p < 0.001$. Error bars represent the standard error of the mean

and *MIM159-3m*, but differed for *OsMIM159*. As a transgenic negative control, *35S:GUS* transformants were used.

Strikingly, *MIM* transcript levels were generally inversely correlated to their silencing efficacies. *MIM159* had the lowest RNA level despite having the greatest efficacy, with *MIM159-3m* plants having at least a tenfold higher levels of the *MIM* RNA (Figure 2d). Interestingly, *OsMIM159* plants had the highest *MIM* RNA level, despite this transgene having a weaker efficacy than *MIM159* (Figure 2c). These results argue against lower RNA levels being the cause of the reduced efficacy of the mutated *MIM159* transgenes or the relatively lower efficacy of *OsMIM159*, but rather other factors are in play. In fact, the inverse correlation may argue that if a *MIM* has a strong interaction with its target miRNA, this may promote *MIM* RNA degradation, so a low RNA level maybe reflective of strong miRNA-*MIM* target recognition. Alternatively, the inverse correlation of *MIM* RNA levels to efficacy may suggest that their expression results in lethality and that only plants expressing *MIMs* below a certain level can be generated, where the greater the efficacy, the lower the lethal dose of *MIM* RNA.

3.4 | Predicted RNA stem-loop structures are adjacent to the miR399 binding site of *IPS1*

Recently, we have shown that a RNA stem-loop structure that is adjacent to the miR159 binding site of *MYB33* is required for strong miR159-mediated silencing (Zheng et al., 2017). Interestingly, within the conserved 32-nt region adjacent to the miR399 binding site of *IPS1 Brassicaceae* homologs, a stem-loop structure of high confidence is predicted to form as determined by the Vienna RNAfold WebServer (Figure 1c). Similarly, within the conserved stretch of nucleotides adjacent to the miR399 binding site of *IPS1* genes from monocotyledonous plants, a stem-loop structure is also predicted to form, albeit with less confidence (Figure 1c).

Similar to the case for *MYB33*, we investigated whether this potential RNA stem-loop promoted miRNA-target interaction. We have previously shown that *MIMs* harboring different miRNA binding sites display dramatically different efficacies in inhibiting their target miRNAs (Reichel et al., 2015). Given that changing the miRNA binding site within a given context can alter local RNA secondary structures, we determined the predicted RNA secondary structure for all the *MIMs* whose expression could result in phenotypic defects (Supporting Information Figure S5; Todesco et al., 2010). We found that most *MIMs* contained the identical predicted RNA stem-loop arising from the conserved adjacent region of *IPS1*, although with varying degrees of confidence. However, *MIM165* was not predicted to contain this stem-loop structure (Figure 3a), but instead the stem-loop nucleotides are predicted to base-pair with the *MIM165* binding site, possibly attenuating the binding site's accessibility. As *MIM165* has been previously reported to have poor silencing efficacy (Reichel et al., 2015; Yan et al., 2012), we investigated whether the efficacy of *MIM165* could be increased via restoring the predicted RNA stem-loop structure.

3.5 | Restoring the predicted RNA stem-loop structure in *MIM165* transforms its silencing efficacy

Based on RNA secondary structure prediction using the Vienna RNAfold WebServer, we generated two different *MIM165* variants, designated *MIM165-3M* and *MIM165-5M*, that had three and five nt alterations in the flanking region of the *MIM165* binding site that was predicted to restore the stem-loop structure and also change the secondary structure of the *MIM165* binding site itself (Figure 3a). The idea was to introduce mutations that would strengthen the stem-loop through additional G-C pairs, with *MIM165-5M* having a G-C pair replace an A-U pair in the stem (Figure 3a). The stem-loop was predicted to be present when the entire *MIM165* RNAs were folded (Supporting Information Figure S6). The *MIM165* binding sites of the two variants remained unchanged from the binding site of the parental *MIM165* transgene, and all three binding sites were identical to the STTM165 binding site (Yan et al., 2012), including the three nucleotide bulge (CTA) at position 10–11 (Todesco et al., 2010). The three different *MIM165* constructs were independently transformed into wild-type plants and their efficacies scored based on the frequency and severity of phenotypes of primary transformants. Plants were either classified as having phenotypes with; None (indistinguishable from wild-type), Mild (evidence of adaxialized leaves), or severe (trumpet or cup-shaped leaves) defects (Figure 3b). As has been previously found (Todesco et al., 2010; Yan et al., 2012), *MIM165* failed to result in any transgenic plants displaying a severe phenotype of trumpet or cup-shaped leaves (Figure 3c). By contrast, many *MIM165-3M* and *MIM165-5M* transformants displayed severe phenotypes, with *MIM165-5M* having the greatest proportion of plants with phenotypic defects. Therefore, seemingly minor nucleotide changes to the flanking regions of the *MIM165* binding site have had a strong impact on the efficacy of this *MIM*.

FIGURE 3 Generation and analysis of the *MIM165*, *MIM165-3M* and *MIM165-5M* transgenes. (a) Sequence of the conserved flanking region of original *MIM165* and the two variants, *MIM165-3M* and *MIM165-5M*, with the putative stem-loop in bold nucleotides (also indicated with arrow) and mutations underlined. Black lines in the cartoons of the transgenes also indicate the position and number of mutations made. The predicted secondary structure of *MIM165* and *MIM165* mutants is shown, with the miR165 binding site indicated by a pink line. (b) Phenotypic categories of "None" (indistinguishable from wild-type), "Mild" (evidence of adaxialized leaves), and "Severe" (trumpet or cup-shaped leaves) plant phenotypes. (c) Percentage of plants displaying each category of phenotypes from the three *MIM165* constructs. Significance values are shown, calculated using Pearson Chi-Squared tests. (d) Relative *MIM165* RNA levels in transgenic plants from different phenotypic categories of the different *MIM165* variants. Measurements are from two biological replicates (shown as dots), each being composed of five primary transformant plants

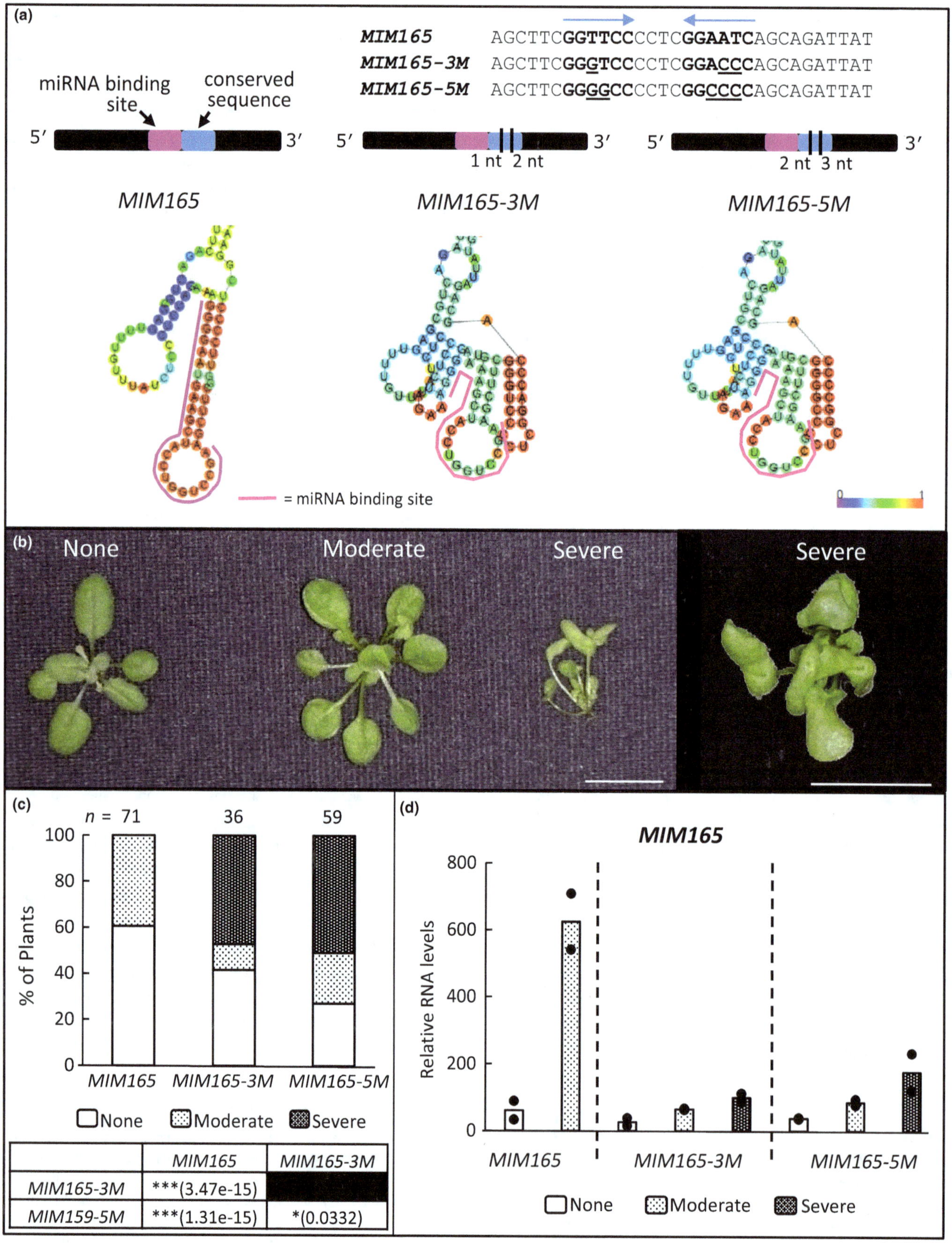

In order to gain insight into the mechanism behind this increased efficacy, we measured *MIM165* RNA levels in the different phenotypic categories for each of the three *MIM165* transgenes. For *MIM165-3M* and *MIM165-5M*, the *MIM165* RNA levels followed a simple trend of directly correlating with phenotypic severity, where the stronger the phenotype, the higher the *MIM165* transcript level (Figure 3d).

However, for the parental *MIM165* transgene, much higher *MIM165* RNA levels were found in plants with Mild defects in comparison to *MIM165-3M* and *MIM165-5M* plants that displayed similar Mild defects (Figure 3d). As these steady-state RNA levels will depend on both transcription and degradation, we hypothesize that only in *MIM165* plants where the transgene is very strongly transcribed can phenotypic defects arise. Additionally, consistent with the different *MIM159* transgenes (Figure 2), *MIMs* that perform poorly have higher transcript levels, supporting the notion that if *MIM*–miRNA interactions are strong, this will promote *MIM* RNA degradation, or that over-expression of *MIMs* with strong efficacy lead to plant death.

3.6 | The predicted RNA stem-loop in *MIM159* does not correlate with strong efficacy

As the *MIM159-1m* mutant no longer contains the predicted RNA stem-loop, we investigated whether this predicted stem-loop correlated with stronger efficacy of *MIM159*. Firstly, two nucleotide mutations were generated in *MIM159* to generate the construct *MIM159-MSL*, which resulted in a construct predicted not to contain the stem-loop (Figure 4a). However, although less *MIM159-MSL* transformants displayed severe phenotypes compared to *MIM159* transformants, this difference was not statistically different (Figure 4b, [p = 0.2195]). Then, a second construct was generated, *MIM159-RES*, in which further mutations were introduced that were predicted to restore the stem-loop, but not the sequence (Figure 4a). However, *MIM159-RES* displayed a weaker efficacy than *MIM159*, as there was significantly less *MIM159-RES* transformants displaying severe phenotypes (Figure 4B, [p = 0.0002063]). This was despite the *MIM159* binding site in *MIM159-RES* having a predicted RNA secondary structure that somewhat resembled the *MIM165-3M* and *MIM165-5M* transgenes (Figure 4a). Therefore, the presence or absence of this predicted stem-loop is not an absolute indicator of the silencing performance of any given *MIM* decoy.

4 | DISCUSSION

Previously, we assessed the capacity of *MIMs*, *STTMs*, and *SPs* regarding their ability to inhibit miRNA activity. Firstly, we found that no one approach guaranteed strong miRNA inhibition, where changing the binding site within the backbone altered their efficacies of miRNA inhibition. Secondly, that identical binding sites within different backbones worked with highly variable efficacies (Reichel et al., 2015). Both of these observations imply an interaction between the binding site and backbone sequences, which can either be favorable or detrimental to the decoy's silencing efficacy. However, the underlying reasons for this could be variable, including the steady-state levels of the RNA decoy transcripts, the accessibility of the miRNA binding site within the decoy or other unknown mechanisms.

Here, we show subtle nucleotide changes adjacent to the miRNA binding site of *MIMs* can dramatically alter their efficacy of inhibition of miRNA function. This was most apparent for *MIM165*, where a three nucleotide change could transform its efficacy from a poor to highly effective miRNA decoy. Interestingly, the design of the alteration was to restore a small predicted stem-loop that is present in a stretch of nucleotides that are conserved immediately adjacent to the miRNA binding site within *IPS1* homologs from the *Brassicaceae.*

From RNA secondary structure predictions, the restoration of this stem-loop in *MIM165-3M* and *MIM165-5M* is predicted to prevent a strong RNA secondary structure that forms in the parental *MIM165* decoy that potentially sequesters much of the *MIM165* binding site (Figure 3a). Such alterations to RNA secondary structure are the most likely explanation underpinning the alteration of efficacy. Here, the mechanism may be, that if sequences flanking a miRNA binding site are in a secondary structure, they are less likely to base-pair with the binding site itself, and hence increase accessibility of that binding site. Such a principle was hypothesized to be behind the strong efficacy of *STTM165/166* (Yan et al., 2012), where this decoy is composed of a 48 nt spacer region, that potentially forms a strong RNA secondary structure, that is flanked by two *MIM165* binding sites. Currently, the analysis of target site accessibility can only be derived from *in silico* prediction programs, where the standard approach is to analyze a region that includes 17 nt upstream and 13 downstream nucleotides of the miRNA binding site (Dai et al., 2018), a region that was found to correlate best with accessibility for animal miRNA-target sites (Kertesz et al., 2007). However, the mutations that we have introduced into the *MIMs* here are outside of this region, as were the flanking mutations that impacted *MYB33* silencing (Zheng et al., 2017). Therefore, the flanking nucleotides included in such an analysis would need to be arbitrarily increased. Given all this uncertainty, it is unlikely that these *in silico* analyses are likely to generate informative predictions.

It is tempting to speculate that the presence of such a stem-loop ensures strong efficacy of the *MIM* decoy, as most reported *MIM* constructs that result in miRNA inhibition have such a structure (Supporting Information Figure S5), as does *MIM165-3M* and *MIM165-5M* (Figure 3). However, mutation of this stem-loop structure within the *MIM159* context, does not significantly attenuate efficacy, and additional compensatory mutations to restore the structure resulted in attenuated efficacy (Figure 4). Given the conflicting results between *MIM165* and *MIM159*, the presence or absence of the predicted stem-loop cannot be regarded as an absolute indicator of efficacy. As mentioned above, the restoration of the stem-loop in *MIM165-3M* and *-5M* may have abolished a competing RNA secondary structure that was predicted to form in *MIM165* (Figure 3a), and which potentially inhibits the accessibility of the miRNA binding site. Therefore, in the *MIM165* context, the stem-loop becomes an important determinant of silencing efficacy. By contrast for *MIM159*, no strong RNA secondary structure was predicted to sequester the miR159 binding site, even when the stem-loop is abolished (Figure 4a). Therefore, the stem-loop is not an important determinant of silencing efficacy in the *MIM159* context. Such a claim will need to be tested with further experimentation. Nevertheless, our observations highlight the complexity of miRNA binding sites, where changing only the miRNA binding site within the *MIM* backbone may not

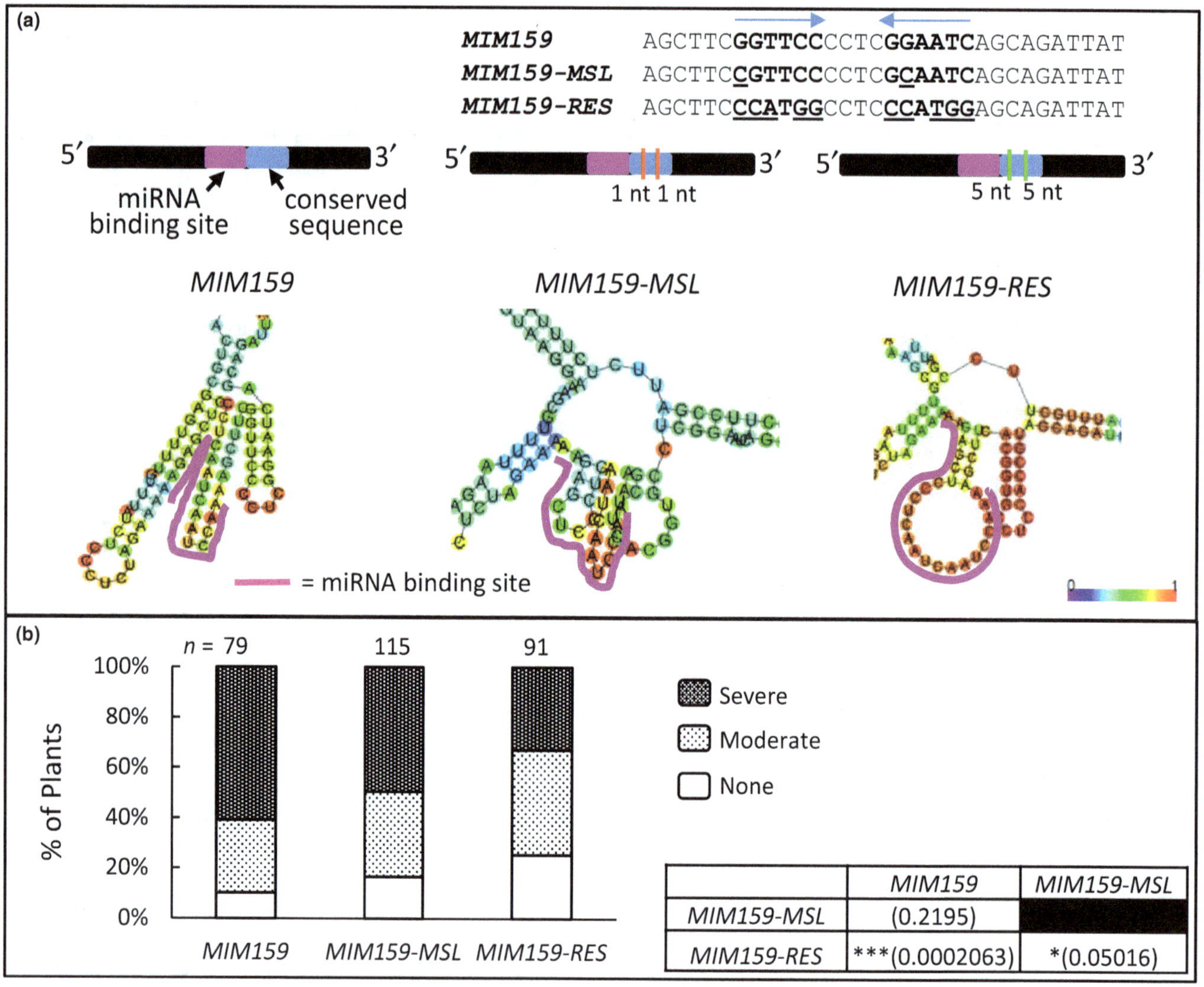

	MIM159	MIM159-MSL
MIM159-MSL	(0.2195)	
MIM159-RES	***(0.0002063)	*(0.05016)

FIGURE 4 Generation and analysis of the *MIM159*, *MIM159-MSL* and *MIM159-RES* transgenes. (a) Sequence of the conserved flanking region of original *MIM159* and the two variants, *MIM159-MSL* and *MIM159-RES*, with the putative stem-loop in bold nucleotides (also indicated with arrow) and mutations underlined. The cartoon depicts the original *MIM159* and the two variants *MIM159-MSL* and *MIM159-RES*. The red lines indicate the position and number of mutations made to destroy the stem-loop in *MIM159-MSL* and the green line for mutations made to restore the stem-loop in *MIM159-RES*. The predicted secondary structure of *MIM159* and *MIM159* mutants. The miR159 binding site is indicated by the pink line (b) Percentage of plants displaying each category of phenotypes from *MIM159* and *MIM159* mutants. The same phenotypic categories were used as for Figure 2b. Significance values are shown, calculated using Pearson Chi-Squared tests

only change the miRNA that it targets, but potentially also the local RNA secondary structure that impacts miRNA-target site accessibility, which ultimately impacts efficacy of the decoy.

Another possibility was that the mutations were altering the abundance of the decoys by altering their RNA stability, hence the more abundant the *MIM* was, the higher its efficacy. However, in all instances, the opposite appeared to be the case, where steady-state levels of the *MIM*s inversely correlated with efficacy. For example, *MIM165-3M* and *MIM165-5M* required much lower steady-state *MIM* transcript levels to induce moderate phenotypes compared to *MIM165* (Figure 3), indicating that the *MIM*s with higher efficacy can produce stronger phenotypes with lower expression levels. This would argue that miRNA-*MIM* interaction is much more favorable with *MIM165-3M* and *MIM165-5M*, than for *MIM165*, as much less *MIM* RNA is required for the same silencing outcome. However, it is clear that the RNA expression level of the *MIM* is important for efficacy, where high enough expression may be able to overcome a weak *MIM*-miRNA interaction. Expression of STTMs via a potato virus-X system was able to strongly inhibit miR159 and miR165 in tobacco, where the STTM159 appeared to work even stronger than STTM165 based on relative miR159 and miR165 levels (Zhao et al., 2016).

In addition to sequences adjacent to the miRNA binding site, altering sequences further away, as is the case for *MIM159-3m*, also altered the performance of the decoy. How the alteration of these sequences impacts the *MIM* efficacy is harder to explain, but as these sequences were more conserved in *IPS1* homologs from the *Brassicaceae*, it could be that they play some important role for inhibition of the miRNA that is yet to be identified.

In summary, we found that subtle mutations in flanking sequences could result in dramatic impacts on the performance of these *MIMs*. We speculate this is most likely due to alterations of RNA secondary structure and associated changes to miRNA binding site accessibility. Although we have analyzed artificial *MIMs* in this study, there is no reason why this principle will not apply to endogenous target genes or endogenous targets *MIMs*, for which many have been predicted (Dai et al., 2018; : Meng, Shao, Wang, & Jin, 2012), but how many of these interactions are functionally relevant remains a pressing question (Li, Reichel, Li et al., 2014). Currently, *MYB33* is one example in which conserved flanking nucleotides facilitates physiologically relevant silencing (Zheng et al., 2017). How many more natural miRNA targets are subject to such phenomenon remain to be determined. Currently, it appears *in silico* predictions of RNA secondary structure/miRNA binding site accessibility are so limited, they are unable to accurately assist in identifying favorable miRNA-target interactions. Until such issues are resolved, identifying favorable miRNA-target interactions will remain elusive and their discovery will be through laborious experimentation.

ACKNOWLEDGMENTS

We thank Dr Terry Neeman for help with statistical analyses.

AUTHOR CONTRIBUTIONS

G.W designed and performed experiments in Figures 1, 2 and 4. M.A-P. and B.L. designed and performed experiments in Figure 3. J.L. and A.A.M supervised the project.

REFERENCES

Alonso-Peral, M. M., Li, J., Li, Y., Allen, R. S., Schnippenkoetter, W., Ohms, S., … Millar, A. A. (2010). The microR159 regulated *GAMYB-like* genes inhibit growth and promote programmed cell death in Arabidopsis. *Plant Physiology*, *154*, 757–771. https://doi.org/10.1104/pp.110.160630

Clough, S. J., & Bent, A. F. (1998). Floral dip: A simplified method for Agrobacterium-mediated transformation of *Arabidopsis thaliana*. *The Plant Journal*, *16*, 735–743. https://doi.org/10.1046/j.1365-313x.1998.00343.x

Curtis, M. D., & Grossniklaus, U. (2003). A gateway cloning vector set for high-throughput functional analysis of genes in planta. *Plant Physiology*, *133*, 462–469. https://doi.org/10.1104/pp.103.027979

Dai, X., Zhuang, Z., & Zhao, P. X. (2018) psRNATarget: A plant small RNA target analysis server (2017 release). *Nucleic Acids Research*, *46*: W49 to W54. https://doi.org/10.1093/nar/gky316

Deveson, I., Li, J., & Millar, A. A. (2013). MicroRNAs with analogous target complementarities perform with highly variable efficacies in Arabidopsis. *FEBS Letters*, *587*, 3703–3708. https://doi.org/10.1016/j.febslet.2013.09.037

Franco-Zorrilla, J. M., Valli, A., Todesco, M., Mateos, I., Puga, M. I., Rubio-Somoza, I., … PazAres, J. (2007). Target mimicry provides a new mechanism for regulation of microRNA activity. *Nature Genetics*, *39*, 1033–1037. https://doi.org/10.1038/ng2079

Hellens, R., Mullineaux, P., & Klee, H. (2000). Technical Focus: A guide to Agrobacterium binary Ti vectors. *Trends in Plant Science*, *5*, 446–451. https://doi.org/10.1016/S1360-1385(00)01740-4

Hofacker, I. L. (2003). Vienna RNA secondary structure server. *Nucleic Acids Research*, *31*, 3429–3431. https://doi.org/10.1093/nar/gkg599

Huang, C. Y., Shirley, N., Genc, Y., Shi, B., & Langridge, P. (2011). Phosphate utilization efficiency correlates with expression of low-affinity phosphate transporters and noncoding RNA, *IPS1*, in barley. *Plant Physiology*, *156*, 1217–1229. https://doi.org/10.1104/pp.111.178459

Iwakawa, H., & Tomari, Y. (2015). The functions of microRNAs: mRNA decay and translational repression. *Trends in Cell Biology*, *25*, 651–665. https://doi.org/10.1016/j.tcb.2015.07.011

Katoh, K., Misawa, K., Kuma, K., & Miyata, T. (2002). MAFFT: A novel method for rapid multiple sequence alignment based on fast Fourier transform. *Nucleic Acids Research*, *30*, 3059–3066. https://doi.org/10.1093/nar/gkf436

Kertesz, M., Iovino, N., Unnerstall, U., Gaul, U., & Segal, E. (2007). The role of site accessibility in microRNA target recognition. *Nature Genetics*, *39*, 1278–1284. https://doi.org/10.1038/ng2135

van Kouwenhove, M., Kedde, M., & Agami, R. (2011). MicroRNA regulation by RNA-binding proteins and its implications for cancer. *Nature Reviews Cancer*, *11*, 644–656. https://doi.org/10.1038/nrc3107

Li, J.-F., Chung, H. S., Niu, Y., Bush, J., McCormack, M., & Sheen, J. (2013). Comprehensive protein-based artificial microRNA screens for effective gene silencing in plants. *The Plant Cell*, *25*, 1507–1522. https://doi.org/10.1105/tpc.113.112235

Li, J., Reichel, M., Li, Y., & Millar, A. A. (2014). The functional scope of plant microRNA-mediated silencing. *Trends in Plant Science*, *19*, 750–756. https://doi.org/10.1016/j.tplants.2014.08.006

Li, J., Reichel, M., & Millar, A. A. (2014). Determinants beyond both complementarity and cleavage govern microR159 efficacy in Arabidopsis. *PLoS Genetics*, *10*, e1004232. https://doi.org/10.1371/journal.pgen.1004232

Liu, Q., Wang, F., & Axtell, M. J. (2014). Analysis of complementarity requirements for plant microRNA targeting using a *Nicotiana benthamiana* quantitative transient assay. *The Plant Cell*, *26*, 741–753. https://doi.org/10.1105/tpc.113.120972

Mathews, D. H., Disney, M. D., Childs, J. L., Schroeder, S. J., Zuker, M., & Turner, D. H. (2004). Incorporating chemical modification constraints into a dynamic programming algorithm for prediction of RNA secondary structure. *Proceedings of the National Academy of Sciences*, *101*, 7287–7292. https://doi.org/10.1073/pnas.0401799101

Meng, Y., Shao, C., Wang, H., & Jin, Y. (2012). Target mimics: An embedded layer of microRNA-involved gene regulatory networks in plants. *BMC Genomics*, *13*, 197. https://doi.org/10.1186/1471-2164-13-197

Reichel, M., Li, Y., Li, J., & Millar, A. A. (2015). Inhibiting plant microRNA activity: Molecular *SPONGE*s, target *MIMIC*s and STTMs all display variable efficacies against target microRNAs. *Plant Biotechnology Journal*, *13*, 915–926. https://doi.org/10.1111/pbi.12327

Schwab, R., Palatnik, J. F., Riester, M., Schommer, C., Schmid, M., & Weigel, D. (2005). Specific effects of microRNAs on the plant transcriptome. *Developmental Cell*, *8*, 517–527. https://doi.org/10.1016/j.devcel.2005.01.018

Todesco, M., Rubio-Somoza, I., Paz-Ares, J., & Weigel, D. (2010). A collection of target mimics for comprehensive analysis of microRNA function in *Arabidopsis thaliana*. *PLoS Genetics*, *6*, e1001031. https://doi.org/10.1371/journal.pgen.1001031

Yan, J., Gu, Y., Jia, X., Kang, W., Pan, S., Tang, X., … Tang, G. (2012). Effective small RNA destruction by the expression of a short tandem target mimic in Arabidopsis. *The Plant Cell*, *24*, 415–427. https://doi.org/10.1105/tpc.111.094144

Zhang, H., Zhang, J., Yan, J., Gou, F., Mao, Y., Tang, G., … Zhu, J. K. (2017). Short tandem target mimic rice lines uncover functions of miRNAs in regulating important agronomic traits. *Proceedings of the National Academy of Sciences of the United States of America*, *114*, 5277–5282. https://doi.org/10.1073/pnas.1703752114

Zhao, J., Liu, Q., Hu, P., Jia, Q., Liu, N., Yin, K., … Liu, Y. (2016). An efficient potato virus X- based microRNA silencing system in *Nicotiana benthamiana*. *Scientific Reports*, *6*, 20573. https://doi.org/10.1038/srep20573

Effect of δ-aminolevulinic acid dehydratase silencing on the primary and secondary metabolisms of citrus

Nabil Killiny | Faraj Hijaz | Yasser Nehela | Subhas Hajeri | Siddarame Gowda

Department of Plant Pathology, Citrus Research and Education Center, IFAS, University of Florida, Lake Alfred, Florida

Correspondence
Nabil Killiny, Department of Plant Pathology, Citrus Research and Education Center, University of Florida, 700 Experiment Station Rd., Lake Alfred, FL 33850.
Email: nabilkilliny@ufl.edu

Present address
Subhas Hajeri, Citrus Pest Detection Program, Central California Tristeza Eradication Agency, Tulare, California.

Funding information
SCRI-NIFA, USDA, Grant/Award Number: 201500955-04

Abstract

δ-aminolevulinic acid dehydratase (ALAD) is an important enzyme in tetrapyrrole synthesis. ALAD combines two δ-aminolevulinic acid (δ-ALA) molecules to form the pyrrole molecule, porphobilinogen, an important precursor for plant pigments involved in photosynthesis, respiration, and nutrient uptake. In this study, we investigated the effects of silencing of *ALAD* gene on citrus leaf pigments and metabolites. The ALAD enzyme was inhibited using virus-induced gene silencing (VIGS) technology using *citrus tristeza* virus (CTV). δ-ALA accumulated in citrus plants inoculated with the recombinant virus (CTV-tALAD) to silence *ALAD* and resulted in discrete yellow spots (yellow islands) and necrosis in leaves and stems. The levels of chlorophylls, starch, sucrose, *trans*- and *cis*-violaxanthin, and α- and β-cryptoxanthin were reduced in CTV-tALAD plants, whereas zeaxanthin was increased. The increase in zeaxanthin and the decrease in its precursors indicated that the reduction in chlorophylls resulted in light damage. Salicylic acid and jasmonic acid levels, as well as emission of (*E*)-α-bergamotene and (*E*)-β-farnesene, increased in CTV-tALAD plants indicating these plants were under stress. Our results showed that silencing of *ALAD* induces stress in plants and that VIGS using mild CTV strains is a promising technique to study biological function of citrus genes.

KEYWORDS
δ-aminolevulinic acid dehydratase, abscisic acid, citrus, leaf pigments, virus-induced gene silencing

1 | INTRODUCTION

Tetrapyrroles are macrocyclic compounds that contain four pyrrole rings with various structural alternatives and numerous biological functions in prokaryotes and eukaryotes. Higher plants possess four classes of tetrapyrroles that play important roles in many biological processes; chlorophyll is essential for photosynthesis, heme is necessary for respiration, siroheme is necessary for sulfur and nitrogen assimilation, and phytochromobilin is required for light sensing (Tanaka & Tanaka, 2007). Tetrapyrrole biosynthesis is a multistep and a multibranched pathway; the first six steps (from glutamate to uroporphyrinogen III) are required by all classes of tetrapyrroles, whereas the next three steps (from uroporphyrinogen III to protoporphyrin IX) are common to phytochromobilin, heme, and chlorophyll biosynthesis (Tanaka & Tanaka, 2007).

In the first step of tetrapyrrole biosynthesis, glutamyl-tRNA synthetase attaches glutamate to t-RNA to form tRNAGlu complex. In the second step, the carboxyl group of glutamyl-tRNA is converted to a formyl group by glutamyl-tRNA reductase to generate glutamate-1-semialdehyde (GSA). In the third step, GSA is converted into δ-ALA by GSA aminotransferase (GSA-AT). In the fourth step, two δ-ALA molecules are combined together by ALA-dehydratase (ALAD) to form a pyrrole molecule, porphobilinogen (PBG) (Tanaka & Tanaka, 2007). The δ-ALA does not accumulate in developing leaves and is directly converted to PBG by ALAD. However, the addition of levulinic acid, a specific competitive inhibitor of ALAD, can induce the accumulation of δ-ALA in greening tissues (Kumar & Söll, 2000).

Functional genomics using mutant plants have been extensively used to investigate the regulatory mechanism of tetrapyrrole synthesis and the role of various enzymes in this pathway. Most of these studies were performed on *Arabidopsis thaliana* (Espineda, Linford, Devine, & Brusslan, 1999; Frick, Su, Apel, & Armstrong, 2003; Hirono & Rédei, 1963; Ishikawa, Okamoto, Iwasaki, & Asahi, 2001; Kumar & Söll, 2000; Molina et al., 1999; Nagata, Tanaka, Satoh, & Tanaka, 2005; Tanaka & Tanaka, 2005) and *Nicotiana tabacum* (Alawady & Grimm, 2005; Höfgen et al., 1994; Kruse, Mock, & Grimm, 1995; Mock, Keetman, Kruse, Rank, & Grimm, 1998; Mock et al., 1999; Papenbrock, Mock, Tanaka, Kruse, & Grimm, 2000; Papenbrock et al., 2001). Other plants species such as *Zea mays* (Hu, Yalpani, Briggs, & Johal, 1998), *Hordeum vulgare* (Jensen et al., 1996; Rzeznicka et al., 2005), *Pisum sativum* (Weller, Terry, Rameau, Reid, & Kendrick, 1996; Weller, Terry, Reid, & Kendrick, 1997), and *Brassica napus* (Tsang et al., 2003) were also investigated. In general, inhibition of δ-ALA formation in mutant plants reduced chlorophyll and heme formation, but did not result in necrotic lesions (Tanaka & Tanaka, 2007). On the other hand, inhibition of later steps in the tetrapyrrole pathway (uroporphyrinogen III decarboxylase to protoporphyrinogen IX oxidase) increased the level of intermediate molecules and resulted in necrotic lesions upon illumination (Tanaka & Tanaka, 2007).

Although the first three enzymes that lead to δ-ALA formation have been extensively targeted (Kim et al., 2005; Kumar & Söll, 2000; Tsang et al., 2003), ALAD rarely has been studied in plants (Chai et al., 2017). In humans, ALAD is necessary for heme biosynthesis and its activity can be inhibited by environmental toxins such as lead (Scinicariello et al., 2007). Lead deactivates ALAD by displacing the zinc from the active site of the enzyme and consequently leads to the accumulation of δ-ALA, which can act as a γ-aminobutyric acid (GABA) receptor agonist in the nervous system resulting in neuropathogenic effects (Brennan & Cantrill, 1979). ALAD enzyme has been isolated and characterized in several organisms, including human, mouse, yeast, bacteria, and many plant species such as pea, soybean, and spinach (Kaczor, Smith, Sangwan, & O'Brian, 1994).

Virus-induced gene silencing (VIGS) technology is a reverse genetic technique that has been widely used to study biological function of plant genes (Romero, Tikunov, & Bovy, 2011). Since its introduction in 1995, VIGS has been applied in about 30 different species, including *A. thaliana*, *N. tabacum*, and *Solanum lycopersicon* (Becker & Lange, 2010). Despite its successful application in many species, VIGS efficiency depends on the distribution and the movement of the virus in plants (Romero et al., 2011). In consequence, it was recommended to use an additional reporter gene such as phytoene desaturase (PDS) to estimate the level of gene silencing (Senthil-Kumar et al., 2007).

VIGS development requires many steps (Becker & Lange, 2010). In the first step, sequences corresponding to the host gene to be silenced are inserted into the viral genome. After creation of the engineered virus, it is inoculated in the host plant. Plant infection with the engineered virus initiates the synthesis of the viral dsRNA. Dicerlike enzymes cleave the dsRNA into short interfering RNAs. These RNAs are recognizable by RNA-induced silencing complex (RISC) and melted into single-stranded RNAs (ssRNAs) which are afterward used as template to target gene degradation.

Because most viruses have a limited host range, numerous virus vectors have been developed for VIGS in plants, including potato virus X (PVX), tobacco mosaic virus (TMV), tomato golden mosaic virus (TGMV), tomato yellow leaf curl virus (TYLCV), and China virus satellite DNA (Igarashi et al., 2009). Although VIGS has been widely used to investigate gene functions in many herbaceous plant species, VIGS has rarely been used in woody plants. Tobacco rattle virus (TRV) was one of the first viruses devolved for VIGS in woody plants such as *Populus euphratica* and *P.* × *canescens* (Shen et al., 2015). In addition, *Citrus leaf blotch virus* (CLBV) has also been used to target the phytoene desaturase (*PDS*) gene in citrus plants (Agüero et al., 2014). Because some of the reliable VIGS vectors can induce similar symptoms to those caused by the silenced gene (Igarashi et al., 2009), it is necessary to develop more VIGS vectors for additional plant species, especially for woody crops. Development of more VIGS vectors will enhance plant genomic studies (Burch-Smith, Anderson, Martin, & Dinesh-Kumar, 2004).

Citrus tristeza virus (CTV) is a closterovirus which is transmitted by the brown citrus aphid (Bar-Joseph, Marcus, & Lee, 1989), and it infects most commercial citrus varieties (Albiach-Marti et al., 2004). T36 and T30 are the most predominant CTV strains in Florida (Folimonova, Robertson, Garnsey, Gowda, & Dawson, 2009; Harper, Cowell, Halbert, Brlansky, & Dawson, 2015). The CTV isolate T36 induces a decline in sweet orange trees grafted onto sour orange rootstock, but it does not induce any appreciable disease syndromes in other common commercial citrus hosts such as grapefruit, and even sweet orange on other rootstocks (Dawson, Robertson, Albiach-Martí, Bar-Joseph, & Garnsey, 2015). Folimonov, Folimonova, Bar-Joseph, and Dawson (2007) developed a CTV strain T36-based vector for transient expression of foreign genes in citrus trees using a green fluorescent protein (GFP) as a reporter. Hajeri, Killiny, El-Mohtar, Dawson, and Gowda (2014) developed CTV strain T36-based vector and demonstrated its capability in silencing transgene *GFP* in *Nicotiana benthamiana* line 16c and citrus endogenous gene *PDS* in citrus. The use of this mild T36 CTV vector offers many advantages: (a) It produces no symptoms in infected plants; (b) has been able to express foreign genes in citrus for many years (Dawson & Folimonova, 2013); (c) can be used for field application to protect against diseases or to treat infected plants (Hajeri et al., 2014); and (d) more than one foreign gene can be inserted into the CTV genome at the same time (Dawson & Folimonova, 2013; El-Mohtar & Dawson, 2014).

In a recent study, we showed that wild-type CTV-T36 (CTV-wt) does not produce dramatic effects on the phloem sap composition and released volatiles of *Citrus macrophylla* (Killiny, Hijaz, Harper, & Dawson, 2017). In the current study, we investigated the effects of silencing of the *ALAD* gene, using the CTV-based silencing vector, on citrus leaf pigments, phytohormones, and volatile and non-volatile metabolites. We hypothesize that silencing of *ALAD* will not only affect chlorophyll biosynthesis, but can also affect the synthesis of other pigments and metabolites.

2 | EXPERIMENTAL PROCEDURES

2.1 | Plant material

Alemow (*C. macrophylla*) plants were grown under controlled greenhouse conditions (temperature of 22–24°C, 16-:8-hr day–light cycle and 60% humidity) at the Citrus Research and Education Centre, Lake Alfred, FL. Plants were about 1 year old (approximately two feet tall and the stem of pencil thickness).

2.2 | *Citrus tristeza virus*-based vectors

The infectious cDNA clones of CTV isolate T36; GenBank accession no. AY170468) in the binary vector pCAMBIA-1380 (Figure 1a) was used as a base plasmid for engineering all of the constructs used in this study (Hajeri et al., 2014). To clone the truncated *ALAD* gene (tALAD) and generate the CTV-tALAD vector, primers were designed based on *Citrus sinensis ALAD* gene cDNA sequence (Phytozome Citrus Gene orange1.1g013861m.g). The antisense of the truncated fragment (400 nucleotides) of the *ALAD* gene was amplified using total RNA from *C. macrophylla* as a template by SuperScript® III One-Step RT-PCR System with Platinum® Taq DNA Polymerase (Life Technologies Corp.) and primers ALAD-PacI.

(5′-GAGC*TTAATTAA*TAAAAAGTGGATATACAAAATTTGCCGGTG AAAG-3′) and ALAD-StuI (5′-GACG*AGGCCT*GCTTCGTCGGTATTT CACGCGCCTTGTG-3′). The PCR product was digested with PacI and StuI restriction enzymes and cloned into similarly digested CTV-tGFP by replacing tGFP with tALAD fragment.

Procedures for agroinfiltration of CTV constructs into *N. benthamiana* were followed as described previously (Hajeri et al., 2014). After 4–6 weeks postinfiltration, systemic leaves from *N. benthamiana* that tested positive for CTV by ELISA were harvested and used in isolating CTV virions and for inoculating *C. macrophylla* by the bark-flap method as described previously (Hajeri et al., 2014). Five plants (about 1 year old) were inoculated with CTV-tALAD virion, five with CTV-wt, and five plants were left as controls. Inoculated plants were kept in the greenhouse

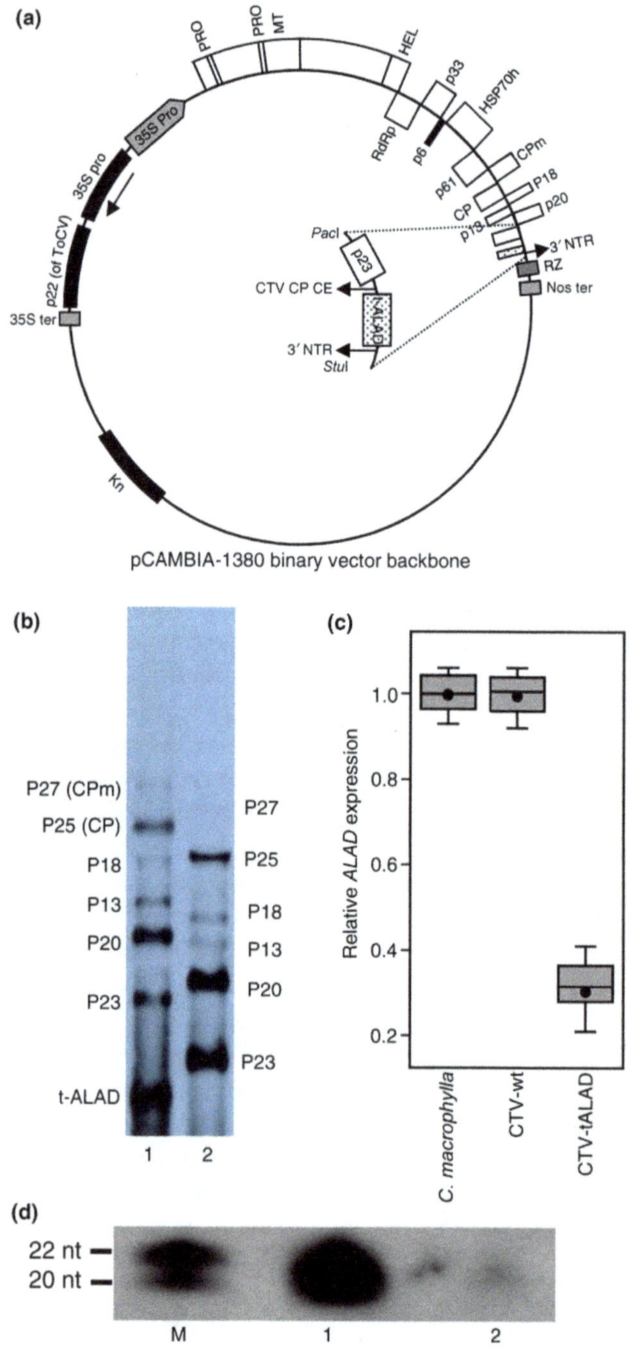

FIGURE 1 Silencing of δ-aminolevulinic acid dehydratase (ALAD) using *Citrus tristeza* virus in a binary vector. (a) Schematic representation of full-length infectious cDNA clones of *C. tristeza* virus (CTV) with its open reading frames (ORF) placed between enhanced 35S promoter of *Cauliflower mosaic* virus at the 5′end, ribozyme (RZ) of Subterranean clover mottle virus satellite RNA and nopaline synthase terminator (Nos ter) at the 3′end in the binary vector pCAMBIA-1380. (b) Northern blot shows the subgenomic RNAs (sgRNAs) of CTV and an extra sgRNA for tALAD, accumulated in CTV-tALAD plants (1) compared to CTV-wt plants (2). The blot was hybridized with digoxigenin labeled minus-sense ribo-probe specific to the 3′nontranslated region of CTV. (c) Gene expression of *ALAD in CTV-tALAD, CTV-wt,* and *Citrus macrophylla.* Elongation factor 1-alpha (*EF1*), F-box/kelch-repeat protein (*F-box*), glyceraldehyde-3-phosphate dehydrogenase GAPC1, cytosolic (*GAPC1*, also known as *GAPDH*), and SAND family protein (*SAND*) were used as endogenous gene (reference gene) to normalize the data of gene expression. Horizontal thick lines indicate the medians, black dots indicate the means, boxes show the interquartile ranges including 25%–75% of the values, and whiskers reflect the highest and the lowest value of data. Experiments in triplicate are presented. (d) Accumulation of ALAD-specific small interfering RNAs (siRNAs) in CTV-tALAD plants (1) compared to CTV-wt (2)

until development of clear genotype (about 6 months). Large RNA northern blot hybridization, small RNA isolation, small RNA northern blot hybridization, reverse transcription quantitative PCR (RT-qPCR) for plant tissue were performed as described previously (Hajeri et al., 2014).

2.3 | High-performance liquid chromatography analysis of citrus leaf pigments

At least three leaves were collected from each plant (five plants were sampled from each treatment, and each sample was analyzed twice). Leaves collected from each plant were placed together in a mortar, mixed with liquid nitrogen, ground into fine powder using a pestle, and citrus leaf pigments were extracted as described previously (Killiny & Nehela, 2017a). In short, about 0.1 g ground material was mixed with 400 μl of acetone 80% and 240 μl of ethyl acetate and vortexed for 30 s. Samples were left in the dark on ice for 10 min and vortexed twice followed by addition of 280 μl of water, and the mixture was centrifuged at 8,500 *g* for 5 min at 4°C. The organic upper layer was removed and dried under a nitrogen stream. The extracted pigments were resuspended in 200 μl ethyl acetate and analyzed immediately by high-performance liquid chromatography (HPLC) as described previously by Wei et al. (2014a,b). The HPLC system consisted of an Agilent 1200 system with photodiode array detector (Agilent Technologies, Santa Clara, CA, USA). Chromatographic separation was performed using a C30 YMC carotenoid column, 250 × 4.6 mm I.D., S-5 μm (YMC America, Allentown, PA, USA). The mobile phase composition and the gradient profile were as described by Mouly, Gaydou, Lapierre, and Corsetti (1999). The column temperature was set at 25°C, mobile phase flow rate was 1 ml/min, and injection volume was 20 μl. Absorbance of the various citrus leaf pigments and peak responses were monitored at several spectral wavelengths (230, 278, 350, 430 and 486 nm). The HPLC output data were analyzed using ChemStation software, B.03.02 (Agilent Technologies, CA, USA). Pigments were identified by comparing experimental retention times and UV–visible spectra with that of published literature (Mouly et al., 1999), as well as with authentic standards. Chlorophyll *a*, chlorophyll *b*, α-carotene, β-carotene, lutein and zeaxanthin were obtained from Sigma-Aldrich (USA). A set of concentrations (1, 5, 10, 25, 50, and 100 ppm) were prepared from each standard. A 20 μl aliquot from each standard was injected into the HPLC under the same condition described above to establish calibration curves and calculate the concentrations of pigments in citrus leaves. The concentrations of the rest of pigments were estimated using the lutein calibration curve.

2.4 | Analysis of leaf pigment by thin-layer chromatography

A 10 μl aliquot of the extracted pigments was spotted on the silica thin-layer chromatography (TLC) plate, and the pigments were separated using a mixture of n-hexane-acetone (70:30 v/v) as described by Zeb and Murkovic (2010).

2.5 | Gene expression analysis

Total RNA was extracted using TRIzol® reagent (Ambion®, Life Technologies, NY, USA). NanoDrop 2000 spectrophotometer (Thermo Scientific, USA) was used to estimate the quantity and quality of isolated RNA. For synthesizing cDNA, SuperScript first-strand synthesis system (Invitrogen) with random hexamer primers was used as described in the manufacturer's instructions. SYBR® Green PCR master mix (Applied Biosystems) was used to perform quantitative PCR (qPCR) on an ABI 7500 Real-Time PCR System (Applied Biosystems). For each treatment, two technical replicates per biological replicate and five biological replicates per treatment were analyzed in triplicate. Primers for 48 genes were used to measure the gene expression. The $2^{-\Delta\Delta CT}$ method was used to determine the relative expression of genes (Livak & Schmittgen, 2001). Elongation factor 1-alpha (*EF1*), F-box/kelch-repeat protein (*F-box*), glyceraldehyde-3-phosphate dehydrogenase GAPC1, cytosolic (*GAPC1*, also known as *GAPDH*), and SAND family protein (*SAND*) were used as endogenous genes (reference genes) to normalize the data of gene expression (Mafra et al., 2012; Wei et al., 2014a,b).

2.6 | Leaves incubation in levulinic acid

Three young leaves from each plant (five plants from each treatment) were harvested, and each leaf was incubated separately by immersing its petiole in 1.5 ml of 100 mM levulinic in 10% dimethylsulfoxide solution (Beale & Castelfranco, 1974) for 0 and 6 hr at room temperature (20°C).

2.7 | Extraction of δ-ALA from citrus leaves

The δ-ALA was extracted using 0.1 N HCl as described by Hijaz and Killiny (2016). In short, leaves were ground in liquid nitrogen and 0.2 g of the ground tissues was transferred to 2-ml centrifuge tube. One millilitre of 0.1 N HCl was added, and the sample was vortexed for 2 min. After 10 min of incubation, samples were further vortexed. The procedure was repeated three times, and the sample was centrifuged at 22,000 *g* for 6 min at room temperature.

2.8 | Colorimetric determination of δ-ALA

The concentration of δ-ALA in the 0.1 N HCl extract was determined colorimetrically as described by Mauzerall and Granick (1956). In short, 0.5 ml of the supernatant was transferred to a new centrifuge tube and was mixed with 0.5 ml of phosphate buffer (equal volume of 0.5 M NaH_2PO_4 and 0.5 Na_2HPO_4; pH: 6.8) and 0.1 ml ethyl acetoacetate. The mixture was placed in a boiling water bath for 10 min. Then, the sample was cooled to room temperature and centrifuged at 14,000 rpm for 6 min at room temperature. After centrifugation, 0.7 ml of the supernatant was mixed with an equal volume of modified Ehrlich's reagent containing 2N perchloric acid. The density of the pink color produced from the reaction of δ-ALA-pyrrole with Ehrlich's reagent was measured at 553 nm using a

Shimadzu UV-Vis spectrophotometer (UV-1700 PharmaSpec (Torrance, CA, USA). Each biological sample was analyzed twice (two technical replicates). A set of δ-ALA standards (10.0, 5.0, 2.5, 1.2, 0.6, and 0.0 ppm in water) was reacted with ethyl acetoacetate and Ehrlich's reagent as mentioned above, and results were used to construct the standard curve. The spectra of the standard and the samples were also scanned between 400 and 700 nm.

2.9 | δ-ALA derivatization for GC-MS

The δ-ALA-pyrrole derivative was reacted with N-methyl-(N-trimethylsilyl) trifluoracetamide (MSTFA) as described by Hijaz and Killiny (2016). In brief, a 0.4 ml of the hydrochloric acid extract was reacted with ethyl acetoacetate as mentioned above and δ-ALA-pyrrole derivative was extracted with 3 × 600 μl of ethyl acetate (Tomokuni & Ogata, 1972). Then, the ethyl acetate was evaporated under a nitrogen stream. A 40 μl of N-methyl-(N-trimethylsilyl) trifluoroacetamide (MSTFA) reagent was added to the dried sample and was incubated at 85°C for 35 min. A 400 μl of 20 ppm δ-ALA standard was derivatized in the same way and 1 μl of the derivatized standard or sample was injected into the GC-MS running in the full scan mode as described by Hijaz and Killiny (2016).

2.10 | Colorimetric determination of starch and sucrose

Starch and sucrose were extracted according to Cimò et al. (2013) with slight modifications. In short, about 100 mg of the same ground leaf tissues was mixed with 500 μl distilled water and vortexed for 30 s. The water extract was boiled for 10 min, vortexed for 10 s, and centrifuged at 650 *g* for 5 min. A 300 μl aliquot of the supernatant was mixed with 900 μl of 100% ethanol, vortexed for 10 s, and centrifuged at 18,000 *g* for 10 min. The supernatant was discarded, and the pellet was suspended in 1 ml distilled water and mixed with 50 μl of iodine solution. Starch was determined by monitoring the absorbance at 595 nm as described by Cimò et al. (2013) using rice starch as a standard. The absorbance was measured using a microplate spectrophotometer (Model 680, Bio-Rad Laboratories, CA, USA). Sucrose determination was accomplished at 620 nm as described by van Handel (1968) using PharmaSpec ultraviolet 1700 spectrophotometer (Shimadzu Corporation, Japan). In short, a 100 μl aliquot of the supernatant was mixed with 100 μl 30% KOH and the mixture was boiled for 10 min. When samples cooled to room temperature, 3 ml of anthrone reagent was added and the samples were incubated at 40°C for 10 min for color development.

2.11 | Analysis of carboxylic compounds and phytohormones in leaves by GC-MS

Phytohormones and carboxylic compounds were extracted, from the same ground samples, as described previously (Nehela, Hijaz, Elzaawely, El-Zahaby, & Killiny, 2016), and then derivatized with methyl chloroformate after spiking of each sample with 5 μl aliquot of 200 ppm heptadecanoic acid, which is not found in citrus leaves, as internal standard. Phytohormones and carboxylic acids were analyzed as previously reported (Killiny & Nehela, 2017b; Nehela et al., 2016). Identification of metabolites was further confirmed by comparing their retention time, linear retention indices (LRIs), and mass spectra with authentic standards. Compound peak areas were normalized to the internal standards (heptadecanoic acid). Quantification of metabolites was based on the peak areas obtained from a series of reference standards derivatized and injected under the same conditions as samples. Calibration curves were constructed from the linear regressions obtained by plotting the concentration versus peak area for each standard.

2.12 | Collection of volatile organic compounds from intact citrus leaves

Released leaf volatiles from intact, young intermediate, and mature leaves (hardened) from branches of *C. macrophylla* were collected on a mixed SPME fiber (50/30 μm divinylbenzene/carboxen/polydimethylsiloxane, 1-cm fiber, Supelco) and analyzed using GC-MS as described by Killiny and Jones (2017). Wiley 9th ed., NIST 2011, and Wiley Flavor and Fragrance mass spectral libraries were used for volatile identification as well as by comparing spectra to standards when available. Volatile standard was prepared in n-hexane and was run using the same GC temperature program as samples. Five plants were sampled from each treatment, and three different types of leaves (young, intermediate, and mature) were taken from each plant.

2.13 | Statistical analysis

For HPLC and GC-MS, five biological and two technical replicates per treatment were analyzed. Comparison of pigment concentrations among the three different treatments was performed using the analysis of variance (ANOVA), followed by post hoc pairwise comparisons using a Tukey–Kramer honestly significant difference test (Tukey's HSD). Statistical significance was established as p-value <0.05. The level of other metabolites in CTV-tALAD plants was compared to that of CTV-wt (T36) plants using two-tailed t test, and statistical significance was established as p-value <0.05. Two-way hierarchical cluster analysis (HCA) was also performed with the means of the gene expressions for each treatment. Distance and linkage were measured using the Bray–Curtis similarity measure method (Michie, 1982). 3D-surface plots were performed among the three pigment groups (chlorophylls, carotenes, and xanthophylls) using the total concentration of each group.

3 | RESULTS

3.1 | CTV-tALAD induced gene silencing and produced yellow islands and necrosis in citrus

Northern blot analysis of RNA showed a of the extra subgenomic RNA in CTV-tALAD plants compared to CTV-wt plants (Figure 1b).

In addition, RT-qPCR showed a twofold to threefold downregulation of δ-ALAD mRNA in CTV-tALAD plants compared to CTV-wt (Figure 1c). The northern blot also showed high accumulation of ALAD-specific small interfering RNAs (siRNAs) in CTV-tALAD plants compared to CTV-wt (Figure 1d).

Citrus macrophylla plants inoculated with CTV-tALAD virions displayed yellow islands and necrosis in leaves, thorns and stems, and the apical meristem of citrus plants (Figure 2c–k). No yellow islands or necrosis was observed in CTV-wt plants (Figure 2a). The symptoms start as a few dots in developing young leaves and the number of dots and their intensity grows until they cover most of the leaf surface (Figure 2f–k). The ALAD phenotype was also different from the PDS-silenced plants (Figure 2b). The PDS plants showed a photobleaching phenotype (Figure 2b), whereas the CTV-tALAD plants showed yellow islands and necrosis in leaves, stems, and the apical meristem (Figure 2c–f).

3.2 | Silencing of *ALAD* reduced chlorophyll content and altered the levels of carotenoids

Fifteen pigments were identified in the leaf extract using HPLC-PDA (Supporting Information Figure S1a), whereas only eight of these

FIGURE 2 Phenotype observed in CTV-tALAD plants and CTV-tPDS plants. (a) CTV-wt. (b) Photobleaching in CTV-tPDS plants. (c–e) Yellow islands and necrosis in apical growing and stem of CTV-tALAD plants. (f–k) Development of yellow islands and necrosis in leaves of CTV-tALAD

pigments were identified using TLC (Supporting Information Figure S1b). The CTV-tALAD significantly reduced the levels of chlorophyll *a*, chlorophyll *b*, chlorophyllide *a*, and pheophytin *a* compared to the controls (Figure 3). The reduction in chlorophyll *a*, chlorophyll *b*, and pheophytin in CTV-tALAD plants was observed in the TLC results (Supporting Information Figure S1b).

While α- and β-cryptoxanthin, *cis*- and *trans*-violaxanthin, and neoxanthin were reduced in CTV-tALAD plants, zeaxanthin was increased (Figure 3). The β-carotene and the *cis*-β-carotene were also significantly reduced in CTV-tALAD plants (Figure 3). Lutein, isolutein, and α-carotene were not significantly affected in CTV-tALAD plants (Figure 3). The TLC also showed that violaxanthin and carotenes were reduced in CTV-tALAD plants (Supporting Information Figure S1b).

To investigate the effect of silencing of the *ALAD* gene on the other genes implicated in the biosynthesis pathway of leaf pigments, the transcription levels of 46 genes involved in biosynthetic pathways of chlorophylls (17 genes) and carotenoids (29 genes) were investigated in *C. macrophylla*, CTV-wt, and CTV-tALAD plants (Figure 4). Gene expression data were normalized using four reference genes (*EF1*, *F-box*, *GAPDH*, and *SAND*). These genes showed high stability and were used for transcript normalization in different citrus organs under biotic stress (Mafra et al., 2012; Wei et al., 2014a,b). The normalizing expression levels using the four reference genes were very similar to each other.

The presented heatmap and the hierarchical clustering dendrogram (HCD) between chlorophyll biosynthetic genes in Figure 4a show that all chlorophyll biosynthetic genes were separated into two major clusters. The first cluster (C1; about 0.55 distance) includes only two genes, glutamyl-tRNA reductase 1 (*CitGluTR*) and glutamate-1-semialdehyde 2,1-aminomutase (*CitGSAT*), which were not affected in CTV-tALAD plants compared to the two controls (*C. macrophylla* and *C. macrophylla* with CTV-wt). The second cluster (C2; about 0.60 distance) contains 15 genes, which were downregulated in CTV-tALAD plants compared to the controls. These include δ-aminolevulinic acid dehydratase 1 (*CitALAD*), chlorophyllide *a* oxygenase (*CitCAO*), protochlorophyllide reductase (*CitPOR*), chlorophyllase (*CitChlases*, also known as *CLHs*), chlorophyll(ide) *b* reductase (*CitCBRs*), including chlorophyll(ide) *b* reductase -NON-YELLOW COLORING 1 (*CitNYC1*) and chlorophyll(ide) *b* reductase -NYC1-Like (*CitNOL*), protoporphyrinogen oxidase (*CitPPO*), chlorophyll synthase (*CitChlG*), and divinyl chlorophyllide a 8-vinyl-reductase (*CitDVR*) (Figure 4a). Furthermore, HCD among treatments indicated that while CTV-wt plants were very similar to *C. macrophylla* plants (less than 0.1 distance), CTV-tALAD plants clustered separately (about 1.65 distance) (Figure 4a).

In addition, silencing of the *ALAD* gene altered the expression of several genes implicated in carotenoids biosynthesis pathways. Of the twenty-nine selected genes, ten genes were downregulated, whereas only four genes were upregulated compared to the controls (Figure 4b). Likewise, the HCA revealed that all carotenoid biosynthetic genes were clustered in three main clusters. The first cluster (C1; about 2.45 distance) included 15 unaffected genes, including phytoene synthase (*CitPYSs*), phytoene desaturase (*CitPDSs*), lycopene ε-cyclase (*CitLCYes*), lycopene β-cyclase (*CitLCYb*), and ζ-carotene desaturase (*CitZDSs*) (Figure 4b). The second cluster (C2; about 2.25 distance) included ten downregulated genes implicated in the ABA-biosynthetic pathway, including zeaxanthin epoxidase (*CitZEPs*), short-chain alcohol dehydrogenase (*CitABA2*), violaxanthin de-epoxidase (*CitVDE*), putative 9-*cis*-epoxycarotenoid dioxygenase 3 (*CitNCED*), capsanthin/capsorubin synthase (*CitCCS*), abscisic aldehyde oxidase (*CitAAO3*), and neoxanthin synthase (*CitNSY*) (Figure 4b). The third cluster included four isoforms of carotenoid hydroxylase β-ring (*CitCHYbs*; converts β-carotene to β-cryptoxanthin and then zeaxanthin), which were upregulated in CTV-tALAD plants compared to the controls (Figure 4b). Furthermore, the HCA among treatments in Figure 4b shows that CTV-wt plants were closer to *C. macrophylla* (less than 0.3 distance) than to CTV-tALAD (about 2.4 distance).

As no differences were found in the concentration of studied pigments and their biosynthetic genes between the *C. macrophylla* and *C. macrophylla* infected with CTV-wt, we chose to use CTV-wt as the control in all subsequent analyses. The 3D surface plots were obtained using the whole data matrix to understand the relationships between chlorophyll, xanthophyll, and carotene groups in CTV-wt and CTV-tALAD plant leaves (Figure 5a–f). Overall, the silencing of the *ALAD* gene affected the total net profiles (TNPs) of other pigments, causing a complex relationship among different pigments groups. The effect of xanthophyll and chlorophyll (as two input parameters) on carotene content (as an associated performance metric) is presented in Figure 5a,b. The carotenes TNP of CTV-tALAD was totally different compared to CTV-wt plants. While the carotenes TNP of CTV-wt plants was flatter and had a clear peak in high chlorophylls and high xanthophylls (Figure 5a), the carotenes TNP of CTV-tALAD plants had more plateaus and curvatures without any clear peaks (Figure 5a). Furthermore, the effect of carotene and chlorophyll content on xanthophyll content is presented in Figure 5c,d. The xanthophylls TNP of CTV-tALAD plants was very different compared to CTV-wt, which was flatter with only one small peak under the high-carotene and high-chlorophyll conditions (Figure 5c). The xanthophylls TNP of CTV-tALAD plants had a clear curvature at relatively low levels of carotenes and moderate levels of chlorophylls with a very sharp peak at relatively low-carotene and moderate-chlorophyll conditions (Figure 5d). In addition, the effect of xanthophylls and carotenes on chlorophyll content presented in Figure 5e,f. As observed, the TNP of chlorophyll content in CTV-tALAD plants appeared very different from that of CTV-wt plants, which was flatter with a small plateau (Figure 5e), while the chlorophylls TNP of CTV-tALAD plants had two clear curvatures: the first one at relatively low levels of carotenes and moderate levels of xanthophylls, and the second one at low levels of carotenes and high levels of xanthophylls. In addition, it had a clear peak at high-carotene and high-xanthophyll conditions (Figure 5f).

3.3 | CTV-tALAD plants accumulated δ-ALA in developing leaves

The colorimetric assay showed that δ-ALA accumulated in the developing leaves from CTV-tALAD (6.6 ± 0.5 mg/g FW) plants, whereas

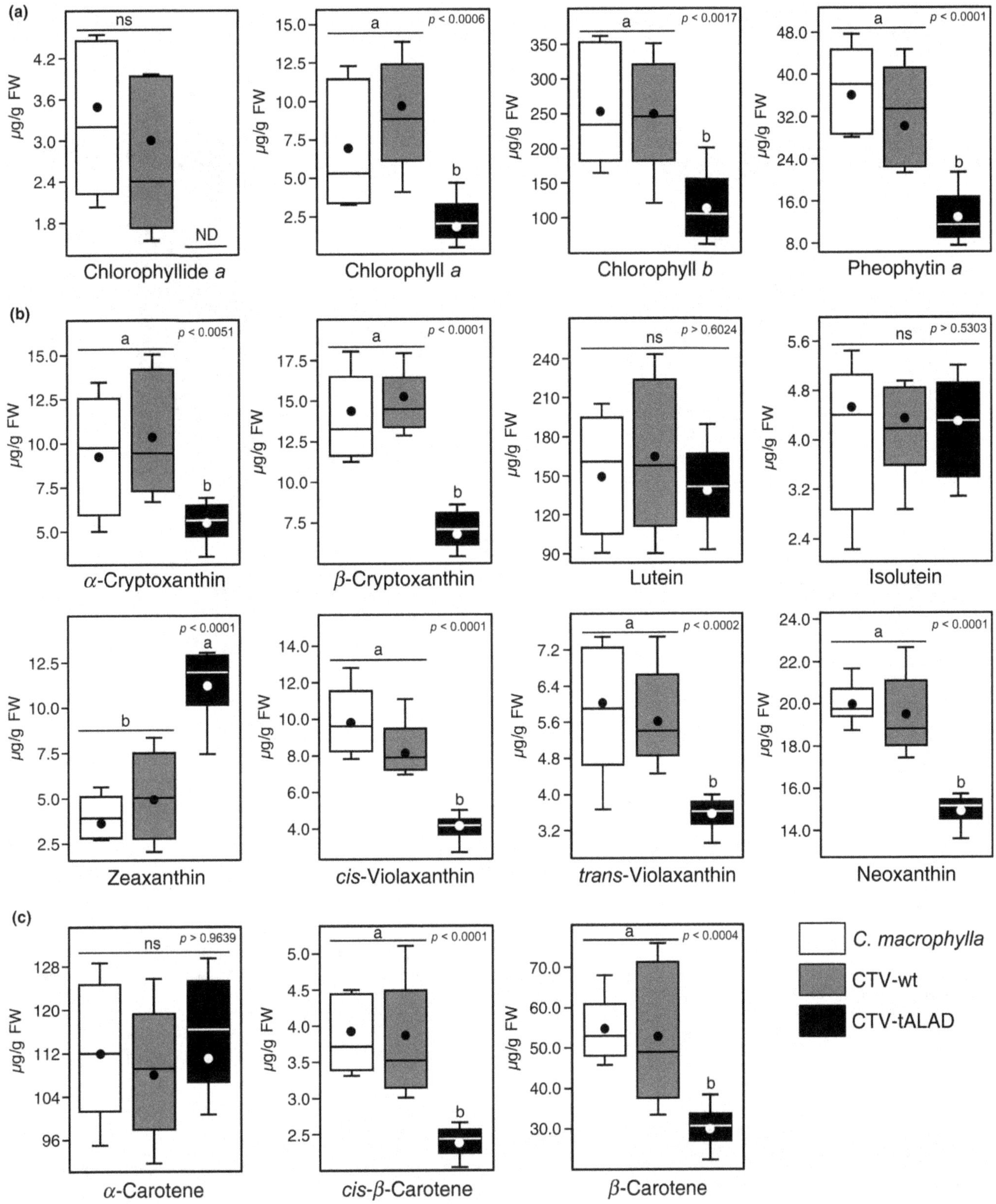

FIGURE 3 Concentrations (μg/g FW) of leaf pigments from control, CTV-wt, and CTV-tALAD *Citrus macrophylla* plants using HPLC (n = 10). Horizontal thick lines indicate the medians, black dots indicate the means, boxes show the interquartile ranges including 25%–75% of the values, whiskers reflect the highest and the lowest value of data. Different letters indicate statistically significant differences among treatments ($p < 0.05$), while "ns" indicates no significant differences among treatments using a Tukey–Kramer honestly significant difference test (Tukey's HSD)

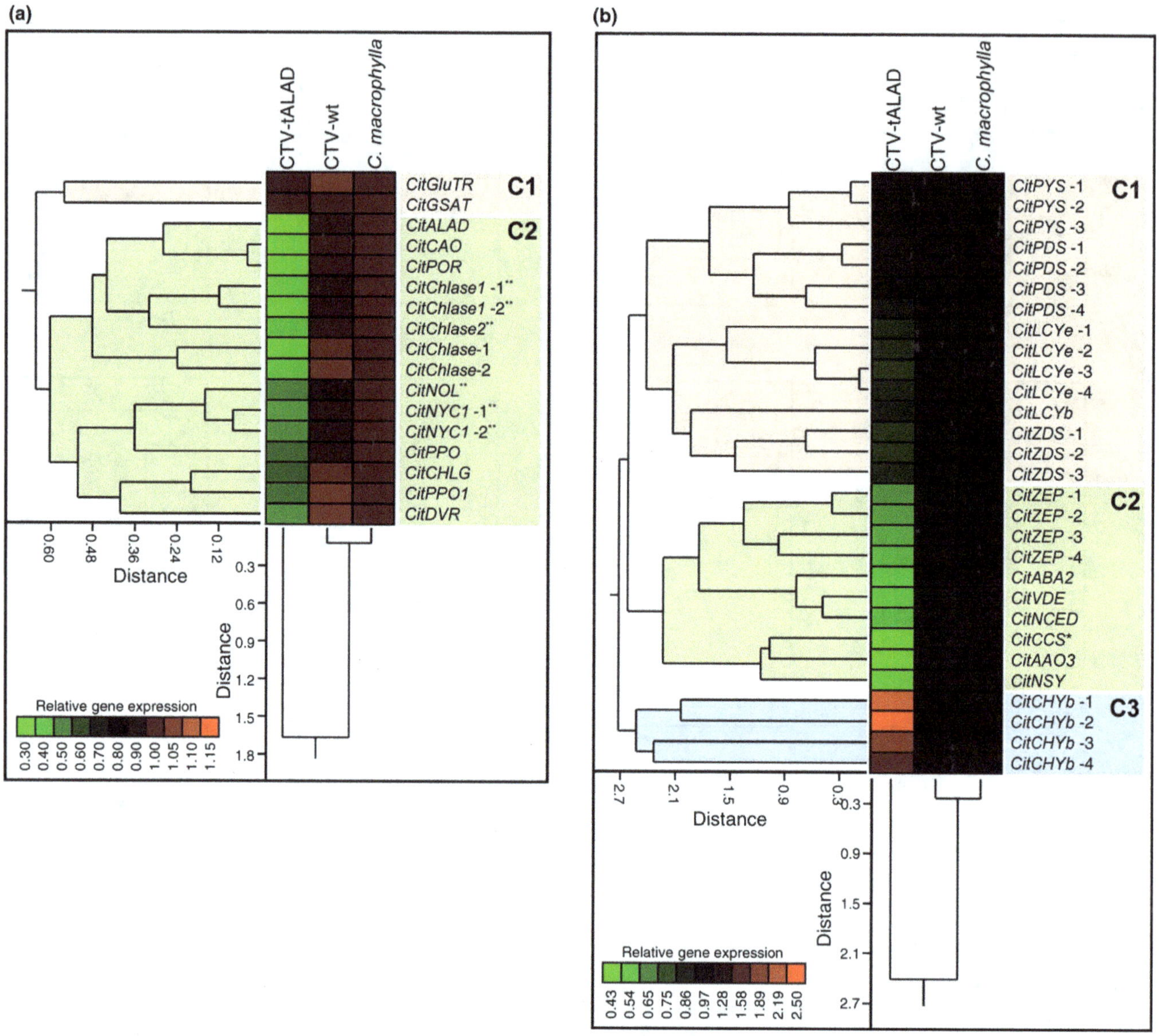

FIGURE 4 Differential biosynthetic gene expression patterns of chlorophylls (a) and carotenoids (b) and detected in leaves of CTV-tALAD, CTV-wt, and *Citrus macrophylla*. Rows represent the genes while columns represent the treatments (n = 30). Lower expressions levels are colored green, and higher expressions are colored red. Treatments and genes were organized using two-way HCA based on similarities in autoscaled values and correlations, respectively. The full lists of expressed genes, accession numbers, and primers are available in our previous study (Killiny & Nehela, 2017a). See text for full names

no δ-ALA was detected in the CTV-wt (Figure 6). No δ-ALA accumulation was detected in mature leaves from CTV-tALAD or CTV-wt plants. In addition, the amount of δ-ALA detected in CTV-tALAD plants after incubation with levulinic acid was higher (138 ± 51 mg/g FW) than that (11 ± 5 mg/g FW) detected in the CTV-wt after 6-hr incubation with levulinic acid (Figure 6). The spectra of the pink color developed in the leaf extracts from CTV-tALAD plants (without incubation with levulinic acid) were similar to those obtained using δ-ALA standard and leaf extracts from leaves incubated with levulinic acid (Supporting Information Figure S2a). The GC-MS results also showed that CTV-tALAD plants accumulate more δ-ALA when incubated with levulinic acid than the CTV-wt (Supporting Information Figure S2b). The mass spectra of the δ-ALA detected in CTV-tALAD plants and in the controls after incubation in levulinic acid were similar to those obtained with authentic standard (Supporting Information Figure S2c).

3.4 | Silencing of *ALAD* decreased the level of photosynthates and affected citrus leaf metabolites and phytohormones

The starch level was significantly reduced (about fivefold) in CTV-tALAD plants compared to the controls (Figure 7a). The sucrose level was also reduced (about threefold) by CTV-tALAD infection (Figure 7b). No effects were observed on the level of sucrose or starch in CTV-wt.

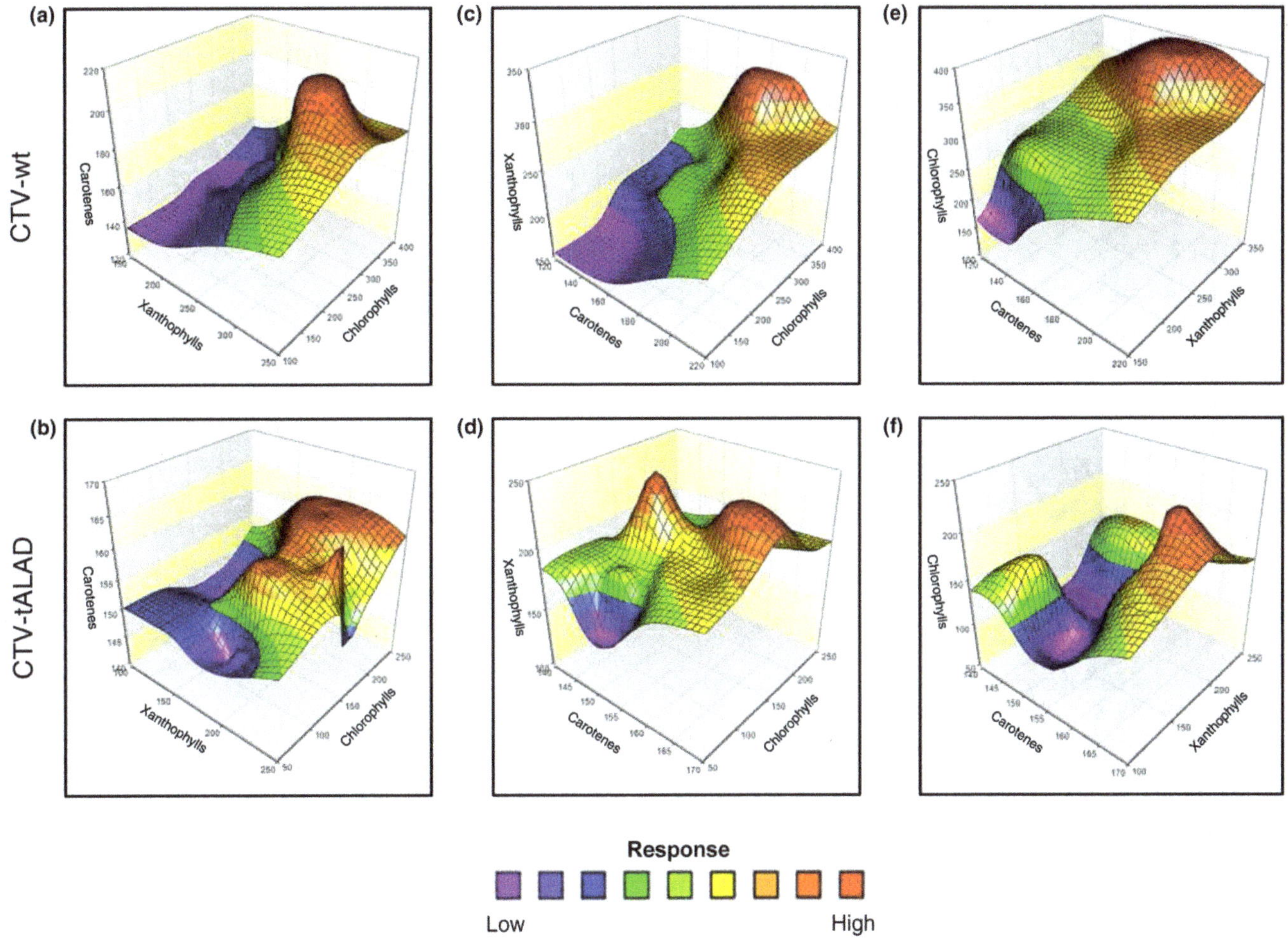

FIGURE 5 Three-dimensional surface plots of leaf pigments groups detected using HPLC in leaves from CTV-wt and CTV-tALAD plants. (a,b) The reciprocal interactions of chlorophylls and xanthophylls on carotene content. (c,d) The reciprocal interactions of carotenes and chlorophylls on xanthophyll content in different treatments. (e,f) The reciprocal interaction of carotenes and xanthophylls on chlorophyll content

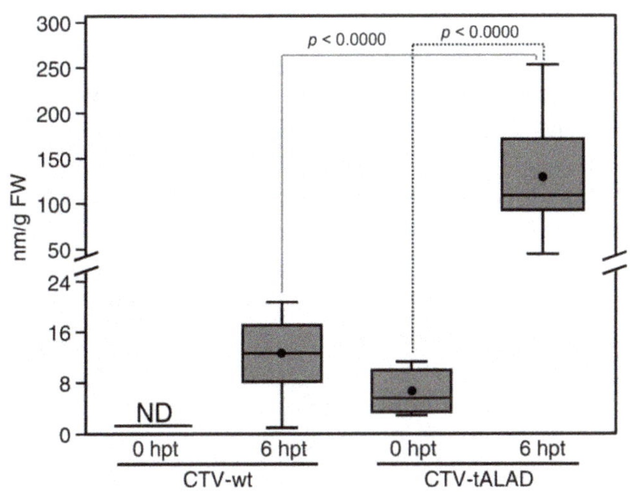

FIGURE 6 Colorimetric determination of δ-ALA. Level of δ-ALA in developing young leaves from CTV-wt and CTV-tALAD plants (n = 5) after 0- and 6-hr incubation with 100 mM levulinic acid. p-value <0.05 indicates statistically significant differences among treatments using two-tailed t test

A few amino acids including L-alanine, *gamma*-aminobutyric acid (GABA), L-threonine, L-glutamic, and L-phenylalanine significantly decreased in leaves of CTV-tALAD plants, whereas L-asparagine increased (Table 1). None of the detected fatty acids was affected in the leaves of CTV-tALAD plants, except linoleic acid, which was significantly increased (Table 1). Fumaric and succinic acid significantly increased in leaves of CTV-tALAD plants (Table 1). In a similar manner, GABA, L-threonine, and L-phenylalanine decreased significantly in the apical meristems of CTV-tALAD plants, whereas L-asparagine increased significantly (Table 1). Linoleic and linolenic acid significantly increased in meristems of CTV-tALAD plants, whereas oleic acid decreased significantly. Fumaric, succinic, and malic also increased significantly in the apical meristems of CTV-tALAD plants (Table 1).

Levels of phytohormones and their precursors were affected in leaves of CT-tALAD plants. BA, *t*CA, IBA, and ABA decreased in leaves from CTV-tALAD plants, whereas SA, JA, *t*JA, IAA, and IPA were not affected (Figure 8a–d). The levels of BA, *t*CA, IBA, and ABA were also reduced in the apical meristems of CTV-tALAD plants, whereas SA, *t*JA, and IPA increased significantly (Figure 8a–d).

The emission of volatile organic compounds was mostly unaffected in juvenile leaves of CTV-tALAD plants. Likewise,

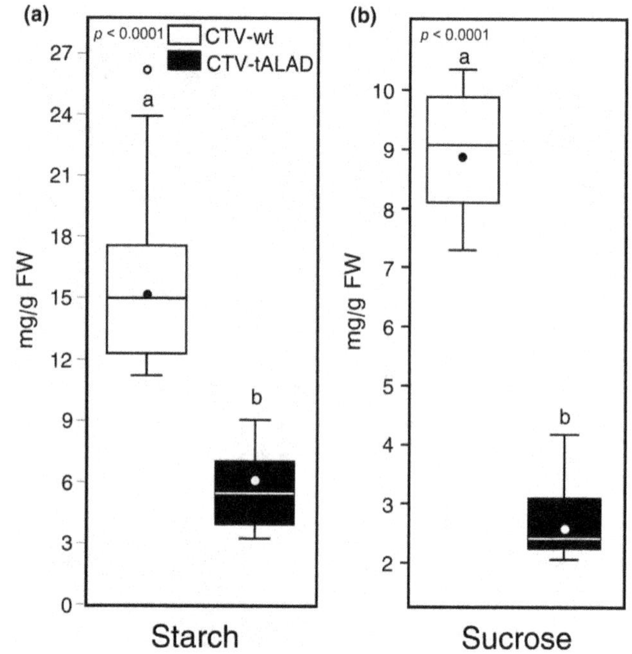

FIGURE 7 Concentrations (mg/g FW) of starch (a) and sucrose (b) in leaves of CTV-wt and CTV-tALA, plants using HPLC (n = 10). Horizontal thick lines indicate the medians, black dots indicate the means, boxes show the interquartile ranges including 25%–75% of the values, and whiskers reflect the highest and the lowest value of data. p-value <0.05 indicates statistically significant differences among treatments using two-tailed t test

intermediate-aged leaves were relatively unaffected, except for (*E*)-β-ocimene, (*E*)-α-bergamotene, (*E*)-β-farnesene, and valencene which increased dramatically (Table 2). Most of the observed differences between the mature leaves of the CTV-wt and the CTV-tALAD plants were small but significant, including increases in β-myrcene, (*E*)-β-ocimene, allo-ocimene, and (*Z*)-β-caryophyllene (Table 2).

4 | DISCUSSION

δ-aminolevulinic acid dehydratase is a key enzyme in plants and is required for the synthesis of all classes of tetrapyrroles, which play important roles in many biological processes, including photosynthesis and respiration. In consequence, we hypothesized that inhibition of this enzyme would not only have an impact on chlorophyll biosynthesis and photosynthesis, but also affect other pigments and could lead to changes in both primary and secondary metabolites. Therefore, we carried out targeted and nontargeted metabolic analysis on CTV-tALAD plants. The level of δ-ALA was determined to confirm the successful silencing of the ALAD gene. We evaluated chlorophyll content as accumulation of δ-ALA was expected to reduce chlorophyll biosynthesis. In addition, we analyzed carotenoid content because carotenoids are functionally linked to chlorophyll in photosynthetic light harvesting. Reduction in chlorophyll biosynthesis was expected to decrease the rate of photosynthesis; thus, we determined the level of the main photosynthates (sucrose and starch). CTV-tALAD plants showed clear symptoms of necrosis, which indicated they were under stress; therefore, four different groups of phytohormones including the main growth regulators and stress-associated groups (auxins, salicylates, jasmonates, and ABA) were quantified. At last, we performed nontargeted metabolomic analysis on volatile and nonvolatile compounds to get more insights on the metabolic status of CTV-tALAD plants. Our results are discussed in the following paragraphs.

4.1 | Silencing of *ALAD* accumulated δ-ALA and reduced chlorophyll and photosynthates levels in citrus leaves

Silencing of *ALAD* led to the accumulation of δ-ALA and reduction in chlorophyll and photosynthates in citrus leaves by the CTV-based silencing vector. These results showed that ALAD is a critical enzyme in chlorophyll synthesis as it is required for the synthesis of all classes of tetrapyrroles (Tanaka & Tanaka, 2007). The δ-ALA that is formed in developing tissues does not accumulate and is directly converted to PBG by ALAD (Kumar & Söll, 2000). Beale and Castelfranco (1974) showed that the rate of δ-ALA accumulation in cucumber cotyledons treated with levulinic acid paralleled the rate of chlorophyll formation, and accumulation of δ-ALA ceased when cucumber cotyledons were returned to dark (Beale & Castelfranco, 1974). In agreement with the previous results, our results also showed that the level of δ-ALA in young leaves from CTV-tALAD was higher than that in mature leaves, which is in agreement with the high rate of chlorophyll synthesis in young leaves.

Inhibition of δ-ALA formation in mutant plants such as tobacco and Arabidopsis reduced chlorophyll and heme biosynthesis, but did not result in necrotic lesions (Tanaka & Tanaka, 2007). In agreement with these observations, our current study showed that silencing of *ALAD* also reduced chlorophyll *a*, chlorophyll *b*, chlorophyllide *a*, and pheophytin *a* synthesis and resulted in yellow islands. However, these changes were accompanied by necrosis in citrus plants. The gene expression results confirmed the reduced levels of pigments and showed that all selected genes in the chlorophyll biosynthesis pathway downstream of the silenced *ALAD* gene were downregulated. Our results indicated that accumulation of δ-ALA could be more toxic to the plants than inhibition of its formation.

The reduction in chlorophyll indicated that CTV-tALAD plants have a lower photosynthesis rate than control plants. In fact, our results showed that sucrose and starch levels (photosynthates) were reduced in CTV-tALAD plants. The reduction in chlorophylls reduces the amount of absorbed light and harvested energy (ATP and NADPH); consequently, carbon fixation is reduced and results in a decrease in photosynthates in leaves.

4.2 | Silencing of *ALAD* increased zeaxanthin level in citrus leaves

Interactions between chlorophylls and carotenoids play a major role in photosynthetic light harvesting. Besides absorbing light and harvesting energy, carotenoids, such as xanthophylls and carotenes,

TABLE 1 Concentrations (ng/g FW) of different amino acids, organic acids, and fatty acids compounds detected in CTV-wt and CTV-tALAD leaves and apical meristems using GC-MS (n = 5)

	Mature leaves		t test	Apical meristems		t test
	CTV-wt	CTV-tALA	p-value	CTV-wt	CTV-tALA	p-value
Amino acids						
Glycine	22.4 ± 4.9	21.4 ± 1.7	0.6819	43.2 ± 2.2	43.3 ± 2.3	0.9442
L-Alanine	612.5 ± 83.0	336.8 ± 85.2	0.0008	174.1 ± 70.6	165.6 ± 77.3	0.8605
L-Valine	91.6 ± 7.4	97.7 ± 1.6	0.1417	85.8 ± 1.4	89.7 ± 3.1	0.0441
L-Leucine	41.1 ± 0.9	42.6 ± 0.9	0.0303	37.3 ± 0.4	39.3 ± 0.8	0.0028
GABA	1288.7 ± 171.9	655.0 ± 172.2	0.0004	973.6 ± 57.6	799.0 ± 141.7	0.0486
L-Isoleucine	54.7 ± 7.4	50.1 ± 2.4	0.2524	43.2 ± 0.6	42.7 ± 1.5	0.5286
L-Threonine	142.6 ± 24.5	103.5 ± 12.2	0.0193	84.7 ± 10.2	67.8 ± 11.1	0.0372
L-Proline	6787.8 ± 1362.0	6237.5 ± 1322.9	0.5351	4531.5 ± 972.5	4974.8 ± 473.5	0.3960
L-Asparagine	3657.2 ± 1717.9	4164.1 ± 580.6	0.5599	947.6 ± 173.7	2260.2 ± 941.0	0.0343
L-Aspartic acid	659.8 ± 60.8	526.8 ± 31.7	0.0049	242.1 ± 21.1	317.5 ± 35.5	0.0054
L-Pyroglutamic acid	410.3 ± 85.2	400.7 ± 18.8	0.8166	14.3 ± 3.2	14.3 ± 10.8	0.9913
L-Serine	1588.4 ± 239.5	1692.8 ± 201.7	0.4780	869.8 ± 151.7	717.3 ± 252.5	0.2874
L-Glutamic acid	2115.4 ± 345.7	1417.7 ± 160.3	0.0073	469.3 ± 263.0	462.8 ± 194.5	0.9659
L-Phenylalanine	164.1 ± 20.2	113.6 ± 14.2	0.0024	151.9 ± 14.8	75.3 ± 27.5	0.0014
L-Lysine	216.1 ± 9.9	214.1 ± 9.2	0.7580	208.5 ± 2.0	209.2 ± 2.8	0.6629
L-Tyrosine	157.2 ± 72.4	122.9 ± 0.5	0.3495	120.8 ± 0.7	121.9 ± 3.0	0.4661
L-Tryptophan	49.5 ± 1.1	48.2 ± 0.2	0.0452	57.2 ± 2.2	48.9 ± 2.4	0.0004
Total amino acids	18059.2 ± 3142.7	16245.3 ± 1958.6	0.3112	9054.9 ± 840.9	10449.8 ± 2033.5	0.2121
Fatty acids						
Palmitic acid (C16:0)	1172.6 ± 62.3	1235.6 ± 112.0	0.3121	1009.6 ± 99.0	1254.3 ± 256.6	0.1015
α-Linolenic acid (C18:3)	10653.0 ± 2161.0	8767.2 ± 831.7	0.1264	5342.0 ± 331.6	6535.1 ± 588.8	0.0068
Linoleic acid (C18:2)	1780.6 ± 238.0	2276.9 ± 200.8	0.0077	684.6 ± 68.4	1647.2 ± 208.8	0.0002
Oleic acid (C18:1)	125.4 ± 55.2	88.3 ± 38.1	0.2550	121.6 ± 18.3	52.6 ± 28.7	0.0029
Stearic acid (C18:0)	279.9 ± 68.6	342.3 ± 50.3	0.1430	227.8 ± 17.9	368.0 ± 187.1	0.1694
Total fatty acids	14011.6 ± 2459.0	12710.3 ± 1079.1	0.3239	7385.6 ± 390.5	9857.1 ± 888.0	0.0017
Organic acids						
Fumaric acid	918.8 ± 30.6	981.2 ± 25.8	0.0086	817.7 ± 17.2	913.8 ± 39.7	0.0033
Succinic acid	506.1 ± 77.0	659.4 ± 61.3	0.0089	405.5 ± 10.0	628.0 ± 75.8	0.0025
Malic acid	2817.3 ± 319.0	3117.9 ± 283.7	0.1544	1574.9 ± 168.7	2202.5 ± 228.9	0.0015
Quinic acid	1548.1 ± 101.2	1497.5 ± 292.0	0.7296	1639.6 ± 322.6	1488.9 ± 745.3	0.6939
Citric acid	1846.7 ± 42.7	1851.8 ± 189.8	0.9554	1759.1 ± 52.3	1724.9 ± 60.3	0.3668
Ferulic acid	390.5 ± 189.7	503.5 ± 84.8	0.2732	747.6 ± 225.7	149.6 ± 30.0	0.0038
Total organic acids	8027.4 ± 376.9	8611.4 ± 678.2	0.1413	6944.4 ± 744.2	7107.7 ± 960.1	0.7718

Note. Means were considered significantly different if p-value was less than 0.05.

protect plants from light damage. In fact, absence of carotenoids in plants causes severe photo-oxidation and leads to plant death (Bartley & Scolnik, 1995). Carotenoids reduce photo-oxidation by quenching of chlorophyll triplets to prevent the formation of singlet oxygen that can oxidize lipids, proteins, and other components in plant cells (Bartley & Scolnik, 1995). The xanthophyll cycle plays an important role in photoprotection in plants (Demmig-Adams, 1990). When the sunlight level exceeds the maximum that can be used by chlorophylls, plants can increase their zeaxanthin levels for photoprotection which in turn reduces chlorophyll fluorescence (Demmig-Adams, 1990).

As the levels of chlorophylls were reduced in leaves from CTV-tALAD plants, the amount of light that can be absorbed is expected to decrease. To prevent light damage, the CTV-tALAD plants increased the level of zeaxanthin. These results were in agreement with the increase in the gene expression of the *CitLCHYB* genes, which are implicated in zeaxanthin synthesis. An increase in zeaxanthin and antheraxanthin and a decrease in violaxanthin were reported in *A. thaliana* after treatment with norflurazon, another specific competitive inhibitor of ALAD (Jung, 2004). The increase in the zeaxanthin was accompanied by an increase in

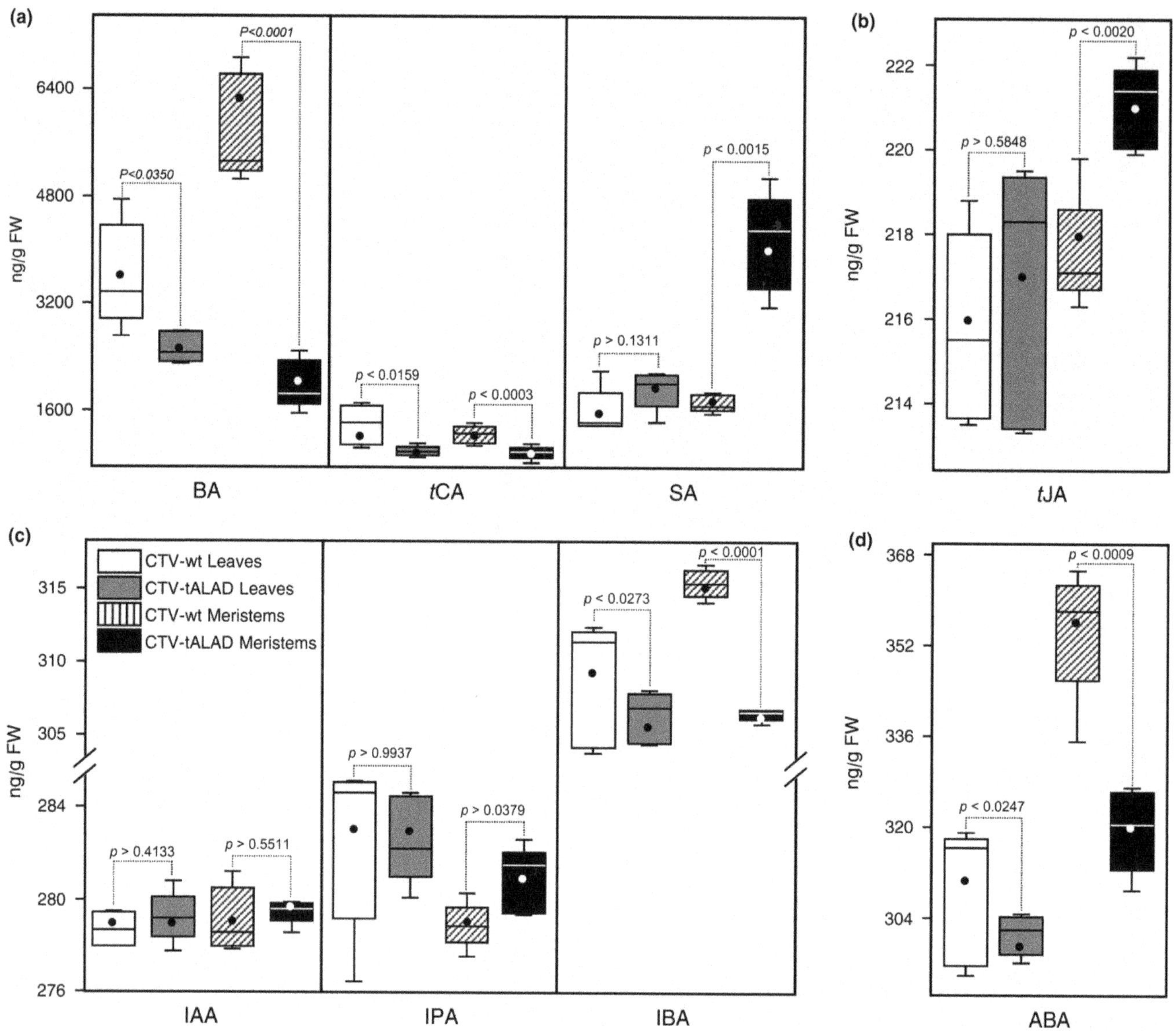

FIGURE 8 Concentrations of different phytohormones and BA, *t*CA in leaves and apical meristems of CTV-wt and CTV-tALA plants a using GC-MS-SIM (n = 10). (a) BA, *t*CA, and SA. (b) *t*JA, (c) auxins. (d) ABA. Horizontal thick lines indicate the medians, black/white dots indicate the means, boxes show the interquartile ranges including 25%–75% of the values, and whiskers reflect the highest and the lowest number of data. p-value <0.05 indicates statistically significant differences among treatments using two-tailed t test. ABA: abscisic acid; BA: benzoic acid; IAA: indole-3-acetic acid; IBA: indole-3-butyric acid; IPA: indole-3-propionic acid; SA: salicylic acid; *t*CA: *trans*-cinnamic acid; *t*JA: *trans*-jasmonic acid

nonphotochemical chlorophyll fluorescence quenching in norflurazon-treated plants (Jung, 2004). In recent times, we also found that the reduction in chlorophyll levels in leaves from *C*Las-infected sweet orange plants was also accompanied by an increase in zeaxanthin (Killiny & Nehela, 2017a). The increase in zeaxanthin indicated that reduction in chlorophyll as a result of silencing of *ALAD* induced light stress in CTV-tALAD plants.

4.3 | The phytohormonal profile indicated that CTV-tALAD plants were under stress

Accumulation of δ-ALA resulted in necrosis in CTV-tALAD, indicating that these plants were under stress. In consequence, we investigated the level of the main phytohormones, which are associated with plant response to biotic and abiotic stresses. Our results showed that several phytohormones were affected by silencing of the *ALAD* gene. SA increased in the apical meristems of CTV-tALAD plants. The *t*JA also increased in the apical meristems of CTV-tALAD plants, and this increase was accompanied by an increase in its precursor, linolenic acid. An increase in *t*JA was observed in citrus plant after *Diaphorina citri* attack and this increase was also accompanied by an increase in linolenic acid and an upregulation of allene oxide synthase (AOS, a JA-biosynthetic enzyme) (Nehela et al., 2018). SA and *t*JA are stress-associated phytohormones in plants (Bari & Jones, 2009; (Nehela et al., 2018)). The SA pathway is associated with plant responses to biotrophic and hemibiotrophic pathogens and plays an

TABLE 2 The relative amounts of volatiles collected from leaves of CTV-tALA plants compared to the CTV-wt (T-36), (n = 5)

		Young leaves		Intermediate leaves		Mature leaves	
Compound	RT	ALAD/T36	*t* test	ALAD/T36	*t* test	ALAD/T36	*t* test
α-Pinene[a]	3.9	0.6	0.4451	1.3	0.5812	0.5	0.0974
β-Myrcene[a]	4.5	0.6	0.4034	1.4	0.4190	0.5	0.0268
β-Phellandrene[a]	5.1	0.4	0.2111	0.3	0.2879	0.3	0.1136
Limonene[a]	5.2	0.6	0.1294	0.7	0.3473	0.8	0.2355
(*E*)-β-Ocimene[a]	5.3	14.4	0.0535	24.2	0.0002	11.0	0.0213
γ-Terpinene[a]	5.4	0.7	0.4730	1.0	0.9415	0.5	0.1025
α-Terpinolene[a]	5.8	0.5	0.3827	1.2	0.5650	0.3	0.0608
Linalool[a]	6.0	1.5	0.5709	0.7	0.6556	0.6	0.2731
Perillene	6.2	0.5	0.3189	2.3	0.3729	1.2	0.8031
allo-Ocimene	6.4	0.1	0.0831	0.1	0.0439	0.0	0.0105
Limonene oxide	6.6	6.5	0.2807	0.4	0.5352	5.3	0.3174
Citronellal[a]	6.7	0.9	0.8148	0.3	0.2484	0.6	0.4206
Methyl salicylate[a]	7.4	1.1	0.8225	0.4	0.4109	0.5	0.3032
Decanal[a]	7.5	0.7	0.3810	0.3	0.1617	0.7	0.4465
Nerol[a]	7.6	0.4	0.2950	0.6	0.5175	0.3	0.1433
Neral	7.7	4.4	0.2855	2.2	0.2510	4.3	0.1757
Geraniol[a]	7.9	1.3	0.6924	0.5	0.5468	0.7	0.4871
Geranial[a]	8.3	0.8	0.6808	0.4	0.4089	0.7	0.4392
δ-Elemene[a]	9.3	5.5	0.0561	8.8	0.0508	7.1	0.1312
Citronellyl acetate	9.4	0.0	0.2659	0.3	0.3007	2.4	0.4045
α-Cubebene	9.5	10.3	0.1283	3.6	0.0543	10.9	0.1176
Geranyl acetate[a]	9.8	1.4	0.6686	0.6	0.6370	1.8	0.3424
β-Cubebene	9.9	0.3	0.1275	0.2	0.2761	0.2	0.0725
β-Selinene	10.0	1.2	0.8040	0.8	0.8353	0.9	0.7858
(*Z*)-β-Caryophyllene	10.4	0.1	0.0617	0.1	0.2308	0.1	0.0177
(*E*)-β-Caryophyllene	10.5	1.0	0.9743	1.0	0.9567	0.7	0.4982
(*E*)-α-Bergamotene	10.6	207.4	0.1002	104.9	0.0000	82.7	0.0616
(*E*)-β-Farnesene[a]	10.7	5.5	0.2186	8.7	0.0006	9.9	0.1174
α-Humulene[a]	11.0	1.2	0.8717	0.6	0.5649	3.7	0.4121
α-Selinene	11.3	1.1	0.8653	0.5	0.4726	0.4	0.1386
Valencene[a]	11.5	3.3	0.0386	11.2	0.0018	4.8	0.1414
β-Bisabolene	11.5	1.1	0.8789	1.2	0.7336	1.2	0.7178
Cadinene	11.7	3.4	0.3375	2.2	0.1478	2.1	0.3136
Elemol	12.0	2.4	0.1686	3.6	0.1237	3.3	0.2017
γ-Gurjunene	12.2	0.2	0.2141	4.5	0.1795	0.4	0.1831
Caryophyllene oxide	12.5	2.1	0.4266	2.6	0.0145	0.9	0.7639
Methyl jasmonate[a]	13.1	2.9	0.2930	1.2	0.7969	0.8	0.6841

Notes. The relative amounts were calculated by dividing the peak area of each compound in the CTV-tALAD plants on the peak area of the same compound in the CTV-wt (T-36).

[a]Compounds were confirmed by comparing their retention time and mass spectra with authentic standards. Means were considered significantly different if p-value was less than 0.05.

important role in systemic acquired resistance (SAR) (Bari & Jones, 2009). On the contrary, the *t*JA pathway is associated with plant responses to necrotrophic pathogens and insect herbivory (Bari & Jones, 2009). In addition, it is believed that both SA and *t*JA are implicated in plant responses to abiotic stresses (Ahmad et al., 2016; Horvàth, Szalai, & Janda, 2007). The increase in SA and *t*JA indicated that CTV-tALAD plants were under stress.

In agreement with our results, Chai et al. (2017) reported an increase in the level of ROS and H_2O_2 in *Gossypium hirsutum* (cotton) lesion mimic mutant (*Ghlmm*), which exhibited an accumulation

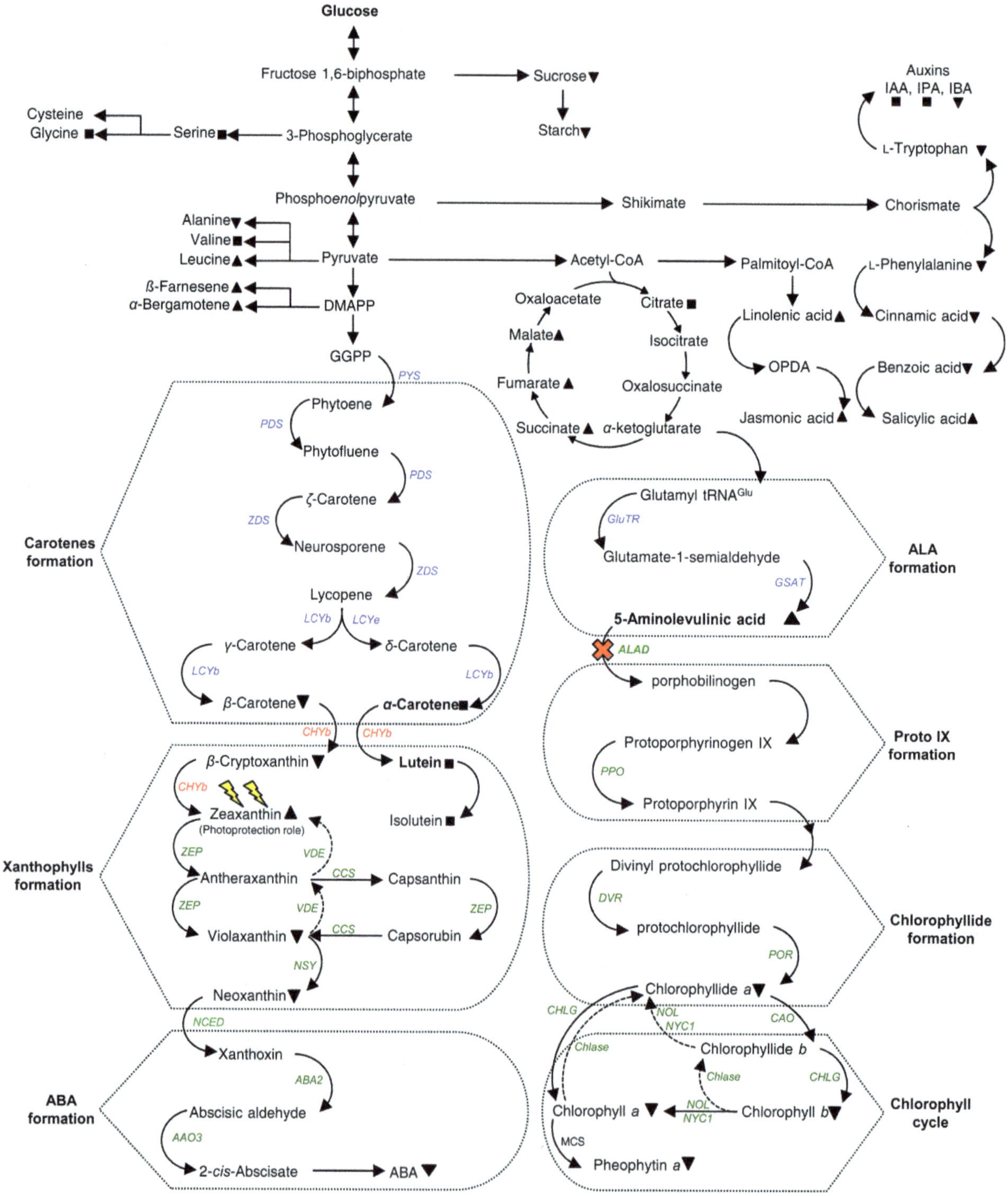

FIGURE 9 Schematic representation of proposed model for the effect of silencing of *ALAD* gene on *Citrus macrophylla* leaf pigments. Silencing of *ALAD* gene (red cross) results in accumulation of δ-ALA and consequently reduces the rate of chlorophylls biosynthesis, photosynthates levels, and induces light stress in CTV-tALAD plants. To prevent light damage, the CTV-tALAD plant increases the rate of zeaxanthin accumulation. The increase in rate of zeaxanthin synthesis decreases the levels of its precursors, upstream pigments, and ABA. Silencing of *ALAD* gene also affected other phytohormones (BA, SA, *t*JA, IBA) and their precursors. In addition, silencing of *ALAD* gene affected the tricarboxylic acid cycle and enhanced the emission of several leaf volatiles. The decrease in upstream pigments reduces ABA synthesis. The up-arrow (▲) indicates increasing, down-arrow (▼) indicates decreasing, and square sign (■) indicates no changes in compound levels. The dotted lines with arrows represent hypothetical mechanisms or uncharacterized elements. Genes in red were upregulated, genes in green were downregulated, whereas those in blue were not affected. ABA: abscisic acid; DMAPP: dimethylallyl diphosphate; GGPP: trans-geranyl-geranyl diphosphate; IAA: indole-3-acetic acid; IBA: indole-3-butaric acid; IPA: indole-3-propionic acid; OPDA: 12-oxo-cis-10,15-phytodienoate

of δ-ALA as a result of low activity of the ALAD enzyme. An increase in methanedicarboxylic aldehyde (MDA) level was also observed in mutant leaves, indicating peroxidation of unsaturated membrane lipids in cells. Silencing of *ALAD* in *G. hirsutum* by the TRV-based VIGS also resulted in the same phenotype observed in mutant plants and showed an accumulation of δ-ALA and H_2O_2 (Chai et al., 2017). The mutant plant also accumulated SA and showed high expression levels of pathogenesis-related genes (PR) and high resistance to *Verticillium dahliae* infection (Chai et al., 2017). In addition, modulation of ALAD activity was found to control programmed cell death (PCD) and PR expression (Chai et al., 2017). Our results, together with the previous results, showed that silencing of *ALAD* induced stress in plants.

The level of ABA decreased in leaves and apical meristems of CTV-tALAD, and this decrease was consistent with the decrease in its precursor (violaxanthin). In agreement with the HPLC results, the gene expression results showed that ten genes implicated in the ABA-biosynthetic pathway were downregulated in CTV-tALAD plants. Previous studies showed that ABA deficiency reduces plant development and elongation (Nitsch et al., 2012). The decrease in ABA indicated that silencing of *ALAD* could affect plant development.

4.4 | Silencing of *ALAD* altered the metabolite profile of citrus leaves

We performed a nontargeted metabolomic analysis on the volatile and nonvolatile profiles of CTV-tALAD plants to gather more information about the effects of silencing of *ALAD* gene in plants. Our results showed that fumaric, succinic, and malic acids were increased in CTV-tALAD plants. The increase in these metabolites could result from the increase in respiration rate, which can be increased to provide more energy during plant responses to stresses (Adi et al., 2012). Citric acid and fumaric acid also increased in CLas-infected citrus plants (Albrecht, Fiehn, & Bowman, 2016). The accumulation of citric acid and fumaric acid was explained as an attempt to enhance nutrient uptake in response CLas-infection (Albrecht et al., 2016).

The level of L-asparagine increased in CTV-tALAD plants compared with CTV-wt plants. L-Asparagine is synthesized in plants by amidation of aspartate using either ammonium or glutamine as ammonium donor (Gaufichon, Reisdorf-Cren, Rothstein, Chardon, & Suzuki, 2010). L-Asparagine is implicated in nitrogen transport, metabolism, recycling and its flow in response to biotic and abiotic stresses (Gaufichon et al., 2010). Light, nitrogen and carbon availability control the three-asparagine synthetase genes in sunflower (*Helianthus annuus*), *HAS1*, *HAS1.1*, and *HAS2* (Herrera-Rodríguez, Maldonado, & Pérez-Vicente, 2004). Light and ample carbon supply induce *HAS2* to supply asparagine to support growth. On the other hand, darkness and a low carbon/nitrogen ratio activate *HAS1* and *HAS1.1* genes to convert excess nitrogen to L-asparagine. Ammonium activates all three genes to enhance its detoxification. The increase in L-asparagine level in CTV-tALAD plants could result from the increase in ammonium, which occurs upon abiotic stresses (Herrera-Rodríguez, Pérez-Vicente, & Maldonado, 2007). It is also possible that the decrease in sucrose and starch decreased the carbon/nitrogen ratio and activated *HAS1* and *HAS1.1* genes to produce more L-asparagine using glutamine as an ammonium donor. The increase in L-asparagine is another indication that accumulation of δ-ALA induces stress in plants.

4.5 | Silencing of *ALAD* enhanced emission of leaf volatiles

Previous studies showed that plant volatiles are triggered by both biotic and abiotic stress (Arimura, Matsui, & Takabayashi, 2009). In agreement with the previous findings, our results showed that the emission of (*E*)-β-ocimene, (*E*)-α-bergamotene and (*E*)-β-farnesene was highly increased in the CTV-tALAD plants compared to the CTV-wt plants. The increased emission of these volatiles indicated that silencing of *ALAD* induced stress in CTV-tALAD. (*E*)-β-farnesene and (*E*)-α-bergamotene emission were increased in *Pinus taeda* upon exposure to high light and temperature (Helmig, Ortega, Guenther, Herrick, & Geron, 2006). In addition, the emission of these two volatiles was also increased in *Z. mays* and *Nicotiana attenuata* upon wounding, herbivory, or application of methyl jasmonate (Gaquerel, Weinhold, & Baldwin, 2009). Terpenes can relieve abiotic stress either by enhancing membrane stability during transient heat stress or by quenching reactive oxygen species (Palmer-Young et al., 2015). Accumulation of ROS upon silencing of *ALAD* (Chai et al., 2017) could be the direct cause of the observed increase in volatile emission (Vickers, Gershenzon, Lerdau, & Loreto, 2009).

5 | CONCLUSION

Silencing of *ALAD* in citrus plants using CTV-tALAD vector resulted in accumulation of the toxic tetrapyrrole intermediate (δ-ALA), decreased chlorophylls and photosynthate levels, altered the levels of carotenoids, phytohormones, volatiles, and many other metabolites and produced yellow islands and necrosis in citrus leaves. These effects are summarized in Figure 9. These changes showed that silencing of *ALAD* gene induces stress in plants.

ACKNOWLEDGMENTS

The authors acknowledge our CREC colleagues for the helpful discussion. We thank Shelley E. Jones for the technical assistance and Laina Lindsey and Floyd Butz for maintaining the trees in the greenhouse. This project is funded by the grant no. 201500955-04 from SCRI-NIFA, USDA.

AUTHOR CONTRIBUTIONS

N.K. and S.G. planned and designed the research. N.K., F.H., Y.N., S.H., and S.G. performed experiments and analyzed data. N.K., F.H., Y.N., S.H., and S.G. wrote the manuscript.

REFERENCES

Adi, M., Jens, P., Brotman, Y., Mikhail, K., Iris, S., Henryk, C., & Rena, G. (2012). Stress Responses to *Tomato Yellow Leaf Curl Virus* (TYLCV) infection of resistant and susceptible tomato plants are different. *Metabolomics*, *S1*(006), 1–13.

Agüero, J., del Carmen Vives, M., Velázquez, K., Pina, J. A., Navarro, L., Moreno, P., & Guerri, J. (2014). Effectiveness of gene silencing induced by viral vectors based on *Citrus leaf blotch virus* is different in *Nicotiana benthamiana* and citrus plants. *Virology*, *460–461*, 154–164. https://doi.org/10.1016/j.virol.2014.04.017

Ahmad, P., Rasool, S., Gul, A., Sheikh, S. A., Akram, N. A., Ashraf, M., ... Gucel, S. (2016). Jasmonates: Multifunctional roles in stress tolerance. *Frontiers in Plant Science*, *7*, 813.

Alawady, A. E., & Grimm, B. (2005). Tobacco Mg protoporphyrin IX methyltransferase is involved in inverse activation of Mg porphyrin and protoheme synthesis. *The Plant Journal*, *41*, 282–290.

Albiach-Marti, M. R., Grosser, J. W., Gowda, S., Mawassi, M., Satyanarayana, T., Garnsey, S. M., & Dawson, W. O. (2004). *Citrus tristeza virus* replicates and forms infectious virions in protoplasts of resistant citrus relatives. *Molecular Breeding*, *14*, 117–128. https://doi.org/10.1023/B:MOLB.0000038000.51218.a7

Albrecht, U., Fiehn, O., & Bowman, K. D. (2016). Metabolic variations in different citrus rootstock cultivars associated with different responses to Huanglongbing. *Plant Physiology and Biochemistry*, *107*, 33–44. https://doi.org/10.1016/j.plaphy.2016.05.030

Arimura, G., Matsui, K., & Takabayashi, J. (2009). Chemical and molecular ecology of herbivore-induced plant volatiles: Proximate factors and their ultimate functions. *Plant and Cell Physiology*, *50*, 911–923. https://doi.org/10.1093/pcp/pcp030

Bari, R., & Jones, J. D. G. (2009). Role of plant hormones in plant defence responses. *Plant Molecular Biology*, *69*, 473–488. https://doi.org/10.1007/s11103-008-9435-0

Bar-Joseph, M., Marcus, R., & Lee, R. F. (1989). The continuous challenge of *Citrus tristeza virus* control. *Annual Review of Phytopathology*, *27*, 291–316. https://doi.org/10.1146/annurev.py.27.090189.001451

Bartley, G. E., & Scolnik, P. A. (1995). Plant carotenoids: Pigments for photoprotection, visual attraction, and human health. *The Plant Cell*, *7*, 1027–1038. https://doi.org/10.1105/tpc.7.7.1027

Beale, S. I., & Castelfranco, P. A. (1974). The biosynthesis of δ-aminolevulinic acid in higher plants. *Plant Physiology*, *53*, 291–296. https://doi.org/10.1104/pp.53.2.291

Becker, A., & Lange, M. (2010). VIGS – Genomics goes functional. *Trends in Plant Science*, *15*, 1–4. https://doi.org/10.1016/j.tplants.2009.09.002

Brennan, M. J. W., & Cantrill, R. C. (1979). δ-Aminolaevulinic acid is a potent agonist for GABA autoreceptors. *Nature*, *280*, 514–515. https://doi.org/10.1038/280514a0

Burch-Smith, T. M., Anderson, J. C., Martin, G. B., & Dinesh-Kumar, S. P. (2004). Applications and advantages of virus-induced gene silencing for gene function studies in plants. *The Plant Journal*, *39*, 734–746. https://doi.org/10.1111/j.1365-313X.2004.02158.x

Chai, Q., Shang, X., Wu, S., Zhu, G., Cheng, C., Cai, C., ... Guo, W. (2017). 5-Aminolevulinic acid dehydratase gene dosage affects programmed cell death and immunity. *Plant Physiology*, *175*, 511–528. https://doi.org/10.1104/pp.17.00816

Cimò, G., Bianco, R. L., Gonzalez, P., Bandaranayake, W., Etxeberria, E., & Syvertsen, J. P. (2013). Carbohydrate and nutritional responses to stem girdling and drought stress with respect to understanding symptoms of huanglongbing in citrus. *HortScience*, *48*, 920–928.

Dawson, W. O., & Folimonova, S. Y. (2013). Virus-based transient expression vectors for woody crops: A new frontier for vector design and use. *Annual Review of Phytopathology*, *51*, 321–337. https://doi.org/10.1146/annurev-phyto-082712-102329

Dawson, W. O., Robertson, C. J., Albiach-Martí, M. R., Bar-Joseph, M., & Garnsey, S. M. (2015). Mapping sequences involved in induction of decline by *Citrus tristeza virus* T36 on the sour orange rootstock. *Journal of Citrus Pathology*, *2*, 1–6.

Demmig-Adams, B. (1990). Carotenoids and photoprotection in plants: A role for the xanthophyll zeaxanthin. *Biochimica et Biophysica Acta (BBA)-Bioenergetics*, *1020*, 1–24. https://doi.org/10.1016/0005-2728(90)90088-L

El-Mohtar, C., & Dawson, W. O. (2014). Exploring the limits of vector construction based on *Citrus tristeza virus*. *Virology*, *448*, 274–283. https://doi.org/10.1016/j.virol.2013.10.017

Espineda, C. E., Linford, A. S., Devine, D., & Brusslan, J. A. (1999). The *AtCAO* gene, encoding chlorophyll a oxygenase, is required for chlorophyll b synthesis in *Arabidopsis thaliana*. *Proceedings of the National Academy of Sciences of the United States of America*, *96*, 10507–10511. https://doi.org/10.1073/pnas.96.18.10507

Folimonov, A. S., Folimonova, S. Y., Bar-Joseph, M., & Dawson, W. O. (2007). A stable RNA virus-based vector for citrus trees. *Virology*, *368*, 205–216. https://doi.org/10.1016/j.virol.2007.06.038

Folimonova, S. Y., Robertson, C. J., Garnsey, S. M., Gowda, S., & Dawson, W. O. (2009). Examination of the responses of different genotypes of citrus to huanglongbing (citrus greening) under different conditions. *Phytopathology*, *99*, 1346–1354. https://doi.org/10.1094/PHYTO-99-12-1346

Frick, G., Su, Q., Apel, K., & Armstrong, G. A. (2003). An *Arabidopsis porB porC* double mutant lacking light-dependent NADPH:protochlorophyllide oxidoreductases B and C is highly chlorophyll-deficient and developmentally arrested. *The Plant Journal*, *35*, 141–153. https://doi.org/10.1046/j.1365-313X.2003.01798.x

Gaquerel, E., Weinhold, A., & Baldwin, I. T. (2009). Molecular interactions between the specialist herbivore *Manduca sexta* (Lepidoptera, Sphigidae) and its natural host *Nicotiana attenuata*. VIII. An unbiased GCxGC-ToFMS analysis of the plant's elicited volatile emissions. *Plant Physiology*, *149*, 1408–1423. https://doi.org/10.1104/pp.108.130799

Gaufichon, L., Reisdorf-Cren, M., Rothstein, S. J., Chardon, F., & Suzuki, A. (2010). Biological functions of asparagine synthetase in plants. *Plant Science*, *179*, 141–153. https://doi.org/10.1016/j.plantsci.2010.04.010

Hajeri, S., Killiny, N., El-Mohtar, C., Dawson, W. O., & Gowda, S. (2014). *Citrus tristeza virus*-based RNAi in citrus plants induces gene silencing in *Diaphorina citri*, a phloem-sap sucking insect vector of citrus greening disease (Huanglongbing). *Journal of Biotechnology*, *176*, 42–49. https://doi.org/10.1016/j.jbiotec.2014.02.010

Harper, S. J., Cowell, S. J., Halbert, S. E., Brlansky, R. H., & Dawson, W. O. (2015). CTV status in Florida. *Citrus Industry*, 8–14.

Helmig, D., Ortega, J., Guenther, A., Herrick, J. D., & Geron, C. (2006). Sesquiterpene emissions from loblolly pine and their potential contribution to biogenic aerosol formation in the Southeastern US. *Atmospheric Environment*, *40*, 4150–4157. https://doi.org/10.1016/j.atmosenv.2006.02.035

Herrera-Rodríguez, M. B., Maldonado, J. M., & Pérez-Vicente, R. (2004). Light and metabolic regulation of *HAS1*, *HAS1.1* and *HAS2*, three asparagine synthetase genes in *Helianthus annuus*. *Plant Physiology and Biochemistry*, *42*, 511–518. https://doi.org/10.1016/j.plaphy.2004.05.001

Herrera-Rodríguez, M. B., Pérez-Vicente, R., & Maldonado, J. M. (2007). Expression of asparagine synthetase genes in sunflower (*Helianthus annuus*) under various environmental stresses. *Plant Physiology and Biochemistry*, *45*, 33–38. https://doi.org/10.1016/j.plaphy.2006.12.002

Hijaz, F., & Killiny, N. (2016). A gas chromatography–mass spectrometry method for the determination of delta-aminolevulinic acid in plant leaves. *Journal of Chromatography A*, *1447*, 57–63. https://doi.org/10.1016/j.chroma.2016.04.041

Hirono, Y., & Rédei, G. P. (1963). Multiple allelic control of chlorophyll *b* level in *Arabidopsis thaliana*. *Nature, 197*, 1324–1325. https://doi.org/10.1038/1971324a0

Höfgen, R., Axelsen, K. B., Kannangara, C. G., Schüttke, I., Pohlenz, H. D., Willmitzer, L., ... von Wettstein, D. (1994). A visible marker for antisense mRNA expression in plants: Inhibition of chlorophyll synthesis with a glutamate-1-semialdehyde aminotransferase antisense gene. *Proceedings of the National Academy of Sciences of the United States of America, 91*, 1726–1730. https://doi.org/10.1073/pnas.91.5.1726

Horvàth, E., Szalai, G., & Janda, T. (2007). Induction of abiotic stress tolerance by salicylic acid signaling. *Journal of Plant Growth Regulation, 26*, 290–300. https://doi.org/10.1007/s00344-007-9017-4

Hu, G., Yalpani, N., Briggs, S. P., & Johal, G. S. (1998). A porphyrin pathway impairment is responsible for the phenotype of a dominant disease lesion mimic mutant of maize. *Plant Cell, 10*, 1095–1105. https://doi.org/10.1105/tpc.10.7.1095

Igarashi, A., Yamagata, K., Sugai, T., Takahashi, Y., Sugawara, E., Tamura, A., ... Yoshikawa, N. (2009). Apple latent spherical virus vectors for reliable and effective virus-induced gene silencing among a broad range of plants including tobacco, tomato, *Arabidopsis thaliana*, cucurbits, and legumes. *Virology, 386*, 407–416. https://doi.org/10.1016/j.virol.2009.01.039

Ishikawa, A., Okamoto, H., Iwasaki, Y., & Asahi, T. (2001). A deficiency of coproporphyrinogen III oxidase causes lesion formation in Arabidopsis. *The Plant Journal, 27*, 89–99. https://doi.org/10.1046/j.1365-313x.2001.01058.x

Jensen, P. E., Willows, R. D., Petersen, B. L., Vothknecht, U. C., Stummann, B. M., Kannangara, C. G., ... Henningsen, K. W. (1996). Structural genes for Mg-chelatase subunits in barley: *Xantha-f*, *-g* and *-h*. *Molecular and General Genetics, 250*, 383–394.

Jung, S. (2004). Effect of chlorophyll reduction in *Arabidopsis thaliana* by methyl jasmonate or norflurazon on antioxidant systems. *Plant Physiology and Biochemistry, 42*, 225–231. https://doi.org/10.1016/j.plaphy.2004.01.001

Kaczor, C. M., Smith, M. W., Sangwan, I., & O'Brian, M. R. (1994). Plant δ -Aminolevulinic acid dehydratase: Expression in soybean root nodules and evidence for a bacterial lineage of the *Alad* gene. *Plant Physiology, 104*, 1411–1417. https://doi.org/10.1104/pp.104.4.1411

Killiny, N., Hijaz, F., Harper, S. J., & Dawson, W. O. (2017). Effects of *Citrus tristeza closterovirus* infection on phloem sap and released volatile organic compounds in *Citrus macrophylla*. *Physiological and Molecular Plant Pathology, 98*, 25–36. https://doi.org/10.1016/j.pmpp.2017.03.003

Killiny, N., & Jones, S. E. (2017). Profiling of volatile organic compounds released from individual intact juvenile and mature citrus leaves. *Journal of Plant Physiology, 208*, 47–51. https://doi.org/10.1016/j.jplph.2016.11.001

Killiny, N., & Nehela, Y. (2017a). Metabolomic response to Huanglongbing: Role of carboxylic compounds in *Citrus sinensis* response to "*Candidatus* Liberibacter asiaticus" and its vector, *Diaphorina citri*. *Molecular Plant-Microbe Interactions, 30*, 666–678. https://doi.org/10.1094/MPMI-05-17-0106-R

Killiny, N., & Nehela, Y. (2017b). One target, two mechanisms: The impact of "*Candidatus* Liberibacter asiaticus" and its vector, *Diaphorina citri*, on citrus leaf pigments. *Molecular Plant-Microbe Interactions, 30*, 543–556. https://doi.org/10.1094/MPMI-02-17-0045-R

Kim, Y.-K., Lee, J.-Y., Cho, H. S., Lee, S. S., Ha, H. J., Kim, S., ... Pai, H.-S. (2005). Inactivation of organellar glutamyl- and seryl-tRNA synthetases leads to developmental arrest of chloroplasts and mitochondria in higher plants. *Journal of Biological Chemistry, 280*, 37098–37106. https://doi.org/10.1074/jbc.M504805200

Kruse, E., Mock, H. P., & Grimm, B. (1995). Reduction of coproporphyrinogen oxidase level by antisense RNA synthesis leads to deregulated gene expression of plastid proteins and affects the oxidative defense system. *EMBO Journal, 14*, 3712–3720.

Kumar, A. M., & Söll, D. (2000). Antisense HEMA1 RNA expression inhibits heme and chlorophyll biosynthesis in Arabidopsis. *Plant Physiology, 122*, 49–56. https://doi.org/10.1104/pp.122.1.49

Livak, K. J., & Schmittgen, T. D. (2001). Analysis of relative gene expression data using real-time quantitative PCR and the $2^{-\Delta\Delta CT}$ method. *Methods, 25*, 402–408. https://doi.org/10.1006/meth.2001.1262

Mafra, V., Kubo, K. S., Alves-Ferreira, M., Ribeiro-Alves, M., Stuart, R. M., Boava, L. P., ... Machado, M. A. (2012). Reference genes for accurate transcript normalization in citrus genotypes under different experimental conditions. *PLoS One, 7*, e31263. https://doi.org/10.1371/journal.pone.0031263

Mauzerall, D., & Granick, S. (1956). The occurrence and determination of delta-amino-levulinic acid and porphobilinogen in urine. *Journal of Biological Chemistry, 219*, 435–446.

Michie, M. G. (1982). Use of the Bray-Curtis similarity measure in cluster analysis of foraminiferal data. *Journal of the International Association for Mathematical Geology, 14*, 661–667. https://doi.org/10.1007/BF01033886

Mock, H. P., Heller, W., Molina, A., Neubohn, B., Sandermann, H., & Grimm, B. (1999). Expression of uroporphyrinogen decarboxylase or coproporphyrinogen oxidase antisense RNA in tobacco induces pathogen defense responses conferring increased resistance to *tobacco mosaic virus*. *Journal of Biological Chemistry, 274*, 4231–4238. https://doi.org/10.1074/jbc.274.7.4231

Mock, H.-P., Keetman, U., Kruse, E., Rank, B., & Grimm, B. (1998). Defense responses to tetrapyrrole-induced oxidative stress in transgenic plants with reduced uroporphyrinogen decarboxylase or coproporphyrinogen oxidase activity. *Plant Physiology, 116*, 107–116. https://doi.org/10.1104/pp.116.1.107

Molina, A., Volrath, S., Guyer, D., Maleck, K., Ryals, J., & Ward, E. (1999). Inhibition of protoporphyrinogen oxidase expression in *Arabidopsis* causes a lesion-mimic phenotype that induces systemic acquired resistance. *The Plant Journal, 17*, 667–678. https://doi.org/10.1046/j.1365-313X.1999.00420.x

Mouly, P. P., Gaydou, E. M., Lapierre, L., & Corsetti, J. (1999). Differentiation of several geographical origins in single-strength Valencia orange juices using quantitative comparison of carotenoid profiles. *Journal of Agriculture and Food Chemistry, 47*, 4038–4045. https://doi.org/10.1021/jf990025l

Nagata, N., Tanaka, R., Satoh, S., & Tanaka, A. (2005). Identification of a vinyl reductase gene for chlorophyll synthesis in *Arabidopsis thaliana* and implications for the evolution of prochlorococcus species. *Plant Cell, 17*, 233–240. https://doi.org/10.1105/tpc.104.027276

Nehela, Y., Hijaz, F., Elzaawely, A. A., El-Zahaby, H. M., & Killiny, N. (2016). Phytohormone profiling of the sweet orange (*Citrus sinensis* (L.) Osbeck) leaves and roots using GC-MS-based method. *Journal of Plant Physiology, 199*, 12–17. https://doi.org/10.1016/j.jplph.2016.04.005

Nehela, Y., Hijaz, F., Elzaawely, A. A., El-Zahaby, H. M., & Killiny, N. (2018). Citrus phytohormonal response to *Candidatus* Liberibacter asiaticus and its vector *Diaphorina citri*. *Physiological and Molecular Plant Pathology, 102*, 24–35. https://doi.org/10.1016/j.pmpp.2017.11.004

Nitsch, L., Kohlen, W., Oplaat, C., Charnikhova, T., Cristescu, S., Michieli, P., ... Rieu, I. (2012). ABA-deficiency results in reduced plant and fruit size in tomato. *Journal of Plant Physiology, 169*, 878–883. https://doi.org/10.1016/j.jplph.2012.02.004

Palmer-Young, E. C., Veit, D., Gershenzon, J., Schuman, M. C., Baldwin, I., & Degenhardt, J. (2015). The sesquiterpenes(*e*)-ß-farnesene and (*e*)-α-bergamotene quench ozone but fail to protect the wild tobacco nicotiana attenuata from ozone, UVB, and drought stresses. *PLoS One, 10*, e0127296. https://doi.org/10.1371/journal.pone.0127296

Papenbrock, J., Mishra, S., Mock, H.-P., Kruse, E., Schmidt, E.-K., Petersmann, A., ... Grimm, B. (2001). Impaired expression of the plastidic ferrochelatase by antisense RNA synthesis leads to a necrotic phenotype of transformed tobacco plants. *The Plant Journal, 28*, 41–50. https://doi.org/10.1046/j.1365-313X.2001.01126.x

Papenbrock, J., Mock, H.-P., Tanaka, R., Kruse, E., & Grimm, B. (2000). Role of magnesium chelatase activity in the early steps of the tetrapyrrole biosynthetic pathway. *Plant Physiology*, *122*, 1161–1169. https://doi.org/10.1104/pp.122.4.1161

Romero, I., Tikunov, Y., & Bovy, A. (2011). Virus-induced gene silencing in detached tomatoes and biochemical effects of phytoene desaturase gene silencing. *Journal of Plant Physiology*, *168*, 1129–1135. https://doi.org/10.1016/j.jplph.2010.12.020

Rzeznicka, K., Walker, C. J., Westergren, T., Kannangara, C. G., von Wettstein, D., Merchant, S., ... Hansson, M. (2005). *Xantha-l* encodes a membrane subunit of the aerobic Mg-protoporphyrin IX monomethyl ester cyclase involved in chlorophyll biosynthesis. *Proceedings of the National Academy of Sciences of the United States of America*, *102*, 5886–5891. https://doi.org/10.1073/pnas.0501784102

Scinicariello, F., Murray, H. E., Moffett, D. B., Abadin, H. G., Sexton, M. J., & Fowler, B. A. (2007). Lead and delta-aminolevulinic acid dehydratase polymorphism: Where does it lead? A meta-analysis *Environmental Health Perspectives*, *115*, 35–41.

Senthil-Kumar, M., Hema, R., Anand, A., Kang, L., Udayakumar, M., & Mysore, K. S. (2007). A systematic study to determine the extent of gene silencing in *Nicotiana benthamiana* and other Solanaceae species when heterologous gene sequences are used for virus-induced gene silencing. *New Phytologist*, *176*, 782–791. https://doi.org/10.1111/j.1469-8137.2007.02225.x

Shen, Z., Sun, J., Yao, J., Wang, S., Ding, M., Zhang, H., ... Chen, S. (2015). High rates of virus-induced gene silencing by tobacco rattle virus in *Populus*. *Tree Physiology*, *35*, 1016–1029. https://doi.org/10.1093/treephys/tpv064

Tanaka, R., & Tanaka, A. (2005). Effects of chlorophyllide *a* oxygenase overexpression on light acclimation in *Arabidopsis thaliana*. *Photosynthesis Research*, *85*, 327–340. https://doi.org/10.1007/s11120-005-6807-z

Tanaka, R., & Tanaka, A. (2007). Tetrapyrrole biosynthesis in higher plants. *Annual Review of Plant Biology*, *58*, 321–346. https://doi.org/10.1146/annurev.arplant.57.032905.105448

Tomokuni, K., & Ogata, M. (1972). Simple method for determination of urinary -aminolevulinic acid as an index of lead exposure. *Clinical Chemistry*, *18*, 1534–1538.

Tsang, E. W. T., Yang, J., Chang, Q., Nowak, G., Kolenovsky, A., McGregor, D. I., & Keller, W. A. (2003). Chlorophyll reduction in the seed of *Brassica napus* with a glutamate 1-semialdehyde aminotransferase antisense gene. *Plant Molecular Biology*, *51*, 191–201. https://doi.org/10.1023/A:1021102118801

van Handel, E. (1968). Direct microdetermination of sucrose. *Analytical Biochemistry*, *22*, 280–283. https://doi.org/10.1016/0003-2697(68)90317-5

Vickers, C. E., Gershenzon, J., Lerdau, M. T., & Loreto, F. (2009). A unified mechanism of action for volatile isoprenoids in plant abiotic stress. *Nature Chemical Biology*, *5*, 283–291. https://doi.org/10.1038/nchembio.158

Wei, X., Chen, C., Yu, Q., Gady, A., Yu, Y., Liang, G., & Gmitter, F. G. (2014a). Comparison of carotenoid accumulation and biosynthetic gene expression between Valencia and Rohde Red Valencia sweet oranges. *Plant Science*, *227*, 28–36. https://doi.org/10.1016/j.plantsci.2014.06.016

Wei, X., Chen, C., Yu, Q., Gady, A., Yu, Y., Liang, G., & Gmitter, F. G. (2014b). Novel expression patterns of carotenoid pathway-related genes in citrus leaves and maturing fruits. *Tree Genetics and Genomes*, *10*, 439–448. https://doi.org/10.1007/s11295-013-0688-7

Weller, J. L., Terry, M. J., Rameau, C., Reid, J. B., & Kendrick, R. E. (1996). The phytochrome-deficient *pcd1* mutant of pea is unable to convert heme to biliverdin IX[alpha]. *Plant Cell*, *8*, 55–67.

Weller, J. L., Terry, M. J., Reid, J. B., & Kendrick, R. E. (1997). The phytochrome-deficient *pcd2* mutant of pea is unable to convert biliverdin IXalpha to 3(*Z*)-phytochromobilin. *The Plant Journal*, *11*, 1177–1186. https://doi.org/10.1046/j.1365-313X.1997.11061177.x

Zeb, A., & Murkovic, M. (2010). Thin-layer chromatographic analysis of carotenoids in plant and animal samples. *Journal of Planar Chromatography-Modern TLC*, *23*, 94–103. https://doi.org/10.1556/JPC.23.2010.2.1

Permissions

All chapters in this book were first published in PD, by John Wiley & Sons Ltd.; hereby published with permission under the Creative Commons Attribution License or equivalent. Every chapter published in this book has been scrutinized by our experts. Their significance has been extensively debated. The topics covered herein carry significant findings which will fuel the growth of the discipline. They may even be implemented as practical applications or may be referred to as a beginning point for another development.

The contributors of this book come from diverse backgrounds, making this book a truly international effort. This book will bring forth new frontiers with its revolutionizing research information and detailed analysis of the nascent developments around the world.

We would like to thank all the contributing authors for lending their expertise to make the book truly unique. They have played a crucial role in the development of this book. Without their invaluable contributions this book wouldn't have been possible. They have made vital efforts to compile up to date information on the varied aspects of this subject to make this book a valuable addition to the collection of many professionals and students.

This book was conceptualized with the vision of imparting up-to-date information and advanced data in this field. To ensure the same, a matchless editorial board was set up. Every individual on the board went through rigorous rounds of assessment to prove their worth. After which they invested a large part of their time researching and compiling the most relevant data for our readers.

The editorial board has been involved in producing this book since its inception. They have spent rigorous hours researching and exploring the diverse topics which have resulted in the successful publishing of this book. They have passed on their knowledge of decades through this book. To expedite this challenging task, the publisher supported the team at every step. A small team of assistant editors was also appointed to further simplify the editing procedure and attain best results for the readers.

Apart from the editorial board, the designing team has also invested a significant amount of their time in understanding the subject and creating the most relevant covers. They scrutinized every image to scout for the most suitable representation of the subject and create an appropriate cover for the book.

The publishing team has been an ardent support to the editorial, designing and production team. Their endless efforts to recruit the best for this project, has resulted in the accomplishment of this book. They are a veteran in the field of academics and their pool of knowledge is as vast as their experience in printing. Their expertise and guidance has proved useful at every step. Their uncompromising quality standards have made this book an exceptional effort. Their encouragement from time to time has been an inspiration for everyone.

The publisher and the editorial board hope that this book will prove to be a valuable piece of knowledge for researchers, students, practitioners and scholars across the globe.

List of Contributors

Prashant Kumar Pandey, Thiago Alexandre Moraes, Regina Feil, John E. Lunn and Mark Stitt
Max Planck Institute of Molecular Plant Physiology, Potsdam, Golm, Germany

Letícia dos Anjos
Max Planck Institute of Molecular Plant Physiology, Potsdam, Golm, Germany Universidade Federal do Ceará, Fortaleza, Brazil

Lauri Nikkanen, Jouni Toivola, Andrea Trotta, Manuel Guinea Diaz, Mikko Tikkanen, Eva-Mari Aro and Eevi Rintamäki
Molecular Plant Biology, Department of Biochemistry, University of Turku, Turku, Finland

Pirko Jalakas, Ebe Merilo and Hannes Kollist
Institute of Technology, University of Tartu, Tartu, Estonia

Mikael Brosché
Institute of Technology, University of Tartu, Tartu, Estonia
Viikki Plant Science Centre, Organismal and Evolutionary Biology Research Programme, Faculty of Biological and Environmental Sciences, University of Helsinki, Helsinki, Finland

Kin H. Lau and Emily Crisovan
Department of Plant Biology, Michigan State University, East Lansing, Michigan

María del Rosario Herrera, Muhammad Awais Khan and Dorcus C. Gemenet
International Potato Center, Lima, Peru

Shan Wu and Zhangjun Fei
Boyce Thompson Institute, Cornell University, Ithaca, New York

Carol Robin Buell
Department of Plant Biology, Michigan State University, East Lansing, Michigan
Plant Resilience Institute, Michigan State University, East Lansing, Michigan

Anna M. Locke
Soybean and Nitrogen Fixation Research Unit, USDA-ARS, Raleigh, North Carolina
Department of Crop and Soil Sciences, North Carolina State University, Raleigh, North Carolina

Rebecca A. Slattery
Carl R. Woese Institute for Genomic Biology, University of Illinois, Urbana, Illinois
Global Change and Photosynthesis Research Unit, USDA-ARS, Urbana, Illinois

Donald R. Ort
Carl R. Woese Institute for Genomic Biology, University of Illinois, Urbana, Illinois
Global Change and Photosynthesis Research Unit, USDA-ARS, Urbana, Illinois
Department of Plant Biology, University of Illinois, Urbana, Illinois

Angela C. Burnett, Mark Rees and Colin P. Osborne
Department of Animal and Plant Sciences, University of Sheffield, Sheffield, UK

Alistair Rogers
Environmental and Climate Sciences Department, Brookhaven National Laboratory, Upton, New York

Sunil Kumar Kenchanmane Raju, Mon-Ray Shao, Yashitola Wamboldt and Sally Mackenzie
Department of Agronomy and Horticulture, University of Nebraska-Lincoln, Lincoln, Nebraska

Masato Shindo, Koichiro Shimomura, Shinjiro Yamaguchi and Mikihisa Umehara
Graduate School of Life Sciences, Toyo University, Gunma, Japan
Graduate School of Life Sciences, Tohoku University, Sendai, Japan

Nischal Karki and Philip D. Bates
Department of Chemistry and Biochemistry, The University of Southern Mississippi, Hattiesburg, Mississippi

Brad M. Binder
Department of Biochemistry and Cellular & Molecular Biology, University of Tennessee, Knoxville, Tennessee

G. Eric Schaller
Department of Biological Sciences, Dartmouth College, Hanover, New Hampshire

Hyo Jung Kim
Department of Biological Sciences, Dartmouth College, Hanover, New Hampshire
Center for Plant Aging Research, Institute for Basic Science (IBS), Daegu, Korea

Dennis E. Mathews
Department of Molecular, Cellular, and Biomedical Sciences, University of New Hampshire, Durham, New Hampshire

Claire E. Hutchison and Joseph J. Kieber
Department of Biology, University of North Carolina, Chapel Hill, North Carolina

Mirai Okuhara, Ruriko Katayama, Noboru Isayama and Isao Enami
Department of Biology, Tokyo University of Science, Tokyo, Japan

Tatsuya Tomo
Department of Biology, Tokyo University of Science, Tokyo, Japan
Graduate School of Science, Tokyo University of Science, Tokyo, Japan

Akinori Okumura
Department of Integrated Sciences in Physics and Biology, College of Humanities and Sciences, Nihon University, Tokyo, Japan

Takehiro Suzuki and Naoshi Dohmae
Biomolecular Characterization Unit, RIKEN Center for Sustainable Resource Science, Saitama, Japan

Ryo Nagao
Research Institute for Interdisciplinary Science, Okayama University, Okayama, Japan

Masako Iwai
School of Life Science and Technology, Tokyo Institute of Technology, Yokohama, Japan

Aaron Kusmec
Department of Agronomy, Iowa State University, Ames, IA, USA

Patrick S. Schnable
Department of Agronomy, Iowa State University, Ames, IA, USA
Plant Sciences Institute, Iowa State University, Ames, IA, USA

Maja Cohen and Robert Fluhr
Department of Plant Sciences, Weizmann Institute of Science, Rehovot, Israel

Brian A. McKinley, Anna L. Casto and John E. Mullet
Department of Biochemistry and Biophysics, Texas A&M University, College Station, Texas

William L. Rooney
Department of Soil and Crop Sciences, Texas A&M University, College Station, Texas

Joseph Lee Hill Jr, Ashley Nicole Hill and Ming Tien
Department of Biochemistry and Molecular Biology, The Center for Lignocellulose Structure and Formation, Pennsylvania State University, University Park, Pennsylvania

Alison W. Roberts
Department of Biological Sciences, University of Rhode Island, Kingston, Rhode Island

Candace H. Haigler
Department of Crop and Soil Sciences and Department of Plant and Microbial Biology, North Carolina State University, Raleigh, North Carolina

Kristina Unland and Christian Schulze Gronover
Fraunhofer Institute for Molecular Biology and Applied Ecology (IME), Muenster, Germany

Katharina M. Pütter, Kirsten Vorwerk and Nicole van Deenen
Institute of Plant Biology and Biotechnology, University of Muenster, Muenster, Germany

Richard M. Twyman
TRM Ltd, Scarborough, UK

Gigi Wong, Maria Alonso-Peral, Bingjun Li, Junyan Li and Anthony A. Millar
Division of Plant Science, Research School of Biology, The Australian National University, Canberra, Australian Capital Territory, Australia

Index